MICROWAVE AND WIRELESS SYNTHESIZERS

MICROWAVE AND WIRELESS SYNTHESIZERS

Theory and Design

Ulrich L. Rohde

Synergy Microwave, Paterson, New Jersey

A Wiley-Interscience Publication

JOHN WILEY & SONS, INC.

New York / Chichester / Weinheim / Brisbane / Singapore / Toronto

This text is printed on acid-free paper.

Copyright © 1997 by John Wiley & Sons, Inc.

All rights reserved. Published simultaneously in Canada.

Reproduction or translation of any part of this work beyond that permitted by Section 107 or 108 of the 1976 United States Copyright Act without the permission of the copyright owner is unlawful. Requests for permission or further information should be addressed to the Permissions Department, John Wiley & Sons, Inc., 605 Third Avenue, New York, NY 10158-0012

Library of Congress Cataloging in Publication Data:
Rohde, Ulrich L.
 RF and microwave digital frequency synthesizers : theory and design / Ulrich L. Rohde.
   p. cm.
   "A Wiley-Interscience publication."
   Includes bibliographical references and index.
   ISBN 0-471-52019-5 (cloth : alk. paper)
   1. Frequency synthesizers—Design and construction. 2. Phase-locked loops. 3. Digital electronics. 4. Microwave circuits--Design and construction. 5. Radio frequency. I. Title.
TK7872.F73R62 1997
621.3815'486—dc20   96-2841

Printed in the United States of America
10 9 8 7 6 5 4 3 2 1

CONTENTS

Preface   xi

Important Notations   xv

1 Loop Fundamentals   1

   1-1 Introduction to Linear Loops / 1
   1-2 Characteristics of a Loop / 4
   1-3 Digital Loops / 9
   1-4 Type 1 First-Order Loops / 11
   1-5 Type 1 Second-Order Loops / 15
   1-6 Type 2 Second-Order Loop / 24
      1-6-1 Transient Behavior of Digital Loops Using Tri-state Phase Detectors / 27
   1-7 Type 2 Third-Order Loop / 34
      1-7-1 Transfer Function of Type 2 Third-Order Loop / 35
      1-7-2 FM Noise Suppression / 43
   1-8 Higher-Order Loops / 44
      1-8-1 Fifth-Order Loop Transient Response / 44
   1-9 Digital Loops with Mixers / 48
   1-10 Acquisition / 53
      1-10-1 Pull-in Performance of the Digital Loop / 60
      1-10-2 Coarse Steering of the VCO as an Acquisition Aid / 62
      1-10-3 Loop Stability / 64
   References and Suggested Reading / 75

2 Noise and Spurious Response of Loops   79

   2-1 Introduction to Sideband Noise / 79
   2-2 Spectral Density of Frequency Fluctuations, Related to $S_{\Delta\theta}$ and $\angle$ / 82
   2-3 Residual FM Related to $\angle(f_m)$ / 83
   2-4 Allan Variance Related to $\angle(f_m)$ / 84
   2-5 Linear Approach for the Calculation of Oscillator Phase Noise / 86
   2-6 Noise Contributions in Phase-Locked Systems / 99

vi CONTENTS

      2-6-1 Phase Noise Characteristics of Amplifiers / 101
      2-6-2 Phase Noise Characteristics of Dividers / 102
      2-6-3 Phase Noise Characteristics of Phase/Frequency Comparators / 105
      2-6-4 Phase Noise Characteristics of Multipliers / 106
      2-6-5 Noise Contribution from Power Supplies / 110
  2-7 Overall Phase Noise Performance of a System / 111
      2-7-1 Practical Results for Noise Contributions / 117
  2-8 Measurement of Phase Noise / 118
      2-8-1 Heterodyne Frequency Measurement Technique / 119
      2-8-2 Phase Noise Measurement with Spectrum Analyzer / 120
      2-8-3 Phase Noise Measurement with Frequency Discriminator / 121
      2-8-4 Delay Line and Mixer as Frequency Comparator / 122
      2-8-5 Phase Noise Measurement with Two Sources and Phase Comparator / 123
  References and Suggested Reading / 132

3 Special Loops   136

  3-1 Direct Digital Synthesis Techniques / 136
      3-1-1 A First Look at Fractional N / 137
      3-1-2 Digital Waveform Synthesizers / 139
      3-1-3 Signal Quality / 153
      3-1-4 Future Prospects / 169
  3-2 Multiple Sampler Loops / 170
  3-3 Loops with Delay Line as Phase Comparators / 171
  3-4 Fractional Division N Synthesizers / 172
      3-4-1 Special Patents for Fractional Division N Synthesizers / 191
  References / 193

4 Loop Components   197

  4-1 Oscillator Design / 197
      4-1-1 Basics of Oscillators / 197
      4-1-2 Low-Noise LC Oscillators / 213
      4-1-3 Switchable/Tunable LC Oscillators / 216
      4-1-4 Use of Tuning Diodes / 226
      4-1-5 Use of Diode Switches / 233
      4-1-6 Use of Diodes for Frequency Multiplication / 236
  4-2 Reference Frequency Standards / 236
      4-2-1 Requirements / 237
      4-2-2 Specifying Oscillators / 239
      4-2-3 Typical Examples of Crystal Oscillator Specifications / 240

4-2-4 Crystal Resonators / 242
4-2-5 Crystal Specifications / 252
4-2-6 Crystal Oscillators / 252
4-2-7 Effect of External Influences on Oscillator Stability / 276
4-2-8 High-Performance Oscillator Capabilities / 277
4-2-9 Surface Acoustic Wave (SAW) Oscillators / 277
4-3 Mixer Applications / 284
4-4 Phase/Frequency Comparators / 288
    4-4-1 Diode Rings / 289
    4-4-2 Exclusive ORs / 291
    4-4-3 Sample/Hold Detectors / 294
    4-4-4 Edge-Triggered JK Master/Slave Flip-Flops / 302
    4-4-5 Digital Tri-state Comparators / 305
4-5 Wideband High-Gain Amplifiers / 319
    4-5-1 Summation Amplifiers / 319
    4-5-2 Differential Limiters / 323
    4-5-3 Isolation Amplifiers / 324
4-6 Programmable Dividers / 330
    4-6-1 Asynchronous Counters / 330
    4-6-2 Programmable Synchronous Up/Down-Counters / 333
    4-6-3 Swallow Counters/Dual-Modulus Counters / 352
    4-6-4 Look-Ahead and Delay Compensation / 364
4-7 Loop Filters / 375
    4-7-1 Passive RC Filters / 375
    4-7-2 Active RC Filters / 376
    4-7-3 Active Second-Order Low-Pass Filters / 378
    4-7-4 Passive LC Filters / 382
4-8 Microwave Oscillator Design / 382
    4-8-1 Introduction / 382
    4-8-2 The Compressed Smith Chart / 387
    4-8-3 Series or Parallel Resonance / 388
    4-8-4 Two-Port Oscillator Design / 390
4-9 Microwave Resonators / 397
    4-9-1 SAW Oscillators / 398
    4-9-2 Dielectric Resonators / 399
    4-9-3 YIG Oscillators / 402
    4-9-4 Varactor Resonators / 403
    4-9-5 Ceramic Resonators / 405
References / 409

5 Digital PLL Synthesizers   419

5-1 Multiloop Synthesizers Using Different Techniques / 419
    5-1-1 Direct Frequency Synthesis / 419
    5-1-2 Multiple Loops / 422

viii   CONTENTS

    5-2   System Analysis / 427
    5-3   Low-Noise Microwave Synthesizer / 437

        5-3-1   Building Bocks / 437
        5-3-2   Output Loop Response / 443
        5-3-3   Low Phase Noise References: Frequency Standards / 444
        5-3-4   Critical Stages / 447
        5-3-5   Time Domain Analysis / 461
        5-3-6   Summary / 462
        5-3-7   Two Commercial Synthesizer Examples / 467

    5-4   Microprocessor Applications in Synthesizers / 471
    5-5   Transceiver Applications / 481
    References / 484

6   High-Performance Hybrid Synthesizer   489

    6-1   Basic Synthesizer Approach / 490
    6-2   Loop Filter Design / 496
    6-3   Summary / 504
    References / 504

APPENDICES   505

A   Mathematical Review   507

    A-1   Functions of a Complex Variable / 507
    A-2   Complex Planes / 513
    A-3   Bode Diagram / 518
    A-4   Laplace Transformation / 524
    A-5   Low-Noise Oscillator Design / 533
    A-6   Oscillator Amplitude Stabilization / 538
    A-7   Very Low Phase Noise VCO for 800 MHz / 544
    References / 549

B   A General-Purpose Nonlinear Approach to the Computation of Sideband Phase Noise in Free-Running Microwave and RF Oscillators   551

    B-1   Introduction / 551
    B-2   Noise Generation in Oscillators / 552
    B-3   Bias-Dependent Noise Model / 552
    B-4   General Concept of Noisy Circuits / 562
    B-5   Noise Figure of Mixer Circuits / 565
    B-6   Oscillator Noise Analysis / 567
    B-7   Limitations of the Frequency-Conversion Approach / 569
    B-8   Summary of the Phase Noise Spectrum of the Oscillator   571
    B-9   Verification Examples for the Calculation of Phase Noise in Oscillators Using Nonlinear Techniques / 573
    B-10   Summary / 585
    References / 585

C	Example of Wireless Synthesizers Using Commercial ICs	587
D	MMIC-Based Synthesizers	625
Index		631

PREFACE

During the last few years, because of the many microwave/RF and wireless activities, interest in frequency synthesizers has grown rapidly. Synthesizers are found in test and measurement equipment, as well as in communication equipment. Many publications have emerged that look at some special aspects of synthesizers, but they do not cover both theoretical and practical aspects. This book is based on theoretical work I have done, courses I have given at George Washington University, and very recent developments set in motion at Compact Software, Inc., supported by a variety of government contracts. I have used software, predominantly from Compact Software, Inc., as a design aid. There are also other suppliers of similar software; however, we tend to have less access to these.

For individuals who are getting acquainted with oscillators and synthesizers, I strongly recommend getting the *ARRL Handbook For Radio Amateurs*, published by the American Radio Relay League, 225 Main Street, Newington, CT 06111. Chapter 14 on oscillators and synthesizers by David Stockton is a really nice first time introduction to this topic.

The book is divided into six chapters beginning with Chapter 1 loop fundamentals, which provides great detailed insight into settling time and other characteristics of the loop. The clear differentiation between analog and digital loops has proven to be quite useful, and topics like pull-in performance and acquisition are discussed in great detail.

Chapter 2 outlines noise and spurious responses of the loops. The linear approach of the oscillator phase noise is very detailed and walks the reader through all the important steps and contributions both inside and outside the loop. We also look at the noise contribution of the various parts of the loop, such as frequency dividers, phase detectors, and even power supplies. Finally, the noise analysis of the entire system and its measurements are covered.

In Chapter 3 we look at special loops. Here the digital direct synthesizer (DDS) technique should prove most interesting to the reader. The fractional division N synthesizer technology competes with DDS. Details regarding a mixed approach are also shown in the appendices.

The fractional division N synthesis principle is more complex. This is best seen from the various patents listed in the references at the end of the chapter. Companies such as Rohde & Schwarz, Marconi Instruments, and Hewlett-Packard, to mention a few, use this technique. The now digital implementation of the accumulator and its compensating network has the greatest influence on its performance. This area is very exciting for me, but only big houses will be able to afford the large-scale custom integrated circuits (LSIs) and gate arrays that will solve these problems. The appearence of the first fractional-N synthesizer chips for cellular phones further illustrates this trend.

Chapter 4 provides a detailed overview of loop components. Many practical circuit details are found in this chapter as it addresses low noise oscillator design, including the use of linear CAD tools. The section on reference frequency standards provides a thorough insight into designing crystal oscillators, which are a vital part of synthesizers and which must provide both low aging and optimum phase noise. Other important components are mixers, phase/frequency discriminators, wideband high-gain amplifiers, programmable dividers, and loop filters. The microwave oscillator design section in Chapter 4 is unique because it is the first systematic evaluation of all aspects of microwave design techniques, including different resonators, such as tuning diodes and SAW oscillators, dielectric resonators, and ceramic resonators. It also uses a nonlinear CAD tool to predict oscillator performance.

Chapter 5 provides in-depth details about multiloop synthesizers. The section on microwave synthesizers deals with analysis, different architectures, and trade-offs. Another unique section is the survey of critical stages and the analysis of their behavior. Finally, we look at the applications of microprocessors to optimize the synthesizer architecture, then show an application of the various techniques in a synthesizer for military communication equipment.

Chapter 6 is dedicated to a practical synthesizer example which combines the techniques outlined in the previous chapters. This design of a high performance hybrid synthesizer, including its measured performance, allows the engineer to follow the various design steps and design rules. It teaches the reader to understand the approaches that must be used for success in the first go-round of a design.

The appendices consist of four different sections. Appendix A provides a mathematical overview and is useful for individuals who want to write their own CAD programs. Also, a very low phase noise oscillator design is provided. Appendix B is a mathematical treatment of the nonlinear approach to calculating phase noise in a free running oscillator. This may be the first complete treatment of its kind in a book and is based on my presentation at the Frequency Control Symposium in 1994. Appendix C is a reprint of application notes provided by Motorola and Philips. This is useful in order to get a feel for the state-of-the-art synthesizer chips available. Finally, Appendix D shows a MMIC-based microwave/millimeter wave oscillator/synthesizer, its design procedures, and its performance.

This collection of practical and theoretical information came from many contributors. I am very grateful to have received the permission of Rohde & Schwarz, Munich, Hewlett-Packard, and MMIC houses such as Texas Instruments and Raytheon to use some of their most recent materials in order to illustrate the most modern trends. I am also grateful to Roger Clark of the Dover Corporation for his considerable enhancement of my material on Reference Frequency Standards.

A note of caution. Currently, it appears that the cleanest and best single or dual-loop synthesizers can be built by using custom made fractional-division N ASIC chips. While many companies are competing for the PLL chip market, at the time of publication of this book, it appears that the performance of

LMX233XA family member PLL chips by National Semiconductors are leading the market. As technology changes, a different company may be leading tomorrow. I strongly advise a thorough evaluation of all available chips before committing to a final design.

Since one needs to test a whole range of parameters for the VCO/PLLs including VCO tuning characteristics, RF power flatness, PLL transients, both spurious and harmonic, the recently introduced HP4352S signal test system, can handle all that and more at a frequency range of 10 MHz–3 GHz and higher. For example, it can measure the phase noise of a 1.8-GHz VCO/PLL in 8.4 seconds covering 204 frequency offset points, normally a 10 minute job.

The completion of this project took significantly longer than I had first anticipated and I am very grateful to the various individuals who supported me throughout this project, including Carol Paolucci and Lynne Peterson, and, of course, my publisher, John Wiley & Sons, Inc., for their continued support of this project.

Ulrich L. Rohde

Upper Saddle River, New Jersey

IMPORTANT NOTATIONS

Symbol	Meaning
$a_i (i = 1, 2, \ldots, n)$	Loop parameters of nth-order PLL
a_n, a_k	Digital data values
A	Amplifier gain
$A(s)$	Open-loop gain
$A(\omega)$	Amplitude of transfer function
	Amplitude response of network
B_i	Bandwidth of input bandpass filter (Hz)
B_L	Noise bandwidth of PLL (Hz)
$B(s)$	Closed-loop gain of PLL
$E(s)$	Error function
F	Noise figure
f, f_i	Frequency (Hz)
f_c	Corner frequency of flicker noise
f_m	Fourier frequency (sideband, offset, modulation, baseband frequency)
f_o	Carrier frequency
$f(t)$	Instantaneous frequency
Δf	Peak frequency deviation (Hz)
$\Delta f(t)$	Instantaneous frequency fluctuation
Δf_{peak}	Peak deviation of sinusoidal frequency modulation
Δf_{res}	Residual FM
$F(s), F(j\omega)$	Transfer function of loop filter
$G(s), G(j\omega)$	Feedforward function (rad/s)
$G_n(s)$	Transfer characteristic of a divider
$H(s), H(j\omega)$	Feedback transfer function
K	Loop gain (rad/s)
K_d'	Phase detector gain before lock (V/rad)
K_θ	Phase detector gain factor (V/rad)
K_m	Multiplier gain
K_o	VCO gain factor (rad/s V)
K_s	Shaper constant
K_v	dc Gain of PLL or Velocity error coefficient (rad/s)
k	An integer or an integer index on a sequence
k	1.4×10^{-23} W s/K
$L(x)$	Laplace transform of x

xvi IMPORTANT NOTATIONS

$\mathcal{L}(f_m)$	Single-sideband phase noise to total signal power in a 1-Hz bandwidth
m	An integer
$m(t)$	Modulation waveshape
M	An integer denoting frequency multiplication or division
n	An integer
n	Loop order
$n(t)$	Noise voltage (V)
$\overline{n(t)}$	Time average of noise
N, N_i	An integer representing frequency division or multiplication
$P(j\omega)$	Fourier transform of pulse waveshape
P_s	Signal power (W)
P_{ssB}	Power of single sideband
P_{sav}	Available signal power
Q_{unl}	Quality factor of unloaded resonator
$s = \sigma + j\omega$	Laplace transform complex variable
SNR	Signal-to-noise ratio
SNR_L	Signal-to-noise ratio in loop bandwidth $2B_L$
S_o	One-sided spectral density of white noise (dB/Hz)
$S_{\Delta f}(f_m)$	Spectral density of frequency fluctuations
$S_y(f_m)$	Spectral density of fractional frequency fluctuations
$S_{\Delta\theta}(f_m)$	Spectral density of phase perturbation
$S_{\Delta\phi}(f_m)$	Spectral density of phase noise
t	Time (s)
t_{acq}	Acquisition time of a loop
t_{lock}	Lock-in time for phase lock
T	Symbol interval of digital data stream (s)
T_{AV}	Average time to first cycle slip (s)
T_o	Temperature (kelvin, K)
T_p	Pull-in time (s)
v_c, V_c	VCO control voltage (V)
v_d, V_d	Phase detector output voltage (V)
$v(t)$	Instantaneous voltage
V_o	Peak amplitude of VCO voltage (V)
V_s	Peak amplitude of signal voltage (V)
V_{sL}	Peak amplitude of sinusoidal signal at limiting port
$V_{n\,rms}$	Equivalent noise voltage (1-Hz bandwidth)
V_{sav}	Available signal voltage
W_e	Maximum energy stored in capacitor
W_i	Spectral density of white noise (W/Hz)
$y(t)$	Instantaneous fractional frequency offset from nominal frequency
θ	Phase angle (rad)
θ_i	Phase angle of input signal (rad)
$\theta_\varepsilon = \theta_i - \theta_o$	Phase error between input signal and VCO (rad)
θ_o	VCO phase (rad)
θ_p	Loop phase error caused by oscillator noise (rad)

θ_v	Steady-state phase error (static phase error, loop stress) due to offset of input frequency (rad)
$\Delta\theta$	Phase deviation (rad)
$\Delta\theta$	Amplitude of phase step (rad)
$\Delta\theta$	Peak deviation of phase modulation (rad)
$\Delta\theta(t)$	Instantaneous fluctuation of phase perturbation
ζ	Damping factor of second-order loop
ρ	Signal-to-noise ratio
ρ_{sav}	Input signal-to-noise ratio
σ_n	Standard deviation (rms value) of noise $n(t)$ (V)
$\sigma_y^2(\tau)$	Allan variance
τ	Time constant (s)
τ_1, τ_2, τ_L	Time constants in loop filter (s)
τ_p	Pull-in time constant
$\Delta\phi(t)$	Instantaneous phase fluctuation
$\Delta\phi_{\text{peak}}$	Peak deviation of sinusoidal phase modulation, also modulation index
ϕ	Loop phase error reduced modulo 2π (rad)
ϕ_{N_o}	Phase fluctuation internal to an oscillator
$\phi(\omega)$	Phase of transfer function of a network
ψ	Phase of a transfer function (rad)
$\omega = 2\pi f$	Angular frequency (rad/s)
$j\omega$	Fourier transform variable
ω_i	Radian frequency of input signal (rad/s)
ω_m	Modulating frequency (rad/s)
ω_n	Natural frequency of second-order loop (rad/s)
$\Delta\omega$	Amplitude of frequency step or of frequency offset (rad/s)
$\Delta\omega$	Peak deviation of frequency modulation (rad/s)
$\Delta\dot{\omega}$	Rate of change of frequency (rad/s^2)
$\Delta\omega_H$	Hold-in limit of PLL (rad/s)
$\Delta\omega_L$	Lock-in limit of PLL (rad/s)
$\Delta\omega_P$	Pull-in limit of PLL (rad/s)
$\Omega(s)$	Laplace transform $L[\Delta\omega(t)]$

1

LOOP FUNDAMENTALS

1-1 INTRODUCTION TO LINEAR LOOPS

The majority of the new frequency synthesizers utilize the phase-locked loop (PLL). Indeed, it was the realization of the PLL in an integrated circuit that led to the inexpensive frequency synthesizer. Because an understanding of PLLs is necessary for the design of frequency synthesizers, they are discussed in detail in this chapter. The emphasis here is on the PLL as used in a frequency synthesizer rather than on the PLL as used for signal detection. For the latter problem, the PLL input is a relatively low-level signal embedded in noise, and the PLL serves to detect the noisy signal. For PLL applications in frequency synthesizers, the input signal-to-noise ratio is high, and the PLL serves to lock out the output frequency on a multiple of the input frequency.

Although the PLL is a nonlinear device, it can be modeled as a linear device over most of its operating range. This chapter first presents the linearized analysis of the PLL, including its stability characteristics. The design of compensating filters to improve PLL performance is then discussed.

PLLs include a phase detector, low-pass filter, and voltage-controlled oscillator, as illustrated in Figure 1-1. The phase detector is a nonlinear device, and its characteristics determine loop performance. The various types of phase detectors are described in Chapter 3. The loop transient performance is discussed in Section 1-10. No generalized results are available for transient performance, but the discussion illustrates one analysis approach that can be used.

Several books have been published on this matter, and for those involved in research or interest in a more theoretical approach of the PLL principle, the following books are recommended:

Best, Roland E., *Phase-Locked Loops: Theory, Design and Applications*, McGraw-Hill, New York, 1989.

LOOP FUNDAMENTALS

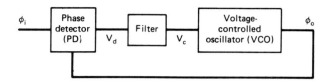

Figure 1-1 Block diagram of a PLL.

Crawford, James A., *Frequency Synthesizer Design Handbook*, Artech House, Boston–London, 1994.

Egan, William F., *Frequency Synthesis by Phase Lock*, Wiley, New York, 1981.

Gardner, Floyd M., *Phaselock Techniques*, 2nd ed., Wiley, New York, 1980.

Gorsky-Popiel, Jerzy, *Frequency Synthesis Techniques and Applications*, IEEE Press, New York, 1975.

Klapper, Jacob, and Frankle, John T., *Phase-Locked and Frequency Feedback Systems*, Academic Press, New York, 1972.

Kroupa, Venceslav F., *Frequency Synthesis Theory, Design and Applications*, Griffin, London, 1973.

Lindsey, William C., and Chie, Chak M., *Phase-Locked Loops & Their Applications*, IEEE Press, New York, 1985.

Lindsey, William C., and Simon, Marvin K., *Phase-Locked Loops & Their Applications*, IEEE Press, New York, 1977.

Manassewitsch, Vadim, *Frequency Synthesizers Theory and Design*, 3rd ed., Wiley, New York, 1987.

Robins, W.P., *Phase Noise in Signal Sources*, IEE Telecommunications Series 9, Peter Peregrinus Ltd. on behalf of the Institution of Electrical Engineers.

Stirling, Ronald C., *Microwave Frequency Synthesizers*, Prentice-Hall, Englewood Cliffs, NJ, 1987.

The term *phase-locked loop* refers to a feedback loop in which the input and feedback parameters of interest are the relative phases of the waveforms. The function of a PLL is to track small differences in phase between the input and feedback signal. The phase detector measures the phase difference between its two inputs. The phase detector output is then filtered by the low-pass filter and applied to the voltage-controlled oscillator (VCO). The VCO input voltage changes the VCO frequency in a direction that reduces the phase difference between the input signal and the local oscillator. The loop is said to be in *phase lock or locked* when the phase difference is reduced to zero.

Although the PLL is nonlinear since the phase detector is nonlinear, it can accurately be modeled as a linear device when the loop is in lock. When the loop is locked, it is assumed that the phase detector output voltage is proportional to the difference in phase between itw inputs; that is,

$$V_d = K_\theta(\theta_t - \theta_o) \qquad (1\text{-}1)$$

where θ_t and θ_o are the phases of the input and VCO output signals, respectively.

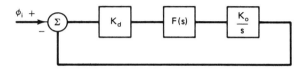

Figure 1-2 Block diagram of a PLL using a linearized model.

K_θ is the phase detector gain factor and has the dimensions of volts per radian. It will also be assured that the VCO can be modeled as a linear device whose output frequency deviates from its free-running frequency by an increment of frequency,

$$\Delta\omega = K_o V_e \qquad (1\text{-}2)$$

where V_e is the voltage at the output of the low-pass filter and K_o is the VCO gain factor, with the dimensions of rad/s per volt. Since frequency is the time derivative of phase, the VCO operation can be described as

$$\Delta\omega = \frac{d\theta_o}{dt} = K_o V_e \qquad (1\text{-}3)$$

With these assumptions, the PLL can be represented by the linear model shown in Figure 1-2. $F(s)$ is the transfer function of the low-pass filter. The linear transfer function relating $\theta_o(s)$ and $\theta_i(s)$ is

$$B(s) = \frac{\theta_o(s)}{\theta_i(s)} = \frac{K_\theta K_o F(s)/s}{1 + K_\theta K_o F(s)/s} \qquad (1\text{-}4)$$

If no low-pass filter is used, the transfer function is

$$B(s) = \frac{\theta_o}{\theta_i} = \frac{K_\theta K_o}{s + K_\theta K_o} = \frac{K}{s + K} \qquad (1\text{-}5)$$

which is equivalent to the transfer function of a simple low-pass filter with unity dc gain and bandwidth equal to K.

This is really the minimum configuration of a PLL. Since there is no divider in the chain, the output frequency and the reference frequency are the same. The first PLL built probably used a ring modulator as a phase detector. The ring modulator or diode bridge is electrically the same as a four-quadrant multiplier and operates from $-\pi$ to $+\pi$ of phase range.

Since the VCO probably has a sine-wave output and the reference frequency also has a sine wave, it is referred to as a *sinusoidal phase detector*. This does not really mean that the phase detector is sinusoidal; it means that the waves applied to the phase detector are sinusoidal.

Since there are no digital components in this basic loop, it is correctly called an *analog phase-locked loop* and, as stated above, is the minimum form of a

4 LOOP FUNDAMENTALS

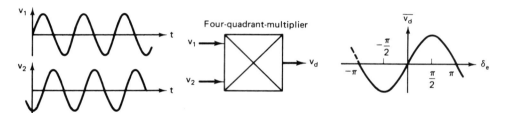

Figure 1-3 Waveform and transient characteristic of a linear phase detector.

phase-locked loop. To model it correctly, a number of assumptions are required.

We have already stated that, for our initial consideration, the loop is locked and that the transfer characteristic of the phase frequency detector is linear in the area of operation. A four-quadrant multiplier or diode quad has a sinusoidal output voltage, as shown in Figure 1-3, and is only piecewise linear for $\theta = 0$ or in the center of operation.

This minimum configuration of a PLL has several drawbacks. The absence of a filter does not allow one to choose parameters for optimized performance; the diode ring has only several hundred millivolts output, and an additional loop amplifier will add noise to the system. Therefore, for frequency synthesizer applications, a simple analog loop without any filter is rarely used. Such a loop would be called a first-order type 1 loop, and we will deal with it later.

There are some applications for a 1:1 loop, as we will see. The 1:1 loop is used to clean up an existing frequency, whereby the loop bandwidth is kept narrow enough to allow fast locking, and wide enough to permit fast acquisition. However, the loop bandwidth is more narrow than the spurious frequencies present at the reference input. The attenuation of the loop filter of such a loop, which should be a second-order loop, will clean up the output signal relative to the reference.

1-2 CHARACTERISTICS OF A LOOP

We have met the minimum-configuration analog PLL and already learned that there are several limitations to this loop. The first step in increasing the output voltage delivered from the diode quad and avoiding an operational amplifier is to use a different phase detector.

If the diode bridge arrangement is changed and the diodes are overdriven by the reference signal and the input signal from the VCO remains sufficiently small relative to the reference, the phase detector is still in a linear operation mode, and the output signal of the phase detector is no longer a sinusoidal curve but rather has a linear sawtooth form. The range over which the phase detector operates is still from $-\pi$ to $+\pi$, and the circuit is called *quasi-digital*. Later, when dealing with ring modulators, we will learn that it is a major requirement that the reference or LO signal must have a specified range, with a minimum typically 0.5 V for hot carrier diodes, and the VCO voltage should be substantially smaller for linear operation.

As the amplitude of the VCO signal is increased, the diode bridge or double-balanced modulator is overdriven, and a large number of harmonics occur. In a 1:1 loop, this is not very dangerous because all harmonics are phase and frequency coherent and do not generate unwanted signals. If a double-balanced mixer is used in a conversion scheme inside a loop, as we will see later, it is of utmost importance to operate the double-balanced mixer in its linear range and use a double-balanced mixer that has sufficient isolation between all ports.

In the case of the double-balanced mixer, whether overdriven from the LO or not, we refer to it as an analog linear phase-locked loop. If the VCO signal also overdrives the double-balanced mixer, we call it quasi-digital.

The digital equivalent of a double-balanced mixer is the exclusive-OR gate. The exclusive-OR gate, when built in CMOS logic, can be operated at 12 V, and therefore the dc output voltage can now be from almost 0 to +12 V, obtained from the integrator. We now have found a way to increase our dc control voltage without the noise sacrifice of an additional operational amplifier.

Both the double-balanced mixer and the exclusive-OR gate are only phase sensitive. The exclusive-OR gate also operates from $-\pi$ to $+\pi$, and the VCO has to be pretuned to be within the capture range.

Both phase detectors can be used for harmonic locking.

When we analyze the various loop components in Chapter 4, we will find some drawbacks that limit the use of the exclusive-OR gate.

The edge-triggered JK master/slave flip-flop can be used as a phase/frequency comparator. IT operates from -2π to $+2\pi$. The edge-triggered JK master/slave flip-flop has two outputs. One output supplies pulses to charge a capacitor, and the other can be used to discharge a capacitor.

To use this phase/frequency detector requires an active loop filter or active integrator, commonly referred to as a *charge pump*. This type of phase detector generates a beat frequency at the output, and the average dc voltage generated in the integrator is either negative or positive, relative to half the power supply voltage, depending on whether the signal frequency is higher or lower than the reference. This is a useful feature and explains why we can state that this circuit is not only phase sensitive but also frequency sensitive. Unfortunately, the frequency sensitivity for this type of phase/frequency discriminator is useful only for large differences. For very small differences in frequency, there is very little advantage in choosing this circuit over the exclusive-OR gate or the double-balanced mixer. We will learn more about this circuit in Chapter 4.

The most important circuit for phase/frequency comparators is probably an arrangement of two flip-flops and several gates. This particular circuit will be called a *tri-state phase/frequency comparator*, for reasons we will see later. It has several advantages:

1. The operating range is linear, from -2π to -2π.
2. It has the best possible locking performance and best frequency and phase difference detection.
3. Regardless of the amount of frequency error, the average output voltage is always above or below half the operating voltage. This results in good locking.

6 LOOP FUNDAMENTALS

Table 1-1 Comparison of various phase/frequency comparators

Type	Operating Range	Sensitivity
Diode ring	$-\pi$ to π	Phase only
Exclusive-OR gate	$-\pi$ to π	Phase only
Edge-triggered JK flip-flop	-2π to 2π ($-\pi$ to π); See Table 1-2	Phase frequency; undefined for small errors
Tri-state phase/frequency comparator	-2π to 2π	Phase frequency

The last two types of phase/frequency detectors, because of their charge/discharge capability, require active filters or summation circuits. It is theoretically possible to avoid the amplifier and use purely resistive circuits, but as we will soon learn, this has disadvantages. So far, we still have only used a loop that has no frequency divider.

The introduction of a frequency divider requires that the input to the divider be a square wave (TTL), and we will now define all loops that use digital dividers and phase/frequency comparators as digital loops. This should not be confused with digital synthesizers. A digital frequency synthesizer is most likely a direct synthesizer in which the output frequency is digitally generated with the help of a computer and is not available anywhere in analog or sine-wave form. The output frequency is then the summation of several digital signals in a quasi-sine wave at the output generated only by means of digital circuitry. There are several methods besides the one currently used, which employs a lookup table for sine or cosine functions. For very low frequency application, a complicated arrangement of diodes can be used to generate sine waves out of triangular waveshapes. Table 1-1 shows a comparison of the various phase/frequency detectors.

We have learned that because of its wider operating range, the phase-locked loop using digital phase/frequency comparators offers significant advantages over the analog PLL, and that is the main reason it is used in frequency synthesizers.

A closed loop is really a feedback system, and the various rules for feedback systems apply. We have already written down without further justification the formula that applies for a closed loop:

$$B(s) = \frac{\text{forward gain}}{1 + (\text{open-loop gain})} \qquad (1\text{-}6)$$

Before we continue, let us take a look at some of the abbreviations and definitions that are used in feedback control systems. Figure 1-4 shows the equivalent block diagram of a feedback system. The gain is equal to the multiplication of the VCO gain K_o times the phase/frequency comparator K_θ and will be abbreviated by K:

$$K = K_o K_\theta \; \frac{\text{rad} \cdot \text{s}^{-1}}{\text{V}} \frac{\text{V}}{\text{rad}} \qquad (1\text{-}7)$$

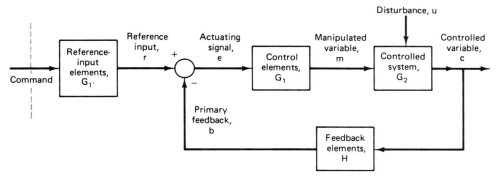

Figure 1-4 Equivalent diagram of a PLL using feedback control system analogy.

The feedforward function

$$G(s) = \frac{K}{s} F(s) \tag{1-8}$$

takes a filter function $F(s)$ into consideration, and the fact that the VCO itself is a perfect integrator is also taken into consideration.

To close the loop, we have to describe the feedback transfer function

$$H(s) = \frac{1}{N} \tag{1-9}$$

which will describe the divider ratio N and assumes that there is no delay in the divider. If there is delay in the divider, $H(s)$ is expressed in the form

$$H(s)^* = \frac{e^{-T_n s}}{N} \tag{1-10}$$

Therefore, the open-loop gain of the system is written

$$A(s) = G(s)H(s) \tag{1-11}$$

This definition of open-loop gain must not be confused with the open-loop gain K or the feedforward gain. From our definition above, the closed-loop gain is

$$\begin{aligned} B(s) &= \frac{\text{forward gain}}{1 + (\text{open-loop gain})} \\ &= \frac{\theta_o(s)}{\theta_i(s)} \\ &= \frac{G(s)}{1 + G(s)H(s)} \\ &= \frac{K(s)F(s)}{s + K(s)F(s)/N} \end{aligned} \tag{1-12}$$

8 LOOP FUNDAMENTALS

It has become customary to incorporate the divider ratio N in the K, which means that K in most cases can be said to equal

$$\frac{K_o K_\theta}{N}$$

If this substitution is made, the formulas are generally valid, provided that the correct factor of K is selected, and the formulas are generally usable regardless of the actual division ratio. In synthesizer design it is interesting to determine the system's noise bandwidth, B_n, which is defined as

$$B_n = \frac{1}{2\pi} \int_0^\infty B(j\omega)\,d\omega \tag{1-13}$$

The 3-dB bandwidth can be determined by solving the equation substituting $|B_n(j\omega)| = 0.707$; the resulting $j\omega$ would then be the 3-dB bandwidth, the complex variable j deleted. Another important piece of information in feedback control system performance is the *steady-state error*, that is, the error remaining after all transients have died out. The equation for the error function is

$$E(s) = \frac{\theta_e}{\theta_i} = 1 - B(s) = \frac{1}{1 + G(s)H(s)} \tag{1-14}$$

We will compute the steady-state error as a function of the various systems in Section 1-10 and make some predictions on loops of various orders. I have previously mentioned that we have classified the loops to be either analog or digital, referring to the phase/frequency detector, and in writing the initial equations, I have already indicated that the loop may or may not have a filter incorporated.

In analyzing the loop, we will find that to describe a loop, we can express it in terms of both the order of the loop and the type of loop. The expression "type of loop" refers to the number of integrators used. A PLL using no active integrator can really only be a type 1 loop. The term "order" refers to the order of the polynomial that is required to express the loop transfer characteristic.

The absolute minimum in a loop is a phase/frequency detector, a VCO, and no filter. By this definition, this would be a first-order type 1 loop. Another way of explaining this is by saying that the type of a system refers to the number of poles of the loop transfer function locked at the origin. The order of a system refers to the highest degree of the polynomial expression of the denominator that can be expanded into a polynomial expression. In this book we deal with loops of types 1 and 2 from first order up to fifth.

We have now learned that there are two classifications for loops:

1. Classification by phase detector, characterizing the loop to be analog or digital.
2. Characterizing the loop by the type of loop filter and number of integrators.

1-3 DIGITAL LOOPS

Using our previous definition, a digital PLL is a PLL system in which the phase/frequency comparator is built from digital components such as gates or flip-flops to form either an exclusive-OR gate, an edge-triggered JK master/slave flip-flop, or what I call a tri-state phase/frequency comparator. In addition, digital PLLs use frequency dividers, and although some circuits using the principle of subharmonic locking for dividers are known, this generally refers to the use of asynchronous or synchronous dividers. Asynchronous dividers are usually ripple counters, and synchronous dividers are counters that are being clocked by a common reset line.

The basic difference between an analog and a digital phase/frequency loop is in the possible delay introduced by the frequency divider and the nonlinear effects of the phase/frequency comparator, and the question of ultimate resolution of the phase/frequency comparator. The phase/frequency comparator using active filters shows some highly nonlinear performance during zero crossings at the output or under perfectly locked conditions. As there is no output from a tri-state phase/frequency comparator under locked conditions, the gain of the loop is zero until there is a requirement to send correcting pulses from the digital phase/frequency comparator, which then results in a jump of loop gain.

We will prove that digital phase/frequency comparators, especially tri-state phase/frequency comparators, have two ranges for acquisition. One is called *pull-in* range, and the other is called *lock-in* range. The acquisition time is the time for both. This total time to acquire both frequency and phase lock is sometimes called *capture time* or *digital acquisition time*. Depending on the loop filter and the phase/frequency comparator, we will have different time constants.

For reasons of convenience and linearity, we have, so far, assumed that the loop is in locked condition. Initially, when the loop is switched on for the first time, it is far from being locked, and the VCO frequency can be anywhere within tuning range. *Tuning range* is defined as the frequency range over which the voltage-controlled oscillator (VCO) can be tuned with the available control voltage.

There are, however, limitations because of the tuning diodes. The minimum voltage that can be applied is determined by the threshold voltage of the diode itself before it becomes conductive. This voltage is typically $0.7\,\text{V}$, and the maximum voltage is the voltage determined by the breakdown voltage of the tuning diode. Even in the case where the familiar back-to-back diode arrangement is used, these are the two limits for the voltage range. In practice, however, this range is even more narrow because the voltage sensitivity of the tuning diode is excessive at the very low end and very small at the extreme high end. Even before the breakdown voltage is reached, the noise contribution from the diode already increases because of some zener effects.

As the loop currently is not in locked condition, we have to help it to acquire lock. Very few loops acquire locking by themselves, a process called *self-acquisition*. Generally, the tuning range is larger than the acquisition range. Self-acquisition is a slow, unreliable process. If the loop is closed for the first time, the process called "pull-in" will occur. The oscillator frequency, together with the reference frequency, will generate a beat note and a dc control voltage of such phases that the VCO is pulled in a direction of frequency lock. As the oscillator itself generates noise in the form of a residual FM, the oscillator is constantly trying to break out of lock, and the loop is constantly monitoring the state and reassuring lock. This results, under normal circumstances, in constant charging and discharging of the holding capacitor responsible for the averaging process.

There is one other phase/frequency comparator that is really more a switch than anything else; it is called a *sample/hold comparator*. The sample/hold comparator, which we will deal with later, has the advantage of very good reference frequency suppression, introduces a phase shift that reduces the phase margin, and is really useful only up to several hundred kilohertz of frequency. For frequencies higher than this, there is too much leakage. The sample/hold comparator, which has been very popular for several years, is described in Chapter 4. Modern frequency synthesizers, however, prefer digital phase/frequency comparators because the sample/hold comparator is only a phase comparator and does not recognize frequency offsets. It is too slow to be used for harmonic sampling and, in my opinion, has only limited use.

The sample/hold comparator is used mostly with T networks for additional reference suppression, and although these circuits provide good reference suppression, the phase margin has to be so high that the loops are generally slow in their response.

The switching speed of the loop and its general performance to noise are covered in detail in Section 1-10.

Now let us take a look at a numerical example. Consider a frequency synthesizer using a PLL to synthesize a 1-MHz signal from a 25-kHz reference frequency. To realize an output frequency of 1 MHz, a division of 1 MHz/25 kHz = 40 is necessary.

Let us assume that there is no filtering included, and therefore the closed-loop transfer function will be

$$B(s) = \frac{K}{s + K/N} \qquad (1\text{-}15)$$

A typical value for K_θ is 2 V/rad and a typical value for the VCO gain factor K_o is 1000 Hz/V. With these values the closed-loop transfer function is

$$B(s) = \frac{K_\theta K_o}{s + K_\theta K_o/N} = \frac{1000 \times 2\pi \times 2}{s + [(1000 \times 2\pi \times 2)/40]} = \frac{4000\pi}{s + 100\pi} \qquad (1\text{-}16)$$

The 3-dB frequency of the system by definition is

$$\omega_{3db} = \frac{K_\theta K_o}{N} \qquad (1\text{-}17)$$

and therefore

$$\omega_n = 100\pi$$

If this is solved to determine f, we obtain $f = 50$ Hz.

As the reference frequency is 25 kHz, the reference suppression of the simple system can be determined from

$$A = 20 \log_{10}\left(\frac{25{,}000}{50}\right) = 20 \log_{10}(500)$$
$$= 54 \text{ dB}$$

The loop bandwidth of the system by itself is 50 Hz.

We have, with very little effort, calculated a first-order type 1 loop. We deal more with these loops in Chapter 2.

Table 1-2 shows the input waveforms and the output average voltage of:

1. A four-quadrant multiplier or double-balanced mixer being driven either by a sine wave or a square wave.
2. The input and output voltages for an exclusive-OR gate.
3. The input and output voltages after the integrator of an edge-triggered JK master/slave flip-flop.
4. The input and output voltages of a tri-state phase/frequency comparator after the integrator. Notice that the extended operating range is linear from -2π to $+2\pi$.

1-4 TYPE 1 FIRST-ORDER LOOPS

The type 1 first-order loop contains a digital phase/frequency comparator and a digital divider, and throughout the rest of this book we will deal only with digital loops. This is done on the assumption that the phase/frequency comparator can be modeled as a linear device over the operating range, which is certainly not true. However, as most of the formulas and deviations that deal with PLLs have certain assumptions that have finite accuracy, it is permissible to do so. I realize that a purist will be offended by this statement. However, as many PLLs have been designed by rule of thumb and the final results were within a few percent accurate, I assume that the reader will permit this simplification.

The analog loop, as mentioned before, does not provide enough dc output, and for reasons of sideband noise, there is no advantage in using the linear loop in digital frequency synthesizers. Therefore, the phase-locked loop without a loop filter, $F(s) = 1$, is called a type 1 first-order loop because it has only one integrator, and the highest power (s) in the denominator of the system transfer function is 1.

The open-loop gain of a type 1 first-order PLL is equal to the forward gain divided by N, $(K_\theta K_o/s)/N$. The transfer function is

$$B(s) = \frac{N}{1 + s[1/(K_\theta K_o/N)]} \tag{1-18}$$

The loop noise bandwidth can be determined from the integration to be

$$B_n = \frac{K_\theta K_o}{4N} \tag{1-19}$$

Table 1-2 Circuit diagrams and input and output waveforms of various phase/frequency comparators[a]

Input signals	Circuit	$V_{out} = f(\theta)$
v_1, v_2 sine waves	Four-quadrant-multiplier	sinusoidal, zeros at $-\pi, 0, \pi$; peaks at $\pm\pi/2$
v_1, v_2 square waves	—	triangular, range $-\pi$ to π
v_1, v_2, Q pulse trains	Exclusive OR	triangular, range $-\pi$ to π
v_1, v_2, Q pulse trains	JK master/slave FF, Q = up, $\bar{Q}$ = down	sawtooth, range -2π to 2π
Case 1: v_1, v_2, Up, Down; Case 2: v_1, v_2, Up, Down	G_1, G_2, G_3, G_4 with S-R FFs; Up and Down outputs	sawtooth, range -2π to 2π

[a]Courtesy of Fachschriftenverlag, Aargauer Tagblatt AG, Aarau, Switzerland.

It should be noted that the noise bandwidth changes as a function of the division ratio N. If there is a large change in the division ratio, the noise bandwidth will change substantially. This is another reason why the type 1 first-order loop is not very popular.

Let us assume that the loop is not in locked condition. The phase/frequency comparator receives two different frequencies at the two inputs. For digital phase/frequency comparators we will use an exclusive-OR gate, which requires an active integrator.

Because of the limits of the operating range, presteering is required for the VCO to be within the range $-\pi$ to $+\pi$ for a locking condition. It is known that the maximum difference between the VCO free-running frequency and the desired final frequency at which phase lock is possible can be equal to

$$\Delta\omega_{\text{capture}} = \Delta\omega_H = \frac{K_\theta K_o}{N} \quad \text{rad/s} \tag{1-20}$$

It is important to keep the steady-state phase error small; therefore, high dc loop gain is required. As the increase in loop gain would require an amplifier, this makes the loop noisy and eventually unstable.

I have explained previously that the VCO gain is limited because of the tuning range of the diodes, and we will learn later that it is desirable to keep the VCO gain as low as possible. The input line to the VCO is a high-impedance line, and pick-up on this line will result in spurious output at the VCO. To keep the spurious frequencies small, the VCO gain must be kept as small as possible.

The phase detector sensitivity depends mainly on the operating voltage. We recall that the double-balanced modulator with four diodes supplies only several hundred millivolts; instead, the exclusive-OR gate was chosen, as this can be operated from 12 V or even higher if CMOS logic is used.

There are several estimates regarding the acquisition time for the type 1 first-order PLL. Using the exclusive-OR gate, the acquisition time is approximately

$$T_A = \frac{2}{K_\theta K_o/N} \ln \frac{2}{\theta_\epsilon} \tag{1-21}$$

where ln refers to the natural logarithm and θ_ϵ refers to the final phase error in radians.

From our previous example, the acquisition time T_A would be determined to be

$$T_A = \frac{2}{50} \ln \frac{2}{0.2}$$
$$= 9.2 \times 10^{-2} \, \text{s}$$

Assuming that the initial offset was less than 50 Hz, the loop locks in frequency without skipping cycles. In practice, it is impossible to presteer the loop within 50 Hz; therefore, this formula has only limited use.

We have just determined the acquisition time, but we have to ask ourselves: What does it really mean? Does it mean that the frequency from the initial offset is now the same as the reference frequency? Does it mean that we have reached a certain percentage of final frequency or gotten very close to the final frequency or final phase? Will we ever reach the final value, or is there a residual error?

14 LOOP FUNDAMENTALS

The *error function*, which we met earlier, provides us with information and insight.

$$E(s) = \frac{\theta_\epsilon}{\theta_i} = 1 - B(s) = \frac{1}{1 - G(s)H(s)} \tag{1-14}$$

It has been shown that, depending on the type of change of input, we get different results. These results are determined by the use of a transformation from the frequency into the time domain. Section A-4 of Appendix A presents the mathematical background for the Laplace transform and discusses how it is applied. Here, we use only the results. Inserting the known factors into Eq. (1-14), we obtain

$$E(s) = \frac{s\theta_\epsilon(s)}{s + K} \tag{1-22}$$

with $K = K_\theta K_o/N$. It is customary to analyze the performance of the loop for three different conditions:

1. To apply a step to the input and see what the output response is.
2. A ramp voltage.
3. A parabolic input.

Case 1, the step input, means an instantaneous jump with zero rise time to which the output will respond with a delay. The steady-state phase error resulting from this step change of input phase of magnitude $\Delta\theta$ for a ramp,

$$\theta(s) = \frac{\Delta\theta}{s} \tag{1-23}$$

is

$$\epsilon = \lim_{s \to 0} \frac{s\,\Delta\theta}{s + K} = 0 \tag{1-24}$$

Case 2, the steady-state error resulting from a ramp of input phase that is the same as a step change in reference frequency in the amount $\Delta\omega [\theta(s) \times \Delta\omega/s^2]$, is

$$\epsilon = \lim_{s \to 0} \frac{\Delta\omega}{s + K} = \frac{\Delta\omega}{K} \tag{1-25}$$

It is apparent from those two equations that after a certain time, a type 1 first-order loop will track out any step change in input phase within the system hold-in range and will follow a step change in frequency with a phase error that is proportional to the magnitude of the frequency steps and inversely proportional to the dc loop gain. The loop will show the same performance if phase or frequency of the VCO changes rather than the reference.

In case 3, for the type 1 first-order PLL, it is of interest to examine a ramp change in frequency, the case in which the reference frequency is linearly changed with a time rate of $\Delta\omega/dt$ rad/s^2 and $\theta(s) = (2\Delta\omega/dt)/s^3$. Why is this so important? Let us assume that we sweep a PLL at a constant rate, as is done in some modern spectrum analyzers, and we have to find the final condition for the steady-state phase error. The final phase error is

$$\epsilon = \lim_{s \to 0} \frac{2\Delta\omega/dt}{s^2 + sK} = \infty \tag{1-26}$$

What does this mean for us? It means that, as there is no infinite value for K in a type 1 first-order loop, this loop is not very attractive for tracking, and it also means that above a certain and critical rate of change of reference frequency or VCO frequency, the loop will no longer stay in locked condition. Therefore, if the loop is swept above a certain rate, it will not maintain lock.

1-5 TYPE 1 SECOND-ORDER LOOPS (see Figure 1-5)

If we insert in our PLL a simple low-pass filter

$$F(s) = \frac{1}{\tau_s + 1} \tag{1-27}$$

the closed-loop transfer function using

$$K = \frac{K_o K_\theta}{N}$$

is
$$\tag{1-28}$$

$$B(s) = \frac{\theta_o(s)}{\theta_i(s)} = \frac{NK}{s(\tau s + 1) + K} = \frac{N}{(s^2/\omega_n^2) + (2\zeta/\omega_n)s + 1}$$

where

$$\omega_n = \sqrt{\frac{K}{\tau}} \tag{1-29}$$

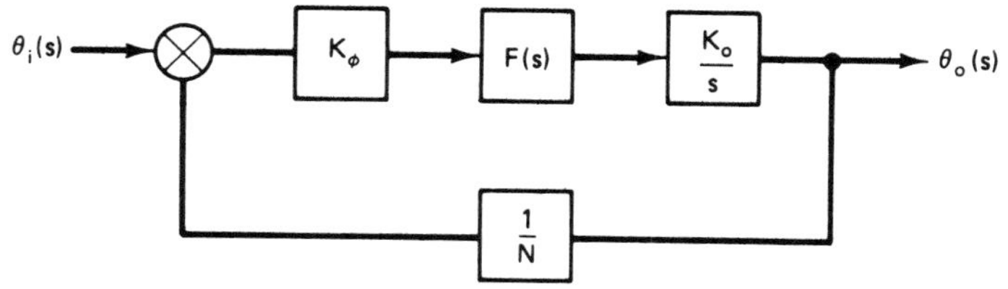

Figure 1-5 Block diagram of the second-order PLL.

16   LOOP FUNDAMENTALS

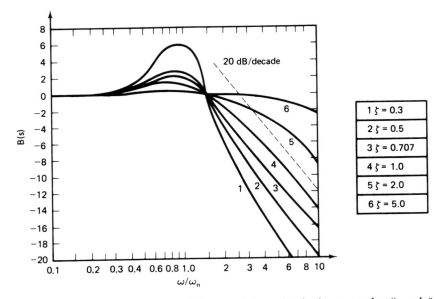

Figure 1-6 Frequency response of the type 1 second-order loop as a function of ζ.

and

$$2\zeta = \frac{\omega_n}{K} = \sqrt{\frac{1}{\tau K}} \qquad (1\text{-}30)$$

The magnitude of the steady-state frequency response is

$$|B(s)| = \left|\frac{\theta_o}{\theta_i}(j\omega)\right| = \frac{1}{[(1 - \omega^2/\omega_n^2)^2 + (2\zeta\omega/\omega_n)^2]^{1/2}} \qquad (1\text{-}31)$$

and the phase shift is

$$\arg\frac{\theta_o}{\theta_i}(j\omega) = \arctan\frac{2\zeta\omega}{\omega_n(1 - \omega^2/\omega_n^2)^2} \qquad (1\text{-}32)$$

The frequency response of this second-order transfer function determined in Eq. (1-28) is plotted in Figure 1-6 for selected values of ζ. For $\zeta = 0.707$, the transfer function becomes the second-order "maximally flat" Butterworth response. For values of $\zeta < 0.707$, the gain exhibits peaking in the frequency domain. The maximum value of the frequency response M_p can be found by differentiating the magnitude of Eq. (1-28) (with $s = j\omega$). M_p is found to be

$$M_p = \frac{1}{2\zeta\sqrt{1 - \zeta^2}} \qquad (1\text{-}33)$$

and the frequency ω_p at which the maximum occurs is

$$\omega_p = \omega_n \sqrt{1 - 2\zeta^2} \qquad (1\text{-}34)$$

The 3-dB bandwidth B can be derived by solving for the frequency ω_h at which the magnitude of Eq. (1-28) (with $s = j\omega$) is equal to 0.707. B is found to be

$$B = \omega_n(1 - 2\zeta^2 + \sqrt{2 - 4\zeta^2 + 4\zeta^4})^{1/2} \qquad (1\text{-}35)$$

The time it takes for the output to rise from 10% to 90% of its final value is called the *rise time* t_r. Rise time is approximately related to the system bandwidth by the relation

$$t_r = \frac{2.2}{B} \qquad (1\text{-}36)$$

which is exact only for the first-order system described by Eq. (1-5).

The error signal θ_ϵ, defined as $\theta_i - \theta_o$, can be expressed (in unity feedback systems) as

$$\theta_\epsilon(s) = \frac{\theta_i(s)}{1 + KG(s)} = \frac{\theta_i(s)}{1 + KF(s)/s} \qquad (1\text{-}37)$$

If the system is stable, the steady-state error for polynomial inputs $\theta_i(t) = t^n$ can be obtained from the final value theorem,

$$\lim_{t \to \infty} \theta_\epsilon(t) = \lim_{s \to 0} s\theta_\epsilon(s)$$
$$= \lim_{s \to 0} \frac{2\theta_i}{KF(s)} \qquad (1\text{-}38)$$

If $\theta_i(t)$ is a step function representing a sudden increase in phase, $\theta_i(s) = 1/s$ and

$$\lim_{t \to \infty} \theta_\epsilon(t) = \lim_{s \to 0} \frac{s}{KF(s)} \qquad (1\text{-}39)$$

$F(s)$ is either a constant or a low-pass filter that may include poles at the origin. That is,

$$\lim_{s \to 0} F(s) = \frac{K^*}{s^n} \neq 0 \qquad (1\text{-}40)$$

Therefore, Eq. (1-39) can be written

$$\lim_{t \to \infty} \theta_\epsilon(t) = \lim_{s \to 0} \frac{s^{n+1} K^*}{KK^*} = 0 \qquad (1\text{-}41)$$

That is, a PLL will track step changes in phase with zero steady-state error.

18 LOOP FUNDAMENTALS

If there is a constant-amplitude change in the input frequency of A rad/s,

$$\theta_i(s) = \frac{A}{s^2} \tag{1-42}$$

Equation (1-39) becomes

$$\lim_{t \to \infty} \theta_\epsilon(t) = \lim_{s \to 0} \frac{A}{KF(s)} = \frac{A}{KF(0)} \tag{1-43}$$

If $F(0) = 1$, the steady-state phase error will be inversely proportional to the loop gain K. Recall that the larger the K is, the larger will be the closed-loop bandwidth and thus the faster the loop response. To increase the response speed and reduce the tracking error, the loop gain should be as large as possible. If $F(0)$ is finite, there will be a finite steady-state phase error. The frequency error,

$$f_\epsilon(t) = \frac{d}{dt} \theta_\epsilon(t) \tag{1-44}$$

will be zero in the steady state. That is, the input and VCO frequencies will be equal ($\omega_i = \omega_o$).

Table 1-3 shows the popular loop filters for the type 2 second-order loop. We have now dealt with case 1, a simple RC filter. The performance obtained with this loop filter is relatively restricted, mainly because the advantage over the loop with no filter was that we only got one additional parameter, a time constant τ_1.

Let us look at the table and the various filters. The passive filter type 2 uses two resistors and one capacitor, which allows compensation of phase.

The active filter, with which we will be dealing shortly, will add an additional integrator and therefore change this loop from a type 1 second-order to a type 2 second-order. The second-order loops we are currently dealing with are of type 1 because there is only one integrator involved, the VCO.

As we have only the time constant available as the additional parameter, which as we saw previously determines both the natural loop frequency ω_n and the damping factor ζ, we have not made much progress toward improving the loop and choosing independent parameters. If we add a resistor in series with the capacitor and obtain the loop filter shown in Table 1-3, type 2, the transfer function $F(s)$ is

$$F(s) = \frac{1 + \tau_2 s}{1 + \tau_1 s} \tag{1-45}$$

or as it is sometimes defined,

$$F(s) = \frac{1 + j\omega\tau_2}{1 + j\omega(\tau_1 + \tau_2)} \tag{1-46}$$

Table 1-3 Circuit and transfer characteristics of several PLL filters

Type	Passive		Active	
	1	2	3	4
Circuit	R₁—C to ground	R₁—R₂—C	Op-amp with R₁ input, R₂ and C feedback	Op-amp with R₁ input, C feedback
Transfer characteristic $F(j\omega) =$	$\dfrac{1}{1+j\omega\tau_1}$	$\dfrac{1+j\omega\tau_2}{1+j\omega(\tau_1+\tau_2)}$	$\dfrac{1+j\omega\tau_2}{j\omega\tau_1}$	$\dfrac{1}{j\omega\tau_1}$

$\tau_1 = R_1 C \qquad \tau_2 = R_2 C$

20 LOOP FUNDAMENTALS

What is the difference? In the first case, we use the abbreviations

$$\tau_1 = (R_1 + R_2)C \quad (1\text{-}47)$$

and

$$\tau_2 = R_2 C \quad (1\text{-}48)$$

whereas in the second case, and as listed in Table 1-3,

$$\tau_1 = R_1 C \quad (1\text{-}49)$$

and

$$\tau_2 = R_2 C \quad (1\text{-}50)$$

This fact should be pointed out, as it may cause confusion to the reader. This results in the transfer function of the type 1 second-order PLL using the first-case definition,

$$B(s) = \frac{K(1 + \tau_2 s/1 + \tau_1 s)}{s + K(1 + \tau_2 s/1 + \tau_1 s)} \\ = \frac{K(1/\tau_1)(1 + \tau_2 s)}{s^2 + (1/\tau_1)(1 + K\tau_2)s + (K/\tau_1)} \quad (1\text{-}51)$$

To be consistent with our previous abbreviations, we now insert the terms of the loop damping factor ζ and the natural frequency ω_n and obtain

$$B(s) = \frac{s\omega_n[2\zeta - \omega_n/K] + \omega_n^2}{s^2 + 2\zeta\omega_n s + \omega_n^2} \quad (1\text{-}52)$$

where

$$\omega_n = \sqrt{\frac{K}{\tau_1}} \quad \text{rad/s} \quad (1\text{-}53)$$

and

$$\zeta = \frac{1}{2}\sqrt{\frac{1}{\tau_1 K}}(1 + \tau_2 K) \quad (1\text{-}54)$$

We remember our abbreviation used previously, $K = K_\theta K_o/N$. The magnitude of the transfer function of the phase-lag filter magnitude is

$$|F(j\omega)| = \sqrt{\frac{1 + (\omega R_2 C)^2}{1 + [\omega C(R_1 + R_2)]^2}} \quad (1\text{-}55)$$

and the phase is

$$\theta(j\omega) = \arctan(\omega\tau_2) - \arctan(\omega\tau_1) \quad \text{degrees}$$

When we use the other definition of τ_1 and τ_2 and insert the abbreviations

$$\omega_n = \sqrt{\frac{K}{\tau_1 + \tau_2}} \tag{1-56}$$

$$\zeta = \frac{1}{2}\sqrt{\frac{K}{\tau_1 + \tau_2}}\left(\tau_2 + \frac{1}{K}\right) \tag{1-57}$$

we obtain the expression

$$B(s) = \frac{s\omega_n(2\zeta - \omega_n/K) + \omega_n^2}{s^2 + 2\zeta(\omega_n^s + \omega_n^2)} \tag{1-58}$$

which turns out to give the same result.

As we were interested in the 3-dB bandwidth of the type 1 first-order loop, we now determine the 3-dB bandwidth of the type 1 second-order loop to be

$$B_{3\text{dB}} = \frac{\omega_n}{2\pi}(a + \sqrt{a^2 + 1})^{1/2} \quad \text{Hz} \tag{1-59}$$

with the substitution

$$a = 2\zeta^2 + 1 - \frac{\omega_n}{K}\left(4\zeta - \frac{\omega_n}{K}\right) \tag{1-60}$$

The noise bandwidth of the type 1 second-order loop is

$$B_n = \frac{\omega_n}{2}\left(\zeta + \frac{1}{4\zeta}\right) \quad \text{Hz} \tag{1-61}$$

Again, we are interested in the final phase error and information we can gain from the phase error function

$$E(s) = \frac{s(1 + \tau_1 s)\theta(s)}{\tau_1 s^2 + (1 + K\tau_2)s + K} \tag{1-62}$$

As we are still dealing with a type 1 system, we obtain zero steady phase error, for a step in phase, and constant error for a ramp input in phase. One of these cases is shown in Figure 1-7, where the transient phase error due to a step in phase has been plotted with a computer.

For the loop to stay in lock, the following critical values have to be considered. The maximum rate of change of reference frequency $d\Delta\omega/dt$ should satisfy the equation

$$\left(\frac{d\Delta\omega}{dt}\right)_{\text{max}} = \omega_n^2 \tag{1-63}$$

22 LOOP FUNDAMENTALS

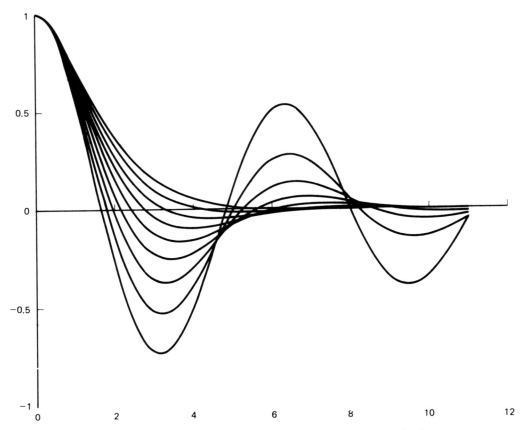

Figure 1-7 Transient phase step response for type 1 second-order loop.

The maximum rate at which the VCO can be swept must satisfy the condition

$$\left(\frac{d\,\Delta\omega}{dt}\right)_{\max} < \frac{\omega_n^2}{2}$$

to achieve lock. The hold-in range of this loop is

$$\Delta\omega_H = K \quad \text{rad/s} \tag{1-64}$$

and the capture range is

$$\Delta\omega_C = K\left(\frac{\tau_1}{\tau_2}\right) \quad \text{rad/s} \tag{1-65}$$

An approximate time required for this type 1 second-order loop to obtain frequency lock is

$$T_{\text{acq}} \approx \frac{4(\Delta f)^2}{B_n^3} \quad \text{seconds} \tag{1-66}$$

TYPE 1 SECOND-ORDER LOOPS

A simple numerical example may give some additional insight. Let us assume that we have an oscillator operating at 45 MHz, using a reference frequency of 1 kHz. The tuning range of the oscillator is 5 MHz, which means that the frequency of the VCO in the beginning can be either at 40 or 50 MHz, as an extreme value. If we take the Δf offset worst-cast condition of 5 MHz, and to get adequate reference suppression we use a 50-Hz natural loop frequency and a damping factor ζ of 0.7, the noise bandwidth is

$$B_n = \frac{50}{2}\left(0.7 + \frac{1}{0.7 \times 4}\right)$$
$$= 25(0.7 + 0.3571)$$
$$= 25 \times 1.0571$$
$$= 26.4 \text{ Hz}$$

For a 5-MHz offset,

$$T_{acq} = \frac{4(5 \times 10^6)^2}{26.4^3} = 5.4 \times 10^9 \text{ s}$$

This is a ridiculous value. Unless the loop gets acquisition help, we would lose patience before it has ever acquired lock. Let us assume for a moment that we have a Δf of only 1 kHz or one step size. This still results in an acquisition time of 380 s, much too long for practical values. In Section 1-10 we will learn some acquisition aids that can speed up the otherwise lengthy procedure. For now it is sufficient to know that in cases where we do not use a phase/frequency comparator, it generates a beat note that is capable of switching and, therefore, aiding acquisition. Externally switched oscillators can be used, or automatic circuits can be incorporated that change the loop bandwidth before acquisition occurs and therefore can speed up the circuit substantially.

We end our discussion of the type 1 second-order loop here and concentrate on the more popular type 2 second-order loop. Why is the type 2 second-order loop so much more popular?

We found out previously that the freedom of choice of parameters was limited.

1. In the type 1 first-order loop with no filter, K determined everything.
2. In the type 1 second-order loop, we had one time constant (τ_1) available, which restricted us in the choice of ω_n and ζ, as these values were related. The type 1 second-order loop has finite dc gain and therefore it is questionable whether the term "PLL" is really justified. By this definition we really should not call it a true PLL system because, from the assumptions made previously, zero phase error requires infinite dc gain.

How do we accomplish the infinite dc gain, and how do we accomplish zero phase error?

24 LOOP FUNDAMENTALS

If it is necessary to have zero phase error in response to step changes in the input frequency, $\lim_{s \to 0} F(s)$ must be infinite. That is, the dc gain of the low-pass filter must be infinite. This can be realized by including in $F(s)$ a pole at the origin. In this case $F(s)$ will be of the form

$$F(s) = \frac{1}{s} \frac{\tau_2 s + 1}{\tau_1} \tag{1-67}$$

The addition of the pole at the origin creates difficulties with the loop stability. In fact, the system will now be unstable unless a lead network is included in $F(s)$. With a passive filter, therefore resulting in a type 1 second-order loop, the condition is generally that we start with specified values for ω_n and ζ and want to determine the time constants τ_1 and τ_2. This has rarely been shown in the literature, and for those interested, here is the result. We start off with $K_\theta K_o/N = K$ and

$$\omega_n = \sqrt{\frac{K}{\tau_1 + \tau_2}} \tag{1-56}$$

$$\zeta = \frac{1}{2} \left(K + \frac{1}{\tau_1 + \tau_2} \right)^{1/2} \left(\tau_2 + \frac{1}{K} \right) \tag{1-68}$$

By squaring ζ and inserting the value for ω_n, after some manipulation we obtain

$$\tau_2^2 K^2 + \tau_2 2K + 1 - 4\zeta^2 \frac{K^2}{\omega_n^2} = 0 \tag{1-69}$$

This equation can be solved with

$$\tau_1 = \frac{K}{\omega_n^2} - \tau_2 \tag{1-70}$$

$$\tau_2 = \frac{-2K + \sqrt{4K^2 - 4K^2[1 - 4\zeta^2(K^2/\omega_n^2)]}}{2K^2} \tag{1-71}$$

and

$$R_1 = \frac{\tau_1}{C} \tag{1-72}$$

and

$$R_2 = \frac{\tau_2}{C} \tag{1-73}$$

1-6 TYPE 2 SECOND-ORDER LOOP

The type 2 second-order loop uses a loop filter in the form

$$F(s) = \frac{1}{s} \frac{\tau_2 s + 1}{\tau_1} \tag{1-67}$$

The multiplier 1/s indicates a second integrator, which is generated by the active amplifier. In Table 1-3, this is the type 3 filter. The type 4 filter is mentioned there as a possible configuration but is not recommended because, as stated previously, the addition of the pole of the origin creates difficulties with loop stability and, in most cases, requires a change from the type 4 to the type 3 filter. One can consider the type 4 filter as a special case of the type 3 filter, and therefore it does not have to be treated separately. Another possible transfer function is

$$F(s) = \frac{1}{R_1 C} \frac{1 + \tau_2 s}{s} \tag{1-74}$$

with

$$\tau_2 = R_2 C \tag{1-50}$$

Under these conditions, the magnitude of the transfer function is

$$|F(j\omega)| = \frac{1}{R_1 C \omega} \sqrt{1 + (\omega R_2 C)^2} \tag{1-75}$$

and the phase is

$$\theta = \arctan(\omega \tau_2) - 90 \text{ degrees}$$

Again, as if for a practical case, we start off with the design values ω_n and ζ, and we have to determine τ_1 and τ_2. Taking an approach similar to that for the type 1 second-order loop, the results are

$$\tau_1 = \frac{K}{\omega_n} \tag{1-}$$

and

$$\tau_2 = \frac{2\zeta}{\omega_n}$$

and

$$R_1 = \frac{\tau_1}{C}$$

and

$$R_2 = \frac{\tau_2}{C}$$

The closed-loop transfer function of a type 2 second-order integrator is

$$B(s) = \frac{K(R_2/R_1)[s + (1/\tau_2)]}{s^2 + K(R_2/R_1)s + (K/\tau_2)(R_2/}$$

LOOP FUNDAMENTALS

By introducing the terms ζ and ω_n, the transfer function now becomes

$$B(s) = \frac{2\zeta\omega_n s + \omega_n^2}{s^2 + 2\zeta\omega_n s + \omega_n^2} \qquad (1\text{-}79)$$

with the abbreviations

$$\omega_n = \left(\frac{K}{\tau_2}\frac{R_2}{R_1}\right)^{1/2} \quad \text{rad/s} \qquad (1\text{-}80)$$

and

$$\zeta = \frac{1}{2}\left(K\tau_2\frac{R_2}{R_1}\right)^{1/2} \qquad (1\text{-}81)$$

and $K = K_\theta K_o/N$.

The 3-dB bandwidth of the type 2 second-order loop is

$$B_{3\text{dB}} = \frac{\omega_n}{2\pi}[2\zeta^2 + 1 + \sqrt{(2\zeta^2 + 1)^2 + 1}]^{1/2} \quad \text{Hz} \qquad (1\text{-}82)$$

and the noise bandwidth is

$$B_n = \frac{K(R_2/R_1) + 1/\tau_2}{4} \quad \text{Hz} \qquad (1\text{-}83)$$

Again, we ask the question of the final error and use the previous error function,

$$E(s) = \frac{s\theta(s)}{s + K(R_2/R_1)\{[s + (1/\tau_2)]/s\}} \qquad (1\text{-}84)$$

or

$$E(s) = \frac{s^2\theta(s)}{s^2 + K(R_2/R_1)s + (K/\tau_2)(R_2/R_1)} \qquad (1\text{-}85)$$

esult of the perfect integrator, the steady-state error resulting from a step in input phase or change of magnitude of frequency is zero.
input frequency is swept with a constant range changae of input frequency or $\theta(s) = (2\Delta\omega/dt)/s^3$, the steady-state phase error is

$$E(s) = \frac{R_1}{R_2}\frac{\tau_2(2\Delta\omega/dt)}{K} \quad \text{rad} \qquad (1\text{-}86)$$

The maximum rate at which the VCO frequency can be swept for maintaining lock is

$$\frac{2\Delta\omega}{dt} = \frac{N}{2\tau_2}\left(4B_n - \frac{1}{\tau_2}\right) \quad \text{rad/s} \tag{1-87}$$

The introduction of N indicates that this is referred to the VCO rather than to the phase/frequency comparator. In the previous example of the type 1 first-order loop, we referred it only to the phase/frequency comparator rather than the VCO.

1-6-1 Transient Behavior of Digital Loops Using Tri-state Phase Detectors

Pull-in Characteristic. The type 2 second-order loop is used with either a sample/hold comparator or a tri-state phase/frequency comparator.

We will now determine the transient behavior of this loop. Figure 1-8 shows the block diagram.

Very rarely in literature is a clear distinction between pull-in and lock-in characteristics or frequency and phase acquisition made as a function of the digital phase/frequency detector. Somehow, all the approximations or linearizations refer to a sinusoidal phase/frequency comparator or its digital equivalent, the exclusive-OR gate.

The tri-state phase/frequency comparator, which seems to be the most popular one and will be explored in greater detail in Chapter 4, follows slightly different mathematical principles.

The phase detector gain

$$K'_d = \frac{V_d}{\omega_0} = \frac{\text{phase detector supply voltage}}{\text{loop idling frequency}}$$

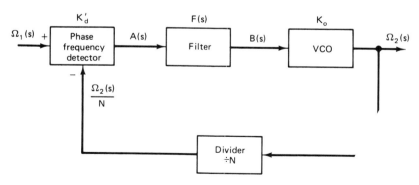

Note: The frequency transfer const. of the VCO = K_o
(not $\dfrac{K_o}{s}$, which is valid for phase transfer only.)

Figure 1-8 Block diagram of a digital PLL before lock

is explained fully in Chapter 4, and we only use the result here. This phase detector gain is valid only in the out-of-lock state and is a somewhat coarse approximation to the real gain, which, due to nonlinear differential equations, is very difficult to calculate. However, practical tests show that this approximation is still fairly accurate.

Definitions:

$$\Omega_1(s) = \mathcal{L}[\Delta\omega_1(t)] \quad \text{Reference input to } \delta/\omega \text{ detector}$$
$$\Omega_2(s) = \mathcal{L}[\Delta\omega_2(t)] \quad \text{Signal VCO output frequency}$$
$$\Omega_e(s) = \mathcal{L}[\omega_e(t)] \quad \text{Error frequency at } \delta/\omega \text{ detector}$$

$$\Omega_e(s) = \Omega_1(s) - \frac{\Omega_2(s)}{N}$$

$$\Omega_2(s) = [\Omega_1(s) - \Omega_e(s)]N$$

From the circuit above,

$$A(s) = \Omega_e(s) K'_d$$
$$B(s) = A(s) F(s)$$
$$\Omega_2(s) = B(s) K_o$$

The error frequency at the detector is

$$\Omega_e(s) = \Omega_1(s) N \frac{1}{N + K_o K'_d F(s)} \tag{1-88}$$

The signal is stepped in frequency:

$$\Omega_1(s) = \frac{\Delta\omega_1}{s} \quad (\Delta\omega_1 = \text{magnitude of frequency step}) \tag{1-89}$$

Active Filter of First Order. If we use an active filter

$$F(s) = \frac{1 + s\tau_2}{s\tau_1} \tag{1-90}$$

d insert this in Eq. (1-88), the error frequency is

$$\Omega_e(s) = \Delta\omega_1 N \frac{1}{s\left(N + K_o K'_d \frac{\tau_2}{\tau_1}\right) + \frac{K_o K'_d}{\tau_1}} \tag{1-91}$$

the Laplase transformation, we obtain

$$\iota_e(t) = \Delta\omega_1 \frac{1}{1 + K_o K'_d(\tau_2/\tau_1)(1/N)} \exp\left[-\frac{t}{(\tau_1 N/K_o K'_d) + \tau_2}\right] \tag{1-92}$$

and

$$\lim_{t \to 0} \omega_e(t) = \frac{\Delta\omega_1 N}{N + K_o K'_d(\tau_2/\tau_1)} \qquad (1\text{-}93)$$

$$\lim_{t \to \infty} \omega_e(t) = 0 \qquad (1\text{-}94)$$

Passive Filter of First Order. If we use a passive filter

$$F(s) = \frac{1 + s\tau_2}{1 + s(\tau_1 + \tau_2)} \qquad (1\text{-}95)$$

for the frequency step

$$\Omega_1(s) = \frac{\Delta\omega_1}{s} \qquad (1\text{-}96)$$

the error frequency at the input becomes

$$\Omega_e(s) = \Delta\omega_1 N \left\{ \frac{1}{s} \frac{1}{s[N(\tau_1 + \tau_2) + K_o K'_d \tau_2] + (N + K_o K'_d)} + \frac{\tau_1 + \tau_2}{s[N(\tau_1 + \tau_2) + K_o K'_d \tau_2] + (N + K_o K'_d)} \right\} \qquad (1\text{-}97)$$

For the first term we will use the abbreviation A, and for the second term we will use the abbreviation B.

$$A = \frac{1/[N(\tau_1 + \tau_2) + K_o K'_d \tau_2]}{s \left[s + \dfrac{N + K_o K'_d}{N(\tau_1 + \tau_2) + K_o K'_d \tau_2} \right]} \qquad (1\text{-}98)$$

$$B = \frac{\dfrac{\tau_1 + \tau_2}{N(\tau_1 + \tau_2) + K_o K'_d \tau_2}}{s + \dfrac{N + K_o K'_d}{N(\tau_1 + \tau_2) + K_o K'_d \tau_2}} \qquad (1\text{-}99)$$

After the inverse Laplace transformation, our final result becomes

$$\mathscr{L}^{-1}(A) = \frac{1}{N + K_o K'_d} \left\{ 1 - \exp\left[-t \frac{N + K_o K'_d}{N(\tau_1 + \tau_2) + K_o K'_d \tau_2} \right] \right\} \qquad (1\text{-}100)$$

$$\mathscr{L}^{-1}(B) = \frac{\tau_1 + \tau_2}{N(\tau_1 + \tau_2) + K_o K'_d \tau_2} \exp\left(-t \frac{N + K_o K'_d}{N(\tau_1 + \tau_2) + K_o K'_d \tau_2} \right) \qquad (1\text{-}101)$$

and finally

$$\omega_e(t) = \Delta\omega_1 N[\mathscr{L}^{-1}(A) + (\tau_1 + \tau_2)\mathscr{L}^{-1}(B)] \qquad (1\text{-}102)$$

30 LOOP FUNDAMENTALS

What does the equation mean? We really want to know how long it takes to pull the VCO frequency to the reference. Therefore, we want to know the value of t, the time it takes to be within 2π or less of lock-in range.

The PLL can, at the beginning, have a phase error from -2π to $+2\pi$, and the loop, by accomplishing lock, then takes care of this phase error.

We can make the reverse assumption for a moment and ask ourselves, as we have done earlier, how long the loop stays in phase lock. This is called the *pull-out range*. Again, we apply signals to the input of the PLL as long as the loop can follow and the phase error does not become larger than 2π. Once the error is larger than 2π, the loop jumps out of lock.

We will learn more about this in Section 1-10, but as already mentioned with regard to the condition where the loop is out of lock, a beat note occurs at the output of the loop filter following the phase/frequency detector.

The tri-state phase/frequency comparator, however, works on a different principle, and the pulses generated and supplied to the charge pump do not allow the generation of an ac voltage. The output of such a phase/frequency detector is always unipolar, but, relative to the value of $V_{\text{batt}}/2$, the integrator voltage can be either positive or negative. If we assume for a moment that this voltage should be the final voltage under a locked condition, we will observe that the resulting dc voltage is either more negative or more positive relative to this value, and because of this, the VCO will be "pulled in" to this final frequency rather than swept in, which had been mentioned previously. The swept-in technique applies only in cases of phase/frequency comparators, where this beat note is being generated. A typical case would be the exclusive-OR gate or even a sample/hold comparator. This phenomenon is rarely covered in the literature and is probably discussed in detail for the first time in the book by Roland Best [1].

Let us assume now that the VCO has been pulled in to final frequency to be within 2π of the final frequency, and the time t is known. The next step is to determine the lock-in characteristic.

Lock-in Characteristic. We will now determine lock-in characteristic, and this requires the use of a different block diagram. Figure 1-8 shows the familiar block diagram of the PLL, and we will use the following definitions:

$$\theta_1(s) = \mathcal{L}[\Delta\delta_1(t)] \quad \text{Reference input to } \delta/\omega \text{ detector}$$

$$\theta_2(s) = \mathcal{L}[\Delta\delta_2(t)] \quad \text{Signal VCO output phase}$$

$$\theta_e(s) = \mathcal{L}[\delta_e(t)] \quad \text{Phase error at } \delta/\omega \text{ detector}$$

$$\theta_e(s) = \theta_1(s) - \frac{\theta_2(s)}{N}$$

From the block diagram, the following is apparent:

$$A(s) = \theta_e(s) K_d$$

$$B(s) = A(s) F(s)$$

$$\theta_2(s) = B(s) \frac{K_o}{s}$$

The phase error at the detector is

$$\theta_e(s) = \theta_1(s) \frac{sN}{K_o K_d F(s) + sN} \qquad (1\text{-}103)$$

In Section A-4 of Appendix A we will see that a step in phase at the input, with the worst-case error being 2π, results in

$$\theta_1(s) = 2\pi \frac{1}{s} \qquad (1\text{-}104)$$

We will now treat the two cases using an active or a passive filter.

Active Filter. The transfer characteristic of the active filter is

$$F(s) = \frac{1 + s\tau_2}{s\tau_1} \qquad (1\text{-}90)$$

This results in the formula for the phase error at the detector,

$$\theta_e(s) = 2\pi \frac{s}{s^2 + (sK_o K_d \tau_2/\tau_1)/N + (K_o K_d/\tau_1)/N} \qquad (1\text{-}105)$$

The polynomial coefficients for the denominator are

$$a_2 = 1$$
$$a_1 = (K_o K_d \tau_2/\tau_1)/N$$
$$a_0 = (K_o K_d/\tau_1)/N$$

and we have to find the roots W_1 and W_2. Expressed in the form of a polynomial coefficient, the phase error is

$$\theta_e(s) = 2\pi \frac{s}{(s + W_1)(s + W_2)} \qquad (1\text{-}106)$$

After the Laplace transformation has been performed, the result can be written in the form

$$\delta_e(t) = 2\pi \frac{W_1 e^{-W_1 t} - W_2 e^{-W_2 t}}{W_1 - W_2} \qquad (1\text{-}107)$$

with

$$\lim_{t \to 0} \delta_e(t) = 2\pi$$

32 LOOP FUNDAMENTALS

and

$$\lim_{t \to \infty} \delta_e(t) = 0$$

The same can be done using a passive filter.

Passive Filter. The transfer function of the passive filter is

$$F(s) = \frac{1 + s\tau_2}{1 + s(\tau_1 + \tau_2)} \qquad (1\text{-}95)$$

If we apply the same phase step of 2π as before, the resulting phase error is

$$\theta_e(s) = 2\pi \frac{[1/(\tau_1 + \tau_2)] + s}{s^2 + s\frac{N + K_o K_d \tau_2}{N(\tau_1 + \tau_2)} + \frac{K_o K_d}{N(\tau_1 + \tau_2)}} \qquad (1\text{-}108)$$

Again, we have to find the polynomial coefficients, which are

$$a_2 = 1$$

$$a_1 = \frac{N + K_o K_d \tau_2}{N(\tau_1 + \tau_2)}$$

$$a_0 = \frac{K_o K_d}{N(\tau_1 + \tau_2)}$$

and finally find the roots for W_1 and W_2. This can be written in the form

$$\theta_e(s) = 2\pi \left[\frac{1}{\tau_1 + \tau_2} \frac{1}{(s + W_1)(s + W_2)} + \frac{s}{(s + W_1)(s + W_2)} \right] \qquad (1\text{-}109)$$

Now we perform the Laplace transformation and obtain our result:

$$\delta_e(t) = 2\pi \left(\frac{1}{\tau_1 + \tau_2} \frac{e^{-W_1 t} - e^{-W_2 t}}{W_2 - W_1} + \frac{W_1 e^{-W_1 t} - W_2 e^{-W_2 t}}{W_1 - W_2} \right) \qquad (1\text{-}110)$$

with

$$\lim_{t \to 0} \delta_e(t) = 2\pi$$

and

$$\lim_{t \to \infty} \delta_e(t) = 0$$

I will show with the type 2 third-order loop how these roots are being determined, as the roots are going to be fifth order. I assume that determining the roots of a cubic equation is known and easy.

As a result of the last equation for the active as well as for the passive filter, Eqs. (1-107) and (1-110) have the dimension of radians. Although mathematically speaking it is not strictly accurate, it is permissible to multiply these values with the division ratio N and the reference frequency to obtain a final error in the dimension frequency. It has been shown in practical experiments that, if this final error is less than 0.1 Hz, the time (t) it takes to get there can be taken as the lock-in time. In Chapter 2, we will learn that any VCO has a certain residual FM. That means that even under locked condition, the output frequency moves within certain boundaries. Medium-quality synthesizers show a residual FM of 3 Hz, whereas a loop with dividers at the output and low division ratio can have a residual FM as low as 0.1 Hz or better. This is, at times, also called *incidental FM*, and similar expressions have been found in the literature.

In Table 1-2 we indicated that the exclusive-OR gate and the edge-triggered JK master/slave flip-flop have a different operation mode than that of the tri-state phase/frequency comparator.

We will go into the details in Section 1-10, since the tri-state phase/frequency comparator does not require any acquisition aid.

Let us take a look at a numerical example. We have a PLL with the following parameters:

Reference frequency	5000 Hz
K_o	$2\pi \times 1$ MHz
K_θ	2.1 V/rad
N	1000
ω_n	500 Hz
Phase margin	45°
ζ	0.7
R_1	1336 Ω
R_2	445 Ω
C_1	1E-6 F
Reference suppression	20 dB
Lockup time	8 ms

These values were determined for the type 2 second-order loop with a program.

In Section 1-7 we will find that the type 2 third-order loop, although initially somewhat more difficult to treat mathematically, will show better reference suppression, faster lockup time, and really is better as far as reproducibility is concerned. This is due to the fact that there are always stray capacitors and some other elements in the circuit, which can be incorporated in the type 2 third-order loop, whereas the type 2 second-order loop really does not exist in its pure form.

Some of the dynamics of the type 2 second-order loop and of the type 2 third-order loop are dealt with in Seciton 1-10. For reasons of consistency, however, the transient behavior of the type 2 second-order loop has already been treated, and the equivalent performance of the type 2 third-order loop will be discussed on the following pages.

1-7 TYPE 2 THIRD-ORDER LOOP

I have stated several times previously that low-order loops really do not exist. The reasons for this are the introduction of phase shift by the operational amplifier, stray capacitors, and other things in the loop.

The type 2 third-order loop is a very good approximation of what is actually happening and can easily be developed from the type 2 second-order loop by adding one more *RC* filter at the output. Most likely, the operational amplifier as part of the active filter has to drive a feed-through capacitor, and for reasons of spike decoupling or additional filtering, one resistor is put in series.

Figure 1-9 shows a loop filter for a type 2 third-order loop. Just a reminder: this loop has two integrators, one being the VCO and one being the operational amplifier and three time constants.

This filter can be redrawn as shown in Figure 1-10. The transfer function for this filter is

$$F(s) = -\frac{1}{s\tau_1}\frac{1+s\tau_2}{1+s\tau_3} \tag{1-111}$$

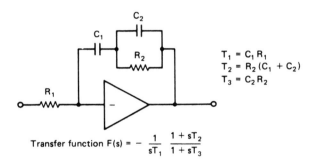

Figure 1-9 Circuit diagram of the loop filter for the third-order loop.

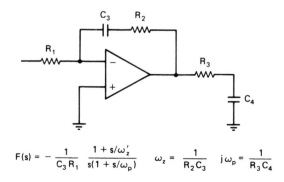

Figure 1-10 Circuit diagram of redrawn Figure 1-9. Note that this is the same type of loop filter as in the type 2 second-order loop with an additional *RC* time constant.

with

$$\tau_1 = C_1 R_1 \tag{1-112}$$

$$\tau_2 = R_2(C_1 + C_2) \tag{1-113}$$

$$\tau_3 = C_2 R_2 \tag{1-114}$$

Let us now determine the transfer function for the type 2 third-order loop.

1-7-1 Transfer Function of Type 2 Third-Order Loop

The forward gain is

$$K = K_\theta F(s) \frac{K_o}{s} \tag{1-115}$$

and for

$$H(s) = \frac{1}{N} \qquad N = \text{division ratio} \tag{1-116}$$

the open-loop gain is

$$A(s) = K_\theta F(s) \frac{K_o}{s} \frac{1}{N} \tag{1-117}$$

and the system transfer function is

$$\begin{aligned} B(s) = \frac{\theta_o(s)}{\theta_i(s)} &= \frac{\text{forward gain}}{1 + (\text{open-loop gain})} \\ &= \frac{K_\theta F(s)(K_o/s)}{1 + \dfrac{K_\theta F(s)(K_o/s)}{N}} \\ &= \frac{K_\theta F(s) K_o}{s + \dfrac{K_\theta F(s) K_o}{N}} \end{aligned} \tag{1-118}$$

If we insert our time constants τ_1, τ_2, and τ_3, we obtain

$$B(s) = \frac{\theta_o(s)}{\theta_i(s)} = \frac{1 + s\tau_2}{s\tau_1 + s^2 \tau_1 \tau_3} \frac{K_\theta K_o}{s + \dfrac{K_\theta K_o}{N} \dfrac{1 + s\tau_2}{s\tau_1 + s^2 \tau_1 \tau_3}} \tag{1-119}$$

and

$$B(s) = N K_\theta K_o (1 + s\tau_2) \frac{1}{s^3 N \tau_1 \tau_3 + s^2 N \tau_1 + s K_\theta K_o \tau_2 + K_\theta K_o} \tag{1-120}$$

$$\text{polynomial } P = s^3 N \tau_1 \tau_3 + s^2 N \tau_1 + s K_\theta K_o \tau_2 + K_\theta K_o \tag{1-121}$$

36 LOOP FUNDAMENTALS

As we have done before, we have to determine the roots:

$$\frac{P}{N\tau_1\tau_3} = s^3 + s^2\frac{1}{\tau_3} + sK_\theta K_o\frac{\tau_2}{N\tau_1\tau_3} + K_\theta K_o\frac{1}{n\tau_1\tau_3} = 0 \quad (1\text{-}122)$$

Therefore, the coefficients are

$$a_4 = 1 \quad (1\text{-}123)$$

$$a_3 = \frac{1}{\tau_3} \quad (1\text{-}124)$$

$$a_2 = K_\theta K_o \frac{\tau_2}{N\tau_1\tau_3} \quad (1\text{-}125)$$

$$a_1 = K_\theta K_o \frac{1}{N\tau_1\tau_3} = \frac{a_2}{\tau_2} \quad (1\text{-}126)$$

The polynomial is of the order of 3, and we can use a calculator routine to find the roots b_1, b_2, and b_3; finally,

$$s_1 = 0 \quad (1\text{-}127)$$

$$s_2 = -b_1 \quad (1\text{-}128)$$

$$s_3 = -b_2 \quad (1\text{-}129)$$

$$s_4 = -b_3 \quad (1\text{-}130)$$

The next step is partial fraction forming:

$$\frac{1}{a_4 s^4 + a_3 s^3 + a_2 s^2 + a_1 s}$$

$$= \frac{1}{a_4 s(s + b_1)(s + b_2)(s + b_3)} \quad (1\text{-}131)$$

$$= \left(\frac{c_1}{s + b_1} + \frac{c_2}{s + b_2} + \frac{c_3}{s + b_3} + \frac{c_4}{s}\right)\frac{1}{a_4}$$

Now we have to determine C_i, for $i = 1$ to 4, and multiply the equation above,

$$\frac{c_1(s + b_2)(s + b_3)s + c_2(s + b_1)(s + b_3)s + c_3(s + b_1)(s + b_2)s + c_4(s + b_1)(s + b_2)(s + b_3)}{(s + b_1)(s + b_2)(s + b_3)s}$$

which is equal to

$$\frac{1}{(s + b_1)(s + b_2)(s + b_3)s} \quad (\text{numerator} = 1)$$

Rearranging yields

$$s^3(c_1 + c_2 + c_3 + c_4)$$
$$+ s^2[c_1(b_2 + b_3) + c_2(b_1 + b_3) + c_3(b_1 + b_2) + c_4(b_1 + b_2 + b_3)]$$
$$+ s[c_1 b_2 b_3 + c_2 b_1 b_3 + c_3 b_1 b_2 + c_4(b_1 b_2 + b_1 b_3 + b_2 b_3)]$$
$$+ c_4 b_1 b_2 b_3$$

and since

$$c_1 + c_2 + c_3 = 0 \tag{1-132}$$

$$c_1(b_2 + b_3) + c_2(b_1 + b_3) + c_3(b_1 + b_2) = 0 \tag{1-133}$$

$$c_1 b_2 b_3 + c_2 b_1 b_3 + c_3 b_1 b_2 = \frac{1}{N\tau_1 \tau_3} \tag{1-134}$$

our final results are

$$c_3 = -c_1 - c_2 \tag{1-135}$$

$$c_2 = -c_1 \frac{b_3 - b_1}{b_3 - b_2} \tag{1-136}$$

$$c_1 = \frac{1}{N\tau_1 \tau_3} \frac{1}{(b_3 - b_1)(b_2 - b_1)} \tag{1-137}$$

Let us test this equation with a step function

$$\theta_i(s) = \frac{1}{s} \tag{1-138}$$

We obtain

$$\theta_o(s) = NK_\theta K_o \frac{1 + s\tau_2}{s} \frac{1/N\tau_1 \tau_3}{s^3 + \frac{1}{\tau_3}s^2 + \frac{\tau_2 K_\theta K_o}{N\tau_1 \tau_3}s + \frac{K_\theta K_o}{N\tau_1 \tau_3}}$$

$$= NK_\theta K_o \left[\tau_2 \left(\frac{c_1}{s + b_1} + \frac{c_2}{s + b_2} + \frac{c_3}{s + b_3} \right) \right.$$

$$+ \frac{c_1}{b_1}\left(\frac{1}{s} - \frac{1}{s + b_1}\right) + \frac{c_2}{b_2}\left(\frac{1}{s} - \frac{1}{s + b_2}\right) + \frac{c_3}{b_3}\left(\frac{1}{s} - \frac{1}{s + b_1}\right) \right] \tag{1-139}$$

$$= NK_\theta K_o \left[\frac{(\tau_2 - 1/b_1)c_1}{s + b_1} + \frac{(\tau_2 - 1/b_2)c_2}{s + b_2} + \frac{(\tau_2 - 1/b_3)c_3}{s + b_3} \right.$$

$$+ \left. \frac{(c_1/b_1) + (c_2/b_2) + (c_3/b_3)}{s} \right]$$

38 LOOP FUNDAMENTALS

If we perform a Laplace transformation, our final result is

$$\theta_o(t) = NK_\theta K_o \left[\left(\tau_2 - \frac{1}{b_1}\right) c_1 e^{-b_1 t} + \left(\tau_2 - \frac{1}{b_2}\right) c_2 e^{-b_2 t} \right.$$
$$\left. + \left(\tau_2 - \frac{1}{b_3}\right) c_3 e^{-b_3 t} + \left(\frac{c_1}{b_1} + \frac{c_2}{b_2} + \frac{c_3}{b_3}\right) \right] \quad (1\text{-}140)$$

To plot the Bode diagram, the open-loop gain equation for $A(j\omega)$ must be determined and plotted in magnitude and phase. We obtain

$$A(j\omega) = -\frac{K_\theta K_o}{Nj\omega}\left(\frac{1}{j\omega\tau_1}\right)\frac{1+j\omega\tau_2}{1+j\omega\tau_3}$$
$$= \frac{K_\theta K_o}{N\omega^2}\left(\frac{1+j\omega\tau_2}{1+j\omega\tau_3}\right)\frac{1}{\tau_1} \quad (1\text{-}141)$$

We will then abbreviate

$$\frac{K_\theta K_o}{N\omega^2} = \tau_9 \quad (1\text{-}142)$$

The phase is determined from

$$\frac{(1+j\omega\tau_2)(1-j\omega\tau_3)}{1+\omega^2\tau_3^2} = \frac{1+\omega^2\tau_2\tau_3 + j\omega(\tau_2-\tau_3)}{1+\omega^2\tau_3^2} \quad (1\text{-}143)$$

and

$$\tan\phi = \frac{\omega\tau_2}{1+\omega^2\tau_2\tau_3} - \frac{\omega\tau_3}{1+\omega^2\tau_2\tau_3} \quad (1\text{-}144)$$

The magnitude is

$$|A(j\omega)| = \frac{\tau_9}{\tau_1}\frac{\sqrt{1+\omega^2\tau_2^2}}{\sqrt{1+\omega^2\tau_3^2}} \quad (1\text{-}145)$$

with

$$|A(j\omega)| = 1 \quad \text{(crossover point)} \quad (1\text{-}146)$$

We finally obtain

$$\tau_1 = \tau_9 \sqrt{\frac{1+\omega^2\tau_2^2}{1+\omega^2\tau_3^2}} \quad (1\text{-}147)$$

and

$$A_1 = 1 + \omega^2\tau_2^2 \quad (1\text{-}148)$$
$$A_2 = 1 + \omega^2\tau_3^2 \quad (1\text{-}149)$$

The phase margin is

$$\phi = \arctan \omega\tau_2 - \arctan \omega\tau_3 + \pi \qquad (1\text{-}150)$$

assuming that

$$\omega^2 \tau_2 \tau_3 \ll 1$$

Let us determine the natural loop frequency ω_o from the point of zero slope of the phase response,

$$\frac{d\phi(\omega)}{d\omega} = 0 \qquad (1\text{-}151)$$

$$\frac{d\phi}{d\omega} = \frac{\tau_2}{1 + (\omega\tau_2)^2} - \frac{\tau_3}{1 + (\omega\tau_3)^2} = 0 \qquad (1\text{-}152)$$

and therefore

$$\omega_o = \sqrt{\frac{1}{\tau_2 \tau_3}} \qquad (1\text{-}153)$$

If we set

$$\alpha = \arctan \omega\tau_2 \qquad (1\text{-}154)$$
$$\beta = \arctan \omega\tau_3 \qquad (1\text{-}155)$$
$$\phi = \alpha - \beta + \pi \qquad (1\text{-}156)$$

and

$$\tan \phi = \tan[(\alpha - \beta) + \pi]$$
$$= \frac{\tan(\alpha - \beta) + 0}{1 - 0} = \tan(\alpha - \beta) \qquad (1\text{-}157)$$

then

$$\tan(\alpha - \beta) = \frac{\tan \alpha - \tan \beta}{1 + \tan \alpha \tan \beta} = \frac{\omega\tau_2 - \omega\tau_3}{1 + \omega^2 \tau_2 \tau_3} = \tan \phi \qquad (1\text{-}158)$$

If we set

$$\omega = \omega_o = \sqrt{\frac{1}{\tau_2 \tau_3}} \qquad (1\text{-}159)$$

then

$$\tan \phi_o = \frac{(1/\sqrt{\tau_2 \tau_3})(\tau_2 - \tau_3)}{1 + 1} = \frac{\tau_2 - \tau_3}{2\sqrt{\tau_2 \tau_3}} \qquad (1\text{-}160)$$

40   LOOP FUNDAMENTALS

and

$$\sqrt{\tau_2 \tau_3} = \frac{1}{\omega_o} \quad (1\text{-}161)$$

$$\omega_o^2 \tau_2 \tau_3 = 1 \quad (1\text{-}162)$$

$$\tau_2 = \frac{1}{\omega_o^2 \tau_3} \quad (1\text{-}163)$$

Using this value, we can determine the time constant τ_3 from

$$\tan \phi_o = \frac{(1/\omega_o^2 \tau_3) - \tau_3}{2(1/\omega_o)} = \frac{(1/\omega_o^2 \tau_3) - \omega_o \tau_3}{2} \quad (1\text{-}164)$$

and

$$2 \tan \phi_o \omega_o \tau_3 = 1 - \omega_o^2 \tau_3^2 \quad (1\text{-}165)$$

$$\omega_o^2 \tau_3^2 + 2 \tan \phi_o \omega_o \tau_3 - 1 = 0 \quad (1\text{-}166)$$

The time constant τ_3 is determined from

$$\begin{aligned}
\tau_3 &= \frac{-2 \tan \phi_o \omega_o + \sqrt{4 \tan^2 \phi_o \omega_o^2 + 4\omega_o^2}}{2\omega_o^2} \\
&= \frac{-2 \tan \phi_o \omega_o + 2\omega_o \sqrt{\tan^2 \phi_o + 1}}{2\omega_o^2} \\
&= \frac{\tan \phi_o + \sqrt{(\cos^2 \phi_o + \sin^2 \phi_o)/\cos^2 \phi_o}}{\omega_o} \\
&= \frac{-\tan \phi_o + 1/\cos \phi_o}{\omega_o}
\end{aligned} \quad (1\text{-}167)$$

τ_3 is now determined independent of the other parameters, τ_1 and τ_2, by setting the value for ω_o and the phase margin ϕ to begin with. Once τ_3 is determined, the values for τ_2 and τ_1 can be computed by inserting them in the necessary equations.

These equations were somewhat lengthy, but they were spelled out in great detail to show the approach taken. A computer program based on these equations is presented in Section A-3 of Appendix A and can be used for Bode diagram plottings.

It may be useful to compare the normalized output response of the different types of loops. Figure 1-11a shows the step response for a phase margin of 10° and 45° and a type 2 third-order loop. Figure 1-11b shows the result for a type 1 second-order loop similar to the normalized output response for a damping factor of 0.1 and 0.45. Figure 1-11c shows the normalized output response for the same damping factors, 0.1 and 0.45.

These responses can be plotted in the Bode diagram. Figure 1-11d shows the integrated response $F(s)$ of the second- and third-order loops.

TYPE 2 THIRD-ORDER LOOP   41

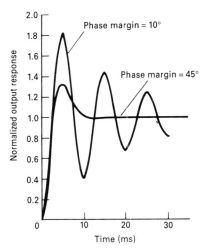

Figure 1-11a Normalized output response of a type 2 third-order loop with a phase margin of 10° and 45°.

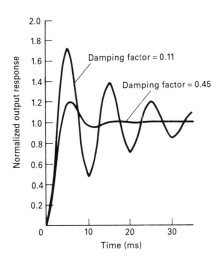

Figure 1-11b Normalized output response of a type 1 second-order loop having a damping factor of 0.1 and 0.45.

The closed-loop response of the type 2, second- and third-order loops is somewhat similar; but as Figure 1-11e shows, we obtain a much better suspension from the type 2 third-order loop than from the type 2 second-order loop.

If the phase margins of the damping factor chosen are inappropriate, we obtain an overshoot that translates into discrete spurious noise at the appropriate offset from the carrier.

42 LOOP FUNDAMENTALS

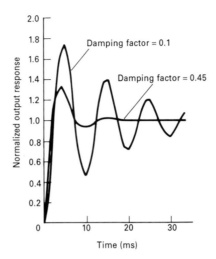

Figure 1-11c Normalized output response of a type 2 second-order loop with a damping factor of 0.1 and 0.05 for $\Omega_n = 0.631$.

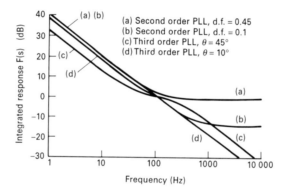

Figure 1-11d Integrated response for various loops as a function of the phase margin.

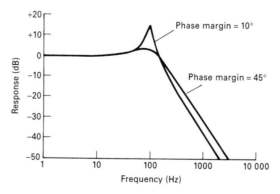

Figure 1-11e Closed-loop response of a type 2 third-order PLL having a phase margin of 10° and 45°.

1-7-2 FM Noise Suppression

In drawing the Bode plot, it is also convenient to show the suppression of noise of the VCO that is provided by the phase-locked loop. Using

$$E_n = \text{VCO noise voltage}$$

and

$$E = \text{noise voltage with loop closed}$$

we can write

$$E(s) = E_n(s) \frac{1}{1 + (\text{open-loop gain})} \quad (1\text{-}168)$$

$$= \frac{1}{1 + G(s)H(s)} = \frac{1}{1 + A(s)}$$

or

$$\left| \frac{E(\omega)}{E_n(\omega)} \right| = \frac{1}{\frac{K_\theta K_o}{N\omega^2 \tau_1} \left| \frac{1 + j\omega\tau_2}{1 + j\omega\tau_3} \right| + 1} \quad (1\text{-}169)$$

The type 2 third-order loop is really the most important but was not used that often in the past. This may be a lack of understanding or not realizing that, by proper combination of the time constants, the unavoidable feed-through capacitors and some series capacitors can be incorporated to obtain this type of loop. The advantages of the third-order loop over the second-order loop are in the higher reference suppression for a given loop frequency, or if a certain loop frequency has to be chosen because of lockup time, the reference suppression is higher than we would find in the case of a type 2 second-order loop.

Let us take a case where we have the following parameters:

Reference frequency	5000 Hz
K_o	$2\pi \times 1$ MHz
K_θ	2.1
Division ratio N	10,000
ω_n	500 Hz
Phase margin	45°

The resulting values for the loop filter are determined from the computer program:

R_1	5600 Ω
R_2	1105 Ω
C_1	5.7E-7 F
C_2	1.19E-7 F
Reference suppression	32 dB
Lockup time	3 ms

44 LOOP FUNDAMENTALS

The same loop in a type 2 second-order system would show reference suppression of 20 dB for the same loop bandwidth and 8-ms lockup time. These last two figures clearly indicate the advantage of the type 2 third-order loop.

Many applications require an even higher reference suppression. In the following analysis, we will deal with higher-order loops that are capable of additional suppression. However, the lockup time and the phase stability now may become a trade-off, as we will soon find out.

1-8 HIGHER-ORDER LOOPS

1-8-1 Fifth-Order Loop Transient Response

The fifth-order loop consists of a type 2 third-order loop with a second-order low-pass filter. The integrator is described by

$$F(s) = -\frac{1}{s\tau_1}\frac{1+s\tau_2}{1+s\tau_3} \tag{1-111}$$

and the second-order low-pass filter is described by

$$K(s) = \frac{1}{s^2(1/\omega_n^2) + s(2d/\omega_n) + 1} \tag{1-170}$$

The transfer function of the filters is

$$T(s) = \frac{-(1+s\tau_2)}{s^2\tau_1\tau_3 + s\tau_1}\frac{1}{s^2(1/\omega_n^2) + s(2d/\omega_n) + 1}$$

$$= -\frac{1+s\tau_2}{s^4\dfrac{\tau_1\tau_3}{\omega_n^2} + s^3\dfrac{\tau_1}{\omega_n}\left(2d\tau_3 + \dfrac{1}{\omega_n}\right) + s^2\tau_1\left(\tau_3 + \dfrac{2d}{\omega_n}\right) + s\tau_1} \tag{1-171}$$

We use the familiar block diagram, Figure 1-12, which shows the phase detector, the low-pass filter, the active integrator [both condensed in $T(s)$], the VCO, and the divider. The forward gain now becomes $(K_\theta K_o/s)T(s)$, and the open-loop gain is

$$A(s) = \frac{K_\theta K_o}{s}\frac{T(s)}{N} \tag{1-172}$$

The closed-loop transfer function is

$$B(s) = \frac{\theta_o(s)}{\theta_i(s)} = \frac{\text{forward gain}}{1 + (\text{open-loop gain})} \tag{1-173}$$

or

$$\frac{\theta_o(s)}{\theta_i(s)} = \frac{(K_\theta K_o/s)T(s)}{1 + (K_\theta K_o/Ns)T(s)} = \frac{K_\theta K_o T(s)}{s + K_\theta K_o T(s)/N} \tag{1-174}$$

HIGHER-ORDER LOOPS

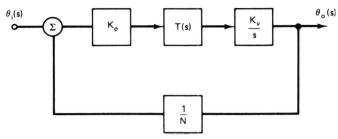

Forward gain: $K_\phi \cdot K_v/s \cdot T(s)$
O.L. gain: $K_\phi \cdot K_v/s \cdot T(s)/N$

Figure 1-12 Noise contributing block in a type 2 fifth-order loop.

Rearranging yields

$$\theta_o(s) = \theta_i(s) K_\theta K_o \frac{T(s)}{s + K_\theta K_o T(s)/N} \quad (1\text{-}175)$$

The output phase, which is the same as the VCO phase, is now assumed to be disturbed by a step of magnitude S_v. The amount S_v would be in the maximum case $N \times 2\pi$, using phase detector operating ranges of $\pm 2\pi$.

If this step is referred to the input of the phase detector, it has to be divided by N, and in Laplace notation, we have

$$\theta_i(s) = \frac{S_v}{Ns} \quad (1\text{-}176)$$

from which results

$$\theta_o(s) = \frac{S_v K_\theta K_o}{N} \frac{T(s)}{s^2 + (K_\theta K_o/N) s T(s)} \quad (1\text{-}177)$$

with $K_\theta K_o/N = U$. Applying a partial fraction, we obtain

$$\theta_o(s) = \frac{S_v}{s} - \frac{S_v s}{s^2 + Us T(s)} = \frac{S_v}{s} - \frac{S_v}{s + UT(s)} \quad (1\text{-}178)$$

After some manipulations, which are deleted, it can be shown that

$$\Delta\theta_o(s) = -S_v \frac{s^4 + s^3 \dfrac{2d\tau_3\omega_n + 1}{\tau_3} + s^2\left(\omega_n^2 + \dfrac{2d\omega_n}{\tau_3}\right) + s\dfrac{\omega_n^2}{\tau_3}}{s^5 + s^4 \dfrac{2d\tau_3\omega_n + 1}{\tau_3} + s^3\left(\omega_n^2 + \dfrac{2d\omega_n}{\tau_3}\right) + s^2 \dfrac{\omega_n^2}{\tau_3} + s\dfrac{U\tau_2\omega_n^2}{\tau_1\tau_3} \dfrac{U\omega_n^2}{\tau_1\tau_3}} \quad (1\text{-}179)$$

We now factorize a denominator polynomial of the form

$$s^5 a_5 + s^4 a_4 + s^3 a_3 + s^2 a_2 + s a_1 + a_o$$

46 LOOP FUNDAMENTALS

and therefore obtain the following coefficients:

$$a_5 = 1 \tag{1-180}$$

$$a_4 = \frac{2d\tau_3\omega_n + 1}{\tau_3} \tag{1-181}$$

$$a_3 = \omega_n^2 + \frac{2d\omega_n}{\tau_3} \tag{1-182}$$

$$a_2 = \frac{\omega_n^2}{\tau_3} \tag{1-183}$$

$$a_1 = U\tau_2 \frac{\omega_n^2}{\tau_1 \tau_3} \tag{1-184}$$

$$a_o = U \frac{\omega_n^2}{\tau_1 \tau_3} \tag{1-185}$$

The next task is to determine the roots W_i and rewrite the denominator polynomial in the form

$$P_{\text{denominator}} = (s - W_5)(s - W_4)(s - W_3)(s - W_2)(s - W_1) \tag{1-186}$$

For the Laplace transform, we need the residues

$$K_s = W_i = e^{ts} \frac{\text{numerator}}{\text{denominator without containing } W_i} \tag{1-187}$$

The next step is to calculate the binomial residues:

$$K_k = e^{tP_k} \frac{A_{k1} + A_{k2} + A_{k3} + A_{k4}}{B_{k1} B_{k2} B_{k3} B_{k4}} \tag{1-188}$$

$$K_1 = e^{tP_1} \frac{A_{16} + A_{15} + A_{14} + A_{13}}{B_{12} B_{13} B_{14} B_{15}} \tag{1-189}$$

$$K_2 = e^{tP_2} \frac{A_{26} + A_{25} + A_{24} + A_{23}}{B_{21} B_{23} B_{24} B_{25}} \tag{1-190}$$

$$K_3 = e^{tP_3} \frac{A_{36} + A_{35} + A_{34} + A_{33}}{B_{31} B_{32} B_{34} B_{35}} \tag{1-191}$$

$$K_4 = e^{tP_4} \frac{A_{46} + A_{45} + A_{44} + A_{43}}{B_{41} B_{42} B_{43} B_{45}} \tag{1-192}$$

$$K_5 = e^{tP_5} \frac{A_{56} + A_{55} + A_{54} + A_{53}}{B_{51} B_{52} B_{53} B_{54}} \tag{1-193}$$

As

$$A_{ij} = P_i \uparrow (j-2) Q_j \tag{1-194}$$

$$i = 1, 2, 3, N_2 \tag{1-195}$$

$$j = 3, 4, 5, N_2 + 1 \tag{1-196}$$

and

$$B_{ij} = P_i - P_j \tag{1-197}$$
$$i = 1, 2, 3, \ldots, N_2 \tag{1-198}$$
$$j = 1, 2, 3, \ldots, N_2 \tag{1-199}$$

it is apparent that

$$A_{1ij} \triangleq \text{real } A_{ij} \tag{1-200}$$
$$A_{2ij} \triangleq \text{imaginary } A_{ij} \tag{1-201}$$
$$B_{1ij} \triangleq \text{real } B_{ij} \tag{1-202}$$
$$B_{2ij} \triangleq \text{imaginary } B_{ij} \tag{1-203}$$
$$K_{1i} \triangleq \text{real } K_i \tag{1-204}$$
$$K_{2i} \triangleq \text{imaginary } K_i \tag{1-205}$$

Using the results, we can rewrite

$$\Delta\theta_o(s) = \frac{s^4 Q_6 + s^3 Q_5 + s^2 Q_4 + s Q_3}{s^5 Q_6 + s^4 Q_5 + s^3 Q_4 + s^2 Q_3 + s Q_2 + Q_1} \tag{1-206}$$

Finally, we obtain the roots P_1 to P_5 in terms of Q_1 to Q_6; then the partial fraction expansion gives the residues:

$$K_1 \bigg|_{s=P_1} = e^{tP_1} \frac{P_1^4 Q_6 + P_1^3 Q_5 + P_1^2 Q_4 + P_1 Q_3}{(P_1 - P_2)(P_1 - P_3)(P_1 - P_4)(P_1 - P_5)} \tag{1-207}$$

$$K_2 \bigg|_{s=P_2} = e^{tP_2} \frac{P_2^4 Q_6 + P_2^3 Q_5 + P_2^2 Q_4 + P_2 Q_3}{(P_2 - P_1)(P_2 - P_3)(P_2 - P_4)(P_2 - P_5)} \tag{1-208}$$

$$K_3 \bigg|_{s=P_3} = e^{tP_3} \frac{P_3^4 Q_6 + P_3^3 Q_5 + P_3^2 Q_4 + P_3 Q_3}{(P_3 - P_1)(P_3 - P_2)(P_3 - P_4)(P_3 - P_5)} \tag{1-209}$$

$$K_4 \bigg|_{s=P_4} = e^{tP_4} \frac{P_4^4 Q_6 + P_4^3 Q_5 + P_4^2 Q_4 + P_4 Q_3}{(P_4 - P_1)(P_4 - P_2)(P_4 - P_3)(P_4 - P_5)} \tag{1-210}$$

$$K_5 \bigg|_{s=P_5} = e^{tP_5} \frac{P_5^4 Q_6 + P_5^3 Q_5 + P_5^2 Q_4 + P_5 Q_3}{(P_5 - P_1)(P_5 - P_2)(P_5 - P_3)(P_5 - P_4)} \tag{1-211}$$

$$\Delta\theta(t) = \left[\sum_{n=1}^{5} K_n\right] \tag{1-212}$$

What is the practical use of higher-order loops? Higher-order loops are really useful only in frequency synthesizers if the reference frequency is substantially higher than the loop frequency. A typical example would be a reference frequency of as high as 5 or 10 MHz down to as low as 25 kHz, which has to be suppressed more than 90 dB. The phase shift introduced by the additional filtering, as can be seen by the computer program in the Appendix and the example shown here,

48 LOOP FUNDAMENTALS

can be allowed only if the cutoff frequency of the loop is small relative to the additional poles of the filter. What is a typical example? Let us assume that we have a reference frequency of 25 kHz, and our loop frequency is set at 1 or 2 kHz. The resulting reference suppression with the simple loop filter would be approximately 20 times $\log(25 \text{ kHz}/2 \text{ kHz})$, or roughly 20 dB. This is totally insufficient, of course.

The third-order loop, because of its deeper filtering, will increase this to roughly 33 dB, but this is still not enough. The insertion of a steep filter, such as an active filter or an LC elliptic filter, with poles at 25 kHz, will have very little phase shift at 2 kHz, the cutoff frequency of the loop filter, and a substantial suppression of reference becomes possible. However, these high-order systems are useful only if there is enough difference between the reference frequency and the loop frequency. The higher the ratio between the two values, the easier it becomes to design a high-order system.

To get high suppression, not too much phase shift, and to be able to work close to the reference frequency, special detectors are generally required. The phase shift introduced by using a sample/hold comparator is fairly small, and it is possible to set the loop filter at about half of the reference. If properly designed, the dual sampler combines good reference suppression with low noise. The phase shift can be adjusted and compensated. The drawback of the sample/hold comparator, as we will see in Chapter 4, is its limits as to what is the highest frequency of operation.

If properly designed, the tri-state phase/frequency comparator, also discussed in Chapter 4, may have 60 to 70 dB inherent reference suppression without the filter and may ease the requirement on the loop filter. Only recent developments, as will be described later, which compensate the spikes at the output of the tri-state phase/frequency comparator, will reduce the phase jitter and incidental FM introduced by this somewhat noisy discriminator. As the tri-state comparator is somewhat noisier than the sample/hold, the particular loop with its cutoff frequency and its noise requirements will determine which sample to use.

It may be useful to do some comparison between the different loops based on their performance. Figure 1-13a shows a simple block diagram of a PLL loop with a 15-MHz reference oscillator and 300-MHz output frequency. The division ratio is therefore 20. An active integrator with the time constant $T_1, \ldots, T_3$ and the open loop gain A_0 is selected.

The plot of Figure 1-13b shows the phase noise of the VCO, the phase noise of the reference oscillator of 15 MHz, and the resulting phase noise when the loop is closed. The overshot is due to the loop band width and the damping factor. In this case, the reduction of the loop and width would have reduced the phase noise as the VCO had better noise properties than the closed loop.

1-9 DIGITAL LOOPS WITH MIXERS

The single-loop synthesizer has a number of restrictions. One immediate restriction is the fact that the step size is the same as the reference frequency, unless special techniques are used, which are described in Chapter 3.

In addition, as we have frequency dividers that work only up to about

DIGITAL LOOPS WITH MIXERS   49

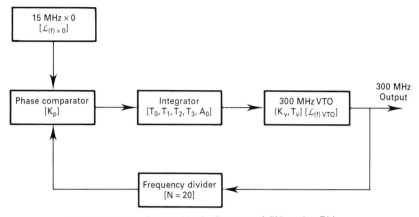

Figure 1-13a Simple block diagram of fifth-order PLL.

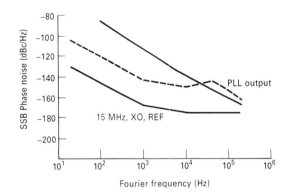

Figure 1-13b Single sideband (SSB) phase noise in dBc/Hz for the 15-MHz crystal oscillator (solid line) 300 MHz L_c VCO (solid line above) and PLL output of total system (curve).

2000 MHz, it becomes difficult, if not impossible, to build a PLL at a much higher frequency unless some mixing techniques are used. These techniques are also sometimes referred to as *heterodyning techniques*. Figure 1-14 shows a single-loop synthesizer that has a mixer incorporated, which heterodynes the VCO frequency down to a lower frequency at which we have dividers available. The auxiliary frequency that is injected in the loop for down-conversion is generated from the reference oscillator, and we will not go into the problems of noise generated because of this up-multiplication, as it is dealt with in Chapter 2. For now we are concerned only with the question of loop stability as a function of the introduction of this mixer, and its possible effects as far as unwanted sidebands are concerned. Let us assume the numerical example shown in Figure 1-15. This synthesizer is intended for use in an amateur transceiver operating from 144 to 148 MHz, a frequency range that is well within the capabilities of current dividers.

However, since we want to minimize power consumption, we find that it requires less power to generate an auxiliary frequency of 140 MHz from our 5-MHz

50 LOOP FUNDAMENTALS

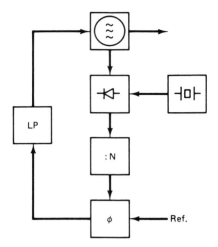

Figure 1-14 Block diagram of a digital PLL using the heterodyne technique.

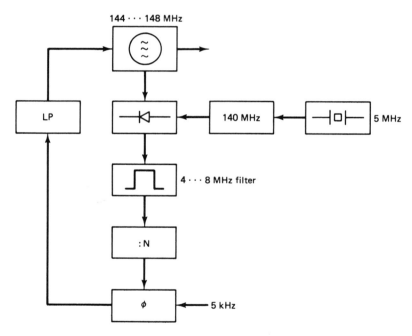

Figure 1-15 Block diagram of a 144- to 148-MHz synthesizer using the heterodyne technique, resulting in an internal IF of 4 to 8 MHz.

standard (synchronizing a 70-MHz crystal and doubling it). The output frequency of this mixer then is 4 to 8 MHz, and this frequency range can be handled by a programmable divider in CMOS. Between the mixer and the divider we will insert a bandpass filter of 4-MHz bandwidth. The divider in CMOS now has to operate between 4 and 8 MHz. For such amateur applications it is frequently required that the same synthesizer be used to transmit and receive.

DIGITAL LOOPS WITH MIXERS

For transmit applications, frequency modulating the synthesizer is required, and the modulation can be inserted in the loop filter, as the loop bandwidth probably is restricted to a few hundred hertz, while the step size will be 5 kHz, in accordance with the channels available. The modulation could also be done if the 140 MHz was generated from a free-running 70-MHz crystal, which is then modulated.

At 70 MHz we will probably have to use a third- or fifth-overtone crystal, which cannot be pulled very well, probably not enough to accommodate a maximum of 3-kHz deviation. More information on pulling of crystals is presented in Chapter 4.

What does the insertion of the mixer do to our system? The division ratio without mixing would have been 28,800 to 29,600, or a ratio of 1:1.0278. After inserting the mixing stage, the division ratio now is 800 to 1600, and the absolute ratio is 1:2.

The loop gain, assuming that the VCO gain is constant over the tuning range, is now changing by the amount 1:2. In some loops we will find that the introduction of such a mixer results in a much larger change of division ratios, and if the frequencies are not selected properly, a change of 1:10 will occur. This will cause two difficulties:

1. If the multiplication changes too much, the sideband of the reference being multiplied at the output will change substantially as a function of frequency setting, while the percentage change of the frequency at the output is very small. Therefore, at the higher frequency, in our case 148 MHz, the noise sideband inside the loop bandwidth will be twice as high as the 144 MHz. In a case where the loop gain changes more dramatically, such as 1:10, the noise will change by this amount.

2. If the loop gain changes with all other parameters remaining constant, this may cause a stability problem. The net result is that, in the case of a type 2 second-order loop, our damping factor ζ can range from 0.1 to 1, and the transient performance of the loop will vary substantially; therefore, stability will become an issue. We have to find a method to compensate for the change of loop gain, and we will show a method of dealing with this phenomenon in Section 1-10. Right now it is only important to know that it does occur and that it can present a problem.

The next effect the mixer produces is the consequence of phase shift of the IF filter following the mixer, in our case a filter ranging from 4 to 8 MHz. As a result, we have to modify our block diagram of the loop, as shown in Figure 1-16. The additional box represents the low-pass equivalent of the bandpass filter, and the delay of this filter has to be determined. There are a number of good books available that provide the necessary computation aids for these filters and information about the delay. The best book, which is by Anatol Zverev [4], will be mentioned herein several times and is really an absolute must for the library of any design engineer who handles RF and filter design. The phase shift of the equivalent low-pass filter of the IF filter is

$$\theta_{\text{low pass}} = \arctan \frac{G_1(f_m) \sin \theta(f_m) - G_1(-f_m) \sin \theta(-f_m)}{G_1(f_m) \cos \theta(f_m) - G_1(-f_m) \cos \theta(-f_m)} \quad (1\text{-}213)$$

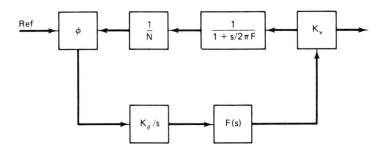

Figure 1-16 Linearized equivalent circuit of a PLL with the heterodyne technique, including delay information introduced by the IF filter.

In some cases, we will have to use loops with conversions, where the resulting IF filter is very narrow. The use of a crystal filter will introduce a group delay in addition to the phase shift. These delays have to be added to the block diagram and added or subtracted from the actual phase. Since this can be done fairly easily and has a substantial impact on the stability, it is recommended that the interested reader analyze such an example mathematically, where it is assumed that the loop bandwidth is 2 kHz, the IF bandwidth is 100 kHz, and the loop bandwidth is increased to 50 kHz. We also have to take into consideration the effect of the phase shift of the low-pass filter equivalent of the bandpass filter. As the output signal from the mixer is passing the IF filter, which will be assumed to be symmetrical, the amplitude of the carrier $G_1(f_m)$ will change relative to the input signal. The losses of the filter and the sidebands will be shifted in phase equal to the phase shift generated by the filter. As is known from the literature, it is possible to convert the bandpass performance of this filter into an equivalent low-pass filter and deal with it. These details and examples are beyond the scope of this book and should be studied in the literature, such as in the book by Zverev already mentioned.

There are a number of unpleasant effects in addition to the one mentioned, such as ringing, as a result of nonequalized group delay, and if the IF filter bandpass is not symmetrical relative to the carrier frequency, AM-to-FM conversion occurs and has to be calculated. These special effects are an interesting topic, and the reader should not only be made aware of them but is encouraged to calculate some numerical examples. In the rest of the book, we will assume that we know how to deal with these effects. The result of the calculations will be that a higher phase margin than the 45° recommended for higher-order loops has to be allowed to compensate for the additional delay. As a result of this higher phase margin, the settling time of the loop will become larger. In analyzing loops and in determining what frequency arrangements to use, one has to be aware of these trade-offs and optimize the loop by calculating a number of examples and finding the best solution by iteration.

The next effect we will analyze is the number of spurious signals introduced by the mixer. The double-balanced mixer, the best for such application, is still a highly nonlinear device that generates harmonics of the two frequencies applied, and those frequencies will mix with each other, resulting in a wide spectrum at the output. To keep the number of unwanted frequencies at the output at an acceptable level, we should observe two design rules:

1. To obtain an IF, the two input frequencies should be as high and as close together as possible; the lower image is then used. A fairly simple low-pass filter at the output minimizes the possibility of feed-through and unwanted products.
2. The power ratio between the RF and the LO should meet certain requirements. A design engineer is well advised to have about a 30-dB difference in level. Let us assume that the LO drive is +17 dB, a typical drive level for a medium-level double-balanced mixer; the RF input should then be −13 dBm or less. A further reduction in RF input may have the disadvantage of requiring too many amplifiers following at the output of the mixer and therefore generating additional noise from the postamplifiers.

In cases such as that of our previous example, where we are mixing a fixed frequency of 140 MHz with a 144- to 148-MHz band, this requirement may be relaxed, and a drive level of 5 or 6 dBm may still be permissible. A decision on this matter depends on the particular case, and the best way of solving it is to take measurements in an actual circuit.

A spectrum analyzer connected at the output of the mixer operating at the range of interest with enough dynamic range should provide the necessary information. If these theoretical guidelines are followed, the design should be fairly trouble-free.

In Chapter 6 you will find several examples where double-balanced mixers are incorporated and IF frequencies inside the synthesizer are being used. In many cases, I have provided level information, and the particular synthesizers shown are spurious-free, at least 90 dB down relative to the carrier.

1-10 ACQUISITION

In order to understand the acquisition performance of the digital PLL, we must first look at the linearized model. As mentioned previously, we have several ranges in which the loop can operate, and Figure 1-17 shows a plot of these ranges. The closest range around the center is the normal operating range and also the capture range, frequently called the *lock-in range*. Once a PLL has acquired lock, it maintains the locked condition within the hold-in range, and the borderline between the hold-in range and instability is fairly narrow. From our previous calculation and Laplace transform, applying

$$\theta_1(s) = \frac{\Delta \omega}{s^2} \tag{1-214}$$

to the input of the PLL, we can determine the maximum error using the final value theorem,

$$\lim_{t \to \infty} \theta_e(t) = \lim_{s \to 0} s\theta_e(s) = \frac{\Delta \omega}{K_o K_\theta F(o)} \tag{1-215}$$

54 LOOP FUNDAMENTALS

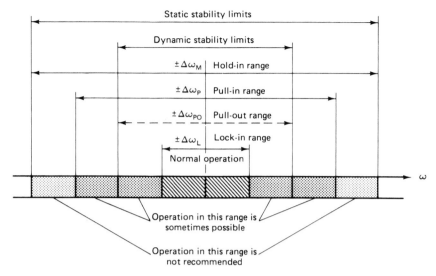

Figure 1-17 Possible operating ranges of a PLL.

With the assumption $\theta \approx \sin \theta$, the maximum amount this can be is 1 and therefore

$$\Delta \omega_H = K_o K_\theta F(o) \qquad (1\text{-}216)$$

For the active filter $F(o)$ is infinite and therefore

$$\Delta \omega_H = \infty \qquad (1\text{-}217)$$

and for the passive filter $F(o) = 1$ and

$$\Delta \omega_H = K_o K_\theta = \langle\!\langle \text{loop gain} \rangle\!\rangle \qquad (1\text{-}218)$$

We have just calculated the hold-in range of the analog linear PLL. It is apparent that the use of an active loop filter guarantees a wider hold-in range.

Now let us go back and assume that the loop has not yet acquired lock. In the beginning we have two different frequencies applied to the phase detector. It is important to understand that, as the loop will acquire lock, we have to deal with two different ranges. One is the area in which we acquire frequency lock, and only after frequency lock has been accomplished can phase lock occur. There are several requirements necessary to make frequency lock possible.

1. The tuning range of the VCO has to be wide enough to cover the desired range of operation.
2. The VCO itself has to be able to oscillate through this required range without disruption of oscillation, a jump phenomenon frequently called "discontinuities."
3. The tuning diodes have to be operated in a range where they do not become conductive (see Chapter 4) since leakage current from the diodes into the phase detector causes difficulty.

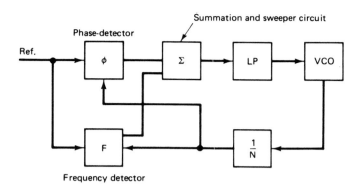

Figure 1-18 Block diagram of a PLL using a phase detector and frequency detector for acquisition aids in addition to the sweeper circuit.

4. Depending on the phase detector used, we either have a pull-in phenomenon, which we will discuss later, or sweeping, where the generated beat note at the output will sweep the oscillator from one end of the range to the other as an aid to acquire lock. In some instances, especially with pure phase detectors, an external frequency lock device may be necessary.

Figure 1-18 shows the two popular techniques used to help obtain frequency lock. The first one is the use of an additional frequency comparator, and an electronic device switches over from this to the phase detector. This technique is also currently used in the Philips HEF4750/51 LSI PLL integrated circuits. To obtain frequency lock, we first use a digital tri-state phase/frequency comparator that has a dead zone in the middle of its operating range where it is locked, and the system then switches over to a sample/hold comparator offering low-noise operation.

Another technique used in the past is an external sweeping device. Figure 1-19 shows the schematic of such an arrangment. We find a phase detector that operates into an NPN transistor. The loop filter is at the output of the phase detector, and the gain of the following dc amplifier is defined as the ratio of R_L/R_E.

A unijunction, or double-base transistor, is used to generate a sweeping signal. As the dc output voltage from this circuit reaches a certain level, the unijunction transistor ignites and starts its sweeping action. The frequency of oscillation of this circuit can be determined with the equation

$$f = \frac{1}{R_T C_T \log 2} \tag{1-219}$$

Note that R_T has to be derived from adding $R_{T1} + R_{T2} + R_{T3}$ in the starting condition, whereby we can assume that the amplifier draws no current. For the values taken here, we obtain a frequency of

$$f = \frac{1}{(18 \text{ k}\Omega + 10 \text{ k}\Omega + 220 \text{ }\Omega) 10^{-6} \log 2} = 118 \text{ Hz}$$

The circuit will be swept with this particular frequency.

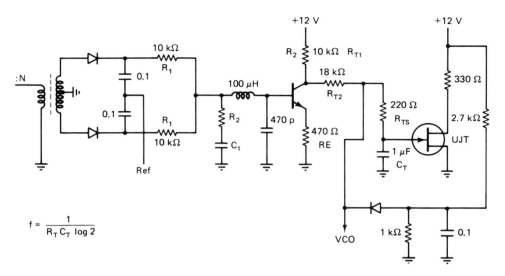

Figure 1-19 Schematic of the phase detector section of a PLL including a dc amplifier and a sweeping circuit.

We must now remember that there is a maximum frequency with which we can sweep the circuit to acquire lock. This is called the *pull-out range*. Pull-out range, defined $\Delta\omega_{PO}$, is determined by the equation

$$\Delta\omega_{PO} = 1.8\omega_n(\zeta + 1) \qquad (1\text{-}220)$$

which is an approximation that can be applied in most cases. In our particular case with the unijunction transistor, the maximum sweep rate for $\zeta = 0.7$ and $\omega_{PO} = 118(2\pi)$ results in 38 Hz for the loop frequency ω_n.

However, our system is currently not yet in lock; we still have to calculate the lock-in range. A similar formula can be found as a function of the type of filter. For the simple *RC* filter with no phase compensation, as shown previously, the lock-in range is

$$\Delta\omega_L \approx \omega_n \qquad (1\text{-}221)$$

and for the filter with phase compensation

$$\Delta\omega_L \approx 2\zeta\omega_n \qquad (1\text{-}222)$$

We are interested in determining how long it takes to acquire lock. The following relation gives a good approximation of the lock-in time:

$$T_L \approx \frac{1}{\omega_n} \qquad (1\text{-}223)$$

This is valid for all type 2 second-order linear PLLs. The calculation that we have just made referred to the phase lock. It becomes apparent from these equations that for $\zeta = 0.7$, the lock-in range is only equal to or slightly larger than the loop bandwidth. To move the VCO frequency within these limits, additional functions are required. This area is called *frequency lock*. We are now dealing with the pull-in range.

The pull-in range is the first range we have to deal with, as the phase-locked loop is about to acquire frequency, and later, phase lock. The explanation given here is somewhat backward, but it is easier to understand if one considers the ranges in this sequence, as the equations indicate the limitations. Pull-in is probably best understood by remembering that the system starts off with an offset in frequencies, and therefore a beat note appears at the output of the phase detector and of the loop filter. In the schematic using the sweeping technique shown previously, the loop bandwidth remains constant. Another way of helping the loop to acquire frequency lock is to widen the bandwidth of the loop filter, enabling phase lock with a larger frequency offset.

Figure 1-20 shows the schematic of a dual-time-constant loop filter, which can be used to explain the pull-in effect. Let us assume that initially we have a difference of several kilohertz between the two frequencies at the phase detector and are currently not considering the effect of a frequency divider.

The beat note generated at the output of the phase detector is an ac voltage together with a dc component. We also have to assume that in the initial condition the beat note is much larger in frequency than the bandwidth of the loop filter. As the output voltage of the loop filter either sweeps the oscillator to a higher

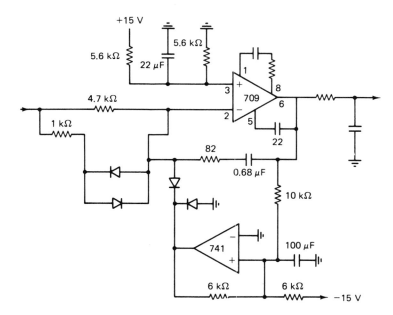

Figure 1-20 Schematic of a loop filter/integrator where the time constant is changed with the help of antiparallel diodes in the loop.

58 LOOP FUNDAMENTALS

or lower frequency, we have to see what the magnitude of the beat note as a function of sweeping is doing. It turns out that the difference in frequency becomes smaller if the VCO is swept toward higher values and becomes larger if the VCO is swept toward smaller frequency values. The output sweep frequency or beat note, therefore, is nonlinear and nonharmonic, and the average frequency at the output is no longer zero. The loop filter acting as an integrator will average the sweep.

The mathematical model for this pull-in is somewhat complicated and is given in the literature [2]. The *pull-in time*, defined as the time required for the average frequency error to decay from the initial condition to the locked limit, is

$$T_p \simeq \frac{(\Delta\omega)^2 \tau_2}{K^2} = \frac{(\Delta\omega)^2}{2\zeta\omega_n^3} \qquad (1\text{-}224)$$

This formula is valid only for the linear type 2 second-order loop. A model for the digital PLL will follow.

Example 1 Let us assume that our initial frequency offset is 1 MHz and that the loop bandwidth ω_n is 10 Hz. The time it requires for the pull-in range is

$$T_p = \frac{(2\pi \times 10^6)^2}{2\zeta(2\pi \times 10)^3} = 113.68 \times 10^6 \text{ s}$$

or 3.6 years. If we use the system that automatically increases the loop bandwidth with the two antiparallel diodes as previously shown, which results in a new loop bandwidth of 10 kHz prior to lock, the formula changes to

$$T_p = \frac{(2\pi \times 10^6)^2}{2\zeta(2\pi \times 10^4)^3} = 0.113 \text{ s}$$

The difference between both is dramatic, and with the simple trick of changing the loop bandwidth, we have speeded up the pull-in substantially.

Immediately after frequency lock has occurred, the switching diodes are no longer conductive. The remaining phase offset is handled by the lock-in function and the time it will now take to phase lock the loop:

$$T_L = \frac{1}{\omega_n} = \frac{1}{10(2\pi)} = 16 \text{ ms}$$

The time required for phase lock, therefore, is much smaller than the time required for frequency lock, and the total lock time would be 0.1296 s.

Lock-in performance of the digital PLL system that uses a tri-state digital phase/frequency comparator as shown in Chapter 4 can no longer be calculated using the linearized model as is frequently done in the literature. The output of the tri-state phase/frequency comparator behaves totally different from the linearized models.

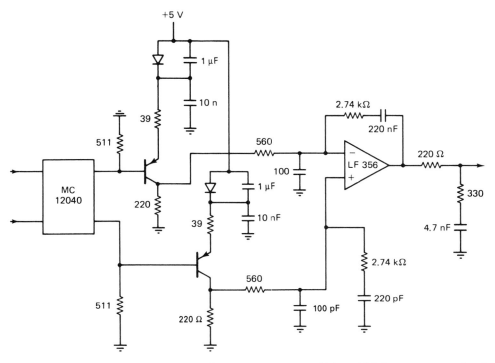

Figure 1-21 Schematic of an ECL phase/frequency comparator driving a high-frequency operational amplifier. The outputs to the inverting and noninverting input of the operational amplifiers are pulses, which are charging and discharging the loop capacitor of 220 nF. Note the additional output filter following the operational amplifier.

We have to take a look at two different examples. First, we will look at a tri-state phase/frequency comparator where the loop filter is placed after the summation stage, and the RCA or Motorola CD4046 fits this description. When the system is switched on, first the two pulses that are combined through the CMOS switches will jam the output voltage up to the power supply voltage and the loop filter will delay this action, depending on the integration time constant. The dc voltage to the VCO, therefore, will slowly rise, the VCO will be swept, and pull-in will be accomplished.

It turns out that this particular example practically behaves as the linear analog phase detector version.

Now let us consider a phase/frequency comparator equal to the MC4044, which has two outputs, where pulses are available to charge or discharage a capacitor.

If the loop filter combination shown in Figure 1-21 is used, whereby each output of the phase/frequency comparator is applied to one input of an operational amplifier provided that the amplifier is fast enough to follow the input frequency, we now have to analyze the statistical average and determine from there what the output of the operational amplifier is showing.

The following mathematical mode, to the best of my knowledge, is the only

60 LOOP FUNDAMENTALS

one that ever assessed this effect correctly and was published by Roland Best in his book *Theory and Application of Phase-Locked Loops* [1]. The following discussion is published here with Dr. Best's permissioin.

1-10-1 Pull-in Performance of the Digital Loop

In the beginning of this chapter, when we were discussing the differences between the analog and digital PLL, we started with the digital phase/frequency comparators. The flip-flop-based digital phase/frequency comparators work on the principle that they analyze the rising edges of the input signals, edge-triggered flip-flops, and are insensitive to the duty cycle.

The output can be used to charge or discharge a capacitor and has to be combined with an active filter to take full advantage of its capabilities. Figure 1-21 showed the typical arrangement using this type of phase/frequency comparator.

Initially, when the loop is not in lock, we can assume that frequencies f_1 and f_2 have a random relation to each other, the phase of one to the other is random, and that the next edge of the following signal within the time interval $0 \le t \le T$ can occur with the same probability in any given time. We define $w(t)\,dt$ as the probability that the next rising edge of the signal f_2 will occur in the time $t \cdots t + dt$. Therefore, we can write

$$w(t) = \frac{1}{T_2} \qquad (t \le T_2) \qquad (1\text{-}225)$$

$$w(t) = 0 \qquad (t > T_2) \qquad (1\text{-}226)$$

We now may have two principal cases:

1. f_2 is smaller than f_1 or $T_2 > T_1 = 1/f_1$, and the negative edge will occur: *case 1*, in the time interval $0 \le t \le T_1$; *case 2*, in the time interval $T_1 < t \le T_2$.
2. In the case of f_1 being smaller than f_2, these conditions are reversed.

As the output signal of the phase/frequency comparator is a chain of pulses that are combined, the duty cycle $\delta(t)$ will change. The average duty cycle $\bar{\delta}$, according to probability theory, can be determined from

$$\bar{\delta} = \int_0^{T_2} w(t)\,\delta(t)\,dt \qquad (1\text{-}227)$$

The integration of this can be done in two steps:

$$\bar{\delta} = \int_0^{T_1} w(t)\,\frac{t}{T_1}\,dt + \int_{T_1}^{T_2} w(t)\,\frac{t}{2T_1}\,dt \qquad (1\text{-}228)$$

with

$$\delta(t) = \frac{t}{T_1} \quad \text{and} \quad \frac{t}{2T_1}$$

depending on the time area, as discussed previously.

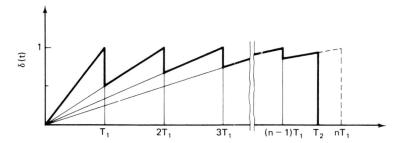

Figure 1-22 Change of duty cycle Δ(t) as a function of T.

In reality, the time interval is not going to lie between T_1 and $2T_1$ but can be between T_1 and ∞. Therefore, our average duty cycle has to be written in the form of several integrals and

$$\bar{\delta} = \int_0^{T_1} w(t) \frac{t}{T_1} dt + \int_1^{T_1} w(t) \frac{t}{2T_1} dt + \cdots$$
$$+ \int_{T_1}^{T_2} w(t) \frac{t}{nT_1} dt \qquad (1\text{-}229)$$

This can be converted into the final equation based on $f_1 > f_2$, and we obtain

$$\bar{\delta} = \frac{f_2}{2f_1} n - \sum_{i=1}^{n} \frac{1}{i} + \frac{f_1}{2nf_2} \qquad (1\text{-}230)$$

with $n = \text{Int}(f_1/f_2) + 1$. Figure 1-22 shows the duty cycle for any combinations of $n\tau_1$, and Figure 1-23 shows the average duty cycle $\bar{\delta}$ as a function of the frequency

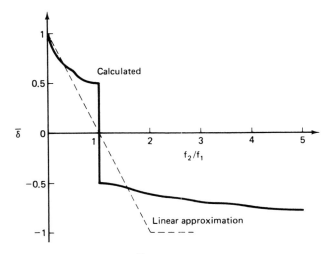

Figure 1-23 Average duty cycle $\bar{\delta}$ as a function of the frequency ratio f_2/f_1.

62 LOOP FUNDAMENTALS

ratio f_2/f_1. The straight-line approximation in this curve can be used to simplify the formula, as the lock-in will occur for the case $f_1 = f_2$. We will then obtain for the average duty cycle

$$\overline{\delta} = \frac{f_1 - f_2}{f_1} = \frac{\omega_1 - \omega_2}{\omega_1} \qquad (1\text{-}231)$$

In this case, the average output voltage is

$$\overline{v}_d = \overline{\delta} V_B = \frac{V_B}{\omega_o}(\omega_1 - \omega_2) \qquad (1\text{-}232)$$

since

$$\overline{v}_d \approx K'_d(\omega_1 - \omega_2) \qquad (1\text{-}233)$$

We finally obtain the previously used gain constant of the phase comparator of this particular type in the out-of-lock condition to be

$$K'_d \approx \frac{V_B}{\omega_o} \qquad (1\text{-}234)$$

1-10-2 Coarse Steering of the VCO as an Acquisition Aid

We have learned so far that we can use a frequency detector or a sweep oscillator to steer or sweep the oscillator close to its final frequency.

In the case of sweeping, we have to make sure that the sweeping speed is not too fast, because if it is, the oscillator will never acquire lock or it will skip cycles several times before its acquires lock. The phenomenon of cycle skipping is explained in Gardner's book [3], but generally not enough information is available about the particular loop to take full advantage of the theoretical evaluation.

Once the transfer characteristic of the VCO is known, it is possible to use a read-only memory (ROM) that receives frequency information and, with the help of a digital-to-analog (D/A) converter within very fine resolution, to coarse steer the oscillator toward its desired final frequency. This method avoids the necessity of the additional external frequency comparator and the sweeping technique. The drawback is that if diodes are changed or the characteristic of the tuning diode as a function of age changes, the lookup table will become incorrect. This is true for extremely fine resolution. Let us assume the case where we have an oscillator operating from 70 to 80 MHz, which we want to coarse steer. If we assume for a moment that the tuning diodes do not produce additional noise or that, under certain circumstances, the additional noise contribution of the coarse-steering tuning diodes can be neglected, it is possible to take an 8-bit D/A converter, as shown in Figure 1-24, that is getting its frequency information from the binary-coded decimal (BCD) commands to the frequency divider, and generate within 100-kHz resolution an output that can be used to coarse steer the tuning diodes. Now the tuning diodes responsible for the fine tuning only have to work over a fairly narrow range, and as a result of this, the VCO gain is very small.

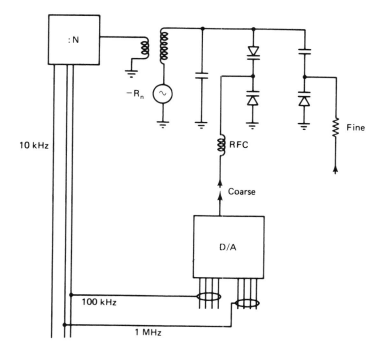

Figure 1-24 VCO coarse steering using a D/A converter.

The output impedance of the D/A converter can be made very low, and because the coarse-tuning diodes are being driven from a low-impedance point rather than the typical high impedance the dc control line has, there is no pickup on the coarse-steering line from any hum of any significant amount. As the fine-control loop now has a voltage gain of 30 to 100 kHz/V at most, the pickup is reduced by at least 20 dB, if not more. Therefore, the amount of spurious signal because of pickup and hum is reduced by the same amount.

This technique has the advantage also that the loop gain for this narrow window remains fairly constant, regardless of the VCO's curvature, as the transfer characteristic in this narrow window does not change very much.

The D/A converter has to generate a dc voltage that is not only nonlinear but rather is the opposite of the transfer characteristic of the tuning diodes used for the wide tuning range. A larger voltage swing will be needed at the higher frequency, whereas less voltage is required at the low end of the VCO. A practical schematic where this technique is used is given in Chapter 6.

In dealing with mixers, we have learned that one of the drawbacks of a heterodyne loop is that the open-loop gain changes more dramatically as the division ratio required becomes much larger. A typical loop without a heterodyne technique may have 30% or 40% variation of loop gain due to change of N, and I have seen cases where the division factor N, as a result of heterodyne technique, has changed by 20:1. How do we cope with this problem? The best way of handling this is either to use a coarse-steering technique with either tuning diodes or switching diodes and allow a very narrow window in which the oscillator will

64   LOOP FUNDAMENTALS

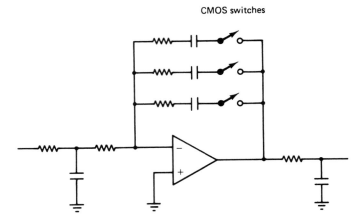

Figure 1-25 Linearizing of loop gain by changing loop components.

operate or change the loop filter dc control gain. Figure 1-25 shows an arrangement where, depending on the frequency setting of the dividers, several CMOS switches change the dc loop gain following the loop filter and, therefore, linearize the loop. The introduction of this amplifier after the loop filter has the drawback that the noise is no longer limited by a following filter, or if such an RC low-pass filter is used after the amplifier, the technique of analyzing high-order loops has to be used, and inside the loop bandwidth, we still find the additional noise contribution.

This approach is typically used in wideband loops where the output oscillator operates from 200 to 300 MHz, as an example, and the reference frequency is between 100 kHz and several megahertz. The loop gain, because of the small division ratio, is fairly high, while the loop gain variation due to some heterodyning may also be very high. In an effort to linearize, either the operational amplifiers are offset with a dc control voltage, or the loop filter is modified with additional capacitors in parallel, or a dc amplifier following the loop filter is used, which changes the dc gain. In some instances, all three techniques are used simultaneously, and it becomes very tricky to avoid additional noise being brought into the loop and to make all systems track without difficulty. Figure 1-26 shows a combination of all these techniques. Table 1-4 shows the most important formulas for digital PLLs.

1-10-3 Loop Stability

The easiest way to analyze the loop stability is to plot the magnitude and phase of the open-loop transfer function $K_v F(s)/s$ as a function of frequency. First, consider the case where $F(s)$ is a simple low-pass filter described by Eq. (1-27). For this case the open-loop frequency response is

$$K_v G(j\omega) = \frac{K_v}{j\omega(j\omega\tau + 1)} \qquad (1\text{-}235)$$

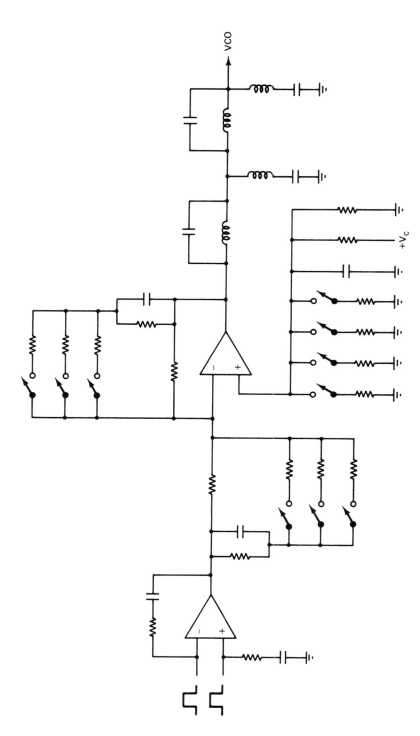

Figure 1-26 Circuit diagram of the loop filter dc amplifier arrangement of a PLL, where the loop gain and dc offset are controlled by CMOS switches to linearize and coarse steer the VCO.

Table 1-4 Most important formulas for digital PLLs (second-order only)

Phase/Frequency Comparator	Exclusive-OR Gate		Edge-Triggered JK Master/Slave Flip-flop	
	Active Filter	Passive Filter	Active Filter	Passive Filter
Hold-in range	$\Delta\omega_H \to \infty$	$\Delta\omega_H = \dfrac{\pi}{2}\dfrac{K_o K_d}{N}$	$\Delta\omega_H \to \infty$	$\Delta\omega_H = \pi\dfrac{K_o K_d}{N}$
Capture range				
$\tau_2 \neq 0$		$\Delta\omega_L \approx \pi\zeta\omega_n$		$\Delta\omega_L \approx 2\pi\zeta\omega_n$
$\tau_2 = 0$		$\Delta\omega_L \approx \dfrac{\pi}{\sqrt{8}}\omega_n$		$\Delta\omega_L \approx \dfrac{\pi}{\sqrt{3}}\omega_n$
Pull-in range	$\Delta\omega_P \approx \dfrac{\pi}{2}\sqrt{\dfrac{2\zeta\omega_n K_o K_d}{N}}$	$\Delta\omega_P \approx \dfrac{\pi}{2}\sqrt{\dfrac{2\zeta\omega_n K_o K_d}{N} - \omega_n^2}$	$\Delta\omega_P \approx \pi\sqrt{\dfrac{2\zeta\omega_n K_o K_d}{N}}$	$\Delta\omega_P \approx \pi\sqrt{\dfrac{2\zeta\omega_n K_o K_d}{N} - \omega_n^2}$
Pull-in time		$T_P \approx \dfrac{4}{\pi^2}\dfrac{\Delta\omega_0^2}{\zeta\omega_n^3}$		$T_P \approx \dfrac{\Delta\omega_0^2}{\pi^2\zeta\omega_n^3}$
Pull-out range				
$\zeta < 1$		$\Delta\omega_{PO} \approx 1.8\omega_n(\zeta + 1)$	$\Delta\omega_{PO} = \pi\omega_n\exp\left(\dfrac{\zeta}{\sqrt{1-\zeta^2}}\arctan\dfrac{\sqrt{1-\zeta^2}}{\zeta}\right)$	
$\zeta > 1$			$\Delta\omega_{PO} = \pi\omega_n\exp\left(\dfrac{\zeta}{\sqrt{\zeta^2-1}}\arctan\dfrac{\sqrt{\zeta^2-1}}{\zeta}\right)$	

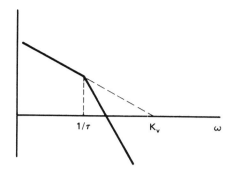

Figure 1-27 Magnitude of the open-loop gain of a PLL system.

The straight-line approximation of the magnitude of this open-loop transfer function is plotted in Figure 1-27. The magnitude of the response decreases at the rate of 6 dB/octave until the frequency is equal to the −3-dB frequency of the low-pass filter ($\frac{1}{\tau}$); for higher frequencies the magnitude decreases at a rate of −12 dB/octave.

Several rules of thumb, developed by Bode for feedback amplifiers, are useful in selecting the loop parameters. The first has to do with selecting the filter bandwidth $\omega_L = 1/\tau$. The approximation is: If the open-loop frequency response crosses the 0-dB line with a slope of −6 dB/octave, the system is stable; if the slope is −12 dB/octave or greater, the system is unstable. The second-order system under consideration is inherently stable, but the model is an approximation to a higher-order system. If the open-loop second-order model crosses the 0-dB line at −12 dB/octave, there is little room left for error. Additional phase shift from the VCO or phase detector could cause the loop to go unstable.

To have the open-loop gain cross the 0-dB line at −6 dB/octave, it is necessary that $\omega_L > K_v$. The larger ω_L is, the better will be the loop stability. From the filtering viewpoint, the smaller ω_L, the smaller the loop bandwidth and the less noise that will reach the VCO. K_v should be as small as possible to minimize the bandwidth. The larger the K_v, the smaller the steady-state error and the faster the loop response. Hence, in PLL design, compromises among noise performance, loop stability, steady-state error, and transient performance must be made.

Another rule of thumb that is helpful in PLL design is that the frequency ω_c at which the magnitude of the open-loop transfer function is unity,

$$\frac{K_v F(j\omega)}{j\omega_c} = 1 \qquad (1\text{-}236)$$

is approximately the closed-loop 3-dB bandwidth. This relation is exact for the case where $F(j\omega) = $ constant.

If $F(s)$ is a simple low-pass filter response and $\omega_L > K_v$, the open-loop frequency response will be as shown in Figure 1-28. In this case, the loop bandwidth is approximately equal to K_v.

If $\omega_L < K_v$, the straight-line approximation will cross the 0-dB line with a slope of −12 dB/octave, which is not good from the standpoint of loop stability. Thus

68 LOOP FUNDAMENTALS

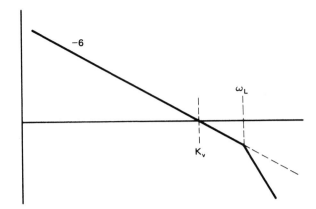

Figure 1-28 Open-loop frequency response in the case $\omega_L = K_v$.

the filter bandwidth should be greater than the open-loop crossover frequency ω_c; ω_c will be approximately equal to the closed-loop bandwidth. Therefore, the filter bandwidth, for good loop stability, should be greater than the loop bandwidth for the simple type 1 system under discussion.

Another parameter that is useful in evaluating the response of second-order and higher systems is the phase margin, which is defined as

$$\phi_m = 180° + \arg KG(j\omega_c) \tag{1-237}$$

That is, the phase margin is equal to 180° plus the phase shift of the open-loop gain (a negative number) at the open-loop crossover frequency ω_c. The greater the phase margin, the more stable the system and the more phase lag from parasitic effects that can be tolerated.

Example 2 Consider a phase-locked loop that has $K_v = 10$ rad/s and that contains a low-pass filter with a corner frequency of 20 rad/s. The magnitude and phase of the open-loop transfer function are plotted in Figure 1-29. The system crossover

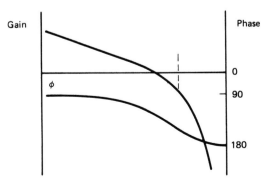

Figure 1-29 Magnitude and phase of the open-loop transfer function for $K_v = 19$ rad/s and $\omega_L = 20$ rad/s.

frequency is approximately 10 rad/s. At this frequency, the phase shift of the open-loop transfer function is −112.5°, so the phase margin is 67.5°.

In this example, the complete phase plot was presented, but once one is familiar with phase plots, they no longer need to be included. One can simply calculate the phase shift after determining the open-loop crossover frequency from the magnitude plot.

Example 3 In Example 2, if the filter corner frequency had been 2 rad/s rather than 20 rad/s, what would have been the system phase margin?

Solution. To determine the phase margin, first plot the magnitude of the open-loop gain and determine the crossover frequency. The straight-line approximation of the magnitude is plotted in Figure 1-30. ω_c is found to be approximately 4.4 rad/s. Thus, the system phase margin is $180° - (90° + \arctan 2.2) = 23.40°$, which is too small for good loop stability. This is in agreement with the rule of thumb, which states that if the magnitude of the open-loop response described crosses the 0-dB line with a slope of −12 dB/octave, the system is unstable. In this example, the straight-line approximation for the gain decreases at −12 dB/octave, but the actual response crosses the 0-dB line with a slope slightly more positive than −12 dB/octave: hence, the small phase margin.

Although the most important frequency-domain design parameters are the closed-loop bandwidth ω_h and the peak value M_p of the closed-loop frequency response, no design techniques exist that allow easy specification of B and M_p. It is relatively easy to design for specified open-loop parameters ω_c and ϕ_m. There are approximations that relate ω_c and ϕ_m to ω_n, M_p, and ζ, and thus to the system rise time and overshoot. Fortunately, the conditions under which these approximations are valid are satisfied by most PLLs.

For the open-loop system (second-order loop),

$$K_v G(s) = \frac{K_v}{s(s/\omega_L + 1)} \qquad (1\text{-}238)$$

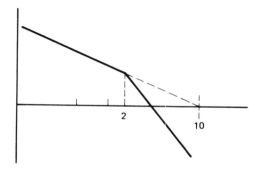

Figure 1-30 Magnitude of the open-loop gain of Example 3.

70 LOOP FUNDAMENTALS

Used with unity feedback, the closed-loop transfer function is given by Eq. (1-28), with

$$\omega_n^2 = K_v \omega_L \tag{1-239}$$

and

$$\zeta = \frac{1}{2}\sqrt{\frac{\omega_L}{K_v}} \tag{1-240}$$

The open-loop unity gain frequency is easily shown to be

$$\omega_c = \omega_L \left[\frac{\sqrt{1 + 4(K_v/\omega_L)^2} - 1}{2} \right]^{1/2} \tag{1-241}$$

Once ω_c is known, the phase margin

$$\phi_m = 90° - \arctan\frac{\omega_c}{\omega_L} = 90° - \arctan\left[\frac{\sqrt{1 + (1/2\zeta)^2} - 1}{2}\right]^{1/2} \tag{1-242}$$

can be calculated. This equation is plotted in Figure 1-31. The closed-loop system parameters of most importance are adequate stability (which is related to phase margin), system bandwidth (which determines the speed of the transient response), and system transient response (rise time and overshoot). For a low-pass transfer

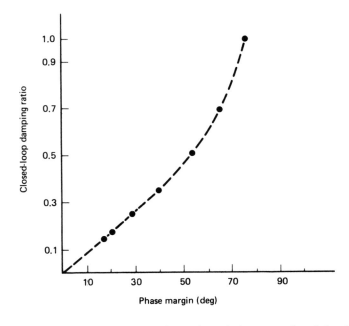

Figure 1-31 Closed-circuit dampling ratio and phase margin relationship.

function, the bandwidth ω_n is defined as the frequency at which the gain is equal to 0.707 of its dc value. The bandwidth of the system represented by Eq. (1-28) is

$$\omega_h = \omega_n(1 - 2\zeta^2 + \sqrt{2 - 4\zeta^2 + 4\zeta^4})^{1/2} \qquad (1\text{-}243)$$

which can be calculated using Eqs. (1-35), (1-239), and (1-240). For the underdamped second-order system given by Eq. (1-28) ($\zeta < 1$), the peak value of the time response to a unit step input can be shown to be

$$P_o = 1 + e^{-\pi\zeta/\sqrt{1-\zeta^2}} \qquad (1\text{-}244)$$

The overshoot is determined solely by ζ. P_o as a function of ζ is plotted in Figure 1-32.

For high-order systems, the overshoot and bandwidth are not readily related to the open-loop system parameters, but a good first approximation is that Eq. (1-241) holds for higher-order systems. It is relatively easy to design a system to have a given phase margin. A design can then be evaluated using computer simulation. If the simulation indicates that the overshoot is too high (or too low) the phase margin can be increased (reduced), but the relations among phase margin, damping, and overshoot are amazingly accurate for higher-order systems. This implies that the response of most feedback systems can be described by a second-order model. Also, the closed-loop bandwidth can be related to the open-loop crossover frequency ω_c and the damping ratio, but it usually suffices to use the rule of thumb that the closed-loop bandwidth of underdamped systems is approximately 50% greater than the open-loop crossover frequency ω_c.

If it is desired to design for a peak transient overshoot, Eq. (1-240) can be used to determine the damping and then Eq. (1-241) is used to determine the required phase margin.

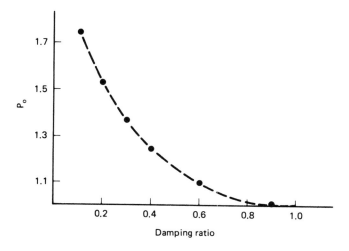

Figure 1-32 Peak overshoot as a function of damping ratio.

72 LOOP FUNDAMENTALS

Example 4 For the phase-locked loop with open-loop transfer function

$$\frac{K_v}{s(s/\omega_L + 1)}$$

($K_v = 1000$), determine the low-pass filter corner frequency ω_L so that the system peak overshoot in response to a step input will be less than 20%.

Solution. Equation (1-242) or Figure 1-32 indicates that for $P_o < 1.2$, the damping ratio ζ must be greater than 0.45. For a ζ of 0.45, the corresponding phase margin is found [using Eq. (1-242)] to be about 50.

The low-pass filter can contribute $-40°$ phase lag 1 and the phase margin will be equal to 50%. Therefore, ω_L must be greater than ω_c (if $\omega_L = \omega_c$, the phase margin would be 45°), so ω_c is approximately 1000 rad/s = K_v. Thus arctan $1000/\omega L = 40°$ or $\omega_L = 1192$. (This is somewhat of an approximation, since adding the low-pass filter will slightly reduce the crossover frequency.) The desired open-loop transfer function becomes

$$\frac{1000}{s\left(\dfrac{s}{1.19 \times 10^3} + 1\right)} = KG(s) \tag{1-245}$$

The step response is plotted in Figure 1-33. The overshoot is 13% and the rise time is 2.1 ms. The overshoot is considerably less than the specified maximum of 20% because of the straight-line approximations used to estimate the gain and crossover frequency. Note that this second-order system is simple enough to be solved analytically since

$$\frac{K_v G}{1 + K_v G} = \frac{1}{(s^2/1000\omega_L) + (s/1000) + 1} = \frac{1}{(s^2/\omega_n^2) + (2\zeta s/\omega_n) + 1}$$

where $\omega_n^2 = 1000\,\omega_L$ and $2\zeta/\omega_n = 1/1000$, or

$$\zeta = \frac{\omega_n}{2000} = \frac{\sqrt{1000\omega_L}}{2000}$$

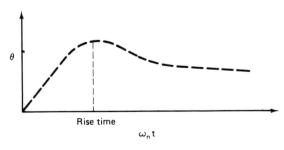

Figure 1-33 Overshoot and rise time of our example.

For $\zeta = 0.45$ (the design value),

$$\omega_L = \frac{(900)^2}{1000} = 810$$

The straight-line approximations resulted in a 32% error in the calculation of the low-pass corner frequency and the overshoot was 13% rather than 20% (for $\omega_L = 810$, the rise time is 2.28 ms). The differences between the two methods could have been reduced by accounting for the fact that the pole of the low-pass filter reduces the crossover frequencies and thus increases the actual phase margin over that estimated with the straight-line approximation.

In some instances it is also necessary to specify the loop bandwidth. In order to control both the loop damping and bandwidth, an amplifier can be added in series with the low-pass filter. If the filter is implemented using active components, the additional gain can be obtained without any additional components.

Example 5 Consider Example 4 with the additional specification that the rise time in response to a unit step input be less than 1 ms. Since the overshoot is to be less than 20%, the phase margin must be approximately 50°. To design for the rise-time specification, it is easiest to use the approximation

$$t_r = \frac{2.2}{B}$$

which is exact only for first-order systems but provides a good design guideline for higher-order systems. Thus, ω_c should be greater than $2.2/t_r = 2.2 \times 10^3$ rad/s.

The previous discussion has shown that

$$\omega_c = K_\theta K_o K = 1000K$$

Thus, for an $\omega_c = 2.2 \times 10^3$ rad/s, an additional amplifier with a gain $K = 2.2$ needs to be added. With the increased ω_c, ω_L will have to be increased from Example 4 in order to meet the phase margin specification. For the second-order systems under discussion,

$$\frac{2\zeta}{\omega_n} = \frac{1}{K_v} \qquad (1\text{-}246)$$

The damping ζ is to be approximately 0.45 to meet the overshoot specification. It suffices to estimate the closed-loop bandwidth by assuming that it is approximately equal to ω_n, which is also approximately equal to the open-loop crossover frequency. Therefore, for $\zeta = 0.45$ and 1 ms rise time,

$$\omega_n = \frac{2.2}{10^{-3}} = 2.2 \times 10^3 = 2\zeta K_v = \sqrt{K_v \omega_L}$$

74 LOOP FUNDAMENTALS

Therefore,

$$K_v = \frac{2.2 \times 10^3}{2 \times 0.45} = 2.44 \times 10^3$$

and

$$\omega_L = \frac{(2.2 \times 10^3)^2}{2.44 \times 10^3} = 1.98 \times 10^3$$

An additional gain K required is

$$K = 2.44$$

The complete open transfer function is then

$$K_v G(s) = \frac{2.44 \times 10^3}{s\left(\dfrac{s}{1.98 \times 10^3} + 1\right)}$$

and the closed-loop transfer function is

$$\frac{K_v G}{1 + K_v G} = \frac{1}{\dfrac{s^2}{(2.2 \times 10^3)^2} + \dfrac{s}{2.44 \times 10^3} + 1}$$

A plot of the step response is shown in Figure 1-34. The peak overshoot is 10% and the rise time is 0.7 ms. The two specifications are now met. In general, two

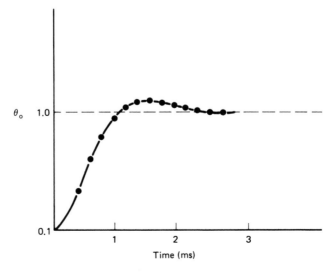

Figure 1-34 Plot of the step response of a 10% overshoot and 0.7-ms rise time.

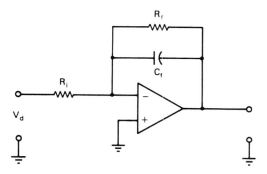

Figure 1-35 Loop filter with a gain of 2.2.

adjustable parameters, such as loop gain and filter bandwidth, are needed to independently specify overshoot and rise time.

An operational amplifier circuit to realize the low-pass filter with a gain of 2.4 is shown in Figure 1-35. Since the feedback impedance is

$$Z_f = \frac{R_f}{R_f C_f s + 1} \qquad (1\text{-}247)$$

the ideal voltage gain is

$$A_v = \frac{-Z_f}{Z_t} = \frac{-R_f/R_i}{R_f C_f s + 1} \qquad (1\text{-}248)$$

which realizes the desired gain and filter provided that

$$\frac{R_f}{R_t} = 2.44 \quad \text{and} \quad R_f C_f = \frac{1}{1.98 \times 10^3}$$

If the phase inversion resulting from this circuit is undesirable, phase inverting at the phase/frequency discriminator can be performed.

REFERENCES AND SUGGESTED READING

1. Roland Best, *Theorie und Anwendungen des Phase-locked Loops*, Fachschriftenverlag, Aargauer Tagblatt AG, Aarau, Switzerland, 1976 (Order No. ISBN-3-85502-011-6).
2. D. Richman, "Color Carrier Reference Phase Synchronization Accuracy in NTSC Color Television," *Proceedings of the IRE*, Vol. 42, January 1954, pp. 106–133.
3. Floyd M. Gardner, *Phaselock Techniques*, 2nd ed., Wiley, New York, 1960.
4. Anatol I. Zverev, *Handbook of Filter Synthesis*, Wiley, New York, 1967.
5. A. Przedpelski, "Analyze, Don't Estimate Phase-Locked Loop Performance of Type 2 Third Order Systems," *Electronic Design*, May 10, 1978.

6. A. Przedpelski, "Optimized Phase Locked Loop to Meet Your Needs or Determine Why You Can't," *Electronic Design*, September 1978.
7. A. Przedpelski, "Suppress Phase Locked Loop Sidebands Without Introducing Instability," *Electronic Design*, September 1978.
8. Alain Blanchard, *Phase-Locked Loops*, Wiley, New York, 1976.
9. C. R. Cahn, "Piecewise Linear Analysis of Phase-Locked Loops," *IRE Transactions on Space Electronics and Telemetry*, Vol. SET-8, No. 1, March 1962, pp. 8–13.
10. J. A. Develet, Jr., "The Influence of Time Delay on Second-Order Phase-Lock Loop Acquisition Range," *Proceedings of the International Telemetering Conference*, Vol. 1, September 23–27, 1963, pp. 432–437.
11. W. F. Egan, "Phase-Locked Loop Simulation Program," *Proceedings of the 1976 GTE Symposium on Computer Aided Design*, Vol. 1, GTE Laboratories, Waltham, MA, June 1976, pp. 239–253.
12. J. Gibbs and R. Temple, "Frequency Domain Yields Its Data to Phase-Locked Synthesizer," *Electronics*, April 27, 1978, pp. 107–111.
13. L. J. Greenstein, "Phase-Locked Loop Pull-in Frequency," *IEEE Transactions on Communications*, Vol. COM-22, August 1974, pp. 1005–1013.
14. U. Mengali, "Acquisition Behavior of Generalized Tracking Systems in the Absence of Noise," *IEEE Transactions on Communications*, Vol. COM-21, July 1973, pp. 820–826.
15. E. N. Protonotarios, "Pull-in Performance of a Piecewise Linear Phase-Locked Loop," *IEEE Transactions on Aerospace and Electronic Systems*, Vol. AES-5, No. 3, May 1969, pp. 376–386.
16. T. J. Rey, "Automatic Phase Control, Theory and Design," *Proceedings of the IRE*, October 1960, pp. 1760–1771.
17. R. G. Robson, "The Pull-in Range of a Phase-locked Loop," Conference on Frequency Generation and Control for Radio Systems, London, *Conference Publication No. 31*, May 1967, pp. 139–143.
18. J. Truxal, *Automatic Feedback Control System Synthesis*, McGraw-Hill, New York, 1955, pp. 38–41.
19. A. Viterbi, *Principles of Coherent Communication*, McGraw-Hill, New York, 1966.
20. C. S. Weaver, "A New Approach to the Linear Design and Analysis of Phase-Locked Loops," *IRE Transactions on Space Electronics and Telemetry*, Vol. SET-5, December 1959, pp. 166–178.
21. D. H. Wolaver, *Phase-Locked Loop Circuit Design*, Prentice-Hall, Englewood Cliffs, NJ, 1991.
22. D. Wulich, "Fast Frequency Synthesis by PLL Using a Continuous Phase Divider," *Proceedings of the IEEE*, Vol. 76, No. 1, January 1988, pp. 85–86.
23. R. C. Halgren et al., "Improved Acquisition in Phase-Locked Loops with Sawtooth Phase Detectors," *IEEE COM*-30, No. 10, October 1982, pp. 2364–2375.
24. W. C. Lindsey and C. M. Chie, *Phase-Locked Loops*, IEEE Press, New York, 1986.
25. P. A. Weisskopf, "Subharmonic Sampling of Microwave Signal Processing Requirements," *Microwaves & RF*, May 1992.
26. P. Vella et al., "Novel Synthesizer Cuts Size, Weight, and Noise Levels," *Microwaves & RF*, May 1991.
27. W. F. Egan, "Sampling Delay—Is It Real?" *RF Design*, February 1991.
28. K. Osafune et al., "High-Speed and Low-Power GaAs Phase Frequency Comparator,"

IEEE Transactions, Vol. MTT-34, January 1986, pp. 142–146.
29. D. T. Gavin and R. M. Hickling, "A PLL Synthesizer Utilizing a New GaAs Phase Frequency Detector," INS EEA 89-000943.
30. A. Hill and J. Surber, "The PLL Dead Zone and How to Avoid It," *RF Design*, March 1992.
31. Analog Devices, "Ultrahigh Speed Phase/Frequency Discriminator," Data Sheet AD9901.
32. M. O'Leary, "Practical Approach Augurs PLL Noise in RF Synthesizers,"*Microwaves & RF*, September 1987.
33. T. J. Endres and J. B. Kirkpatrick, "Sensitivity of Fast Settling PLLs to Differential Loop Filter Component Variations," *IEEE 47th Annual Symposium on Frequency Control*, Hershey, PA, May 27–29, 1992.
34. M. Marlin, "RF Sidebands Caused by DC Power Line Fluctuations," *Microwave Journal*, September 1991.
35. V. Manassewitsch, *Frequency Synthesizers Theory and Design*, 2nd ed., Wiley, New York, 1980, pp. 502–503.
36. D. L. Snyder, "The State Variable Approach to Analog Communication Theory," *IEEE Transactions on Information Theory*, Vol. IT-14, No. 1, January 1986.
37. W. D. Lindsey, *Synchronization Systems in Communication and Control*, Prentice-Hall, Englewood Cliffs, NJ, 1972.
38. J. A. Crawford, "The Phase-Locked Loop Concept for Frequency Synthesis," *M/A COM Linkabit*, February 1987.
39. K. K. Clarke and D. T. Hess, *Communication Circuits*, Addison-Wesley, Reading MA, 1971, Chap. 6.
40. C. H. Houpis et al., "Refined Design Method for Sampled-Data Control Systems: The Pseudo-Continuous Time (PCT) Control System Design," *IEEE Proceedings*, Vol. 132, Part D, No. 2, March 1985, pp. 69–74.
41. W. F. Egan, *Frequency Synthesis by Phase Lock*, Wiley, New York, 1981, pp. 115–123, 126–129.
42. J. Crawford, "The Phase/Frequency Detector," *RF Design*, February 1985, pp. 46–57.
43. B. C. Kuo, *Digital Control Systems*, Holt, Rinehart and Winston, New York, 1980, p. 48.
44. E. Kreyszig, *Advanced Engineering Mathematics*, 3rd ed., Wiley, New York, 1972, p. 604.
45. G. F. Franklin and J. D. Powell, *Digital Control of Dynamic Systems*, Addison-Wesley, Reading, MA, 1980, p. 86.
46. J. E. Marshall, *Control of Time-Delay Systems*, Peter Peregrinus, London, 1982, Chap. 3.
47. B. C. Kuo, *Automatic Control Systems*, 3rd ed., Prentice-Hall, Englewood Cliffs, NJ, 1975, pp. 316–374, 434–444.
48. J. A. Gibson et al., "Transfer Function Models of Sampled Systems," *IEE Proceedings*, Vol. 130, Part G, No. 2, April 1983, pp. 37–44.
49. K. Horiguchi and N. Hamada, "System Theoretical Considerations on N-Point Padé Approximation," *Electronic Communicator Japan*, Vol. 69, Part 1, No. 4, 1986, pp. 10–20.
50. D. Y. Abramovitch, "Analysis and Design of a Third-Order Phase-Locked Loop," Milcom 1988, pp. 455–459.

51. R. I. Ross, "Evaluating the Transient Response of a Network Function," *Proceedings of the IEEE*, May 1967, pp. 693–694.
52. M. S. Corrington, "Simplified Calculation of Transient Response," *Proceedings of the IEEE*, March 1965, pp. 287–292.
53. J. Vlach, *Computerized Approximation and Synthesis of Linear Networks*, Van Nostrand, New York, 1993, pp. 106–112.
54. L. O. Chua and P. Lin, *Computer-aided Analysis of Electronic Circuits: Algorithms & Computational Techniques*, Prentice-Hall, Englewood Cliffs, NJ, 1975, Chap. 12.
55. T. R. Cuthbert, *Circuit Design Using Personal Computer*, Wiley, New York, 1983, Chap. 3.
56. K. W. Henderson and W. H. Kantz, "Transient Response of Conventional Filters," *IRE Transactions on Circuit Theory*, December 1958, pp. 333–347.
57. D. L. Jagerman, "An Inversion Technique for the Laplace Transform," *Bell System Journal*, Vol. 61, No. 8, October 1982, pp. 1995–2003.
58. J. M. Smith, *Mathematical Modeling and Digital Simulation for Engineers and Scientists*, 2nd ed., Wiley, New York, 1987, App. A.
59. H. S. Malvar, "Transform Analog Filters into Digital Equivalents," *Electronic Design*, April 30, 1981, pp. 145–148.
60. M. E. Holder and V. A. Thomason, "Using Time Moments to Determine System Response," *IEEE Transactions on Circuits and Systems*, Vol. CAS-35, No. 9, September 1988, pp. 1193–1195.
61. F. D. Waldauer, *Feedback*, Wiley, New York, 1982.

2

NOISE AND SPURIOUS RESPONSE OF LOOPS

2-1 INTRODUCTION TO SIDEBAND NOISE

In the course of dealing with various synthesizer configurations, we will learn that the output noise of a synthesizer is an important design consideration. The main sources of noise are leakage of the reference frequency in phase-locked loops and incomplete suppression of the unwanted component of mixer output (spurious). Another source of noise is the noise inherent in the oscillator.

If the spectral power density is measured at the output of an oscillator, a curve such as that of Figure 2-1 is observed. Rather than all of the power being concentrated at the oscillator frequency, some is distributed in frequency bands on both sides of the oscillator frequency.

As noise is a form of stability, it is useful to characterize frequency stability in the time domain in several areas. Short-term stability extends between a very small fraction of a second to 1 s, maybe under some considerations up to 1 min, and the value for the stability between 1 s and 1 min will be about the same. For longer time periods, we talk about long-term stability or aging. The aging is typically expressed in forms of how many parts in 10^{-10} or 10^{-11} per day the frequency changes. This information is in the time domain; in the frequency domain, we find terms like "random walk," "flicker," and "wide phase noise," which describe the slope of spectral density. The Fourier frequency, at times labeled f_m, is at times called sideband frequency, offset frequency, modulation frequency, or baseband frequency. In this book we will refer to it as *offset frequency*, describing the signal-to-noise ratio of an oscillator at a certain offset off the center frequency. The most common characterization of phase noise of a source is the frequency power density, and the probable reason for this is that it can be seen only as a spectrum analyzer when the AM noise contribution is insignificant. The spectrum analyzer display is then symmetrical.

Each one takes one side and by looking at sideband noise in a 1-Hz bandwidth

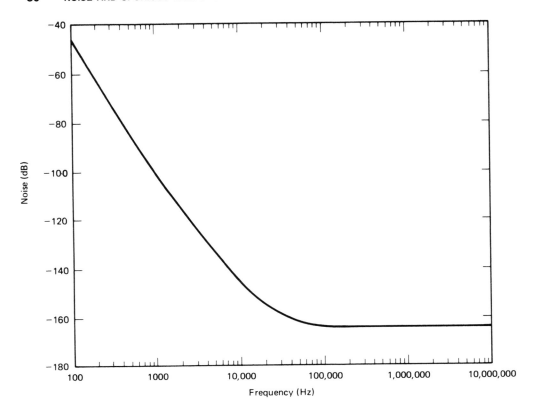

Figure 2-1 Typical noise sideband curve of a free-running oscillator, Rohde & Schwarz signal generator SMDU.

leads to the definition of $\mathscr{L}(f_m)$. $\mathscr{L}(f_m)$ is defined as the ratio of the single sideband power of phase noise in a 1-Hz bandwidth f_m hertz away from the carrier frequency to the total signal power. This is plotted in Figure 2-1.

These unwanted frequency components are now referred to as *oscillator noise*. The oscillator output $S(t)$ can be expressed by the equation

$$S(t) = A(t) \cos[\omega_o t + \theta(t)] \qquad (2\text{-}1)$$

where $A(t)$ describes the amplitude variation as a function of time and $\theta(t)$ is the phase variation. $\theta(t)$ is referred to as *phase noise*. A well-designed, high-quality oscillator is very amplitude stable and $A(t)$ can be considered constant. For a constant-amplitude signal, all oscillator noise is due to $\theta(t)$. Leeson [7] has developed a model that describes the origins of phase noise in oscillators, and since it closely fits experimental data, the model is widely used in describing the phase noise of oscillators and frequency synthesizers. Leeson's model will be described, but first a relation between the observed power spectral density function and $\theta(t)$ will be developed.

A carrier signal of amplitude V that is frequency modulated by a sine wave of frequency f_m can be represented by the equation

$$S(t) = V \cos\left(\omega_o t + \frac{\Delta f}{f_m} \sin \omega_m t\right) \quad (2\text{-}2)$$

where Δf is the peak frequency deviation and $\theta_p = \Delta f / f_m$ is the peak phase deviation, often referred to as the *modulation index* β. Equation (2-2) can be expanded as

$$S(t) = V[\cos(\omega_o t)\cos(\theta_p \sin \omega_m t) \\ - \sin \omega_o t \sin(\theta_p \sin \omega_m t)] \quad (2\text{-}3)$$

If the peak phase deviation is much less than 1 ($\theta_p \ll 1$),

$$\cos(\theta_p \sin \omega_m t) \approx 1$$

and

$$\sin(\theta_p \sin \omega_m t) \approx \theta_p \sin \omega_m t$$

Thus, for $\theta_p \ll 1$, the signal $S(t)$ is approximately equal to

$$\begin{aligned} S(t) &= V[\cos(\omega_o t) - \sin \omega_o t (\theta_p \sin \omega_m t)] \\ &= V\left\{\cos(\omega_o t) - \frac{\theta_p}{2}[\cos(\omega_0 + \omega_m)t - \cos(\omega_0 - \omega_m)t]\right\} \end{aligned} \quad (2\text{-}4)$$

That is, when the peak phase deviation is small, the phase deviation results in frequency components on each side of the carrier of amplitude $\theta_p/2$. This frequency distribution of a narrowband FM signal is useful for interpreting an oscillator's power spectral density as being due to phase noise. The phase noise in a 1-Hz bandwidth has a noise power-to-power ratio of

$$\mathscr{L}(f_m) = \left(\frac{V_n}{V}\right)^2 = \frac{\theta_p^2}{4} = \frac{\theta_{\text{rms}}^2}{2} \quad (2\text{-}5)$$

The total noise is the noise in both sidebands and will be denoted by S_θ. That is,

$$S_\theta = 2\frac{\theta_{\text{rms}}^2}{2} = \theta_{\text{rms}}^2 = 2\mathscr{L}(f_m) \quad (2\text{-}6)$$

With this interpretation of the noise power, the noise can now be described in terms of its origin; see Figure 2-2.

Noise can be expressed in a number of ways; therefore, we want to try to cover the various methods of describing other forms of stability before we analyze the oscillator.

82 NOISE AND SPURIOUS RESPONSE OF LOOPS

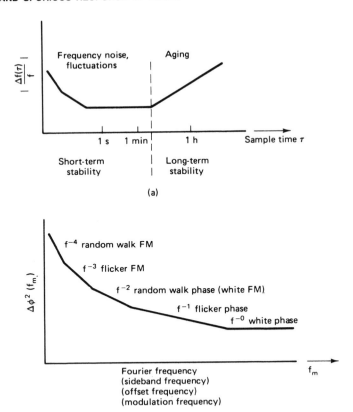

Figure 2-2 Characterization of a noise sideband in the time and frequency domain and its contributions: (a) time domain and (b) frequency domain.

2-2 SPECTRAL DENSITY OF FREQUENCY FLUCTUATIONS, RELATED TO $S_{\Delta\theta}$ AND $\mathscr{L}$

Stability measurements using frequency comparators give the *spectral density of frequency fluctuations*,

$$S_{\Delta f}(f_m) = \Delta f_{rms}^2 \tag{2-7}$$

To relate the spectral density of frequency fluctuations to the spectral density of phase noise, we recall that

$$\Delta f(t) = \frac{1}{2\pi} \frac{d\Delta\theta(t)}{dt} \tag{2-8}$$

Transformed into the frequency domain,

$$\Delta f(f_m) = f_m \Delta\theta(f_m) \tag{2-9}$$

$$S_{\Delta f}(f_m) = \Delta f_{rms}^2(f_m) = f_m^2 S_{\Delta\theta}(f_m) = 2 f_m^2 \mathscr{L}(f_m) \tag{2-10}$$

NBS proposes to standardize the definition of the spectral density of fractional frequency fluctuations. The instantaneous frequency deviation is normalized to the carrier frequency f_o.

$$y(t) = \frac{\Delta f(t)}{f_o} \tag{2-11}$$

$$S_y(f_m) = \frac{1}{f_o^2} S_{\Delta f}(f_m) = \frac{f_m^2}{f_o^2} S_{\Delta\theta}(f_m) = \frac{2f_m^2}{f_o^2} \mathscr{L}(f_m) \tag{2-12}$$

Characterizing fractional frequency fluctuations allows better comparison between sources with different carrier frequencies.

2-3 RESIDUAL FM RELATED TO $\mathscr{L}(f_m)$

Residual FM, the total rms frequency deviation within a specified bandwidth, is another common way to specify the frequency stability of signal generators. Commonly used bandwidths are 50 Hz to 3 kHz, 300 Hz to 3 kHz, and 20 Hz to 15 kHz.

$$\Delta f_{\text{res}} = \sqrt{2} \sqrt{\int_a^b \mathscr{L}(f_m) f_m^2 \, df_m} \tag{2-13}$$

Table 2-1 correlates Δf_{res} and $\mathscr{L}(f_m)$ for specific slopes of $\mathscr{L}(f_m)$ and $\mathscr{L}$ at 1 kHz = -100 dBc.

Table 2-1

$\mathscr{L}^a$ at 1 kHz (dBc)	Slope of $\mathscr{L}(f_m)$		Residual FM Δf_{res}		
	Exponent	dB/oct	50 Hz to 3 kHz	300 Hz to 3 kHz	20 Hz to 15 kHz
-100	0	0	1.34	1.34	15.0
-100	-1	-3	0.95	0.94	4.74
-100	-2	-6	0.77	0.73	1.73
-100	-3	-9	0.90	0.68	1.15

^{a}For any $\mathscr{L}$ at 1 kHz different from -100 dBc, multipoy Δf_{res} of the table by

$$\text{antilog} \frac{100 - |\mathscr{L} \text{ at } 1 \text{ kHz/dBc}|}{20}$$

The table does not take into account any microphonic or spurious sidebands.

Example: $\mathscr{L}$ at 1 kHz = -88 dBc, slope -9 dB. For bandwidth 20 Hz to 15 kHz:

$$\Delta f_{\text{res}} = 1.15 \text{ Hz} \times \text{antilog} \frac{100 - 88}{20} = 4.6 \text{ Hz}$$

2-4 ALLAN VARIANCE RELATED TO $\mathscr{L}(f_m)$

For many applications, such as high-stability crystal oscillators or doppler radar systems, it is more relevant to describe frequency stability in the time domain. The characterization is based on the sample variance of fractional frequency fluctuations. Averaging differences of consecutive sample pairs with no deadtime in between yields the *Allan variance*, $\sigma_y^2(\tau)$, which is the proposed standard measure of frequency stability.

$$\sigma_y^2(\tau) \approx \frac{1}{2(M-1)} \sum_{K=1}^{M-1} (\bar{y}_{k+1} - \bar{y}_k)^2 \qquad (2\text{-}14)$$

$\bar{y}_k$ is the average fractional frequency difference of the *k*th sample measured over sample time τ.

Conversions from frequency- to time-domain data and vice versa are possible but tedious. The power spectrum $\mathscr{L}(f_m)$ needs to be approximated by integer slopes of 0, −1, −2, −3, −4. Then conversion formulas (see Table 2-2) can be applied. A good description of this procedure is given in Refs. 11 and 12.

We have covered the most frequently used measures of phase noise and have interrelated them. Before we take a look at the generation of phase noise in amplifiers and oscillators, let us take a look at the noise-conversion nomograph in Table 2-3. The example given there is self-explanatory.

As most of these relationships, for reasons of convenience, are expressed in decibels rather than absolute values, the following formulas are commonly used:

$$\mathscr{L}(f_m) = 10 \log_{10} \left(\frac{\Delta f_{\text{peak}}}{2 f_m} \right)^2 \qquad (2\text{-}15)$$

Table 2-2 Conversion table[a]

	Slope of $\sigma_y^2(\tau)$	$\sigma_y(\tau) =$	$\mathscr{L}(f) =$	Slope of $\mathscr{L}(f)$
White phase	−2	$\dfrac{\sqrt{\mathscr{L}(f)_h}}{2.565 f_o} \tau^{-1}$	$\dfrac{[\sigma_y(\tau)\tau f_o(2.565)]^2}{f_h} f^0$	0
Flicker phase	−1.9	$\dfrac{\sqrt{\mathscr{L}(f) f [2.184 + \ln(f_h \tau)]}}{2.565 f_o} \tau^{-1}$	$\dfrac{[\sigma_y(\tau)\tau f_o(2.565)]^2}{2.184 + \ln(f_h \tau)} f^{-1}$	−1
White frequency	−1	$\dfrac{\sqrt{\mathscr{L}(f) f^2}}{f_o} \tau^{-1/2}$	$[\sigma_y(\tau)\tau^{1/2} f_o]^2 f^{-2}$	−2
Flicker frequency	0	$\dfrac{1.665 \sqrt{\mathscr{L}(f) f^3}}{f_o} \tau^0$	$0.361 [\sigma_y(\tau) f_o]^2 f^{-3}$	−3
Random walk frequency	+1	$\dfrac{3.63 \sqrt{\mathscr{L}(f) f^4}}{f_o} \tau^{1/2}$	$[(0.276) \sigma_y(\tau) \tau^{-1/2} f_o]^2 f^{-4}$	−4

[a] τ = measurement time, $y = \Delta f_o / f_o$, f_o = carrier, f = sideband frequency, f_h = measurement system bandwidth.

Table 2-3 Noise-conversion nomograph: relationship among modulating frequency (f_m), power spectral density of phase (S_ϕ), modulation index, sideband-to-carrier ratio (dBc), dBmO, and frequency deviation (Δf_{rms})[a]

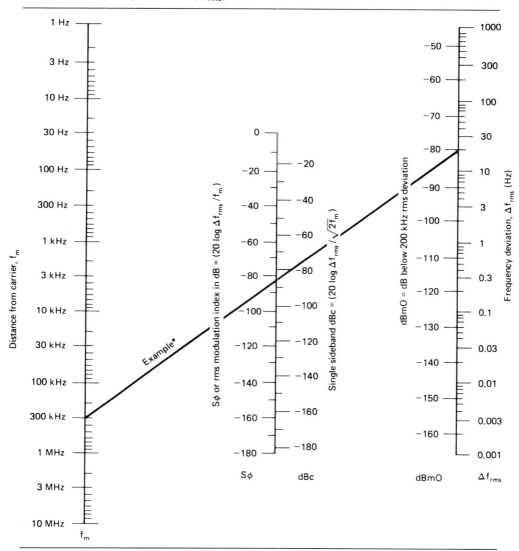

[a] Use consistent measurement bandwidth. *Example:* 20-Hz deviation in a 1-kHz band at 300 kHz from carrier = single-sideband dBc of −87 dB in a 1-kHz band.

$$\mathcal{L}(f_m) = 10\log_{10}\left(\frac{\Delta f_{rms}}{\sqrt{2}f_m}\right)^2 \qquad (2\text{-}16)$$

$$\mathcal{L}(f_m) = 20\log_{10}\frac{\Delta f_{rms}}{\sqrt{2}f_m} \qquad (2\text{-}17)$$

$$\mathcal{L}(f_m) = 20\log_{10}\frac{\theta_d}{2} \qquad (2\text{-}18)$$

2-5 LINEAR APPROACH FOR THE CALCULATION OF OSCILLATOR PHASE NOISE

Since an oscillator can be viewed as an amplifier with feedback (Figure 2-3), it is helpful to examine the phase noise added to an amplifier that has a noise figure F. With F defined by [7]

$$F = \frac{(S/N)_{\text{in}}}{(S/N)_{\text{out}}} = \frac{N_{\text{out}}}{N_{\text{in}} G} = \frac{N_{\text{out}}}{GkTB} \tag{2-19}$$

$$N_{\text{out}} = FGkTB \tag{2-20}$$

$$N_{\text{in}} = kTB \tag{2-21}$$

where N_{in} is the total input noise power to a noise-free amplifier. The input phase noise in a 1-Hz bandwidth at any frequency $f_0 + f_m$ from the carrier produces a phase deviation given by (Figure 2-4)

$$\Delta\theta_{\text{peak}} = \frac{V_{\text{nRMS1}}}{V_{\text{avsRMS}}} = \sqrt{\frac{FkT}{P_{\text{avs}}}} \tag{2-22}$$

$$\Delta\theta_{\text{1RMS}} = \frac{1}{\sqrt{2}} \sqrt{\frac{FkT}{P_{\text{avs}}}} \tag{2-23}$$

Since a correlated random phase relation exists at $f_0 - f_m$, the total phase deviation becomes

$$\Delta\theta_{\text{RMS total}} = \sqrt{FkT/P_{\text{avs}}} \tag{2-24}$$

The spectral density of phase noise becomes

$$S_\theta(f_m) = \Delta\theta_{\text{RMS}}^2 = FkTB/P_{\text{avs}} \tag{2-25}$$

where $B = 1$ for a 1-Hz bandwidth. Using

$$kTB = -174 \text{ dBm/Hz} \quad (B = 1) \tag{2-26}$$

allows a calculation of the spectral density of phase noise that is far removed from

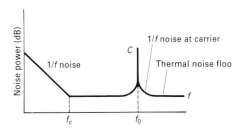

Figure 2-3 Noise power versus frequency of a transistor amplifier with an input signal applied.

the carrier (i.e., at large values of f_m). This noise is the theoretical noise floor of the amplifier. For example, an amplifier with +10 dBm power at the input and a noise figure of 6 dB gives

$$S_\theta(f_m > f_c) = -174 \text{ dBm} + 6 \text{ dB} - 10 \text{ dBm} = -178 \text{ dB}$$

For a modulation frequency close to the carrier, $S_\theta(f_m)$ shows a flicker or $1/f$

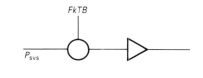

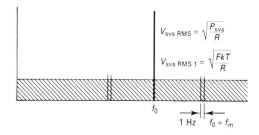

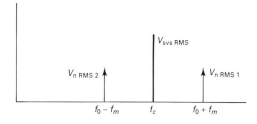

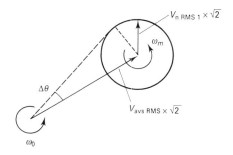

Figure 2-4 Phase noise added to carrier.

88 NOISE AND SPURIOUS RESPONSE OF LOOPS

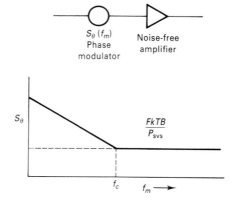

Figure 2-5 Phase noise modeled by a noise-free amplifier and a phase modulator.

component, which is empirically described by the corner frequency f_c. The phase noise can be modeled by a noise-free amplifier and a phase modulator at the input as shown in Figure 2-5. The purity of the signal is degraded by the flicker noise at frequencies close to the carrier. The spectral phase noise can be described by

$$S_\theta(f_m) = \frac{FkTB}{P_{\text{avs}}}\left(1 + \frac{f_c}{f_m}\right) \quad (B = 1) \tag{2-27}$$

The oscillator may be modeled as an amplifier with feedback as shown in Figure 2-6. The phase noise at the input of the amplifier is affected by the bandwidth of the resonator in the oscillator circuit in the following way. The tank circuit or bandpass resonator has a low-pass transfer function

$$L(\omega_m) = \frac{1}{1 + j(2Q_L \omega_m/\omega_0)} \tag{2-28}$$

where

$$\omega_0/2Q_L = B/2 \tag{2-29}$$

is the half-bandwidth of the resonator. These equations describe the amplitude response of the bandpass resonator; the phase noise is transferred unattenuated through the resonator up to the half-bandwidth. The closed-loop response of the phase feedback loop is given by

$$\Delta\theta_{\text{out}}(f_m) = \left(1 + \frac{\omega_0}{j2Q_L\omega_m}\right)\Delta\theta_{\text{in}}(f_m) \tag{2-30}$$

LINEAR APPROACH FOR THE CALCULATION OF OSCILLATOR PHASE NOISE

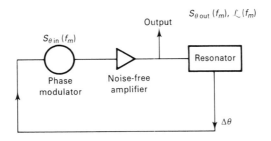

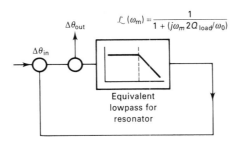

Figure 2-6 Equivalent feedback models of oscillator phase noise.

The power transfer becomes the phase spectral density

$$S_{\theta\text{out}}(f_m) = \left[1 + \frac{1}{f_m^2}\left(\frac{f_0}{2Q_L}\right)^2\right] S_{\theta\text{in}}(f_m) \tag{2-31}$$

where $S_{\theta\text{in}}$ was given by Eq. (2.27). Finally, $\mathscr{L}(f_m)$ is

$$\mathscr{L}(f_m) = \frac{1}{2}\left[1 + \frac{1}{f_m^2}\left(\frac{f_0}{2Q_L}\right)^2\right] S_{\theta\text{in}}(f_m) \tag{2-32}$$

This equation describes the phase noise at the output of the amplifier. The phase perturbation $S_{\theta\text{in}}$ at the input of the amplifier is enhanced by the positive phase feedback within the half-bandwidth of the resonator, $f_0/2Q_L$.

Depending on the relation between f_c and $f_0/2Q_L$, there are two cases of interest, as shown in Figure 2-7. For the low-Q case, the spectral phase noise is unaffected by the Q of the resonator, but the $\mathscr{L}(f_m)$ spectral density will show a $1/f^3$ and $1/f^2$ dependence close to the carrier. For the high-Q case, a region of $1/f^3$ and $1/f$ should be observed near the carrier. Substituting Eq. (2.27) in (2.32) gives an overall noise of

$$\begin{aligned}\mathscr{L}(f_m) &= \frac{1}{2}\left[1 + \frac{1}{f_m^2}\left(\frac{f}{2Q_L}\right)^2\right]\frac{FkT}{P_{\text{avs}}}\left(1 + \frac{f_c}{f_m}\right) \\ &= \frac{FkTB}{2P_{\text{avs}}}\left[\frac{1}{f_m^3}\frac{f^2 f_c}{4Q_L^2} + \frac{1}{f_m^2}\left(\frac{f}{2Q_L}\right)^2 + \frac{f_c}{f_m} + 1\right] \quad \text{dBc/Hz}\end{aligned} \tag{2-33}$$

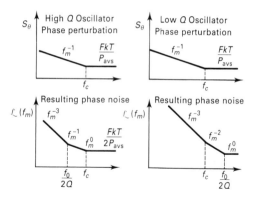

Figure 2-7 Oscillator phase noise for high-Q and low-Q resonator viewed as spectral phase noise and as noise-to-carrier ratio versus frequency from the carrier.

Examining Eq. (2-33) gives the four major causes of oscillator noise: the up-converted $1/f$ noise or flicker FM noise, the thermal FM noise, the flicker phase noise, and the thermal noise floor, respectively.

Q_L (loaded Q) can be expressed as

$$Q_L = \frac{\omega_o W_e}{P_{\text{diss, total}}} = \frac{\omega_o W_e}{P_{\text{in}} + P_{\text{res}} + P_{\text{sig}}} \quad (2\text{-}34)$$

$$= \frac{\text{reactive power}}{\text{total dissipated power}}$$

where W_e is the reactive energy stored in L and C,

$$W_e = \tfrac{1}{2} C V^2 \qquad P_{\text{res}} = \frac{\omega_o W_e}{Q_{\text{unl}}}$$

$$\mathcal{L}(f_m) = \frac{1}{2}\left[1 + \frac{\omega_o^2}{4\omega_m^2}\left(\frac{P_{\text{in}}}{\omega_o W_e} + \frac{1}{Q_{\text{unl}}} + \frac{P_{\text{sig}}}{\omega_o W_e}\right)^2\right]\left(1 + \frac{\omega_c}{\omega_m}\right)\frac{FkT_o}{P_{\text{sav}}} \quad (2\text{-}35)$$

- input power over reactive power
- resonator Q
- signal power over reactive power
- flicker effect
- phase perturbation

This equation is extremely significant because it contains most of the causes of phase noise in oscillators. To minimize the phase noise, the following design rules apply:

1. Maximize the unloaded Q.

2. Maximize the reactive energy by means of a high RF voltage across the resonator and obtain a low LC ratio. The limits are set by breakdown voltages of the active devices and the tuning diodes and the forward-bias condition of the tuning diodes.

3. Avoid saturation at all cost, and try to either have limiting or automatic gain control (AGC) without degradation of Q. Isolate the tuned circuit from the limiter or AGC circuit. Use antiparallel tuning diode connections to avoid forward bias.

4. Choose an active device with the lowest possible noise figure. The noise figure of interest is the noise figure obtained at the actual impedance at which the device is operated. Using field-effect transistors rather than bipolar transistors, it is preferable to deal with the equivalent noise voltage and noise currents rather than with the noise figure, since they are independent of source impedance. The noise figure improves as the ratio between source impedance and equivalent noise resistance increases. In addition, in a tuning circuit, the source impedance changes drastically as a function of the offset frequency, and this effect has to be considered.

The following transistors have the lowest noise figure:

- BFG65/67 by Philips
- BFR106/92; BFP405/1420 by Siemens
- 2SC3358/3356 by NEC
- HXTR4105 by HP
- AT2100 (chip) by Avantek

Among the lowest noise junction, field-effect transistors are U310 and 2N4416/17 and 2N5397.

In designing MMICs, one must resort to MESFETs, which are also referred to as GaAs FETs. They have a fairly high flicker frequency, which is typically around 10 MHz but can go up to as high as 100 MHz.

Table 2-4 shows equivalent circuits and measurement data for the Avantek AT2100 20-GHz NPN silicon bipolar oscillator transistor. (Siemens of Germany is about to come out with 30-GHz f_t transistors, probably closely followed by NEC.)

5. Phase perturbation can be minimized by using high-impedance devices such as field-effect transistors, where the signal-to-noise ratio of the signal voltage relative to the equivalent noise voltage can be made very high. This also indicates that in the case of a limiter, the limited voltage should be as high as possible.

6. Choose an active device with low flicker noise. The effect of flicker noise can be reduced by RF feedback. An unbypassed emitter resistor of 10 to 30 Ω in a bipolar circuit can improve the flicker noise by as much as 40 dB. In Chapter 4 we will study such an oscillator.

Table 2-4 Linear and Spice equivalent circuits for the Avantek AT21400 chip

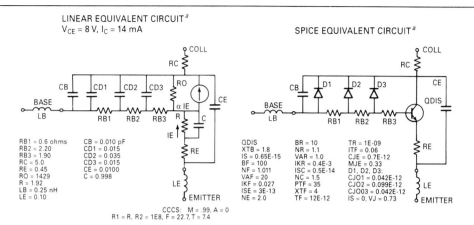

Modeled Scattering Parameters, Common Emitter[b] $T_A = 25°C$, $V_{CE} = 8$ V, $I_C = 14$ mA

Freq. GHz	S_{11} Mag	S_{11} Ang	S_{21} dB	S_{21} Mag	S_{21} Ang	S_{12} dB	S_{12} Mag	S_{12} Ang	S_{22} Mag	S_{22} Ang
2	.67	−168	15.8	6.19	90	−34.0	.02	52	.65	−9
3	.67	−180	12.4	4.17	81	−32.0	.025	59	.65	−10
4	.67	172	10.0	3.15	74	−30.5	.03	64	.65	−11
5	.68	165	8.1	2.53	68	−28.0	.04	68	.65	−12
6	.68	160	6.5	2.12	62	−28.0	.04	70	.65	−14
7	.69	154	5.2	1.82	56	−26.0	.05	72	.65	−16
8	.70	149	4.1	1.60	51	−24.4	.06	73	.65	−18
9	.70	144	3.1	1.43	46	−24.4	.06	74	.65	−20
10	.71	140	2.2	1.29	41	−23.1	.07	74	.65	−22
11	.72	135	1.4	1.18	36	−21.9	.08	74	.65	−24
12	.73	131	0.7	1.08	31	−21.9	.08	75	.66	−26
13	.74	127	0	1.00	27	−20.9	.09	75	.66	−28
14	.75	123	−0.6	0.93	23	−20.0	.10	75	.66	−30
15	.76	120	−1.3	0.86	18	−19.2	.11	74	.67	−32
16	.77	116	−1.9	0.80	14	−19.2	.11	74	.67	−34
18	.79	109	−3.1	0.70	7	−17.7	.13	74	.68	−39
20	.80	103	−4.2	0.62	−1	−17.1	.14	73	.69	−44
22	.82	97	−5.4	0.54	−7	−15.9	.16	72	.70	−48
24	.84	92	−6.6	0.47	−14	−14.9	.18	71	.71	−53
26	.85	87	−8.0	0.40	−20	−14.0	.20	70	.72	−58

[a]These equivalent circuits are provided only as first-order design aids. Their accuracy for critical designs at very high frequencies has not been validated.
[b]S-Parameters are from linear equivalent circuit. Below 10 GHz, they have been fit to measurements of die on a standard carrier with one bond wire to the base and four wires to the emitter.

The proper bias point of the active device is important, and precautions should be taken to prevent modulation of the input and output dynamic capacitance of the active device, which will cause amplitude-to-phase conversion and therefore introduce noise.

7. The energy should be coupled from the resonator rather than from another portion of the active device so that the resonator limits the bandwidth. A crystal oscillator using this principle is described later.

8. For microwave application, the lowest GaAs FET transistors are pseudomorphic HJ FETs. A reproduction of the data sheet of the NEC NE42484A, which is generally the state of the art today, appears on pages 94–95.

Equation (2-35) assumes that the phase perturbation and the flicker effect are the limiting factors, as practical use of such oscillators requires that an isolation amplifier be used.

In the event that the energy is taken directly from the resonator and the oscillator power can be increased, the signal-to-noise ratio can be increased above the theoretical limit of $-174\,\text{dB}$, due to the low-pass filter effect of the tuned resonator. However, since this is mainly a theoretical assumption and does not represent the real world in a system, this noise performance cannot be obtained. In an oscillator stage, even a total noise floor of 170 dB is rarely achieved.

What other influences do we have that cause the noise performance to degrade?

So far, we have assumed that the Q of the tuned circuit is really determined only by the LC network and the loading effect of the transistor. In synthesizer applications, however, we find it necessary to add a tuning diode. The tuning diode has a substantially lower Q than that of a mica capacitor or even a ceramic capacitor. As a result of this, the noise sidebands change as a function of the additional loss. This is best expressed in the form of adjusting the value for the loaded Q in Eq. (2-34).

There seems to be no precise mathematical way of predetermining the noise influence of a tuning diode, but the following approximation seems to give proper results:

$$\frac{1}{Q_{T\text{load}}} = \frac{1}{Q_{\text{load}}} + \frac{1}{Q_{\text{diode}}} \qquad (2\text{-}36)$$

The tuning diode is specified to have a cutoff frequency $f_{\max}$, which is determined from the loss resistor R_s and the value of the junction capacitance as a function of voltage (i.e., measured at 3 V). This means that the voltage determines the Q and, consequently, the noise bandwidth.

We will go into more detail in dealing with the mechanism and influence of tuning diodes in the oscillator section, where we will evaluate the various methods of building voltage tunable oscillators using tuning diodes and switching diodes. In this chapter we limit ourselves to practical results.

The loading effect of the tuning diode is due to losses, and these losses can be described by a resistor parallel to the tuned circuit.

It is possible to define an equivalent noise $R_{a\text{eq}}$ that, inserted in Nyquist's equation,

$$V_n = \sqrt{4KT_o R \Delta f} \qquad (2\text{-}37)$$

where $KT_o = 4.2 \times 10^{-21}$ at about 300 K, R is the equivalent noise resistor, and Δf is the bandwidth, determines an open noise voltage across the tuning diode. Practical values of R equivalent for carefully selected tuning diodes are in the vicinity of $200\,\Omega$ to $50\,\text{k}\Omega$. If we now determine the noise

SUPER LOW NOISE PSEUDOMORPHIC HJ FET | NE42484A

FEATURES
- **VERY LOW NOISE FIGURE:**
  0.8 dB typical at 12 GHz
- **HIGH ASSOCIATED GAIN:**
  10.5 dB Typical at 12 GHz
- $L_G = 0.35$ µm, $W_G = 200$ µm
- **LOW COST METAL CERAMIC PACKAGE**
- **TAPE & REEL PACKAGING OPTION AVAILABLE**

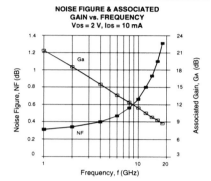

NOISE FIGURE & ASSOCIATED GAIN vs. FREQUENCY
$V_{DS} = 2$ V, $I_{DS} = 10$ mA

DESCRIPTION
The NE42484A is a pseudomorphic Hetero-Junction FET that uses the junction between Si-doped AlGaAs and undoped InGaAs to create very high mobility electrons. The device features mushroom shaped TiAl gates for decreased gate resistance and improved power handling capabilities. The mushroom gate also results in lower noise figure and high associated gain. This device is housed in an epoxy-sealed, metal/ceramic package and is intended for high volume consumer and industrial applications.

NEC's stringent quality assurance and test procedures ensure the highest reliability and performance.

ELECTRICAL CHARACTERISTICS ($T_A = 25°C$)

SYMBOLS	PART NUMBER / PACKAGE OUTLINE / PARAMETERS AND CONDITIONS	UNITS	NE42484A 84AS MIN	TYP	MAX
NF_{OPT}[1]	Optimum Noise Figure, $V_{DS} = 2.0$ V, $I_{DS} = 10$ mA, $f = 12$ GHz	dB		0.8	1.2
G_A[1]	Associated Gain, $V_{DS} = 2.0$ V, $I_{DS} = 10$ mA, $f = 12$ GHz	dB	9.0	10.5	
P_{1dB}	Output Power at 1 dB Gain Compression Point, $f = 12$ GHz $V_{DS} = 2.0$ V, $I_{DS} = 10$ mA $V_{DS} = 2.0$ V, $I_{DS} = 20$ mA	dBm dBm		9.7 10.2	
G_{1dB}	Gain at P_{1dB}, $f = 12$ GHz $V_{DS} = 2.0$ V, $I_{DS} = 10$ mA $V_{DS} = 2.0$ V, $I_{DS} = 20$ mA	dB dB		10.3 10.5	
I_{DSS}	Saturated Drain Current, $V_{DS} = 2.0$ V, $V_{GS} = 0$ V	mA	15	40	70
V_P	Pinch-off Voltage, $V_{DS} = 2.0$ V, $I_{DS} = 0.1$ mA	V	-2.0	-0.8	-0.2
g_m	Transconductance, $V_{DS} = 2.0$ V, $I_D = 10$ mA	mS	45	60	
I_{GSO}	Gate to Source Leakage Current, $V_{GS} = -3.0$ V	µA		0.5	10.0
$R_{TH(CH-A)}$	Thermal Resistance (Channel to Ambient)	°C/W		750	
$R_{TH(CH-C)}$[2]	Thermal Resistance (Channel to Case)	°C/W			350

Notes:
1. Typical values of noise figures and associated gain are those obtained when 50% of the devices from a large number of lots were individually measured in a circuit with the input individually tuned to obtain the minimum value. Maximum values are criteria established on the production line as a "go-no-go" screening tuned for the "generic" type but not for each specimen.
2. R_{TH} (channel to case) for package mounted on an infinite heat sink.

California Eastern Laboratories

CEL/NEC Data Sheet NE42484A, copyright of California Eastern Laboratories, Santa Clara, CA. Used with permission

LINEAR APPROACH FOR THE CALCULATION OF OSCILLATOR PHASE NOISE

NE42484A

ABSOLUTE MAXIMUM RATINGS[1] ($T_A = 25°C$)

SYMBOLS	PARAMETERS	UNITS	RATINGS
V_{DS}	Drain to Source Voltage	V	4.0
V_{GS}	Gate to Source Voltage	V	-3.0
I_{DS}	Drain Current	mA	I_{DSS}
I_{GRF}	Gate Current (RF Drive)	μA	200
P_{IN}	RF Input (CW)	dBm	15
T_{CH}	Channel Temperature	°C	150
T_{STG}	Storage Temperature	°C	-65 to +150
P_T	Total Power Dissipation	mW	165

Note:
1. Operation in excess of any one of these parameters may result in permanent damage.

TYPICAL NOISE PARAMETERS ($T_A = 25°C$)

$V_{DS} = 2$ V, $I_{DS} = 10$ mA

FREQ. (GHz)	NF_{OPT} (dB)	G_A (dB)	Γ_{OPT} MAG	Γ_{OPT} ANG	$R_n/50$
1.0	0.31	21.4	0.78	10	0.43
2.0	0.34	18.5	0.76	28	0.38
4.0	0.40	15.5	0.72	58	0.28
6.0	0.47	13.6	0.65	84	0.21
8.0	0.56	12.4	0.57	113	0.15
10.0	0.66	11.4	0.50	141	0.10
12.0	0.80	10.5	0.44	173	0.09
14.0	0.93	9.8	0.39	-157	0.08
16.0	1.10	9.3	0.36	-125	0.08
18.0	1.31	8.7	0.35	-90	0.08

TYPICAL PERFORMANCE CURVES ($T_A = 25°C$)

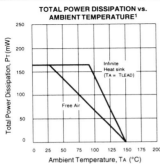

TOTAL POWER DISSIPATION vs. AMBIENT TEMPERATURE[1]

Note
1. If P_T exceeds the Free Air Value, reliable operation can be assured by measuring the worst-case temperature, $T_{(LEAD)}$, at the lead where heat flow is maximum (usually the source lead) and limiting T_A, P_T or $R_{TH\,(CKT)}$.

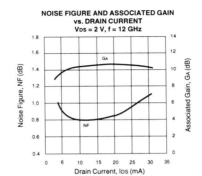

NOISE FIGURE AND ASSOCIATED GAIN vs. DRAIN CURRENT
$V_{DS} = 2$ V, $f = 12$ GHz

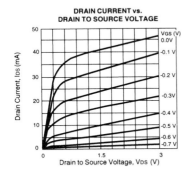

DRAIN CURRENT vs. DRAIN TO SOURCE VOLTAGE

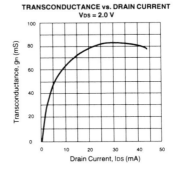

TRANSCONDUCTANCE vs. DRAIN CURRENT
$V_{DS} = 2.0$ V

voltage $V_n = \sqrt{4 \times 4.2 \times 10^{-21} \times 10{,}000}$, the resulting voltage value is $1.265 \times 10^{-8}\,\text{V}\sqrt{\text{Hz}}$.

This noise voltage generated from the tuning diode is now multiplied with the VCO gain, resulting in the rms frequency deviation:

$$(\Delta f_{\text{rms}}) = K_o \times (1.265 \times 10^{-8}\,\text{V}) \text{ in 1-Hz bandwidth} \qquad (2\text{-}38)$$

In order to translate this into the equivalent peak phase deviation,

$$\theta_d = \frac{K_o \sqrt{2}}{f_m}(1.265 \times 10^{-8}\,\text{rad}) \text{ in 1-Hz bandwidth}$$

or for a typical oscillator gain of 100 kHz/V,

$$\theta_d = \frac{0.00179}{f_m} \text{ rad in 1-Hz bandwidth}$$

For $f_m = 25$ kHz (typical spacing for adjacent channel measurements for FM mobile radios), the $\theta_c = 7.17 \times 10^{-8}$. This can be converted now into the SSB signal-to-noise ratio

$$\begin{aligned}\mathcal{L}(f_m) &= 20\log_{10}\frac{\theta_c}{2} \\ &= -149\,\text{dB/Hz}\end{aligned} \qquad (2\text{-}39)$$

This is the value typically achieved in the Rohde & Schwarz SMDU or with the Hewlett-Packard 8640 signal generator and considered state of the art for a free-running oscillator. It should be noted that both signal generators use a slightly different tuned circuit; the Rohde & Schwarz generator uses a helical resonator, whereas the Hewlett-Packard generator uses an electrically shortened quarter-wavelength cavity. Both generators are mechanically pretuned and the tuning diode with a gain of about 100 kHz/V is used for frequency-modulation purposes or for the AFC input. It is apparent that, because of the nonlinearity of the tuning diode, the gain is different for low dc voltages than for high dc voltages. The impact of this is that the noise varies within the tuning range. A detailed discussion of these phenomena is given in Chapter 4.

If this oscillator had to be used for a frequency synthesizer, the 1-MHz tuning range would be insufficient; therefore, a way had to be found to segment the band into the necessary ranges. In VCOs, this is typically done with switching diodes that allow the proper frequency bands to be selected. These switching diodes insert in parallel or series, depending on the circuit, with additional inductors or capacitors, depending on the design.

In low-energy-consuming circuits, the VCO frequently is divided into a coarse-tuning section using tuning diodes and a fine-tuning section with a tuning diode. In the coarse-tuning range, this results in very high gains, such as 1 to 10 MHz/V, for the diodes, and therefore the noise contribution of those diodes is very high and can hardly be compensated by the loop. For low-noise applications, which automatically mean higher power consumption, it is un-

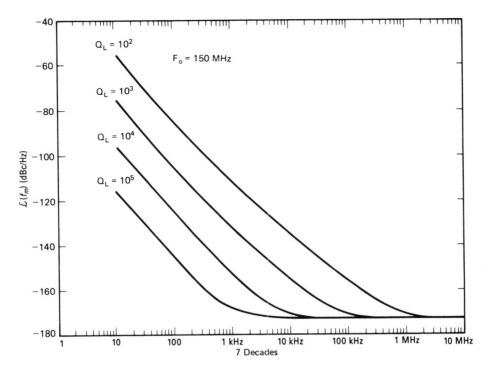

Figure 2-8 Noise sideband of an oscillator at 150 MHz as a function of the loaded Q of the resonator.

avoidable to use switching diodes. More detailed information about the switching diodes and their applications is presented in Chapter 4.

Let us now examine some test results. If we go back to Eq. (2-33), Figure 2-8 shows the noise sideband performance as a function of Q, whereby the top curve with $Q_L = 100$ represents a somewhat poor oscillator and the lowest curve with $Q_L = 100,000$ probably represents a crystal oscillator where the unloaded Q of the crystal was in the vicinity of 3×10^6. Figure 2-9 shows the influence of flicker noise.

Corner frequencies of 10 Hz to 10 kHz have been selected, and it becomes apparent that around 1 kHz the influence is fairly dramatic, whereas the influence at 20 kHz off the carrier is not significant. Finally, Figure 2-10 shows the influence of the tuning diodes on a high-Q oscillator.

Curve A in Figure 2-10 uses a lightly coupled tuning diode with a K_o of 10 kHz/V; the lower curve is the noise performance without any diode. As a result, the two curves are almost identical, which can be seen from the somewhat smeared form of the graph. Curve B shows the influence of a tuning diode at 100 kHz/V and represents a value of 143 dB/Hz from 155 dB/Hz, already some deterioration. Curve C shows the noise if the tuning diode results in a 1-MHz/V VCO gain, and the noise sideband at 25 kHz has now deteriorated to 123 dB/Hz. These curves speak for themselves.

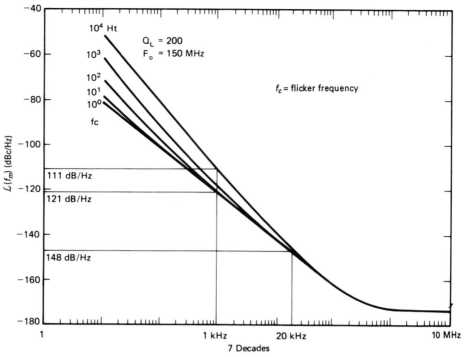

Figure 2-9 Noise sideband performance as a function of the flicker frequency ωC varying from 10 Hz to 10 kHz.

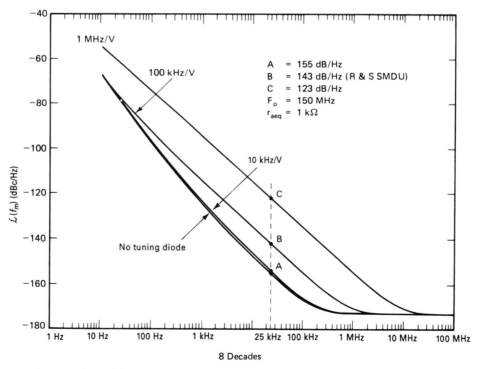

Figure 2-10 Noise sideband performance of an oscillator at 150 MHz, showing the influence of various tuning diodes.

NOISE CONTRIBUTIONS IN PHASE-LOCKED SYSTEMS 99

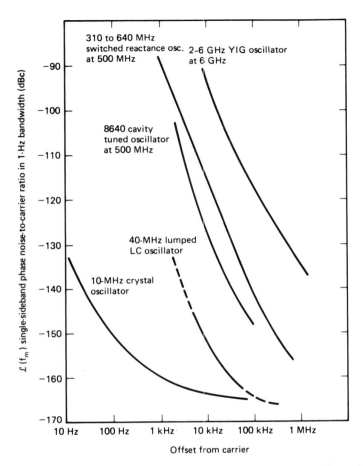

Figure 2-11 Comparison of noise sideband performances of a crystal oscillator, LC oscillator, cavity-tuned oscillator, switched reactance oscillator, and YIG oscillator.

It is of interest to compare various oscillators. Figure 2-11 shows the performance of a 10-MHz crystal oscillator, a 40-MHz LC oscillator, the 8640 cavity tuned oscillator at 500 MHz, the 310- to 640-MHz switched reactance oscillator of the 8662 oscillator, and a 2- to 6-GHz YIG oscillator at 6 GHz.

In the following paragraphs, we deal with the noise influence of other loop components, such as dividers, phase/frequency comparators, and operational amplifiers, as well as examine the influence of the reference frequency on the loop noise sideband performance.

2-6 NOISE CONTRIBUTIONS IN PHASE-LOCKED SYSTEMS

Figure 2-12 shows a block diagram of a phase-locked loop consisting of a phase detector, an integrator, a shaper-pretuned circuit, an additional attenuator as a second loop filter, the VCO, and finally the divider.

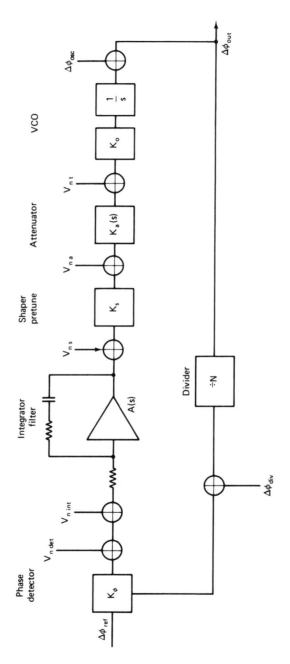

Figure 2-12 Block diagram showing the various noise sources in a phase-locked loop.

The reference has not been included. The reason for this is that we assume for a moment that the reference is ideal and noise-free, and take a look at the influence of the other elements of the loop.

2-6-1 Phase Noise Characteristics of Amplifiers

As mentioned several times previously, it is an absolute necessity to provide an isolation amplifier between the VCO and the following circuits (e.g., the divider chain or others) to minimize any feedback or noise contribution because of periodic loading.

What does this do to our system noise? Unfortunately, if we assume that we generate our initial signal in a noise-free environment or at least start off with the theoretical minimum of -174 dBm/Hz, the spectral density of phase noise that is generated by a resistor at room temperature as a sideband noise floor, the signal will be degraded by the postamplifier. For example, a signal of 0 dBm passes through an amplifier with a 3-dB noise figure. The resulting spectral density of phase noise is

$$S_{\Delta\theta} = -174 \text{ dBm} + 3 \text{ dB} - 0 \text{ dBm} = -171 \text{ dBm/Hz} \qquad (2\text{-}40)$$

This theoretical floor can be observed only at a fairly large offset. With practical transistors, various noise sources have to be taken into consideration. As mentioned previously, flicker noise is the major reason for the noise degradation, and its contribution is very device dependent and can range from a few hundred hertz to 1 MHz. It is caused by low-frequency device noise modulating the phase of the passing signal by modulating the transconductance and the input and output impedances of the amplifier (depletion layer and diffusion capacitance). There is very little one can do to reduce this effect on a large scale. One is limited to:

1. Some negative feedback at low frequency, such as an unbypassed emitter resistor.
2. Some negative feedback at RF frequency to stabilize the transconductance.
3. Designing the RF amplifier for a low noise figure, also at low frequency.

Depending on the author, there are various speculations as to which device is best for low flicker noise. It appears to me that the first decision to make is to determine whether the device is being used in high- or low-power RF application. From recent experiments, it has been possible to prove that, at medium levels, junction field-effect transistors show a significant advantage over bipolar transistors, as the modulation of the input and output impedances is less. The field-effect transistor shows a 10- to 20-dB better performance at drive levels of up to 1 V at the gate electrode in a frequency range from 50 to 500 MHz over the bipolar transistor. A plausible explanation is that the base spreading resistor and other loss resistors in the bipolar transistor have a significant influence at these high drive levels in addition to the effect of gain saturation, a nonlinear phenomenon in bipolar transistors that causes cross-modulation and intermodulation distortion.

102 NOISE AND SPURIOUS RESPONSE OF LOOPS

However, these things are device dependent and may change as new devices are developed, and it is necessary to update this information periodically.

2-6-2 Phase Noise Characteristics of Dividers

Generally, the phase noise at the input of a divider appears at the divider output reduced by N. However, there are some limitations to this effect.

1. In accordance with Figure 2-13, the practical noise limit is in the vicinity of 170 dB for TTL dividers and 155 dB for ECL. CMOS dividers, up to an input frequency of 10 MHz, behave similar to TTL devices. However, the close-in noise or noise between 1 and 10 Hz off the carrier is slightly higher than that of TTL devices. TTL devices require higher shielding and better power supply decoupling to prevent external crosstalk between the various stages, which otherwise results in unwanted spurious outputs. Any unwanted sidebands are also reduced by the same amount of this ratio.

2. Most of these dividers, however, have another unpleasant effect, in the form of internal *crosstalk*. Crosstalk is defined as the amount of input frequency appearing at the output of the divider chain. In high-performance synthesizers it is necessary to use a low-pass filter after the reference or the

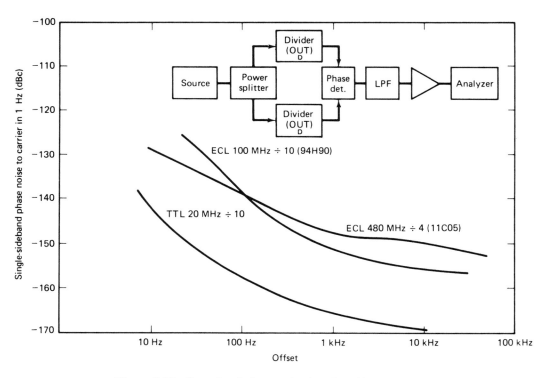

Figure 2-13 Example of phase noise introduced by dividers.

programmable divider and a pulse shaper to translate the resulting sine wave back into a square wave in order to keep the output voltage at the input frequency sufficiently suppressed. Further details are given in Section 4-6.

For microwave applications, other dividers like GaAs FETs or analog dividers are used. Figure 2-14 shows the residual noise for such different dividers while Figure 2-15 shows the phase noise normalized to 10 GHz, which is easier to compare.

Let us take a look at some of the mathematics involved. The instantaneous phase $\theta_i(t)$ of a carrier frequency modulated by a sine wave of frequency f_m is given by

$$\theta_i(t) = \omega_o t + \frac{\Delta f}{f_m} \sin \omega_m t \qquad (2\text{-}41)$$

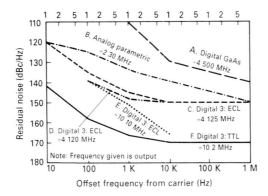

Figure 2-14 Residual phase noise of different dividers as a function of offset from the carrier frequency.

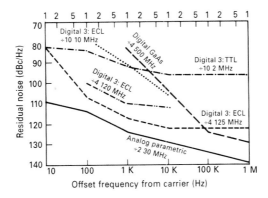

Figure 2-15 Phase noise of different dividers normalized to 10 GHz.

104 NOISE AND SPURIOUS RESPONSE OF LOOPS

Instantaneous frequency is defined as the time rate of change of phase

$$\omega(t) = \frac{d\theta i(t)}{dt} = \omega_o + \frac{\Delta f}{f_m}\omega_m \cos \omega_m t \qquad (2\text{-}42)$$
$$\leq \omega_o + \Delta\omega$$

If this signal is passed through a frequency divider that divides the frequency by N, the output frequency ω' will be given by

$$\omega' = \frac{\omega_o}{N} + \frac{\Delta\omega}{N} \qquad (2\text{-}43)$$

and the output phase by

$$\theta_i(t) = \frac{\omega_o t}{N} + \frac{\Delta f}{Nf_m}\sin \omega_m t \qquad (2.44)$$

The fundamental frequency at the divider output is

$$S(t) = V\cos\left(\frac{\omega_o t}{N} + \frac{\Delta f}{Nf_m}\sin \omega_m t\right) \qquad (2\text{-}45)$$

where V is the input peak voltage. The divider reduces the carrier frequency by N but does not change the frequency of the modulation signal. The peak phase deviation θ_p is reduced by the divider ratio N. Since it was shown that the ratio of the noise power to carrier power is

$$\frac{V_n^2}{V^2} = \frac{\theta_p^2}{4} \qquad (2\text{-}46)$$

frequency division by N reduces the noise power by N^2 for a perfect divider.

Example 1 The indirect frequency synthesizer shown in Figure 2-16 is used to generate a 5-GHz (5×10^9) signal. A 1-kHz reference signal is obtained from a 5-MHz reference oscillator ($M = 5000$), which is specified to have a single-sideband noise power of -140 dBc/Hz at a frequency separation of 0.5 kHz from the oscillator's operating frequency. If the loop bandwidth is assumed to be approximately 1 kHz, the noise from the reference oscillator will not be reduced by the low-pass filtering of the PLL. Although the divider N will reduce the noise power by the factor $1/N^2$, the approximate loop transfer function is

$$\theta_o = \frac{\theta_r[K_v F(s)/s]}{1 + [K_v F(s)/sN]} = N\theta_r \qquad (2\text{-}47)$$

for reference frequencies below the loop bandwidth of 1 kHz. The net effect is that the output noise power is the reference oscillator noise power multiplied by

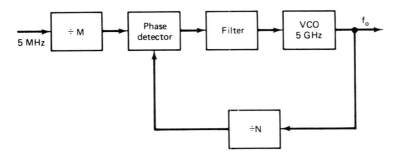

Figure 2-16 A 5-GHz YIG oscillator harmonic stabilized from a 5-MHz reference.

$(N/M)^2$. N must be equal to 5×10^6 to obtain the specified output frequency of 5×10^9 Hz, and the output noise power due to the reference oscillator is

$$N_o = -140 \text{ dB/Hz} + 10 \log \left(\frac{5 \times 10^6}{5 \times 10^3} \right)^2 = -80 \text{ dB/Hz} \qquad (2\text{-}48)$$

at a frequency offset of 0.5 kHz.

Example 1 illustrates a problem inherent in PLL frequency synthesizers used to generate an output frequency much higher than the reference oscillator frequency. Although the reference oscillator noise power may be small, the same noise power appears on the output signal amplified by the factor N^2, where N is the output frequency/reference oscillator frequency ratio.

2-6-3 Phase Noise Characteristics of Phase/Frequency Comparators

With the phase/frequency comparator's simplest form being a two-diode arrangement, a double-balanced mixer, an exclusive-OR gate, a flip-flop, or a tri-state comparator, it is a highly nonlinear device despite the fact that we had linearized its performance for an easier understanding of the PLL performance. This means that radiation into any of the ports of this device is being transferred as a sideband spur, depending on the conversion loss of the system. Any hum reference radiation or outer signal somehow fed into the phase/frequency comparator up to a very high order of harmonics of the input signal can be detected. This has caused grief for many design engineers because it is normally not obvious that high orders of the reference can mix with some RF pickup. Figure 2-17 shows the phase noise of an ECL phase detector compared with a silicon diode mixer and a hot carrier diode mixer (double-balanced mixer).

There is another unpleasant effect related to digital and analog phase/frequency comparators. In an analog phase/frequency comparator, the dc output voltage is limited and has to be amplified up to 10 or 20 V, while the digital phase/frequency comparators, with their up-and-down pulse output, have to use a summation amplifier. In the first case, the additional amplifier introduces flicker noise and other sideband noise that will reduce the maximum signal-to-noise ratio; in the case of the digital phase/frequency comparator under perfect locked condition, there is a range called the *zero gain area*. What does this mean? It means that

106 NOISE AND SPURIOUS RESPONSE OF LOOPS

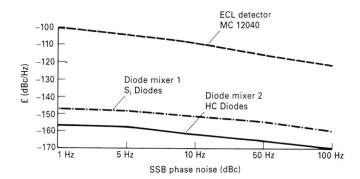

Figure 2-17 Phase noise of an ECL phase detector compared to a silicon diode mixer and hot carrier diode mixer (double-balanced mixer).

if we assume that the loop is in locked condition and requires no pulses to update the holding capacitor, the phase/frequency comparator has zero gain. This causes loop instability at very low frequencies and degrades the close-in noise sideband performance by up to 20 dB. Cures for this dead-zone effect are described in Chapter 4.

2-6-4 Phase Noise Characteristics of Multipliers

In general, the reference oscillator operating at either 5 or 10 MHz is divided to 1 MHz, 100 kHz, or even down to 1 kHz to generate the reference frequency for the various loops.

The dividers have an ultimate noise floor depending on the offset from the center frequency from around 130 dB/Hz at 10 Hz to 160 dB/Hz at 10 kHz or maybe up to 170 dB/Hz for TTL dividers, and, if necessary, shielding is used and ground loops are avoided. The PLL acts as a multiplier, as most likely the output frequency of the VCO is substantially higher than the reference frequency, and later we will analyze these effects inside and outside the loop.

Generally, in a multiplier the reverse result of that found in a frequency divider is encountered: the sideband noise and spurious response are increased by the multiplication factor. Let us assume that a 10-MHz reference frequency is divided down to 1 MHz, first with a noise floor of 170 dB/Hz, wideband, and then used as a reference frequency for a 10-MHz PLL. The noise performance is degraded by 20 dB because of this multiplication.

If the crystal oscillator is replaced by another synthesizer loop with some discrete spurs 80 dB below the carrier, they will maintain their 80-dB level, and if the VCO frequency is then increased to 100 MHz, they will deteriorate to 60 dB below the carrier.

In multiloop synthesizers, there is a frequent requirement for auxiliary frequencies that can be generated by one of the following:

1. Phase-locked loop as a multiplier.
2. Transistor multiplier.
3. Step recovery diode multiplier.

In the case of the PLL multiplier, we have two choices:

1. Use a fixed divider.
2. Use harmonic sampling.

Harmonic sampling is generally used for frequencies above 2000 MHz because there are no dividers available that work reliably at higher frequencies. Attempts to use tunnel diodes for this purpose or parametric effects in tuning diodes have shown up in the literature from time to time but in production have failed to show reliable performance, due to component tolerances in temperature.

A harmonic sampler is typically a balanced modulator that uses hot carrier diodes, which are being driven from a pulse or needle generator with extremely high harmonic contents. A typical application for such a circuit is in spectrum analyzers, where the input frequency and the YIG oscillator can be locked together. A similar application is where a harmonic comb is being generated from a 1-MHz reference, and locking can occur every 1 MHz to several gigahertz. These circuits require a pretuned mechanism to make sure that the desired harmonic is being selected and false locking is being prevented. This type of multiplication is used in systems where the frequency of the VCO is changed frequently and low spurious contents and high signal-to-noise ratio are required.

For fixed-frequency application, fixed-tuned frequency multipliers with transistors or step recovery diodes are used. The transistor multipliers work well up to several hundred megahertz, and the step recovery diodes or snap-off diodes can be used up to several gigahertz.

For higher frequency ranges, impatt diodes or other exotic devices can generate the necessary frequencies, and some of these multipliers are also built as *injection-lock oscillators*. An injection-lock oscillator can be considered as a frequency multiplier with a certain pulling range, where the oscillator somehow locks up with the reference frequency. These are highly nonlinear phenomena, described in the literature from time to time, and the explanations and mathematical models are built primarily around experimental data and are not always very reliable. Low-frequency injection locking is a very convenient way of combining extremely high stability in certain types of crystal oscillators, which are being used as a reference for extremely low noise crystal oscillators operating at the same frequency. In Section 4-2-5 we will see an example of a 10-MHz crystal oscillator being injection locked to an external reference. This method is also used now in many frequency synthesizers and frequency counters for the same reason: there is a major trade-off between the best operating mode for short-term stability versus long-term stability (aging).

For single-frequency applications, we find in synthesizer loops high-frequency crystal oscillators at discrete frequencies between 70 and 150 MHz, which are locked against a frequency standard but with an extremely narrow loop so that the output noise sideband depends only on the crystal oscillator frequency, rather than on the input frequency. These loops have time constants of 1 Hz or less and therefore compensate only for temperature effects or aging.

If higher frequencies are required, such as 600 to 700 MHz, several choices are available. As the wideband noise floor is being multiplied, depending on the type of multiplier, different results can occur. Let us take a look at Figure 2-18, which

108 NOISE AND SPURIOUS RESPONSE OF LOOPS

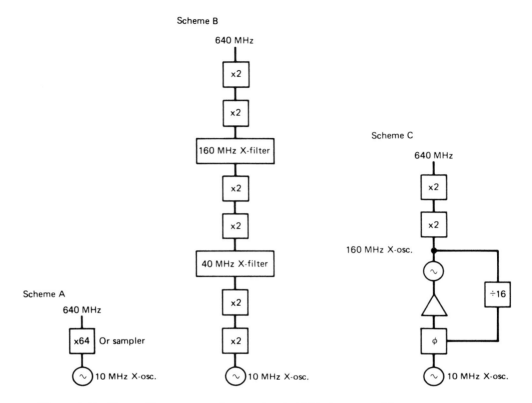

Figure 2-18 Three different ways of generating 640 MHz from a 10-MHz crystal oscillator.

shows three different multiplier chains. In scheme A, a 10-MHz crystal oscillator is multiplied directly up to 640 MHz. This system is bound to show a very high noise floor at 10 kHz and more off the carrier.

Scheme B uses the same multiplication scheme but incorporates two crystal filters, one at 40 MHz and one at 160 MHz. It should be noted that it is very difficult to build narrow-frequency crystal filters at these high frequencies. The 40-MHz crystal most likely is a third overtone crystal, and the 160-MHz crystal filter uses a ninth overtone crystal. These crystal filters are probably single-pole filters, and there is a trade-off between how narrow they can be made and how narrow the designer wants them to be. As aging, production, reproducibility, and temperature effects have to be taken into consideration, these crystal filters cannot be narrower than 1 kHz in a practical circuit.

Scheme C shows a phase-locked crystal oscillator at 160 MHz that is multiplied up to 640 MHz.

Schemes B and C both use a chain of times-2 multipliers. These can be built fairly conveniently with two diodes rather than transistors, and the advantage of these multipliers is the better noise performance, since the close-in noise of fast switching, hot carrier diodes is less than the noise found in bipolar transistors.

There is some merit in using a field-effect transistor as a frequency doubler since junction field-effect transistors, as well as MOS field-effect transistors, are

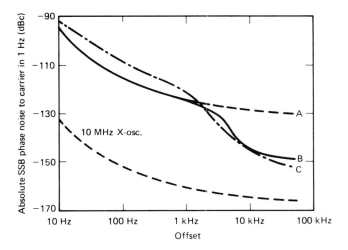

Figure 2-19 Noise sideband performance of the multipliers of Figure 2-18 cases A, B, and C.

square-law devices, and frequency multiplication becomes easy, as these devices have gain and a high output impedance, so they can operate in a tuned circuit for filtering purposes. Again, a decision has to be made regarding the frequency at which these multiplication stages should be operated. Above 300 or 400 MHz the diode multiplier would be preferred.

By now, we are curious to see the result of these different types of multiplication at the output frequency of 640 MHz. In Figure 2-19, we have plotted the noise performance of the 10-MHz crystal oscillator, which is very clean, and have shown the noise curves for cases A, B, and C.

Method A results in straight multiplication of the reference frequency; therefore, the reference oscillator noise is increased by 36 dB. Method B is also a straight multiplication, but the narrowband crystal filters have improved the noise sideband performance above 2 kHz off the carrier. Method C uses a 160-MHz overtone crystal oscillator. Because of the lower Q of this crystal compared to the 10-MHz crystal oscillator, the noise sideband performance is worse. In addition, because of the divider generating noise into the phase/frequency comparator, the close-in noise below 1 kHz of this method is higher than with any other method. It now becomes clear that method B is the preferred choice.

Method B is not without risks, however. The use of these crystal filters also means that the system has to be built mechanically to be extremely stable. Moving components can generate mechanical vibration of the crystal, which in turn causes phase jumps. Several recently developed synthesizers using this scheme ran into difficulties with mechanical vibration from the built-in fan and minute oscillations from the power supply transformer generating line frequency spurs. There are other potential problems in this approach:

1. Additive noise in the first stages of multiplication.
2. Low-frequency device noise and power supply noise, causing phase modulation in amplifiers, most sensitive again in the first stages of multiplication.

3. Doubler noise.
4. Crystal filter noise.
5. Microphonic noise, inducing phase noise in crystal filters (already mentioned).

2-6-5 Noise Contribution from Power Supplies

I have mentioned the effect of line frequency pickup several times so far, the most direct being ripple on the dc supply voltage.

Power supplies can generally be built in one of two ways:

1. Using a monolithic regulator.
2. Using discrete components.

The safe approach is generally to use two cascaded regulator systems, starting with a monolithic regulator, followed by a discrete postregulation.

In synthesizers, it is typical to find the following voltage requirements: +5 V, ±12 V, +9 V, and +24 V. When using a power supply fed from a 110- or 220-V power line, the generation of these auxiliary voltages is fairly easy. As the 5 V probably has the highest current drain, this will be kept totally separate from the other voltages. The current consumption on the ±12 V is on the order of several hundred milliamperes, and the 9 V is probably an auxiliary voltage that can be generated in a postregulator from the +12 V.

The +24 V requirement is generally of low power consumption and is required for the phase/frequency detector stages and the tuning diodes. If a dc amplifier translation stage is used following the phase/frequency comparators to drive the tuning diode, such a high voltage is necessary.

The dynamic regulation found in a regulator is typically 60 or sometimes 70 dB, which reduces the input ripple voltage to about 1 mV. This is insufficient for sensitive lines and a postregulator of at least the same amount must be added. Here a discrete circuit is the proper choice.

There are numerous regulators on the market, but the one with the lowest noise is probably the National LM723. The typical output noise of this regulator is in the vicinity of a few microvolts. Figure 2-20 shows a regulator for extremely low noise output. It is based on the fact that the current generating PNP transistor produces much less noise than its emitter follower equivalent.

In battery-operated synthesizers, especially if they operate from 12 V dc, it is somewhat difficult to generate the higher voltage for the tuning diodes. One of the best approaches is to use a switching dc/dc converter stage that is being driven from the reference oscillator at a rate of 10 kHz to 1 MHz. As the power consumption on the tuning line is very small, no special power transistors are required, and regulators take care of reference suppression. As these stages are being driven from a square wave generated from a regulated power supply, extremely high values of regulation can be obtained, and the tuning voltage is therefore very clean and noise-free. Attempts to generate the auxiliary voltage from asynchronous dc/dc converters have generally resulted in poor performance, and this approach is not recommended.

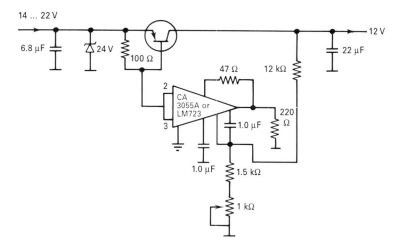

Figure 2-20 Schematic diagram of an extremely low noise output regulator based on the fact that the current generating PNP transistor produces much less noise than its emitter follower equivalent. Also, this type of circuit has a much smaller voltage drop than the source follower. It operates quite well with voltage differences as low as 0.7 V.

2-7 OVERALL PHASE NOISE PERFORMANCE OF A SYSTEM

By now we are curious to see how the various stages of noise contribution affect the system. However, we have not yet taken into consideration the absolute reference frequency relative to the output frequency. It is apparent that a very narrow loop will show a different performance than a wide loop, and therefore the reference frequency plays an important role. From a system point of view, let us now take a look at the effects of reference frequency on loop performance.

The expression for the output frequency of a single loop shows that, in order to obtain fine frequency resolution, the reference frequency must be small, equivalent to the step size. This creates conflicting requirements. One problem is that to cover a broad frequency range requires a large variation in N. Even if the hardware problems can be overcome, some method will normally be needed to compensate for the variations in loop dynamics that occur for widely varying values of N. The linearized loop transfer function is

$$B(s) = \frac{\theta_o(s)}{\theta_i(s)} = \frac{K_v F(s)/s}{1 + K_v F(s)/Ns} \quad (2\text{-}49)$$

If N is to assume a large number of values, say, from 1 to 1000, there will be a 60-dB variation in the open-loop gain and a correspondingly wide variation in the loop dynamics unless some means is used to alter the loop gain for different N values. A second problem encountered with a low reference frequency is that the loop bandwidth must be less than the reference frequency because the low-pass filter must filter out the reference frequency and its harmonics. It was explained in Section 1-10-3 that the loop bandwidth must be less than the filter bandwidth

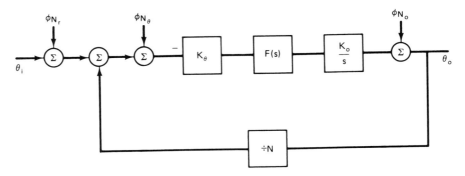

Figure 2-21 Linearized model of a PLL with the three main sources of noise, whose largest components are at the reference frequency; the harmonics of this frequency is the noise introduced by the VCO.

for adequate stability. Therefore, a low reference frequency results in a frequency synthesizer that will be slow to change frequency. Although the transient behavior is difficult to analyze, a rule of thumb often used in digital PLLs is that it takes approximately 25 cycles of the reference signal to change frequency. Thus, if a 1-Hz reference frequency is used, it will take approximately 25 s to switch to a different frequency. Another problem introduced by a low reference frequency is its effect on noise introduced in the VCO. Figure 2-21 shows a linearized model of a PLL with the three main sources of noise. ϕ_{N_r} is the noise on the reference signal. ϕ_{N_θ} is the noise created in the phase detector; its largest components are at the reference frequency and the harmonics of this frequency. ϕ_{N_o} is the noise introduced by the VCO. VCO noise has most of its energy content near the oscillator frequency; in the PLL model it can be interpreted as a low-frequency noise. The noise at the VCO output is given by

$$\phi = \frac{(\phi_{N_v} + \phi_{N_\theta})[K_v F(s)/s]}{1 + K_v F(s)/Ns} + \frac{\phi_{N_o}}{1 + K_v F(s)/Ns}$$
$$= G(s)(\phi_{N_r} + \phi_{N_\theta}) + Gr(s)/\phi_{N_o} \quad (2\text{-}50)$$

$G(s)$ is a low-pass transfer function, and $Gr(s)$ is a high-pass transfer function. Since $F(s)$ is either unity or a low-pass transfer function, the PLL functions as a low-pass filter for phase noise arising in the reference signal and phase detector, and it functions as a high-pass filter for phase noise originating in the VCO. Since the VCO noise is a low-frequency noise, the output noise due to ϕ_{N_o} is minimized by having the loop bandwidth as wide as possible. At the same time, the loop bandwidth should be less than the reference frequency in order to minimize the effect of ϕ_{N_θ}, which is dominated by spurious frequency components at the reference frequency and its harmonics.

Figures 2-22 and 2-23 show the rules for low noise operation.

Therefore, the desire to have a low reference frequency f_r in order to obtain fine frequency resolution is offset by the need to have f_r large in order to reduce the loop settling time and also the amount of noise contributed by the VCO. One

OVERALL PHASE NOISE PERFORMANCE OF A SYSTEM

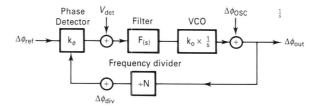

PARAMETERS TO OPTIMIZE FOR MINIMUM OUTPUT PHASE NOISE

- Minimize phase noise of free-running VCO

$$\frac{\Delta\phi_{out}}{\Delta\phi_{osc}} = \frac{1}{1 + G_{ol}(s)}$$

Open loop gain $G_{ol}(s) = k_\phi F_{(s)} k_o \frac{1}{s} \frac{1}{N}$

- Maximize bandwidth and open loop gain

$$\frac{\Delta\phi_{out}}{\Delta\phi_{ref}} = \frac{1}{1 + \frac{1}{G_{ol}(s)}}$$

Constraints: N × Reference phase noise
N × Divider phase noise
N × Phase detector noise
Filtering of f_{ref} and spurious on reference signal
Loop stability

- Avoid dividers if possible

Figure 2-22 Parameters to be optimized for minimum output phase noise for phase-locked sources.

method frequently used to obtain line frequency resolution and fast loop response is to use a multiloop synthesizer.

A *multiloop synthesizer* uses the various stages we have covered in this chapter as building blocks for a complete synthesizer. The particular example we will use is the Rohde & Schwarz SMPC signal generator, and we will look only at the synthesizer itself, not the entire generator. Without the modulation capability, the SMPC is also being sold under the name XPC. Figure 2-24 shows the block diagram of the synthesizer. A 10-MHz frequency standard is used to generate the auxiliary frequencies; it can be synchronized externally.

The main resolution of the synthesizer down to 1 Hz is achieved in a direct synthesizer where the frequencies are generated with the help of a lookup table of a sine-wave function, a principle explained in greater detail in Chapter 3.

This direct synthesizer operating from 2.2 to 2.3 MHz is mixed with the 10 MHz to generate a frequency of 12.2 to 12.3 MHz.

The 100-kHz reference frequency generated from the 10-MHz standard is used for the programmable divider and to phase lock a 135-MHz crystal oscillator in which the crystal acts as a narrow filter. This circuit has the advantage of combining the filter effects of two crystals in the multiplier loop, as mentioned earlier in this chapter. The loop also reduces some of the microphonic disturbances.

A third auxiliary frequency is generated by dividing the 135-MHz crystal oscillator output to 45 MHz.

114 NOISE AND SPURIOUS RESPONSE OF LOOPS

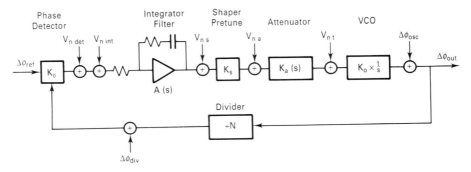

Open Loop Gain: $G_{ol}(s) = K_\phi \, A(s) \, K_a(s) \, K_o \, \dfrac{1}{N} \, \dfrac{1}{s}$

- Minimize integrator, shaper, attenuator noise

- Maximise phase detector gain K_ϕ $\quad \dfrac{\Delta\phi_{out}}{V_{n\,int}} = \dfrac{1}{K_\phi} \dfrac{N}{1 + \dfrac{1}{G_{ol}(s)}}$

- Minimize sensitivity of VCO, K_o $\quad \dfrac{\Delta\phi_{out}}{V_{nt}} = \dfrac{\dfrac{K_o}{s}}{1 + G_{ol}(s)}$

- Employ attenuator and minimize $K_a(s)$

  e.g., effect of $V_{n\,int}$ (outside loop bandwidth): $\Delta\phi_{out} = K_a(s)\, A(s)\, K_s\, K_o \dfrac{1}{s} V_{n\,int}$

Total response due to all noise excitations:

$$\Delta\phi_{out}^2(s) = \left(\dfrac{N}{1 + \dfrac{1}{G_{ol}(s)}}\right)^2 \left[\Delta^2\phi_{ref}(s) + \Delta^2\phi_{div}(s)\right]$$

$$+ \dfrac{1}{K_\phi^2}\left(\dfrac{N}{1 + \dfrac{1}{G_{ol}(s)}}\right)^2 \left[V_{n\,det}^2(s) + V_{n\,int}^2(s) + \dfrac{1}{A(s)^2}V_{ns}^2(s) + \dfrac{1}{A(s)^2}\dfrac{1}{K_s^2}V_{n\,a}^2(s)\right]$$

$$+ \left(\dfrac{1}{1 + G_{ol}(s)}\right)^2 \left[\left(\dfrac{K_o}{s}\right)V_{n\,t}^2(s) + \Delta\phi_{osc}^2(s)\right]$$

Figure 2-23 Calculation of all the noise excitation.

The main oscillator is the 240- to 248-MHz phase-locked-loop oscillator system, which is locked by 240 to 248 MHz. Such a loop can be considered as a "cleanup" loop, where the loop bandwidth is kept narrow enough to clean up all the spurs but still allow a fast-enough switching time.

The output of this loop is then divided to 20 . . . 20.667 MHz and used to drive the output phase-locked loop, which covers 680 to 1360 MHz.

The programmable divider gets its information from the keyboard, and the microprocessor inside the synthesizer generates the necessary voltages to pretune the 680–1360-MHz oscillator. This is also called coarse steering. Therefore, the output oscillator will lock with the proper harmonic of the 20- to 20.667-MHz spectrum in a sampling-type phase/frequency comparator. This can be done without a frequency division under the harmonic lock principle, also described in this chapter.

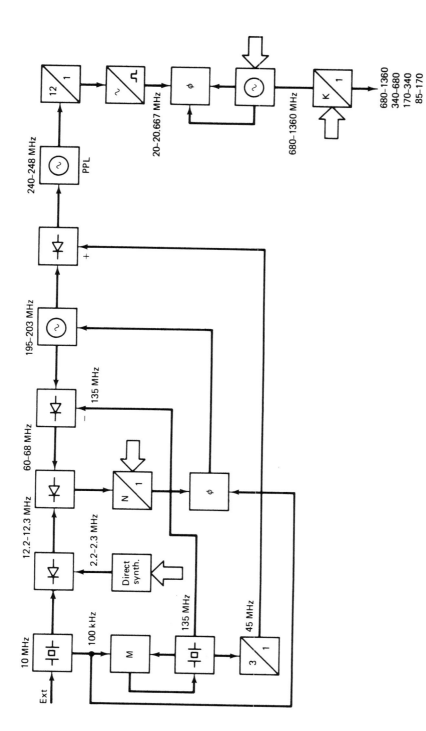

Figure 2-24 Rohde & Schwarz type XPC synthesizer.

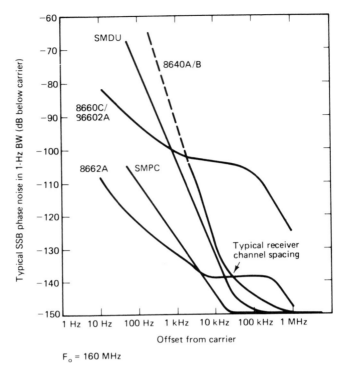

Figure 2-25 Comparison of phase noise of various signal generators and synthesizers. The Rohde & Schwarz SMPC has since been replaced by the SMG/SMH and SMHU synthesized signal generators. They exhibit similar phase noise performance.

In order to obtain the other frequency ranges, the output divider (K) divides the output range 680 to 1360 MHz into the subranges 340 to 680 MHz, 170 to 340 MHz, and 85 to 170 MHz. It is obvious that the noise sideband at the lower range is therefore better than at the higher end; the noise performance therefore improves. The 135-MHz crystal oscillator is being used in an arrangement to generate the range 100 kHz to 100 MHz but is not shown in this block diagram.

As the microprocessor keeps track of all the various division ratios, it becomes apparent that the resolution for the direct synthesizer in the range 2 to 2.3 MHz has to be better than 1 Hz in order to get 0.1-Hz resolution up to 100 MHz at the final output frequency and 1 Hz from 100 to 1300 MHz.

How does this principle compare with other signal generators and synthesizers on the market? Figure 2-25 shows a comparison in which the noise performance of the Hewlett-Packard 8640A/B and the Rohde & Schwarz SMDU free-running signal generators are displayed. In addition, the older model HP8660C, 86602A, and the latest version, 8662A, are compared with the Rohde & Schwarz SMPC.

It becomes apparent that the close-in noise of the HP8662A from 1 Hz to about 3 kHz off the carrier is better than the SMPC, whereas the noise performance from 5 to 800 kHz of the SMPC is better. The reason for this is the different concept;

the 8662A, being practically twice as expensive as the SMPC, is a much faster synthesizer, locking in less than 0.5 ms while the SMPC has about 15 ms switching time. This indicates that the loop bandwidth of the SMPC is narrow, and therefore less cleanup can be achieved from the auxiliary circuits.

However, the far-out noise of the SMPC is better because it relies more heavily on the performance of the 680- to 1360-MHz oscillator that is locked against the 240- to 248-MHz oscillator with a fast divider of 12.

Special divider circuits were created to provide a low enough noise floor so that the total system's noise after the multiplication is not substantially degraded.

We can learn from this analysis that there is a speed and cost trade-off whereby the more complex system, especially using the fractional division synthesizer principle like that in the 8662A, can be made much faster and at higher cost, while in allowing a longer switching time and a somewhat more traditional approach, costs can be kept lower.

2-7-1 Practical Results for Noise Contributions

It was mentioned earlier that when choosing the wrong damping factor or wrong phase margin, we will see "noise ears." The following are some examples.

Let us use a Frequency West microwave source and internal crystal reference oscillators, which get multiplied to 9.543 GHz. When looking at the output spectrum and replacing the crystal oscillator with a different signal source, the lack of purity of the source will become much more obvious. This method is frequently used to expose the performance if the resolution of the spectrum analyzer at the base frequency is not sufficient. Figure 2-26 shows the output signal using the internal crystal oscillator for the microwave source.

Replacing the crystal oscillators by a Fluke 6160B synthesizer at the same frequency, the noise performance gets significantly degraded. The two ears, as seen in Figure 2-27 (left and right), from the carrier are due to the phase margin set.

A similar picture is obtained using a PTS160 source, as seen in Figure 2-28.

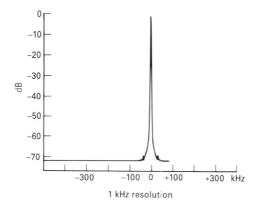

Figure 2-26 Measured SSB phase noise of the Frequency West MS-70XCE-XX Microwave Source. output = 9.543 GHz, internal crystal 98.46875 MHz, measured with HP8565A spectrum analyzer.

118 NOISE AND SPURIOUS RESPONSE OF LOOPS

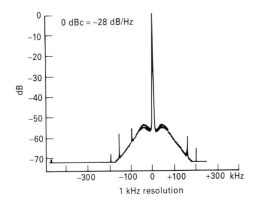

Figure 2-27 Same signal generator as shown in Figure 2-26, driven at 98.46875 MHz from Fluke 6160B. The two ears—left and right—are due to the error in the phase margin from optimum 45°.

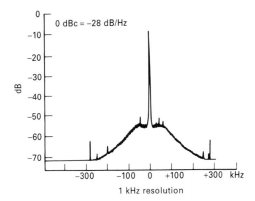

Figure 2-28 Phase noise output of Frequency West MS-70XCE-XX. Microwave source output = 9.543 GHz driven at 98.46875 MHz from PTS160.

Increasing the resolution of the spectrum analyzer and looking much closer around the carrier, we find ac power line spurious and related harmonics, as shown in Figure 2-29.

In order to obtain such fine resolution from the spectrum analyzer, the latest and most modern test equipment is required. While the technology changes constantly, for 1995, it appears that the Rohde & Schwarz spectrum analyzer Series FSEB30/FSEM30 and the HP8566 are the best available tools for this purpose in a similar price range.

2-8 MEASUREMENT OF PHASE NOISE

The emphasis on low phase noise sources also guides the selection of test methods. Phase noise measurement techniques will be compared on the basis of minimum measurable phase noise $\mathcal{L}$.

MEASUREMENT OF PHASE NOISE 119

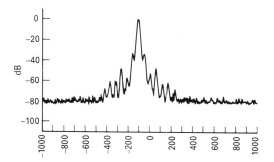

Figure 2-29 Close-in analysis of the Frequency West MS-71XCE-XX Microwave Source. We find ac power line spurious and related harmonics.

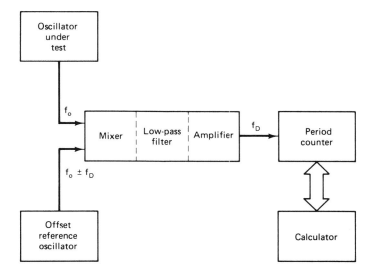

Figure 2-30 Noise measurements based on heterodyne technique.

2-8-1 Heterodyne Frequency Measurement Technique

In the time domain, frequency stability is measured with period counters. Given a stable reference source, the resolution is greatly enhanced by heterodyning (Figure 2-30). Resolution is expressed by

$$\frac{\Delta f}{f_o} = \frac{f_D^2 \Delta \tau}{f_o} \tag{2-51}$$

where $\Delta f/f_o$ = minimum fractional frequency difference
f_D = difference frequency
τ = sample time, minimum, = $1/f_D$
$\Delta \tau$ = least digit of period count

With computing counters the Allan variance σ_y can be obtained conveniently. Desktop computer-based systems such as the HP5390A frequency stability analyzer, convert time-domain data into spectral densities. The system noise floor is given by

$$\mathscr{L}_{\text{system}} = -174 + \log_{10} \frac{f_D^2}{f_m^2} \quad \text{dBc/Hz} \tag{2-52}$$

For example: at $f_m = 1\,\text{Hz}$ with $f_D = 10\,\text{Hz}$, the system can measure down to $-154\,\text{dBc}$.

There are a number of other counters on the market that allow time-interval averaging or period-interval averaging, and the results may be processed by a computer used as a controller on the IEEE bus.

Compared with other methods, this technique loses its advantage quickly above Fourier frequencies greater than 100 Hz.

2-8-2 Phase Noise Measurement with Spectrum Analyzer

RF spectrum analyzers measure the spectral density $\mathscr{L}$ directly, provided that the phase noise of the source under test is significantly above its AM noise.

By down-converting with a clean reference source, AM noise of the source under test can be suppressed if it is used as the high-level LO drive for the mixer.

Limitations of this direct method are phase noise of the spectrum analyzer LO, dynamic range, and resolution.

An RF spectrum analyzer with a YIG oscillator as LO can measure $\mathscr{L}$ at 100 kHz down to approximately $-120\,\text{dBc}$. Spectrum analyzers with synthesized LO allow phase noise measurements closer in. The various companies producing spectrum analyzers have so many new models under development that it is not possible to recommend certain types of analyzers, as new models may be available at the time of publication of this book.

However, the two most powerful spectrum analyzers currently available are the HP Model 8568A from 100 Hz to 1500 MHz, and the Model 8566A from 100 Hz to 22 GHz.

Both of these spectrum analyzers are synthesized rather than using a simple swept YIG oscillator. Hewlett-Packard uses a novel technique in the synthesizer portions of those spectrum analyzers, which had been referred to as "rock-and-roll" technique, an expression borrowed from dance music rather than science. What it means is that at certain frequencies and at certain times the frequency is phase locked for a short period and then swept to the next lock point, a method that apparently combines precise frequency resolution and extremely low noise. The interested reader is referred to a brief discussion of this technique in Section 3-5 and in the details given in the instruction manuals for those analyzers. Figure 2-31 shows the two common test setups.

MEASUREMENT OF PHASE NOISE   121

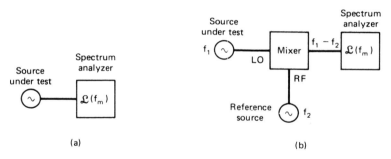

Figure 2-31 Two commonly used arrangements to determine noise sideband of a source.

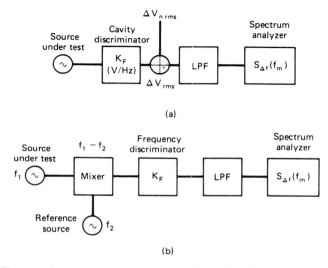

Figure 2-32 Phase noise measurement setup with cavity discriminator or frequency discriminator.

2-8-3 Phase Noise Measurement with Frequency Discriminator

The spectral density of frequency fluctuations $S_{\Delta f}(f_m)$ of the source under test is obtained when the signal is applied to a frequency discriminator either directly or in a heterodyne fashion, as shown in Figure 2-32.

$$\Delta f_{\text{rms}} = \frac{1}{K_F} \Delta V_{\text{rms}}$$

$$S_{\Delta f}(f_m) = \frac{1}{K_F^2} (\Delta V_{\text{rms}})^2 (1 \text{ Hz}) \qquad (2\text{-}53)$$

$\mathcal{L}(f_m)$ is calculated from $S_{\Delta f}$:

$$\mathcal{L}(f_m) = \frac{1}{2} \frac{1}{f_m^2} S_{\Delta f}(f_m) \qquad (2\text{-}54)$$

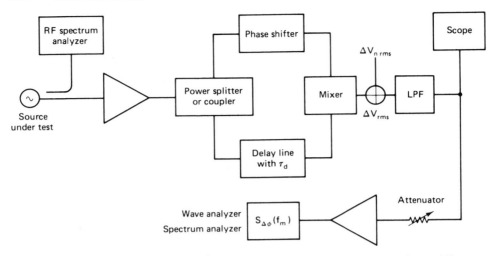

Figure 2-33 Noise sideband evaluation system using a delay line and a phase shifter.

Assuming a noise floor of the discriminator represented by $\Delta V_{n\,\text{rms}}$, the system noise for $\mathscr{L}(f_m)$ is

$$\mathscr{L}_{\text{system}}(f_m) = \frac{1}{2}\frac{1}{K_F^2}\frac{1}{f_m^2}(\Delta V_{n\,\text{rms}})^2(1\text{ Hz}) \tag{2-55}$$

It indicates the basic drawback of the use of the frequency discriminator method in determining the phase noise $\mathscr{L}(f_m)$ of a source. The system's noise floor rises with f_m^{-2} toward low offsets. This assumes a white spectrum of ΔV_n.

Using the 8901A modulation analyzer ($\Delta f_{\text{res}} = 0.5$ Hz), for example, as the frequency discriminator, $\mathscr{L}_{\text{system}}$ at 1 kHz will be -105 dBc.

2-8-4 Delay Line and Mixer as Frequency Comparator

A mixer operating as a phase detector and a delay line has the combined effect of a frequency comparator, again yielding $S_{\Delta f}(f_m)$, as seen in Figure 2-33. Both inputs to the mixer have to be in quadrature to assure maximum phase sensitivity.

The output voltage ΔV of the mixer is proportional to the frequency deviation Δf of the source and to the phase detector constant K_θ and has a periodic, $(\sin x)/x$, dependence on $f_m \tau_d$.

$$\Delta V = K_\theta \Delta \theta$$
$$\Delta V = K_\theta \tau_d \frac{\sin(\omega_m \tau_d/2)}{\omega_m \tau_d/2} \Delta \omega \tag{2-56}$$

where τ_d = delay time
K_θ = phase detector constant, = $V_{\text{beat,peak}}$ for sinusoidal beat signal

For $f_m \ll 1/2\tau_d$,

$$\Delta f_{rms} = \frac{\Delta V_{rms}}{2\pi k_\theta \tau_d}$$

$$\mathscr{L}(f_m) = \frac{1}{2f_m^2} S_{\Delta f}(f_m) = \frac{1}{2} \frac{(\Delta V_{rms})^2 (1\text{ Hz})}{(2\pi)^2 K_\theta^2 \tau_d^2 f_m^2} \quad (2\text{-}57)$$

$$S_{\Delta f}(f_m) = \frac{(\Delta V_{rms})^2 (1\text{ Hz})}{(2\pi)^2 K_\theta^2 \tau_d^2}$$

The sensitivity of the system can again be evaluated by replacing ΔV_{rms}, caused by frequency fluctuations of the source, with $\Delta V_{n\,rms}$, representing mixer noise plus noise of the following amplifier.

$$\mathscr{L}_{system}(f_m) = \frac{1}{2} \frac{(\Delta V_{n\,rms})^2 (1\text{ Hz})}{(2\pi)^2 K_\theta^2 \tau_d^2 f_m^2} \quad (2\text{-}58)$$

With white mixer (plus amplifier) noise, the system sensitivity decreases with f_m^{-2}. The flicker characteristic of the mixer noise causes the noise floor to rise with f_m^{-3} toward low offsets.

This method is also referred to as the *autocorrelation method* [19] and can be optimized toward two different goals. To determine the highest sensitivity for AM or spurious output at the input of the system, the delay should be set at $\tau_d = m(\pi/\omega_m)$, where the delay-line length is an integral number of even half-wavelengths. The system is optimized for maximum sensitivity of AM by varying the delay until a maximum dc level is obtained at the output of a mixer.

The system has the maximum sensitivity to FM noise for $\tau_d = [(2m + 1)/\omega_m]\pi$. This time the system is tuned for maximum sensitivity to FM by adjusting the delay τ_d until a dc null is obtained by the output of the mixer.

This principle can be reversed, which means that the output from the double-balanced mixer can be fed backward into the oscillator and can be used to stabilize the noise of the free-running oscillator.

This method of feedback, described later for a signal generator, has been used in Germany since about 1968, together with a PAL delay line in a 5- to 5.5-MHz VFO. The method has been adapted lately by Fluke in its generator Model 6070A and Model 6071A synthesized signal generator. Reference 20 provides more details.

References 15 and 16 explore this method extensively. $\mathscr{L}_{system}$ at 1 kHz can be as low as -115 dBc.

2-8-5 Phase Noise Measurement with Two Sources and Phase Comparator

The most direct and most sensitive method to measure the spectral density of phase noise $S_{\Delta\theta}(f_m)$ requires two sources—one or both of them may be the device(s) under test—and a double-balanced mixer used as a phase detector. The RF and

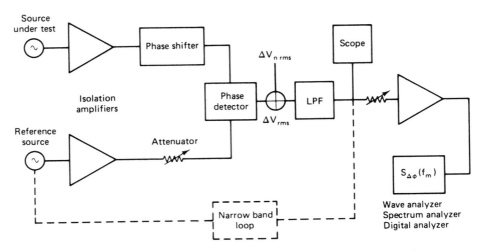

Figure 2-34 Phase noise system with two sources maintaining phase quadrature.

LO input to the mixer should be in phase quadrature, indicated by 0 V dc at the IF port. Good quadrature assures maximum phase sensitivity K_θ and minimum AM sensitivity. With a linearly operating mixer, K_θ equals the peak voltage of the sinusoidal beat signal produced when both sources are frequency offset (see Figure 2-34).

When both signals are set in quadrature, the voltage ΔV at the IF port is proportional to the fluctuating phase difference between the two signals.

$$\Delta \theta_{rms} = \frac{1}{K_\theta} V_{rms}$$

$$S_{\Delta\theta}(f_m) = \frac{(\Delta V_{rms})^2 (1 \text{ Hz})}{V_{B\,peak}^2} \frac{1}{2} \frac{(\Delta V_{rms})^2 (1 \text{ Hz})}{V_{B\,rms}^2} \qquad (2\text{-}59)$$

$$\mathcal{L}(f_m) = \frac{1}{2} S_{\Delta\theta}(f_m) = \frac{1}{4} \frac{(\Delta V_{rms})^2 (1 \text{ Hz})}{V_{B\,rms}^2}$$

where K_θ = phase detector constant, and $V_{B\,peak}$ for sinusoidal beat signal.

The calibration of the wave analyzer or spectrum analyzer can be read from the equations above. For a plot of $\mathcal{L}(f_m)$ the 0-dB reference level is to be set 6 dB above the level of the beat signal. The -6-dB offset has to be corrected by $+1.0$ dB for a wave analyzer and by $+2.5$ dB for a spectrum analyzer with log amplifier and average detector. In addition, noise bandwidth corrections may have to be applied.

Since the phase noise of both sources is measured in this system, the phase noise performance of one of them needs to be known for definite data on the other source. Frequently, it is sufficient to know that the actual phase noise of the dominant source cannot deviate from the measured data by more than 3 dB. If three unknown sources are available, three measurements with three different source combinations yield sufficient data to calculate accurately each individual performance.

Figure 2-34 indicates a narrowband phase-locked loop that maintains phase quadrature for sources that are not sufficiently phase stable over the period of the measurement. The two isolation amplifiers should prevent injection locking of the sources.

Residual phase noise measurements test one or two devices, such as amplifiers, dividers (Figure 2-13), or synthesizers (Figure 2-42), driven by one common source. Since this source is not free of phase noise, it is important to know the degree of cancellation as a function of Fourier frequency.

The noise floor of the system is established by the equivalent noise voltage ΔV_n at the mixer output. It represents mixer noise as well as the equivalent noise voltage of the following amplifier:

$$\mathscr{L}_{\text{system}}(f_m) = \frac{1}{4} \frac{(\Delta V_{n\,\text{rms}})^2 (1\,\text{Hz})}{V_{B\,\text{rms}}^2} \tag{2-60}$$

Noise floors close to -180 dBc can be achieved with a high-level f25mixeand a low-noise port amplifier. The noise floor increases with f_m^{-1} due to the flicker characteristic of ΔV_n. System noise floors of -166 dBc at 1 kHz have been realized.

In measuring low-phase-noise sources, a number of potential problems have to be understood to avoid erroneous data:

- If two sources are phase locked to maintain phase quadrature, it has to be ensured that the lock bandwidth is significantly lower than the lowest Fourier frequency of interest.
- Even with no apparent phase feedback, two scources can be phase locked (injection locked), resulting in suppressed close-in phase noise.
- AM noise of the RF signal can come through if the quadrature setting is not maintained sufficiently.
- Deviation from the quadrature setting will also lower the effective phase detector constant.
- Nonlinear operation of the mixer results in a calibration error.
- A nonsinusoidal RF signal causes K_θ to deviate from $V_{B\,\text{peak}}$.
- The amplifier or spectrum analyzer input can be saturated during calibration or by high spurious signals such as line frequency multiples.
- Closely spaced spurious signals such as multiples of 60 Hz may give the appearance of continuous phase noise when insufficient resolution and averaging are used on the spectrum analyzer.
- Impedance interfaces should remain unchanged when going from calibration to measurement.
- In residual measurement system phase, the noise of the common source might be insufficiently canceled due to improperly high delay-time differences between the two branches.

126   NOISE AND SPURIOUS RESPONSE OF LOOPS

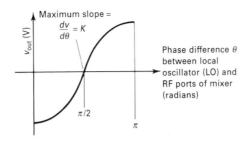

Figure 2-35 Output voltage of double-balanced mixer versus phase difference between local oscillator and RF signal ports.

- Noise from power supplies for devices under test or the narrowband phase-locked loop can be a dominant contributor of phase noise.
- Peripheral instrumentation such as the oscilloscope, analyzer, counter, or DVM can inject noise.
- Microphonic noise might excite significant phase noise in devices.

Mathematical Analysis. The mixer is usually a double-balanced mixer consisting of four diodes. The IF port is dc-coupled to provide the phase-locked dc signal. This phase-locked dc signal is adjusted to be 0 V on the voltmeter, since the sensitivity $dv/d\theta$ is maximum for this condition. This is done by adjusting a line length such that the phases of the two oscillators are 90° apart.

Figure 2-35 shows the typical sensitivity of the mixer. Beyond the loop bandwidth, the output of the mixer may be described as follows:

$$v_1 = V_1 \cos(\omega t + \theta_{n1}) \tag{2-61}$$

$$v_2 = V_2 \cos(\omega t + \theta_{n2} - \pi/2) \tag{2-62}$$

$$v_3 = V_1 V_2 \cos(\omega t + \theta_{n1}) \cos(\omega t + \theta_{n2} - \pi/2)$$
$$= \frac{V_1 V_2}{2} \cos\left(\theta_{n1} - \theta_{n2} + \frac{\pi}{2}\right) \tag{2-63}$$

The θ_{n1} and θ_{n2} terms are rms phase noise, which can be combined as

$$\theta_{nT} = \sqrt{\overline{\theta_{n1}^2} + \overline{\theta_{n2}^2}} \tag{2-64}$$

$$v_3 = \frac{V_1 V_2}{2} \cos\left(\theta_{n1} - \theta_{n2} + \frac{\pi}{2}\right) = \frac{V_1 V_2}{2} \cos\left(\theta_{nT} + \frac{\pi}{2}\right)$$
$$= -\frac{V_1 V_2}{2} \sin \theta_{nT} \tag{2-65}$$

For θ_{nT} very small,

$$\sin \theta_{nT} \simeq \theta_{nT} = \sqrt{\overline{\theta_{n1}^2} + \overline{\theta_{n2}^2}} \tag{2-66}$$

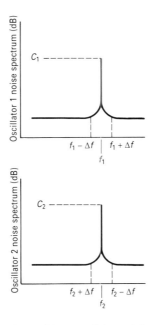

Figure 2-36 Noise spectrum of two oscillators at f_1 and f_2 carrier frequency.

Since the spectrum analyzer displays power, it will show the square of the term,

$$-\frac{V_1 V_2}{2}\sqrt{\overline{\theta_{n1}^2} + \overline{\theta_{n2}^2}} \quad \text{or} \quad \left(\frac{V_1 V_2}{2}\right)^2 (\overline{\theta_{n1}^2} + \overline{\theta_{n2}^2})$$

$$\overline{\theta_{n1}^2} = S_\theta(f_m) \quad \text{of oscillator 1} \tag{2-67}$$

$$\overline{\theta_{n2}^2} = S_\theta(f_m) \quad \text{of oscillator 2} \tag{2-68}$$

If the spectral densities have equal power distribution but are not correlated, the mixer output is 3 dB greater than either one alone. This technique yields the sum of the $S_\theta(f_m)$ for oscillator 1 and oscillator 2.

$S_\theta(f_m)$ can now be related to $\mathcal{L}(f_m)$. $S_\theta(f_m)$ is equal to $\mathcal{L}(f_m)$ folded about itself. Therefore, $S_\theta(f_m) = 2\mathcal{L}(f_m)$ if the noise sidebands about f_1 are correlated and $S_\theta(f_m) = \sqrt{2}\mathcal{L}(f_m)$ if they are not correlated.

In Figure 2-36 the noise below $f_0 - \Delta f$ is assumed to be uncorrelated to the noise above $f_0 + \Delta f$ in oscillator 1 and in oscillator 2. Closer than $f_0 \pm \Delta f$ the assumption is that there is correlation of the noise above and below the carrier in both oscillators. Beyond $\pm\Delta f$ we assume that this is the noise floor of the device. Closer than $\pm\Delta f$ we assume that the noise is caused by phase modulation mechanisms in the device or other components that generate related sidebands above and below the carrier. When these two spectra are mixed together, the following occurs: If $f_1 = f_2$, then (if we ignore the sum frequency components, which are eliminated by the low-pass filter), $f_1 - f_2 = 0$; f_1 then mixes against the noise spectrum of f_2. This causes the noise spectrum of f_2 to fold upon itself. For instance, f_1 mixing against $f_2 \pm \Delta f_x$ will yield two correlated noise components at Δf_x, which add in

128 NOISE AND SPURIOUS RESPONSE OF LOOPS

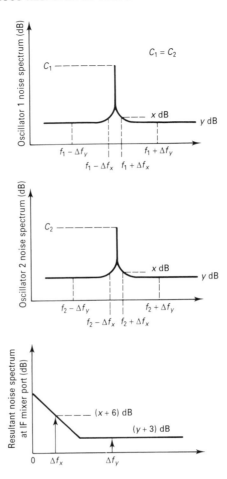

Figure 2-37 Resultant noise spectrum due to foldover of spectrum about the carrier for one oscillator.

power to cause a 6-dB increase, as in Figure 2-37. However, if f_1 mixes against $f_2 \pm \Delta f_y$, there is only a 3-dB increase, since the noise at $f_2 - \Delta f_y$ is not correlated to that at $f_2 + \Delta f_y$.

The reverse also occurs: f_2 can mix with the noise of f_1 at $f_1 \pm \Delta f_x$ and $f_1 \pm \Delta f_y$ to cause an additional 3-dB increase in noise measured at the mixer's output. This increase occurs because this reverse process generates another spectrum identical in amplitude to that in Figure 2-37; however, the noise of the two oscillators is not correlated except within the phase-locked-loop bandwidth. The mixer takes these two uncorrelated spectra and adds them at its output, causing an additional 3-dB increase in noise, as shown in Figure 2-38. $\mathscr{L}(f_m)$ can be obtained from this spectrum by subtracting 9 dB from the part where the upper and lower noise sidebands are correlated and by subtacting 6 dB from the area where no correlation exists.

MEASUREMENT OF PHASE NOISE

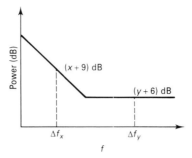

Figure 2-38 Power spectrum of mixer IF port as displayed on a spectrum analyzer due to the combined effects of foldover and addition of 3 dB for noise spectrum of two uncorrelated oscillators.

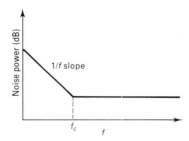

Figure 2-39 Noise power versus frequency of a transistor amplifier.

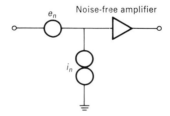

Figure 2-40 Equivalent noise sources at the input of an amplifier.

It is possible to go back to Figure 2-37 before the addition of 3 dB (due to two uncorrelated oscillators) to see how $S_\theta(f_m) = \Delta\theta^2_{\text{rms}}$ is related to $\mathscr{L}(f_m)$. Since

$$\mathscr{L}(f) = \tfrac{1}{2}\Delta\theta^2_{\text{rms}} \tag{2-69}$$

and

$$S_\theta(f_m) = \Delta\theta^2_{\text{rms}} \tag{2-70}$$

we see that the folded-over spectrum of a single oscillator at Δf_x or where the upper and lower sidebands of f_1 are correlated is equal to $S_\theta(f_m)$.

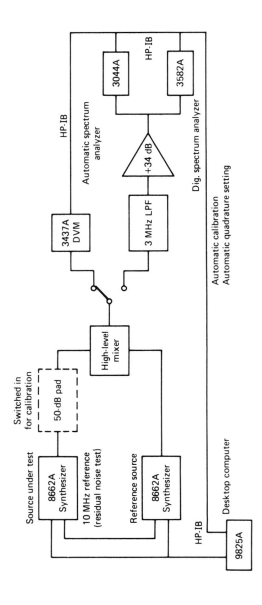

Figure 2-41 Automatic system to measure residual phase noise of two 8662A synthesizers. (Courtesy of Hewlett-Packard Company.)

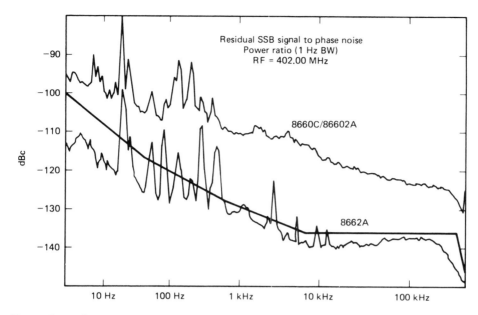

Figure 2-42 Signal-to-noise measurement of an 8660C/86602A and an 8662A Hewlett-Packard frequency synthesizer in an automated system.

The noise spectrum of an amplifier would appear as in Figure 2-39. For a moment, it is of interest to discuss the $1/f$ noise spectrum near dc. Noise in amplifiers is often modeled as in Figure 2-40, which was also discussed in Section 2-3. In bipolar amplifiers, e_n is related to the thermal noise of the base spreading resistance:

$$e_n = \sqrt{4kTr_bB} \qquad (2\text{-}71)$$

This noise source has a relatively flat frequency response. The i_n noise source is associated with the shot noise in the base current:

$$i_n = \sqrt{2qI_bB} \qquad (2\text{-}72)$$

This i_n noise generator has associated with it a $1/f$ noise mechanism.

In FET devices, the situation is reversed. The e_n noise generator has a $1/f$ noise component where i_n shows none. It is interesting to note that, in general, the $1/f$ noise corner of bipolar silicon devices is higher than that of silicon JFETs. Silicon JFETs are less noisy than silicon MOSFETS. GaAs MESFETs usually have the highest $1/f$ corner frequencies, which can extend to several hundred megahertz. Carefully selected bipolar devices can have $1/f$ noise corners below 100 Hz.

There are various instruments that can measure e_n and i_n directly with no carrier signal present. These measurement methods would provide a noise plot as in Figure 2-39. However, if a carrier signal is applied to the amplifier, the noise plot would be modified as in Figure 2-3. The low-frequency noise sources can affect the phase shift through the amplifier, causing the $1/f$ phase noise spectrum about the carrier.

Despite all these hazards, automatic test systems have been developed and operated successfully [15]. Figure 2-41 shows a system that automatically measures the residual phase noise of the 8662A synthesizer. It is a residual test, since both instruments use one common 10-MHz referenced oscillator. Quadrature setting is conveniently controlled by probing the beat signal with a digital voltmeter and stopping the phase advance of one synthesizer when the beat signal voltage is sufficiently close to zero.

The two plots of Figure 2-42 were done with the 3044A automatic spectrum analyzer covering 10 Hz to 13 MHz. The test system also measures spurious signals. On this chart the signals appear to be rather broad due to a limited number of data points per decade. Again, the older 8660C/86602A synthesized signal generator is compared with the new 8662A.

REFERENCES AND SUGGESTED READING

1. Jacques Rutman, "Characterization of Frequency Stability: A Transfer Function Approach and Its Application to Measure via Filtering of Phase Noise," *IEEE Transactions on Instrumentation and Measurement*, Vol. 22, 1974, pp. 40–48.
2. Chuck Reynolds, "Measure Phase Noise," *Electronic Design*, February 15, 1977, pp. 106–108.
3. J. A. Barnes, A. R. Chie, L. S. Cutter, et al., "Characterization of Frequency Stability," *IEEE Transactions on Instrumentation and Measurement*, Vol. IM-20, No. 2, May 1971, pp. 105–120.
4. J. A. Barnes and R. C. Mockler, "The Power Spectrum and Its Importance in Precise Frequency Measurements," *IRE Transactions on Instrumentation*, pp. 149–155.
5. E. J. Baghdady, R. N. Lincoln, and B. D. Nelin, "Short-Term Frequency Stability: Characterization, Theory, and Measurements," *Proceedings of the IEEE*, 1965, pp. 704–722.
6. D. B. Leeson, "Short-Term Stable Microwave Sources," *Microwave Journal*, June 1970, pp. 59–69.
7. D. B. Leeson, "A Simple Model of Feedback Oscillator Noise Spectrum," *Proceedings of the IEEE*, 1966, pp. 329–330.
8. L. S. Cutler and C. L. Searle, "Some Aspects of the Theory and Measurement of Frequency Fluctuations in Frequency Standards," *Proceedings of the IEEE*, Vol. 54, 1966, pp. 136–154.
9. W. A. Edson, "Noise in Oscillators," *Proceedings of the IRE*, 1960, pp. 1454–1466.
10. Erich Hafner, "The Effects of Noise in Oscillators," *Proceedings of the IEEE*, Vol. 54, 1966, pp. 179–198.
11. M. C. Fischer, "Frequency Stability Measurement Procedures," Eighth Annual Precise Time and Time Interval Applications and Planning Meeting, December 1976.
12. D. A. Howe, "Frequency Domain Stability Measurements: A Tutorial Introduction," *NBS Technical Note 679*, March 1976.
13. D. J. Healey III, "Flicker of Frequency and Phase and White Frequency and Phase Fluctuations in Frequency Sources," *Proceedings of the 26th Annual Symposium on Frequency Control*, Fort Monmouth, NJ, June 1972, pp. 43–49.
14. Hewlett-Packard staff, "Understanding and Measuring Phase Noise in the Frequency Domain," *Application Note 207*, October 1976.

REFERENCES AND SUGGESTED READING 133

15. A. L. Lance, W. D. Seal, F. G. Mendozo, and N. W. Hudson, "Automatic Phase Noise Measurements in the Frequency Domain," *Proceedings of the 31st Annual Symposium on Frequency Control*, June 1977.
16. A. L. Lance, W. D. Seal, N. W. Hudson, F. G. Mendozo, and Donald Halford, "Phase Noise Measurements Using Cross-Spectrum Analysis," Conference on Precision Electromagnetic Measurements, Ottawa, June 1978.
17. J. H. Shoaf, D. Halford, and A. S. Risley, "Frequency Stability Specification and Measurement: High Frequency and Microwave Signals," *NBS Technical Note 632*, January 1973.
18. Dieter Scherer, "Design Principles and Test Methods for Low Phase Noise RF and Microwave Sources," RF & Microwave Measurement Symposium and Exhibition, Hewlett-Packard.
19. A. Tykulsky, "Spectral Measurements of Oscillators," *Proceedings of the IEEE*, February 1966, p. 306.
20. Fred Telewski, Kingsley Craft, Eric Drucker, and Joe Martins, "Delay Lines Give RF Generator Spectrum Purity, Programmability," *Electronics*, August 28, 1980, pp. 133–142.
21. Hewlett-Packard staff, "Timekeeping and Frequency Calibration," *Application Note 52-2*, November 1975 (Hewlett-Packard, Palo Alto, CA 94304).
22. David W. Allan, "Report on NBS Dual Mixer Time Difference System (DMTD) Built for Time-Domain Measurements Associated with Phase 1 of GPS," *NBSIR 750827*, January 1976, National Bureau of Standards (U.S.) (NTIS, Springfield, VA 22151).
23. Hewlett-Packard staff, "Measuring Warmup Characteristics and Aging Rates of Crystal Oscillators," *Application Note 174-11*, November 1974 (Hewlett-Packard, Palo Alto, CA 94394).
24. David W. Allan, "The Measurement of Frequency and Frequency Stability of Precision Oscillators," *NBS Technical Note 669*, May 1975, National Bureau of Standards (U.S.) (SD Catalog No. C13.46: 669, U.S. Government Printing Office, Washington, DC 20402).
25. Luiz Peregrino and David W. Ricci, "Phase Noise Measurement Using a High Resolution Counter with On-Line Data Processing," *Proceedings of the 30th Annual Symposium on Frequency Control*, U.S. Army Electronics Command, Fort Monmouth, NJ, 1976. (Copies available from Electronic Industries Association, 2001 I Street, NW, Washington, DC 20006.)
26. Byron E. Blair, ed., *Time and Frequency: Theory and Fundamentals*, NBS Monograph 140, May 1974, National Bureau of Standards (U.S.) (SD Catalog No. C13.44: 140, U.S. Government Printing Office, Washington, DC 20402).
27. Hewlett-Packard staff, "Measuring Fractional Frequency Standard Deviation (sigma) versus Averaging Time (tau)," *Application Note 174-7*, November 1974 (Hewlett-Packard, Palo Alto, CA 94304).
28. James E. Gray and David W. Allan, "A Method for Estimating the Frequency Stability of an Individual Oscillator," *Proceedings of the 28th Annual Symposium on Frequency Control*, U.S. Army Electronics Command, Fort Monmouth, NJ, 1974, pp. 243–246. (Copies available from Electronic Industries Association, 2001 I Street, NW, Washington, DC 20006.)
29. *Reference Data for Radio Engineers*, 5th ed., Howard W. Sams, Indianapolis, IN, 1968, pp. 21–27.
30. Hewlett-Packard staff, "Spectrum Analysis: Noise Measurements," *Application Note 150-4*, January 1973 (Hewlett-Packard, Palo Alto, CA 94304).
31. Hewlett-Packard staff, "Spectrum Analysis: Signal Enhancement," *Application Note 150-7*, June 1975; and "Spectrum Analysis: Accuracy Improvement," *Application Note 150-8*, March 1976 (Hewlett-Packard, Palo Alto, CA 94304).

32. Hewlett-Packard staff, "Spectrum Analysis: Noise Figure Measurement," *Application Note 150–9*, April 1976 (Hewlett-Packard, Palo Alto, CA 94304).
33. Patrick Lesage and Claude Audoin, "Characterization of Frequency Stability: Uncertainty Due to the Finite Number of Measurements," *IEEE Transactions on Instrumentation and Measurement*, Vol. IM-22, No. 2, June 1973, pp. 157–161.
34. K. Kurokawa, "Noise in Synchronized Oscillators," *IEEE Transactions on Microwave Theory and Techniques*, April 1968, pp. 234–240.
35. A. N. Riddle, *Oscillator Noise: Theory and Characterization*, Ph.D. thesis, North Carolina State University, 1986.
36. R. A. Pucel and J. Curtis, "Near-Carrier Noise in FET Oscillators," *IEEE International Microwave Theory and Techniques Science Digest*, 1983, pp. 282–284.
37. H. Rohdin, C.-Y. Su, and C. Stolte, "A Study of the Relation Between Device Low-Frequency Noise and Oscillator Phase Noise for GaAs MESFETs," *IEEE International Microwave Theory and Techniques Science Digest*, 1984, pp. 267–269.
38. J. H. Abeles, S. H. Wemple, W. O. W. Schlosser, and J. P. Beccone, "Third-Order Nonlinearity of GaAs MESFETs," *IEEE International Microwave Theory and Techniques Science Digest*, 1984, pp. 224–226.
39. A. B. Carlson, *Communication Systems*, McGraw-Hill, New York, 1975.
40. K. H. Sann, "The Measurement of Near-Carrier Noise in Microwave Amplifiers," *IEEE Transactions on Microwave Theory and Techniques*, September 1968, pp. 761–766.
41. F. L. Walls and S. R. Stein, "Accurate Measurements of Spectral Density in Devices," *31st Annual Frequency Control Symposium*, 1977.
42. M. Marlin, "RF Sidebands Caused by DC Power Line Fluctuations," *Microwave Journal*, September 1991.
43. "Local Oscillator Phase Noise and Its Effect on Receiver Performance," *Technical Notes, Watkins Johnson Co.*, Vol. 8, No. 6, November/December 1981.
44. J. Gagnepain, J. Groslambert, and R. Brendel, "The Fractal Dimension of Phase and Frequency Noises: Another Approach to Oscillator Characterixation," *IEEE* CH2186-0/85/0000-0113, 1985.
45. W. P. Robins, *Phase Noise in Signal Sources*, Peter Peregrinus, London, 1982.
46. W. A. Gardner, *Introduction to Random Processes with Applications to Signals and Systems*, 2nd ed., McGraw-Hill, New York, 1990.
47. J. Rutman and F. L. Walls, "Characterization of Frequency Stability in Precision Frequency Sources," *Proceedings of the IEEE*, June 1991, pp. 952–960.
48. D. B. Percival, "Characterization of Frequency Stability: Frequency-Domain Estimation of Stability Measures," *Proceedings of the IEEE*, June 1991, pp. 961–972.
49. F. L. Walls and D. W. Allan, "Measurements of Frequency Stability," *Proceedings of the IEEE*, Vol. 74, No. 1, January 1986, pp. 162–168.
50. L. Cohen, "Time-Frequency Distributions—A Review," *Proceedings of the IEEE*, Vol. 77, No. 77, July 1989, pp. 941–981.
51. D. Allan et. al., "Standard Terminology for Fundamental Frequency and Time Metrology," *42nd Annual Frequency Control Symposium*, 1988, pp. 419–425.
52. CCIR Recommendation 686 (1990), "Glossary," in *Standard Frequencies and Time Signals*, Vol. 7, International Telecommunications Union, General Secretariat—Sales Section, Place des Nations, CH-1211, Geneva, Switzerland.
53. R. Gilmore, "Specifying Local Oscillator Phase Noise Performance: How Good Is Good Enough," RF Expo, 1991.
54. W. J. Riley, "Integrate Phase Noise and Obtain Residual FM," *Microwaves*, August 1979.
55. J. Cheah, "Analysis of Phase Noise in Oscillators," *RF Design*, November 1991.
56. M. R. McClure, "Residual Phase Noise of Digital Frequency Dividers," *Microwave Journal*, March 1992, pp. 124–130.

57. F. L. Walls and C. M. Felton, "Low Noise Frequency Synthesis," *41st Annual Frequency Control Symposium*, 1987.
58. "Silicon/GaAs Complementary, Not Competitive," *RF Letters RF Design*, August 1986.
59. M. Bomford, "Selection of Frequency Dividers for Microwave PLL Applications," *Microwave Journal*, November 1990.
60. R. E. Best, *Phase-Locked Loops Theory, Design, & Applications*, McGraw-Hill, New York, 1984.
61. D. H. Wolaver, *Phase-Locked Loop Circuit Design*, Prentice-Hall, Englewood Cliffs, NJ, 1991.
62. T. F. Hock, "Synthesizer Design with Detailed Noise Analysis," *RF Design*, July 1993, pp. 37–48.
63. F. L. Walls et al., "Extending the Range and Accuracy of Phase Noise Measurements," *42nd Annual Frequency Control Symposium*, 1989, pp. 432–441.
64. L. Martin, "Program Optimizes PLL Phase Noise Performance," *Microwaves & RF*, April 1992, pp. 78–91.
65. M. O'Leary, "Practical Approach Augurs PLL Noise in RF Synthesizers," *Microwaves & RF*, September 1987, pp. 185–194.
66. J. A. Mezak and G. D. Vendelin, "CAD Design of YIG Tuned Oscillators," *Microwave Journal*, December 1992.
67. R. Kiefer and L. Ford, "CAD Tool Improves SAW Stabilized Oscillator Design," *Microwaves & RF*, December 1992.

3

SPECIAL LOOPS

Chapters 1 and 2 have familiarized us with the phase-locked loop (PLL), the fundamental building block of all modern frequency synthesizers. We now understand the various types and orders of loops, the performance of the loop, and the evaluation of the loop.

This chapter deals with special loops that are basically one-loop synthesizers. These systems can be combined, as we will see later, in multiloop synthesizers, or some of them can be used as stand-alone systems.

The resolution or step size of the synthesizer, as we have learned, is equal to the reference frequency. There is a conflict between speed and step size, and this chapter deals with ways of minimizing this conflict. First, we will take a look at a system generating frequencies digitally with the help of logic circuitry and/or a digital computer. As today's technology provides us with fast microprocessors, these systems, using microprocessors and lookup tables, are capable of ultrafine-resolution synthesizers.

Then we take a look at multiloop sampler loops, where the various samplers are being used to speed up the response of the very narrow loops commonly required in high-resolution systems. Loops with sequential phase shifters allow increased resolution at the expense of absolute accuracy.

Then we will see how a delay line can be used to improve noise performance. This is almost the reverse technique of what we saw in Chapter 2, where the delay line was used to measure the phase noise.

Finally, we acquaint ourselves with the fractional N phase-locked loop, a spin-off of the digiphase system.

3-1 DIRECT DIGITAL SYNTHESIS TECHNIQUES

This chapter deals with modern digital synthesis concepts and implementations. Rapid advances in digital electronic circuitry, as well as digital to analog converters, have led to some very attractive solutions in frequency synthesis based on a totally digital approach.

Fractional N implementations are essentially phase-locked loop (PLL) solutions to which digital logic has been added to perform certain useful functions. In particular, fractional N is an effective and economic way of increasing the frequency resolution of PLL frequency synthesis, while maintaining an acceptable level of spurious sidebands. The approach still suffers from some inherent constraints. The main ones are relatively long settling times, when the frequency is switched, and limited phase modulation bandwidth. Recent trends toward the use of a spread spectrum in radar and communications are driving the need for faster switching and more phase modulation of bandwidth in synthesizers.

We will examine the potential for increasing the speed of fractional N implementations before reviewing the field of direct digital synthesis. An interesting question arises: Will direct digital synthesis implementations be limited to those applications requiring fast switching or will they find a natural place (dictated by economics) in general-purpose frequency synthesis? It is hoped that this chapter will provide significant insight into this matter, as we discuss direct digital synthesis architectures and their advantages and drawbacks, modulation signal quality, and future prospects.

Particular attention is paid to the quality of the signal being synthesized. It is important that the minimum requirements for the application be met. Digital synthesis has traditionally suffered from high spurious sidebands, precluding its use in many radar and communications applications. It is important to set some realistic standards for spurious sideband levels and to evaluate the potential for various competing approaches.

Contributions to this chapter were made by Albert W. Kovalick and Roland Hassun, of Hewlett-Packard Co., Palo Alto, California.

3-1-1 A First Look at Fractional *N*

This topic is reviewed extensively in Section 3-4 and an analysis of spurious sideband levels is given in Refs. 1 and 2. The method has led to cost effective implementations for slow switching narrowband requirements. What are the prospects for improving the efficacy of fractional N and extending it to faster switching, wider band applications?

The bandwidth of the PLL in fractional N implementations is an important parameter in determining its ability to switch rapidly and to sustain a high rate of phase modulation. An important variation on the traditional fractional N implementation is presented in Ref. 3.

Figure 3-1 shows a traditional fractional N configuration. The frequency divider modulus is set by the overflow indicator or the MSB from the phase accumulator. This allows the creation of fractional divisors, as explained more fully in Section 3-4. Fractional divisors give rise to frequencies at the VCO that are related to the reference frequency, at the phase detector, in a nonintegral way. The spurious sidebands caused by the dithering of the divider are canceled by introducing appropriate phase modulation through a DAC that is connected to the phase accumulator.

This configuration has been implemented in a large number of Hewlett-Packard

138 SPECIAL LOOPS

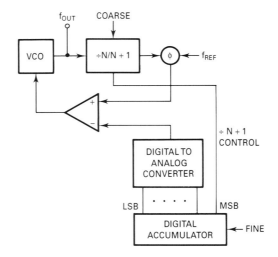

Figure 3-1 Traditional fractional N loop block diagram.

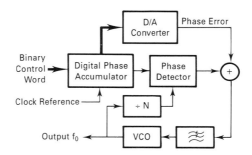

Figure 3-2 Simplified block diagram of configuration developed in Ref. 3.

synthesizers and measuring instruments. The reference frequency has been 100 kHz and was more recently increased to 400 kHz.

A different fractional N configuration is shown in Figure 3-2. It was reported in 1976 [ref. 3]. In this approach, the reference frequency is the overflow or MSB output of the phase accumulator. This signal is not periodic when a noninteger value of binary control word is used. This corresponds to a fractional frequency condition in the traditional fractional N approach. The DAC is used for correction. The main differences between the two configurations are that Figure 3-2 uses a fixed divider in the feedback and Figure 3-1 uses the accumulator output as the reference. A higher VCO frequency is possible with the use of fixed dividers and a higher reference frequency is possible with the use of the accumulator, as shown.

It is interesting to note that fractional N benefits from the same advances in digital and conversion technologies that have propelled direct digital synthesis. The key factor is to increase the reference frequency to the phase detector. This has two important benefits: improvement in phase noise and ability to extend the bandwidth of the PLL, which leads to faster switching.

After reading Section 3-1-2, it will become apparent that the method proposed in Ref. 3 is a combination of direct digital synthesis, frequency multiplication, and filtering by means of a PLL. There is no evidence of this approach being implemented commercially to date.

3-1-2 Digital Waveform Synthesizers

This class of synthesizer uses sampled data methods to produce waveforms. Three methods will be discussed: one is a digital recursion oscillator, another is phase accumulator based, and the third is a direct table lookup method. The block diagram for all three processes is shown in Figure 3-3.

The digital hardware block provides a data stream of K bits per clock cycle for the DAC. Ideally, the DAC is a linear device with glitch-free performance. The practical limits of the DAC will be discussed later in this chapter. The DAC output is the desired signal plus replications of it around the clock frequency and all of the clock's harmonics. Also present in the DAC output signal is a small amount of quantization noise from the effects of finite math in the hardware block. Figure 3-4 shows the frequency spectrum of an ideal DAC output with a digitally sampled sine-wave data stream at its input. Note that the desired signal, F_o (a single line in the frequency domain), is replicated around all clock terms. Figure 3-5 shows the same signal in the time domain.

The DAC performs a sample-and-hold operation as well as converting digital values to analog voltages. The sample occurs on each rising edge of the clock; the hold occurs during the clock period. The transfer function of a sample-and-hold operator is a $(\sin x)/x$ envelope response with linear phase. In this case, $x = (\pi F/F_{\text{clock}})$. (See Ref. 4.)

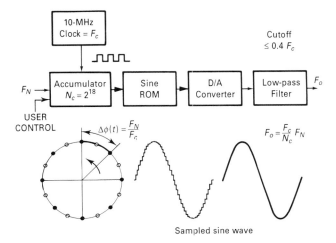

Figure 3-3 Direct digital frequency synthesizer.

140 SPECIAL LOOPS

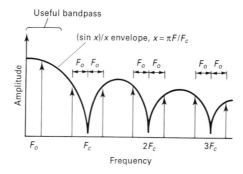

Figure 3-4 Ideal DAC output with F_o, a sampled-and-held sine wave, at its output. Notice the $(\sin x)/x$ envelope rolloff. As F_o moves up in frequency, an aliased component $F_c - F_o$ moves down into the passband.

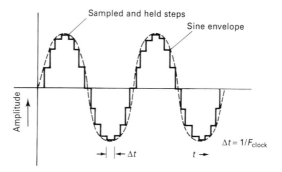

Figure 3-5 Samples/cycle sine wave. This is typical of a single-tone DAC output. $F_o = F_{clock}/16$ after low-pass filtering; only the sine envelope is present. The LPF removes the sampling energy. Each amplitude step is held for a clock period.

It should be noted that the sinc function rolloff affects the passband flatness. A 2.4-dB drop should be expected at 40% of F_{clock}. See Ref. 5 for a solution to this rolloff problem that uses a method called half-hold sampling.

Referring again to Figure 3-3, the output of the DAC is passed through a lowpass filter (LPF). With proper attention to design, an LPF may be realized that has linear phase in a flat passband with width of $0.4F_{clock}$. With this design, the maximum available bandwidth is achieved. For example, with $F_{clock} = 125\,\text{MHz}$, a useful synthesized bandwidth of about 50 MHz is attained. The LPF output is the desired signal without any sampling artifacts. Viewing the LPF strictly as a device to remove sampling energy, it is obvious why the output contains only the desired signal. It is also instructive to view the LPF from the time domain. From this point, the LPF may be seen as the perfect interpolator. It fills the space between time samples with a smooth curve to reconstruct perfectly the desired signal.

DIRECT DIGITAL SYNTHESIS TECHNIQUES   141

In general, the theory of sampled data is based on Nyquist's sampling theorem. For complete coverage, see Ref. 6.

Systems Concerns. If the desired signal, F_o (or the highest component in the desired signal if it is composed of many frequencies), is greater in frequency than $F_{\text{clock}}/2$, then an aliased version of it will appear in the passband at $F_{\text{clock}} - F_o$. See Figure 3-4. Some aliasing will always occur when non-band-limited signals, like square waves, triangular waves, or wideband FM waves, are desired. This is so because the desired signal has energy in it that extends beyond $F_{\text{clock}}/2$. Careful signal design will always guarantee that the aliasing contribution is below some specified minimum. In the case of a pure carrier, that is, a synthesizer without any modulation, aliasing is not a concern as long as $F_{\text{carrier}} < F_{\text{clock}}/2$.

Another general concern is that of finite wordlength effects. The digital hardware block of Figure 3-3 outputs a K-bit value each clock cycle. This word is typically formed through either roundoff or truncation (roundoff with a bias). In any event, the desired signal must be quantized into at most 2^K amplitude levels. All commercial DACs are of fixed point design ranging from about 6 to 20 bits.

The effect of truncation/roundoff error is quantization noise in the final output. Since each sample has an error ranging from $-\text{LSB}/2$ to $+\text{LSB}/2$, a sequence of noisy values is created. Figure 3-6 shows how amplitude quantization noise is formed. This noise may be made as small as desired by selecting the value K. It is a fact of life, however, that as K increases the DAC must be clocked slower. This is the big trade-off in designing digital synthesizers. Another strategy is to contour the effects of the noise. This may be accomplished with the aid of a dither source. Reference 7 gives a complete analysis of this scheme.

Now that the basics have been covered, let's investigate each of the three ways to generate synthesized waveforms. Hybrid combinations are possible too.

Digital Recursion Oscillator. This method is simple in concept. Build a second-order structure with its z transform poles on the unit circle. Ideally, this

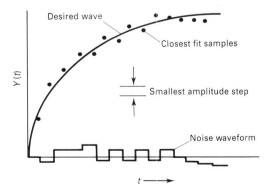

Figure 3-6 Example of amplitude quantization noise.

142 SPECIAL LOOPS

structure will oscillate at the location of the poles with unity amplitude. The following difference equation describes such a second-order structure:

$$Y(n) = Y(n-1)(2\cos\omega t) - Y(n-2) \tag{3-1}$$

with initial conditions

$$Y(n-2) = \cos\phi \tag{3-2}$$

$$Y(n-1) = \cos\phi - \omega t \tag{3-3}$$

where ϕ = starting phase
  ω = desired radial frequency = $2\pi(F_{\text{desired}}/F_{\text{clock}})$
  n = Nth clock pulse
$Y(n-1)$ = the desired sinusoidal output

Figure 3-7 shows the hardware implementation required to compute $Y(n)$. This hardware fills the hardware block in Figure 3-3.

The beauty of this realization is seen in its simplicity. For certain applications it performs as required. For example, for slow rate tones, which may be computed with floating point hardware (or computed in a software loop), this method is ideal. This method has been used successfully with a variety of DSP chips currently on the market (e.g., TMS 320 series from Texas Instruments).

However, the realization of Figure 3-7 in fixed point hardware will not produce a pure sine wave. In most cases, due to recursion, limit-cycle noise will build up. The resulting SNR may be unacceptable. Limit cycles are the nemeses of DSP designs that use recursion. Typically, IIR type filters exhibit limit-cycle noise. See Jackson's classic paper [8]. The resuting $Y(n)$ output will have small amounts of undesired AM and PM due to noise contribution.

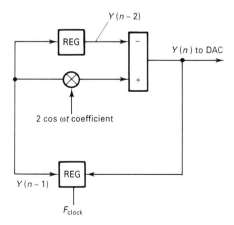

Figure 3-7 Hardware implementation required to computer Y(n). This hardware fills the hardware block in Figure 3-3.

Modulating the recursion oscillator is a painful chore. If the synthesizer is to have FM or PM, the $2\cos\omega t$ coefficient must be computed for each modulation data point. For this reason, other structures should be considered when modulation is needed. Note, too, that the output frequency of $Y(n)$ is determined by the fixed coefficient term $2\cos\omega t$. Representing the coefficient with a finite wordlength means that the computed sinusoidal frequency may not be exactly what is required.

In general, this method is used infrequently for hardware synthesizers. It shows more promise as a method to generate sine waves using software. It is relatively inflexible regarding FM and PM modulation.

Phase Accumulator Method. This method relies on the direct computation of $Y(n) = \sin\omega t$. The computation requires a phase ramp ωt and phase to sine amplitude converter (PAC). Figure 3-8 shows a block diagram of the digital hardware block. In this case, the radial frequency ω is determined by

$$\omega = d\phi/dt$$

where $dt = 1/F_{\text{clock}}$
$d\phi$ = phase increment per clock cycle

The register output is a quantized version of the pure ramp, ωt. The adder is binary and modulo (2^M). By definition, $2^M = 2\pi$ radians. So the adder overflow is exactly at the 2π position. Any overflow remainder phase will foldover into the next cycle of the output sinusoid. This overflow phase is exactly the required amount. For example, if the register is holding a value 350° at clock N and if $d\phi$ is 36° ($F_{\text{sig}} = F_{\text{clock}}/10$), the register will contain 350 + 36 Mod(360) degrees at clock $N + 1$. So, at clock $N + 1$, the register contains 26° as desired.

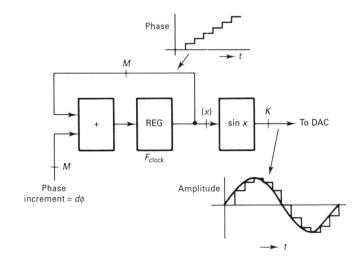

Figure 3-8 Phase accumulator-based, digital synthesizer hardware block.

144 SPECIAL LOOPS

There are four main design variables that affect performance:

M Phase increment bit width
L Truncated phase for PAC
K PAC output width (DAC bit width)
F_{clock} Phase update rate

Let's consider the contribution of each term to the final output performance.

The phase accumulator width M determines the frequency resolution of the synthesizer; that is, $F_{res} = F_{clock}/2^M$. So for $M = 30$ and $F_{clock} = 2^{27} = 134.27$ MHz, the resolution is exactly 0.125 Hz. Using a BCD adder instead of a binary adder would yield a different F_{res}. The advantage of a BCD adder is that F_{clock} may be a clean power of 10 (e.g. 100 MHz) and the resulting F_{res} is also a "nice" frequency. The choice of a BCD adder will cause the entire system design to be affected. When using a binary adder, the clock needs to be a power of 2 if F_{res} is to be a "nice" frequency. If F_{clock} is not a power of 2, then certain cardinal frequencies cannot by synthesized. For example, if $M = 30$ and $F_{clock} = 100$ MHz, then $F_{res} = 0.093132$ Hz. In this case, commonplace frequencies such as 1, 5, or 10 MHz cannot be generated exactly. As usual, the choice must be made depending on end requirements.

The bit width input to the PAC is L. Ideally, $L = M$. This is impractical if M is large. In practice, $L = K + 2$ is a good choice. Let us see why. The L bits represent the truncated phases of the carrier. However, phase truncation causes quantization noise. Also, amplitude quantization noise is present due to K, the DAC width, as mentioned earlier.

So both K and L contribute to the total quantization noise. It would help us to know the SNR of each process so that the K/L trade-off will become more clear. First, let us derive the phase noise SNR assuming that $K = \infty$. In this way there is only phase truncation noise in the output.

$$Y(n) = \sin \omega t + N(t) \tag{3-4}$$

where $N(t)$ is the phase truncation noise. So

$$Y(n) = \sin(\omega t)\cos[N(t)] + \cos(\omega t)\sin[N(T)] \tag{3-5}$$

assuming $\cos[N(t)] = 1$ and $\sin[N(t)] = N(t)$ since $N(t) \ll 1$. Then

$$Y(n) = \sin(\omega t) + \cos(\omega t) \times N(t) \tag{3-6}$$

So

$$\text{SNR} = \frac{\text{power in } \sin(\omega t)}{\text{power in } \cos(\omega t) \times N(t)} \tag{3-7}$$

The normalized power in $\sin(\omega t)$ and $\cos(\omega t)$ is 0.5 watt ($R_{load} = 1\,\Omega$). So

$$\text{SNR} = \frac{1}{\text{power in } N(t)}$$

Assuming a uniform distribution of error states in $N(t)$, the noise power can be derived to be

$$\text{Power}(N(t)) = (Q^2)/12 \tag{3-8}$$

where $Q = 2\pi/2^L$ is the smallest phase step size. So $\text{SNR} = 12/Q^2$ or $\text{SNRdB} = 10\log(12/Q^2)$

$$\text{SNRdB} = 6.02L - 5.17\,\text{dB} \quad \text{(phase quantization noise)} \tag{3-9}$$

The noise energy falls between 0 and F_{clock}. If we assume that only 40% of the noise bandwidth is preserved at the output of the LPF, then the SNR is enhanced by 4 dB. So the $\text{SNRdB}_{\text{fil}} = 6.02L - 1.17\,\text{dB}$. As an example, if $L = 12$, the $\text{SNRdB}_{\text{fil}} = 71.1\,\text{dB}$.

For sufficiently long noise sequences, the spectral distribution is nearly evenly spread across the passband. The SNR value is the *total* power in all the noise spectra and *not* the height of the individual noise lines. If the noise sequence is very long, the noise will be distributed in many lines very close together and low in level. The sum of all the noise lines will always equal the SNR value.

Next, we derive the SNR for a finite DAC width, K. In this derivation we may assume that $L = \infty$ and only K is of finite length. For a sinusoid the quantized signal may be expressed as

$$Y(n) = \sin(\omega t) + N(t) \tag{3-10}$$

where $N(t)$ is the quantization noise due to amplitude truncation. The power in $\sin(\omega t)$ is 1/2 normalized into 1-Ω resistor.

Now, assuming that the noise states in $N(t)$ are uniformly distributed, the noise power is [9]

$$\text{Power}(N(t)) = Q^2/12 \tag{3-11}$$

where $Q = 2/2^K$ is the smallest amplitude step size.

$$\text{SNR} = \frac{\text{power}(\sin(\omega t))}{\text{power}(N(t))} \tag{3-12}$$

$$\text{SNR} = 0.5/(4/2^{2K})/12 = (3/2)(2^{2K}) \tag{3-13}$$

$$\text{SNRdB} = 6.02K + 1.8\,\text{dB} \quad \text{(amplitude quantization noise)} \tag{3-14}$$

Using the reasoning of the previous derivation, the SNR at the LPF output is enhanced by 4 dB. So the final SNR is

$$\text{SNRdB}_{\text{fil}} = 6.02K + 4.8\,\text{dB} \tag{3-15}$$

So for $K = 12$, the SNR is

$$\text{SNRdB}_{\text{fil}}(K = 12) = 78\,\text{dB} \tag{3-16}$$

146   SPECIAL LOOPS

Now if both K and L are of finite length, the total noise comes from a contribution from each noise term. A good design practice is to make one noise source subordinate. As may be seen from the SNR results, for equal values of K and L, the phase noise is about 7 dB higher than the amplitude noise. Typical DAC bit widths are 8, 10, and 12 bits for fast DACs. So if $K = 12$, the phase noise would be subordinate if L were chosen to be 14. Also, a value of $L = 14$ is a reasonable choice for practical hardware design. A value of $L > 14$ would not improve system performance measurably since the amplitude quantization noise would always predominate. So $L = K + 2$ is a reasonable design guideline.

Finally, the designer has a choice of F_{clock}. Due to the Nyquist sampling theorem and practical LPF filter design considerations, the maximum useful output frequency is $0.4F_{\text{clock}}$. The choice will often depend on the DAC speed and the rate at which one can economically generate the required sampled digital data.

Other Considerations. The design of the PAC poses some interesting design challenges. A brute force method uses a ROM (or RAM) with 2^L addresses and an output width of K bits. With $L = 14$, a $16K \times 12$ bit lookup ($K = 12$) table would be needed. An alternative to straight lookup is a structure that uses piecewise interpolation and quadrant logic to form the $\sin x$ output. One design uses 32 segments per quadrant to form the output. It uses only 640 bits of coefficient ROM. See Ref. 10.

Piecewise methods are frequently based on the partitioning of the PAC phase term. One method splits the input phase into upper and lower pieces. Let us call these terms the bottom (B) and top (T). In this case,

$$\text{PAC input phase} = B + (2^J)T$$

Where the top bits are shifted by J bits from the LSB position. So

$$\sin(\text{phase}) = \sin(B)\cos(2^J) + \sin(2^J T)\cos(B) \tag{3-17}$$

For small B,

$$\cos(B) = 1 \quad \text{and} \quad \sin(B) = B \tag{3-18}$$

So

$$\sin(\text{phase}) = B\cos(2^J T) + \sin(2^J T) \tag{3-19}$$

Let us look at an example with $K = 12$ bits. If the top term, T, is only 6 bits wide then the ROM storage needed is 128 words. This is so since you must store 654 points of $\sin(2^J T)$ and 64 points of $\cos(2^J T)$. Also, B is 6 bits, so the required multiplication is 6×12. For $T = B - 6$ bits, the error using this approximation is less than 0.015%, which is better than the required 12-bit resolution. Implementing this architecture yields good results when the synthesizer must be compact and composed of only a few VLSI-type chips.

Incidentally, the piecewise approximation method of sine generation surprisingly is spectrally clean. For only two segments from 0° to 90°, the total harmonic distortion is only 2.3%. See Refs. 11 and 12.

DIRECT DIGITAL SYNTHESIS TECHNIQUES

Another scheme to increase the sample rate is through parallelism. Using multiple channels of computation/PAC yields very favourable results [13].

The phase accumulator structure is flexible and lends itself to modification for implementing AM, FM, and PM. For an example of a commercial product, see the Hewlett-Packard 8791 synthesizer.

Modulation with the Phase Accumulator Synthesizer. Fortunately, this structure is amenable to AM, PM, and FM simultaneously and in real time. Let us investigate a structure with all three modulations. The expression that describes the filtered output is

$$Y(n) = \text{Am}(t) \sin\left[\Sigma F_i(t) + \text{Pm}(t)\right] \qquad (3\text{-}20)$$

where Am(t) = desired AM
Pm(t) = desired PM
Σ = phase accumulator operator (identical to discrete integrator)
$F_i(t)$ = instantaneous frequency

F_i is composed of a carrier term and any desired modulation. So

$$F_i(t) = F_{\text{carrier}} + \text{Fm}(t) \qquad (3\text{-}21)$$

The AM may be added with a real-time hardware multiplier after the PAC. The PM is easily implemented by adding the Pm(t) term to the phase accumulator register output. Likewise, the $F_i(t)$ term is formed with the inclusion of an FM adder block. The output of this adder is the input to the phase accumulator. So, with minor architectural adjustments, real-time modulation is available. The hardware must be designed so that a new point on $Y(n)$ is computed each clock cycle.

From where will the Am(t), Pm(t), and Fm(t) data come? In general, the data come from two sources. One is a real-time user-supplier input. Here, the user must be able to supply high-speed digital data to the synthesizer. Another source of modulation data is the RAM. In this case the AM, PM, and FM data are stored in different dedicated RAMs. The RAMs are addressed and the data are combined to produce $Y(n)$.

Adding modulation to the phase accumulator structure provides a very flexible synthesizer. For signals that may be described by their AM, PM, or FM components, this method provides a completely deterministic approach to signal synthesis.

RAM-Based Synthesis. The third architecture to be discussed is RAM-based synthesis. Figure 3-9 shows a block diagram of such a synthesizer. The major blocks are:

Fast static RAM for waveform storage
Memory address sequencer
DAC/LPF subassembly (as in Figure 3-3)
Waveform development station

148 SPECIAL LOOPS

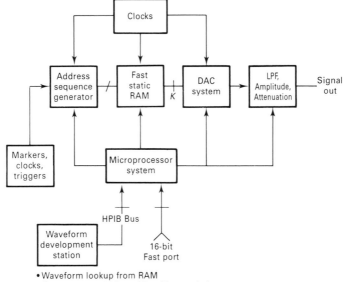

- Waveform lookup from RAM
- Frequency resolution = $(F_{clock})/$(array size)
- Bandwidth = dc to $0.4 F_{clock}$
- Amplitude resolution (norm) = $1/2^K$

Figure 3-9 Block diagram of a RAM-based synthesizer.

In essence, the method uses a sampled image of the desired final waveform. This image is stored in the waveform RAM. The sequencer scans the desired waveform samples and these samples, in turn, are sent to the DAC for conversion to the analog domain.

The theory of sampled data referred to and discussed in the section on the Phase Accumulator Method is the basis for understanding this method as well.

Before the components are discussed, let us consider an analogy to the RAM-based synthesizer, namely, the compact disc (CD) player.

The CD player has a rotating platter that contains the music as sampled data. In the RAM-based synthesizer, a memory contains the sampled data to be played back.

In a CD player, the data are sampled at a 44.1-KHz rate. In a memory-based synthesizer, the rate is selected by the user and is usually much higher.

CD players feature two 16-bit DACs, one per channel. The synthesizer has one DAC per output channel as well. For speed reasons, the DAC width is more likely to be in the 8- to 12-bit range.

Most CD players have a sequencer that lets you play back the tracks in any order. A memory-based synthesizer has a memory sequencer that allows playing back the "tracks" in any order.

An example of a RAM-based synthesizer is the HP8770A waveform synthesizer. See Ref. 5.

Components in a RAM-Based Synthesizer. The RAM stores the sampled data. It is a key component. The RAM must be clocked at F_{clock}, so its access time will be a limiting factor in useful bandwidth. A successful way to increase the RAM output data rate is to form a parallel RAM array and multiplex the individual RAMs to form a very high speed data path. The size of the RAM is a major design variable, as we will see.

The RAM is useless unless it is addressed by a sequencer. In the simplest sense, a sequencer scans a wave segment of data in the RAM. A wave segment is defined as a block of sampled waveform data. The simplest sequencer is an address counter with stop and start address parameters. A more sophisticated sequencer has a mini-program that directs the addressing. In this type, several levels of looping are allowed. In many waveforms there are wave segments that are repeated often. These segments may be scanned by the sequencer to form a complex final output. In effect, the sequencer allows for the RAM data to be compressed.

A simple example will shed some light on the method. It is desired to synthesize an NSTC color bar test pattern. This pattern has redundancy. Many horizontal lines have the same color. A brute force RAM lookup would require about 525,000 points with a 14,317,816-Hz clock (this is exactly four times the color burst frequency). Careful analysis of the signal reveals that there are many wave segments that repeat. By loading the RAM with only the nonredundant data, the RAM size need only be 20K addresses. So the sequencer has given us a data compression ratio of about 26:1.

Another component in a RAM-based synthesizer is the waveform development environment. The user needs a methodology to compose the desired waveform. For simple waves, like pure carriers or simple AM carriers, the user may choose to write dedicated software routines to calculate the sampled data. A more general solution is to provide the user with a waveform design language. Using this language, the user may create any waveform within the limits of creativity, the synthesizer's bandwidth, and amplitude resolution. An example of this is the waveform generation language (WGL) that is a companion product of the HP8770A waveform synthesizer [14].

Understanding the Design Variables in RAM Synthesis. The output spectral purity is limited by the DAC bit width. For random data, the SNR of the system is nearly $6.02K$. However, since the DAC is nonideal (it glitches), the actual limiting performance may come from the DAC produced spurious energy. Another source of spurious spectra comes from the digital data feed-through. The output picks up crosstalk from the digital section of the system.

Besides the value of K, another design parameter is the size of the memory. Even though the RAM can contain the image of any arbitrary time, finite length, or waveform, insight into the method is gained by investigating the simple case of producing a single-frequency tone. Let F_{low} be the lowest frequency that may be produced. Thus

$$F_{\text{low}} = \frac{F_{\text{clock}}}{\text{sequence length}} \qquad (3\text{-}22)$$

This tone would be a single cycle in Q points, assuming that the sequence length is Q points. In general, the single-tone output may be described by

$$Y(n) = \sin[2\pi(P/Q)I + P_{\text{off}}] \qquad (3\text{-}23)$$

where P_{off} = any desired phase offset
I = Ith point in the sequence
P = number of cycles of the desired tone in Q points
Q = number of sampled data points for P cycles

Both P and Q must be integers. Also,

$$F_{out} = F_{clock}(P/Q) \quad \text{Hz} \tag{3-24}$$

Note that $2\pi(P/Q)I$ is just another equivalent way to write ωt, with $\omega = 2\pi(P/Q)$ and I the time index. The Q points of $Y(n)$ are stored in the RAM.

So if F_{clock} is 100 MHz and a 28-MHz tone is desired, then F_{out} = (100 MHz)(P/Q). By inspection with $P = 7$ and $Q = 25$, an F_{out} of 28 MHz would be realized. This tone would only have 25/7 or about 3.5 points per cycle. This is fine as long as the LPF is designed to remove the sampling energy at frequencies greater than and equal to $100 - 28$ MHz.

By adjusting P/Q, many different tones may be generated. For this example, the user observes an analog output with the tone at exactly 28 MHz (actually the only error would be due to F_{clock} not being exactly at the desired frequency). In the tone, each cycle would be identical to all the others. However, the data feeding the DAC are composed of 7 cycles of the desired tone before the sequence repeats. Each cycle has exactly the same frequency. The difference is that each cycle has a different distribution of sample points compared to any of the other 6 cycles. Try computing $\sin[2\pi(7/25)I]$ for all 25 points to see this effect for yourself.

As an aside, well-designed waveforms exhibit closure; that is, the last point in the sampleddata is immediately followed by the first point in the RAM stored sequence. This allows the address counter to return to the first point in the sequence immediately after the last and the final output has no discontinuity. If either P or Q is not an integer, closure will not be maintained. In this case, the spectrum will be salted with unwanted spurious signals. Again, try an example for yourself using a simple software loop and you will see that closure will not be obtained.

Given that the problem is to find P and Q for any desired F_{out}, some interesting results surface. It turns out that for some maximum value of Q (Q_{max}), there is a solution set of P/Q such that there is no better fit to the desired tone. It is true that there will usually be an error in the resulting frequency, but it may be made very small by choosing Q_{max} large enough.

The analysis is complicated by the fact that P and Q may only be whole numbers. To make matters worse, many combinations of P/Q yield identical frequencies. For example, for $P = 21$ and $Q = 75$, we obtain the same 28-MHz tone as with $P = 7$ and $Q = 25$. Only for P and Q relatively prime (no common factors) is F_{out} obtained with a minimum value of Q. Naturally, we want Q to be small since it conserves memory space.

The problem of finding P and Q is the same one mathematicians face when asked to find the best rational approximation to a fraction number like $0.dddddddd$. With the help of Euler's method of continued fractions, P and Q may be found given the desired fraction (P/Q) of F_{clock} that is to be synthesized. For an excellent study in this area, refer to Ref. 15.

Some results of solving for P and Q may be summarized.

The frequency resolution of a RAM-based synthesizer is not a constant. In fact, there is no simple expression that you may use to find the exact resolution versus frequency. However, a typical or expected resolution may be expressed as

$$F_{res} = F_{clock}(2\pi^2/3Q_{max}^2) \qquad (3\text{-}25)$$

This odd expression may be derived by finding how many pairs of relatively prime P/Q fractions (with $P/Q < 1$) are available given an upper limit on Q of Q_{max}. In this analysis, Euler's totient function is used to find the sum of the pairs. On average, a unique tone will be found at a spacing of F_{res}. For $Q_{max} > 32$, F_{res} as computed is accurate to $<1\%$.

Fortunately, F_{res} drops as the square of Q_{max}. Let us consider an example. For $Q_{max} = 1024$ points and $F_{clock} = 10\,\text{MHz}$,

$$F_{res} = (10\,\text{MHz}(19.74)/(3 \times 1048576) = 62\,\text{Hz} \qquad (3\text{-}26)$$

This means that there are about 162,000 unique frequencies spanning dc to 10 MHz spaced approximately by 62 Hz. The available frequencies higher than $0.4 F_{clock}$ are not used, but this does not affect the resolution.

With $Q_{max} = 4096$ points,

$$F_{res} = 62/16 = 4\,\text{Hz} \qquad (3\text{-}27)$$

There is no guarantee, however, that you will get the desired frequency within F_{res}. But it is highly likely.

Figure 3-10 shows a plot of the percentage error in $F_{desired}$ for a 4K RAM size. The worst-case error expected is 100/size. For this case, worst case = 0.024%. Examining the figure, we find almost no errors that are worst case. In fact, 95%

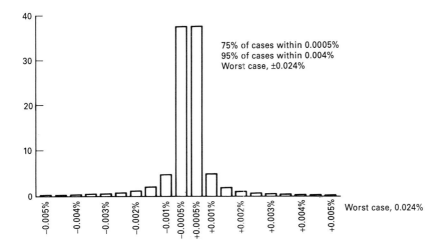

Figure 3-10 Percentage error in desired frequency for a 4 K RAM size.

of all desired tones fall within 0.004% of desired frequency and 75% are within 0.0005% of F_{desired}. The Y axis is the percentage of all possible frequencies (using P/Q) that fall within any given error bin. A curious side effect of the P/Q relationship is that 61% of all possible P/Q values are relatively prime, assuming that $P_{\max} = Q_{\max}$.

It should be mentioned, however, that if Q is fixed and only P is a variable, then $F_{\text{res}} = F_{\text{clock}}/Q$ and is *much* larger than if Q is also a variable.

Another way to analyze the frequency resolution is given in Ref. 5. The method utilizes partial derivatives to solve for F_{res}.

RAM-based synthesis is well suited for the generation of arbitrary time functions. Any desired waveform of finite length may be sampled and stored in the RAM. Generating a carrier with an AM, PM, or FM is just as easy as generating a pure carrier with no modulation. Depending on user needs, this method may be employed with success for generating a large class of waveforms.

Applications. Applications fall into several broad categories. At the low end there are simple single-tone, low-frequency oscillators. With just a few chips such an oscillator can be built. At the high end, a group of synthesizers classified as signal simulators exist. These units can generate very complex signal scenarios with independent AM, PM, FM, and frequency hopping. Such systems find application in radar, multiple satellite signal generation, and communication channel simulation, for example.

RAM-based systems are currently being used to simulate, read, and serve waveforms from a disc drive read head. This technique provides very flexible waveforms for disc drive testing and development. See Ref. 18.

For communications systems that use I/Q modulation techniques, two synthesizers may be paired to generate synchronously the I and Q components. With this method, for example, two 50-MHz bandwidth synthesizers may be used to modulate an I/Q modulator and the resulting bandwidth is 100 MHz. In this application, the I and Q channels must be matched in amplitude and phase to better than 0.01 dB and 0.35°, respectively, for good performance.

With this design requirement, the prudent choice is a digital synthesizer. Achieving this match with two analog instruments is almost impossible. See Ref. 19.

New applications are being discovered every day. By applying digital signal processing techniques to signal problems, designers are discovering the joy of digital synthesis. Increasingly, what used to be done awkwardly with analog methods may now be done with finesse using digital methods.

Summary of Methods. Three architectures for digital synthesizers have been discussed. The major design elements of each method have been analyzed. In brief, now, here are some reasons to choose one architecture over another.

Recursion Method. Simple to implement. Needs floating point multiply to achieve excellent frequency resolution and low limit-cycle noise. Difficult to modulate. Good for software simulation of single tones.

Phase Accumulator. Practical synthesizer using fixed point hardware. Easy to achieve constant, useful F_{res} values. Modulation may be implemented with additional hardware. Allows for real-time modulation of a carrier with user data. Very flexible, long scenario lengths attained. Amount of hardware required depends on the modulation capabilities needed. The system can become large when AM, PM, and FM are required.

RAM-Based Method. Stored sample image lookup. Nonuniform frequency resolution is typical. Any desired waveform may be generated. Scenario time is a direct function of the amount of RAM available and how the sequencer is programmed. No real-time modulation allowed (AM would be easy, however). All waveforms must be precomputed. This may be time-consuming depending on the application. A waveform generation software package will often be needed for all but the simplest signals.

In conclusion, the choice is really between the last two methods for flexible, hardware synthesizers. Each has its advantages/disadvantages depending on the application. As more designers become aware of the advantages of digital synthesis, new applications will emerge.

There are a number of companies producing DDS synthesizers. As an example, the following pages are reproductions of DDS synthesizers produced by Stanford Telecom (Fig. 3.11), which can serve as examples of the capabilities offered by DDS synthesizers. These chips and boards are available in different operating ranges and frequency ranges. The first is an example of a 1-GHz GaAs 32-bit modulated numerically controlled oscillator, STEL-2173, and its descriptive data.

The STEL-2173 is also available as a synthesizer board, model STEL-2273 (Fig. 3.12). Its specifications can be seen from the reproduced data sheets that follow.

A good targeted use of these digital direct synthesizers is application in HF radios and UHF applications.

Figures 3-13 through 3-14 show examples of an HF synthesizer using a DDS for the fine resolution. The 20-MHz temperature-compensated crystal oscillator (TCXO) is used as the frequency standard for the DDS synthesizer. Its output frequency, which operates from 1 to 2 MHz, is then up-converted from 21/22 MHz. The output is then divided by 4 and fed into a phase detector. A second loop, operating from 147 to 410 MHz in 1-MHz steps is also divided by 4 and serves as the auxiliary frequency to mix down the output frequencies into a range of approximately 5 MHz.

3-1-3 Signal Quality

It is important to discuss this subject in some detail since digital synthesis has acquired a reputation for having poor signal quality, especially with regard to spurious sidebands.

Signal quality is a measure of the purity of a desired signal. In the case of sinusoidal signals, a single infinitesimally wide spectral line is the highest form of purity by definition. Broadening of the spectral line is caused by amplitude or phase noise. For a number of reasons outside the scope of this chapter, phase noise is the more important parameter of the two. Nonharmonically related sidebands, commonly referred to as "spurs," are another signal-corrupting factor and are

FEATURES

- **1 GHz CLOCK FREQUENCY**
  - 0 TO OVER 400 MHz OUTPUT FREQUENCY (STANDARD OPERATING CONDITIONS)
- **32-BIT FREQUENCY RESOLUTION**
  - 0.23 Hz @ 1 GHz CLOCK
- **2-BIT PHASE MODULATION (BPSK AND QPSK)**
- **8-BIT PARALLEL SINE OUTPUT**
  - 2 DEVICES CAN BE USED TO GENERATE QUADRATURE OUTPUT SIGNALS
- **PHASE CONTINUOUS INSTANTANEOUS FREQUENCY SWITCHING**
- **UP TO 62.5 MHz FREQUENCY HOPPING**
- **ECL INPUTS AND OUTPUTS FOR CONVENIENT INTERFACING**
- **50Ω OUTPUTS**
- **132-PIN CERAMIC FLATPACK**

TYPICAL APPLICATIONS

- **VHF/UHF FREQUENCY SYNTHESIZERS**
- **CHIRP GENERATORS**
- **VERY HIGH SPEED FREQUENCY HOPPED SOURCES**

FUNCTIONAL DESCRIPTION

The STEL-2173 is a GaAs Modulated Numerically Controlled Oscillator (MNCO) which operates at clock frequencies up to 1 GHz and uses digital techniques to provide a cost-effective solution for precision very high frequency signal sources. This monolithic device is ideal for use in frequency synthesizers, frequency hoppers, and other precision frequency sources. It has a frequency resolution of 32 bits, making it possible to generate signals from 0 to more than 400 MHz with a resolution of 0.23 Hz. The device provides an 8-bit digitized sine wave output, making it possible to generate sine waves with better than 55 dB purity. The device also incorporates a 2-bit phase modulator, allowing it to be used to generate BPSK, and QPSK signals. In addition, two STEL-2173 NCOs can be used to generate quadrature signals by setting the phase modulation to 0° on one device and 90° on the other. Both phase and frequency can be updated as rapidly as every sixteenth clock cycle, making this a very versatile device for hopped frequency and spread-spectrum applications. A lower cost 800 MHz version is also available.

BLOCK DIAGRAM

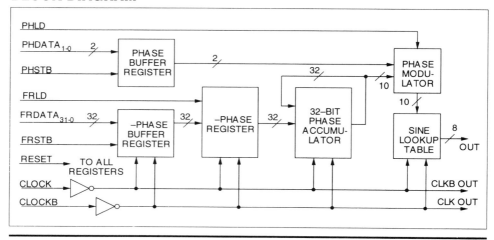

Figure 3-11 Reproduction of the data sheet for the STEL 2173. Copyright Standford Telecommunications, Sunnyvale, CA. Reproduced with permission.

CIRCUIT DESCRIPTION

The NCO maintains a record of phase which is accurate to 32 bits. At each clock cycle the number stored in the 32-bit Δ-Phase Register is added to the previous value of the Phase Accumulator. The number in the Phase Accumulator represents the current phase of the synthesized sine and cosine functions. The number in the Δ-Phase Register represents the phase change for each cycle of the clock. This number is directly related to the output frequency by the following:

$$f_o = \frac{f_c \times \Delta\text{-Phase}}{2^{32}}$$

where: f_o is the frequency of the output signal

and: f_c is the clock frequency.

The sine functions are generated from the ten most significant bits of the phase accumulator.

The NCO generates a sampled sine wave where the sampling function is the clock. The practical upper limit of the NCO output frequency is about 40% of the clock frequency due to spurious components that are created by sampling. Those components are at frequencies greater than half the clock frequency, and become more difficult to remove by filtering as the output frequency approaches half the clock frequency.

The phase noise of the NCO output signal may be determined from the phase noise of the clock signal input and the ratio of the output frequency to the clock frequency. This ratio squared times the phase noise power of the clock specified in a given bandwidth is the phase noise power that may be expected in that same bandwidth relative to the output frequency.

The NCO achieves its high operating frequency by making extensive use of pipelining in its architecture. The pipeline delays within the NCO represent 23 clock cycles. The pipeline delay associated with the phase modulator is only 7 clock cycles, since the phase modulating function is at the output of the accumulator. Note that when a phase or frequency change occurs at the output the change is instantaneous, i.e., it occurs in one clock cycle, with complete phase continuity.

FUNCTION BLOCK DESCRIPTION

Δ-PHASE BUFFER REGISTER BLOCK
The Δ-Phase Buffer Register is used to temporarily store the Δ-Phase data written into the device. This allows the data to be written asynchronously on the rising edge of **FRSTB**. The data is transferred from this register into the Δ-Phase Register after a falling edge on the **FRLD** input.

PHASE BUFFER REGISTER BLOCK
The Phase Buffer Register is used to temporarily store the Phase data written into the device. This allows the data to be written asynchronously on the rising edge of **PHSTB**. The data is transferred from this register into the Phase ALU after a falling edge on the **PHLD** input.

Δ-PHASE REGISTER BLOCK
This block controls the updating of the Δ-Phase word used in the Accumulator. The frequency data from the Δ-Phase Buffer Register is loaded into this block after a falling edge on the **FRLD** input.

PHASE ACCUMULATOR BLOCK
This block forms the core of the NCO function. It is a high-speed, pipelined, 32-bit parallel accumulator, generating a new sum in every clock cycle. The overflow signal is discarded, since the required output is the modulo(2^{32}) sum only. This represents the modulo(2π) phase angle.

PHASE ALU BLOCK
The Phase ALU performs the addition of the PM data to the Phase Accumulator output. The PM data word is 2 bits wide, and this is added to the 2 most significant bits of the Phase Accumulator output to form the modulated phase used to address the lookup table.

SINE LOOKUP TABLE BLOCK
This block is the sine memory. The 10 bits from the Phase ALU are used to address this memory to generate the 8-bit OUT_{7-0} outputs.

Figure 3-11 (Continued).

INPUT SIGNALS

RESET
The **RESET** input is asynchronous and active low, and clears all the registers in the device. When **RESET** goes low, all registers are cleared within 1 nsec, and normal operation will resume after this signal returns high. The data on the OUT_{7-0} bus will then be invalid for 5 clock cycles, and thereafter will remain at the value corresponding to zero phase until new frequency or phase modulation data is loaded with the **FRLD** or **PHLD** inputs after the **RESET** returns high.

CLOCK and CLOCKB
All synchronous functions performed within the NCO are referenced to the rising edge of the **CLOCK** input. The **CLOCK** signal should be nominally a square wave at a maximum frequency of 1 GHz. A non-repetitive **CLOCK** waveform is permissible as long as the minimum duration positive or negative pulse on the waveform is always greater than 400 picoseconds. **CLOCKB** is the inverse phase clock input.

$FRDATA_{31}$ through $FRDATA_0$
The 32-bit $FRDATA_{31-0}$ bus is used to program the 32-bit Δ-Phase Buffer Register. $FRDATA_0$ is the least significant bit of the bus. The data programmed into the Δ-Phase Register in this way determines the output frequency of the NCO.

FRSTB
The **Frequency Strobe** input is used to latch the data on the $FRDATA_{31-0}$ bus into the device. The information on the $FRDATA_{31-0}$ bus is latched into the Δ-Phase Buffer Register on the rising edge of the **FRSTB** input. The buffer register is transparent when **FRSTB** is set low and the device may be used in this mode within the restrictions specified below in the description of the **FRLD** signal.

FRLD
The **Frequency Load** input is used to control the transfer of the data from the Δ-Phase Buffer Register to the Δ-Phase Register. The data at the output of the Δ-Phase Buffer Register must be valid during the clock cycle following the falling edge of **FRLD** and must remain valid for the next 16 clock cycles. The data is then transferred beginning with the subsequent cycle. Since the buffer register is transparent when the **FRSTB** input is set low it is important that the data on the $FRDATA_{31-0}$ bus remains stable during this period if **FRSTB** is held low. The frequency of the NCO output will change 24 clock cycles after the falling edge of **FRLD**, due to pipelining delays.

$PHDATA_1$ through $PHDATA_0$
The 2-bit $PHDATA_{1-0}$ bus is used to program the 2-bit Phase Buffer Register. $PHDATA_0$ is the least significant bit of the bus. The data programmed into the Phase ALU in this way determines the output phase of the NCO relative to the zero offset sine output.

PHSTB
The **Phase Strobe** input is used to latch the data on the $PHDATA_{1-0}$ bus into the device. On the rising edge of the **PHSTB** input, the information on the 2-bit data bus is transferred to the Phase Buffer Register.

PHLD
The **Phase Load** input is used to control the transfer of the data from the Phase Buffer Register to the Phase ALU. The data at the output of the Phase Buffer Register must be valid during the clock cycle following the falling edge of **PHLD**. The phase of the NCO output will change 8 clock cycles after the falling edge of **PHLD**, due to pipelining delays.

OUTPUT SIGNALS

OUT_{7-0}
The signal appearing on the OUT_{7-0} output bus is derived from the 10 most significant bits of the Phase Accumulator. The 8-bit sine function is presented in offset binary format. The value of the output for a given phase value follows the relationship:

$OUT_{7-0} = 127 \times \sin(360 \times (phase+0.5)/1024)° + 128$

The result is accurate to within 1 LSB. When the phase accumulator is zero, e.g., after a reset, the decimal value of the output is 129 (81_H).

CLK OUT and CLKB OUT
The clock signals used to latch the output data into the output registers are brought out on the **CLK OUT** and **CLKB OUT** pins. **CLKB OUT** is the inverse phase clock. The output data changes on the rising edges of **CLK OUT**.

VREF
An internal reference generator provides a –1.3 volt reference for the ECL input comparators. The output of this reference generator is available at the **VREF** pins. Since the generator has a high output impedance this voltage can be modified by connecting a different voltage source to these pins.

Figure 3-11 (Continued).

DIRECT DIGITAL SYNTHESIS TECHNIQUES 157

PIN CONNECTIONS

1	V_{EE}	34	V_{EE}	67	V_{EE}	100	V_{EE}
2	N.C.	35	N.C.	68	N.C.	101	N.C.
3	$FRDATA_8$	36	$FRDATA_{24}$	69	N.C.	102	OUT_5
4	V_{SS}	37	V_{SS}	70	V_{SS}	103	V_{SS}
5	$FRDATA_9$	38	$FRDATA_{25}$	71	N.C.	104	OUT_4
6	$FRDATA_{10}$	39	$FRDATA_{26}$	72	$PHDATA_0$	105	OUT_3
7	V_{SS}	40	V_{SS}	73	V_{SS}	106	V_{SS}
8	$FRDATA_{11}$	41	$FRDATA_{27}$	74	N.C.	107	OUT_2
9	$FRDATA_{12}$	42	$FRDATA_{28}$	75	$PHDATA_1$	108	OUT_1
10	V_{SS}	43	V_{SS}	76	V_{SS}	109	V_{SS}
11	$FRDATA_{13}$	44	$FRDATA_{29}$	77	PHSTB	110	$OUT_{0(LSB)}$
12	$FRDATA_{14}$	45	$FRDATA_{30}$	78	PHLD	111	FRLD
13	V_{SS}	46	V_{SS}	79	V_{SS}	112	V_{SS}
14	$FRDATA_{15}$	47	$FRDATA_{31}$	80	N.C.	113	FRSTB
15	N.C.	48	N.C.	81	N.C.	114	N.C.
16	V_{SS}	49	V_{SS}	82	V_{SS}	115	V_{SS}
17	N.C.	50	N.C.	83	N.C.	116	N.C.
18	V_{SS}	51	V_{SS}	84	V_{SS}	117	V_{SS}
19	N.C.	52	N.C.	85	N.C.	118	N.C.
20	$FRDATA_{16}$	53	N.C.	86	VREF	119	$FRDATA_0$
21	V_{SS}	54	V_{SS}	87	V_{SS}	120	V_{SS}
22	$FRDATA_{17}$	55	VREF	88	CLOCKB	121	$FRDATA_1$
23	$FRDATA_{18}$	56	N.C.	89	CLOCK	122	$FRDATA_2$
24	V_{SS}	57	V_{SS}	90	V_{SS}	123	V_{SS}
25	$FRDATA_{19}$	58	N.C.	91	RESET	124	$FRDATA_3$
26	$FRDATA_{20}$	59	N.C.	92	N.C.	125	$FRDATA_4$
27	V_{SS}	60	V_{SS}	93	V_{SS}	126	V_{SS}
28	$FRDATA_{21}$	61	N.C.	94	CLK OUT	127	$FRDATA_5$
29	$FRDATA_{22}$	62	N.C.	95	CLKB OUT	128	$FRDATA_6$
30	V_{SS}	63	V_{SS}	96	V_{SS}	129	V_{SS}
31	$FRDATA_{23}$	64	N.C.	97	OUT_7	130	$FRDATA_7$
32	N.C.	65	N.C.	98	OUT_6	131	N.C.
33	V_{EE}	66	V_{EE}	99	V_{EE}	132	V_{EE}

Note: N.C. denotes No Connection.

Figure 3-11 *(Continued).*

ELECTRICAL CHARACTERISTICS
ABSOLUTE MAXIMUM RATINGS

Note: Stresses greater than those shown below may cause permanent damage to the device. Exposure of the device to these conditions for extended periods may also affect device reliability. All voltages are referenced to V_{ss}.

Symbol	Parameter	Range	Units
T_{stg}	Storage Temperature	−55 to +125	°C
V_{EEmax}	Supply voltage on V_{EE}	0 to −7	volts
$V_{I(max)}$	Input voltage	0.5 to VEE − 0.5	volts
$V_{O(max)}$	Output voltage	0.5 to VEE − 0.5	volts
I_i	DC input current	± 1	mA
I_O	DC output current	40	mA

RECOMMENDED OPERATING CONDITIONS

Symbol	Parameter	Range	Units	
V_{EE}	Supply Voltage	−5 ± 5%	Volts	(Commercial Conditions)
		−5 ± 10%	Volts	(Military Conditions)
T_c	Operating Temperature (Case)	0 to +70	°C	(Commercial Conditions)
		−55 to +125	°C	(Military Conditions)

D.C. CHARACTERISTICS (Operating Conditions: V_{EE} = −5.0 V ±5%, VSS = 0 V, T_c = 0° to 70° C, Commercial
V_{EE} = −5.0 V ±10%, VSS = 0 V, Tc = −55° to 125° C, Military)

Symbol	Parameter	Min.	Typ.	Max.	Units	Conditions
I_{EE}	Supply Current, Operational		1.2		A	f_{CLK} = 1 GHz
$V_{IH(min)}$	High Level Input Voltage					
	Extended Operating Conditions	−1.1		0	volts	Logic '1'
$V_{IL(max)}$	Low Level Input Voltage	−2.0		−1.5	volts	Logic '0'
$I_{IH(min)}$	High Level Input Current			10	μA	V_I = 0 volts
$I_{IL(max)}$	Low Level Input Current			−10	μA	V_I = −2.0 volts
$V_{OH(min)}$	High Level Output Voltage	−1.0		−0.5	volts	R_L = 50Ω to −2.0 volts
$V_{OL(max)}$	Low Level Output Voltage	−2.0		−1.6	volts	R_L = 50Ω to −2.0 volts
$I_{OH(min)}$	High Level Output Current	20	23	30	mA	R_L = 50Ω to −2.0 volts
$I_{OL(max)}$	Low Level Output Current	0	5	8	mA	R_L = 50Ω to −2.0 volts
	($I_{OH(min)}$ and $I_{OL(max)}$ are not tested.)					
C_{IN}	Input Capacitance		2		pF	All inputs
C_{OUT}	Output Capacitance		4		pF	All outputs

Figure 3-11 *(Continued).*

DIRECT DIGITAL SYNTHESIS TECHNIQUES   159

NCO RESET SEQUENCE

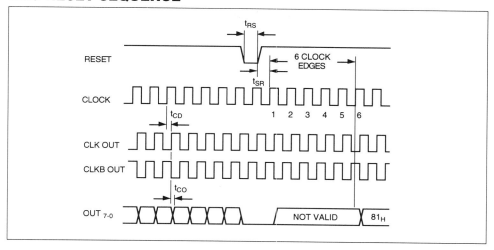

NCO FREQUENCY CHANGE SEQUENCE

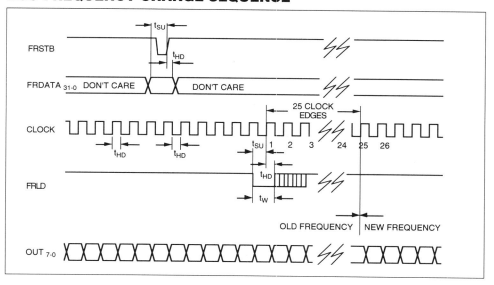

Figure 3-11 *(Continued).*

NCO PHASE CHANGE SEQUENCE

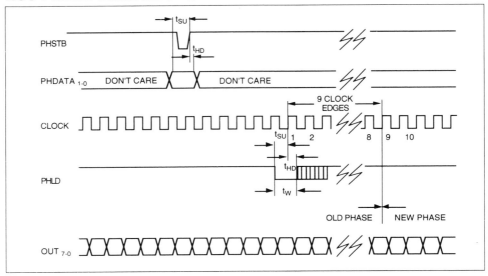

ELECTRICAL CHARACTERISTICS

A.C. CHARACTERISTICS (Operating Conditions: $V_{EE} = -5.0$ V $\pm 5\%$, VSS = 0 V, $T_c = 0°$ to $70°$ C, Commercial
$V_{EE} = -5.0$ V $\pm 10\%$, VSS = 0 V, Tc = $-55°$ to $125°$ C, Military)

Symbol	Parameter	Min.	Typ.	Max.	Units	Conditions
t_{RS}	RESET pulse width	750			psec.	
t_{SR}	RESET to CLOCK Setup	750			psec.	
t_{SU}	FRDATA, or PHDATA to FRSTB or PHSTB Setup, and FRLD or PHLD to CLOCK Setup		100		psec.	
t_{HD}	FRDATA, or PHDATA to FRSTB or PHSTB Hold, and FRLD or PHLD to CLOCK Hold		50		psec.	
t_{CH}	CLOCK high	400			psec.	$f_{CLK} = 1$ GHz
t_{CL}	CLOCK low	400			psec.	$f_{CLK} = 1$ GHz
t_W	FRSTB, PHSTB, FRLD or PHLD pulse width		1000		psec.	
t_{CD}	CLOCK to CLK OUT delay		1500		psec.	Load = 15 pF max.
t_{CO}	CLK OUT to output delay (All outputs)		300		psec.	Load = 15 pF max.

Figure 3-11 *(Continued).*

INTERFACING THE STEL-2173 TO THE TQ6114 DAC

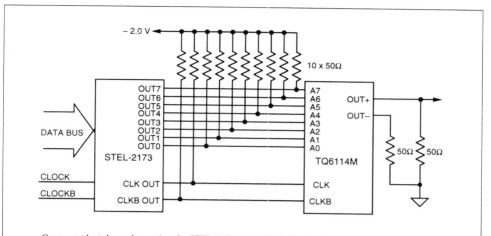

Care must be taken when using the STEL-2173 with a Digital to Analog converter because of the high frequencies involved. Great care must be taken to ensure that the power supplies are properly decoupled, especially the −5 volt supply. The preferred practice is to use a 4 layer printed circuit board, using one layer as a ground plane and another as a −5 volt supply plane. In this way the transients in the ground plane can be minimized, and the analog output purity maximized. It is recommended that four 1000 pF chip capacitors be connected between the supply lines and ground at every one of the supply pins.

Figure 3-11 *(Continued).*

TYPICAL APPLICATION
A FAST SWITCHING DC TO 400 MHz SYNTHESIZER

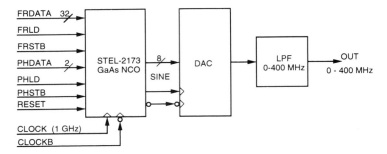

If the output of the STEL-2173 is fed into a high-speed D-to-A converter, such as the TriQuint TQ6122, a phase continuous, fast switching frequency synthesizer may be realized. The spurious components at the output of the low-pass filter will be about 45 dB below the primary output component. If the clock frequency is set to 858.9935 MHz (0.2×2^{32} Hz), the frequency steps obtainable will be exactly 0.2 Hz, although this exceeds the guaranteed speed rating of the STEL-2173+800. The output frequency may be programmed from DC to over 400 MHz, the limit depending on the sharpness of the low-pass filter. In order to keep the sampling components above the Nyquist frequency at a level compatible with the other spurious components, the low-pass filter will need to have at least 45 dB of attenuation above $f_c - f_o$, where f_c is the clock frequency and f_o is the highest output frequency desired.

SPECTRAL PURITY

In many applications, the STEL-2173 GaAs-NCO is used with a digital-to-analog converter to generate an analog waveform which approximates an ideal sinewave. The spectral purity of this synthesized waveform is a function of many variables, including phase quantization, amplitude quantization, the ratio of the clock frequency to output frequency, and the dynamic characteristics of the D/A converter.

The sine function produced by the STEL-2173 has 8-bit amplitude quantization and 10-bit phase quantization which result in spurious levels which are theoretically about –55 dBc. The highest output frequency the NCO can generate is half the clock frequency ($f_c/2$), so spurious components which occur at frequencies greater than $f_c/2$ can be removed by filtering. As the output frequency of the STEL-2173 approaches $f_c/2$, an "image" spurious component also approaches $f_c/2$ from above. If the programmed output frequency is only slightly below $f_c/2$, it is virtually impossible to remove the "image" spurious component just above $f_c/2$ by filtering. For this reason, the maximum practical output frequency of the STEL-2173 should be limited to about 40% of the clock frequency.

Probably the most significant contribution to spectral purity is the dynamic performance of the D/A converter (DAC). To minimize these effects, connections between the DAC and STEL-2173 should be kept equal length using transmission line techniques. The analog output of the DAC should be isolated from the clock signal and other digital signals as much as possible. Grounding and decoupling should be done with the objective of optimizing the step response of the DAC.

Figure 3-11 (Continued).

A spectral plot of the NCO output after conversion with a DAC (TriQuint TQ6122) is shown below. In this case, the clock frequency is 1 GHz and the output frequency is programmed to 234.567 MHz. The maximum non-harmonic spur level observed over the entire useful output frequency range in this case is −44 dBc. The spur levels are limited by the dynamic linearity of the DAC. It is important to remember that when the output frequency exceeds 25% of the clock frequency, the second harmonic frequency will be higher than the Nyquist frequency, 50% of the clock frequency. When this happens, the image of the harmonic at the frequency $f_c - 2f_o$, which is not harmonically related to the output signal, will become intrusive since its frequency falls as the output frequency rises, eventually crossing the fundamental output when its frequency crosses through $f_c/3$. It would be necessary to obtain a DAC with better dynamic linearity to improve the harmonic spur levels. (The dynamic linearity of a DAC is a function of both its static linearity and its dynamic characteristics, such as settling time and slew rates.) At higher output frequencies the waveform produced by the DAC will have large output changes from sample to sample. For this reason, the settling time of the DAC should be short in comparison to the clock period. As a general rule, the DAC used should have the lowest possible glitch energy as well as the shortest possible settling time.

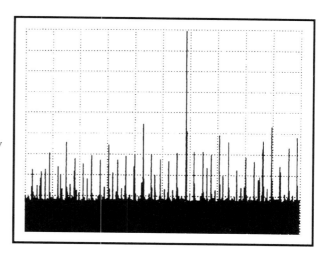

TYPICAL SPECTRUM

Frequency Span:	0 – 400 MHz
Reference Level:	+0 dBm
Resolution Bandwidth:	10 KHz
Scale:	Log, 10 dB/div
Output frequency:	234.567 MHz
Clock frequency:	1 GHz

Figure 3-11 *(Continued).*

FEATURES

- **1 GHz CLOCK FREQUENCY – 0 TO OVER 400 MHz OUTPUT FREQUENCY (COMMERCIAL OPERATING CONDITIONS)**
- **32-BIT FREQUENCY RESOLUTION – 0.23 Hz @ 1 GHz CLOCK**
- **2-BIT PHASE MODULATION (BPSK AND QPSK)**
- **50Ω COMPLEMENTARY SINE (OR COSINE) OUTPUTS – 2 UNITS CAN BE USED TO GENERATE QUADRATURE OUTPUT SIGNALS**
- **PHASE CONTINUOUS INSTANTANEOUS FREQUENCY SWITCHING**
- **UP TO 62.5 MHz FREQUENCY HOPPING**
- **ECL INPUTS FOR CONVENIENT INTERFACING**
- **HIGH-SPEED, LOW GLITCH GaAs DAC**
- **–40 dBc SPURIOUS TYPICAL**
- **3.0" x 5.125", CONTROLLED IMPEDANCE, 4-LAYER PRINTED CIRCUIT BOARD**

DESCRIPTION

The STEL-2273A is a synthesizer board using the STEL-2173 GaAs Numerically Controlled Oscillator (NCO) chip and is a revised version of the STEL-2273. The board uses the STEL-2173 to drive a high-speed 8-bit DAC (TriQuint TQ6114M) to generate complementary output signals which can then be low-pass filtered to give continuous output waveforms. The system operates at clock frequencies up to 1 GHz over the temperature range of 0-70°C, giving an output frequency range of 0 to over 400 MHz (with the appropriate filters) with spurious signal levels typically below –40 dBc. For information on programming the STEL-2173 GaAs-NCO to set the output frequency and other parameters, please refer to the STEL-2173 data sheet.

BLOCK DIAGRAM

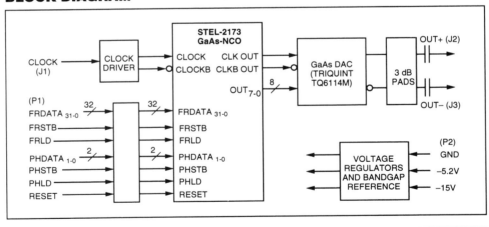

Figure 3-12 Partial reproduction of the data sheet for the STEL 2273 GaAs numerically controlled oscillator chip. Copyright Standford Telecommunications, Sunnyvale, CA. Reproduced with permission.

SPECIFICATIONS

FREQUENCY RANGE:
300 KHz to approximately 400 MHz at a clock frequency of 1 GHz.

RESOLUTION:
0.186 Hz @ f_{CLK} = 1 GHz (32 bits)

SWITCHING SPEED:
Frequency changes 25 clock cycles after strobe. (25 nsec. @ 1 GHz)

Maximum update rate: $f_{clk}/16$

(= 62.5 MHz @ f_{CLK} = 1 GHz)

SPURIOUS PERFORMANCE:
$\leq$ –39.5 dBc, measured at f_{OUT} = 333.3 MHz, f_{CLK} = 1 GHz.

(Includes spurious products up to fiftieth order)

CLOCK INPUT:
J1: –5 to +5 dBm, A.C. coupled, 50Ω
(or ECL level signal)

Maximum frequency, $f_{CLK(max.)}$ = 1 GHz

CONTROL INPUTS:
P1: 2 x 25 pin header

(25 mil square posts, 100 mil centers)

Impedance: 50Ω terminated to –2 volts.

Frequency Control: 32-bit parallel input.

Phase Control: 2-bit parallel input.

OUTPUTS: J2 (OUT+) and J3 (OUT–), 50Ω.
Complementary outputs. SMA connectors.

AC coupled, –3 dB point: Approx. 300 KHz

Output level: Approx.   0 dBm @ f < 0.1 x f_{CLK}
                        –2.4 dBm @ f = 0.4 x f_{CLK}

Note: The output level falls as the frequency rises according to the equation:

$$V_{OUT} = \frac{V_{OUT(DC)} (\sin f')}{f'}$$

where: $f' = \pi f_{OUT} / f_{CLK}$

DIMENSIONS:
3.0" x 5.125". See Mechanical Specifications on Page 6.

COOLING:
Forced air, approx. 400 lin. ft./min.

(Directed over the heat sinks on U1 and U3).

CONTROL INTERFACE (P1)

Pin Connections (see drawing on page 6):

Pin	Signal	Pin	Signal
1	GND	2	$FRDATA_0$
3	$FRDATA_1$	4	$FRDATA_2$
5	GND	6	$FRDATA_3$
7	$FRDATA_4$	8	$FRDATA_5$
9	$FRDATA_6$	10	GND
11	$FRDATA_7$	12	$FRDATA_8$
13	$FRDATA_9$	14	$FRDATA_{10}$
15	$FRDATA_{11}$	16	$FRDATA_{12}$
17	GND	18	$FRDATA_{13}$
19	$FRDATA_{14}$	20	$FRDATA_{15}$
21	$FRDATA_{16}$	22	GND
23	$FRDATA_{17}$	24	$FRDATA_{18}$
25	$FRDATA_{19}$	26	$FRDATA_{20}$
27	$FRDATA_{21}$	28	$FRDATA_{22}$
29	GND	30	$FRDATA_{23}$
31	$FRDATA_{24}$	32	$FRDATA_{25}$
33	$FRDATA_{26}$	34	GND
35	$FRDATA_{27}$	36	$FRDATA_{28}$
37	$FRDATA_{29}$	38	$FRDATA_{30}$
39	$FRDATA_{31}$	40	RESET
41	GND	42	GND
43	FRLD	44	FRSTB
45	GND	46	$PHDATA_0$
47	$PHDATA_1$	48	GND
49	PHLD	50	PHSTB

POWER REQUIREMENTS (P2):

Pin 1 –15   volts ± 10 % @ 0.4 A
    2 GND
    3 –5.2 volts ± 5% @ 1.8 to 2.2 A*
    4 N.C.

* All inputs have 50Ω terminations to –2.0 volts. Consequently, additional current will be drawn from the –5.2 volt supply whenever any of these inputs are connected to a higher voltage level than –2.0 volts. The worst case is when they are all grounded.

SPECTRAL PURITY

In the STEL-2273A, the NCO is used with a digital to analog converter (DAC) to generate an analog waveform which approximates an ideal sinewave. The spectral purity of this synthesized waveform is a function of many variables including the phase and amplitude quantization, which are both fixed, the ratio of the clock frequency to output frequency, and the dynamic characteristics of the DAC. A high speed GaAs DAC is used, minimizing the effects of the DAC's performance on the system performance.

Figure 3-12   *(Continued).*

The signal generated by the STEL-2273A has eight bits of amplitude resolution and ten bits of phase resolution which results in spurious levels which are theoretically about −55 dBc. When operating at clock frequencies over 600 MHz the analog characteristics of the DAC limit this to approximately −40 dBc in practice. The highest output frequency the NCO can generate is half the clock frequency ($f_{CLK}/2$, the Nyquist frequency), and the spurious components at frequencies greater than $f_{CLK}/2$ can then removed by filtering. As the output frequency f_{OUT} of the NCO approaches $f_{CLK}/2$ the "image" spur at $f_{CLK} - f_{OUT}$ also approaches $f_{CLK}/2$ from above. If the programmed output frequency is very close to $f_{CLK}/2$ it is virtually impossible to remove this "image" spur by filtering. For this reason, the maximum practical output frequency of the NCO should be limited to 40 – 45% of the clock frequency, depending on the anti-aliasing filter used.

A spectral plot of the STEL-2273A is shown below. In this case the clock frequency is 1 GHz and the output frequency is programmed to 234.567 MHz. The maximum spur level observed over the output frequency range shown in this case is below −45 dBc. Under other conditions the spurious levels can be greater than this since at higher output frequencies the waveform produced by the DAC will have large output changes from sample to sample, and slew rate limitations will introduce more harmonic distortion. Better results can also be obtained by reducing the clock frequency, when this is feasible.

TYPICAL SPECTRUM

Frequency Span:	0 – 400 MHz
Reference Level:	+0 dBm
Resolution Bandwidth:	10 KHz
Scale:	Log, 10 dB/div
Output frequency:	234.567 MHz
Clock frequency:	1 GHz

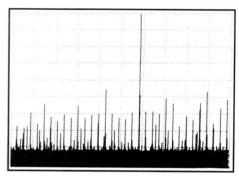

Figure 3-12 *(Continued).*

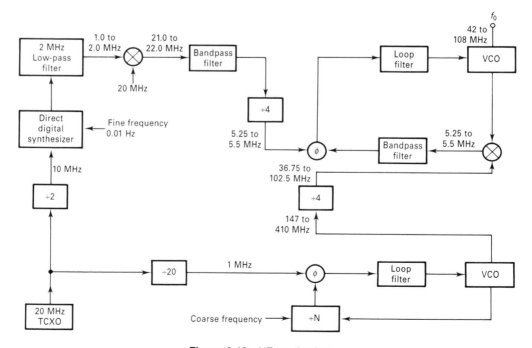

Figure 3-13 HF synthesizer.

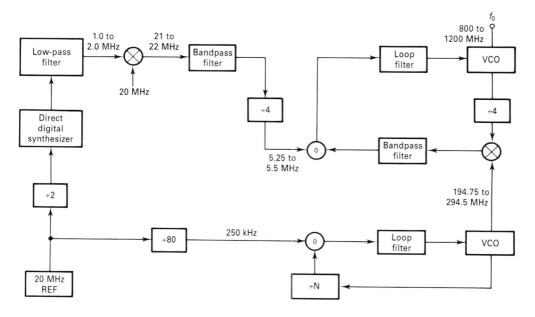

Figure 3-14 Extension to L-band frequencies.

measured with respect to the level of the carrier. Harmonic distortion is the third corrupting factor but is harmless in many applications.

The effect of quantization on signal quality has been treated earlier in this chapter. It represents the best that can be achieved theoretically by digital synthesis. There are a number of practical effects that affect the signal in a more serious way. These important effects will be considered here.

Spurious Sideband Mechanisms. We will examine the various factors that cause spurious sidebands in digital synthesizers.

Quantization Effects. These have been analyzed earlier in the chapter.

Nonlinear Transition Effects in the DAC. This occurs entirely during the transition from one state to the other. The spurs created by this mechanism are determined in the frequency by the same formula that governs quantization spurs, $F_{spur} = MF_{clock} + -NF_{signal}$. The difference is that values of M and N are fairly low. M is usually less than 3 and N is less than 10. If the nonlinear effects are very pronounced, these values could increase. For the values mentioned, the spurs generated are well above those created by the ideal quantization process. It is not unusual to find spurs around -30 dBc when a high conversion rate is attempted. A good deglitching circuit (sample/hold or simple sampler) can reduce these spurs by up to 30 to 40 dB. The deglitcher, of course, introduces its own set of spurs, but these are lower and tend to be limited to values of $M < 2$ and $N < 5$. A good way to evaluate this effect is to generate signals at frequencies close, but not equal, to fractions of the clock. At approximately 1/3 of the clock frequency, one would observe, on a spectrum analyzer in proximity to the signal frequency, spurs caused by $M = 1$ and $N = 2$ among others that satisfy the equation stated earlier. One should repeat this process for 1/4, 1/5, 1/6 of the clock frequency until the level of the spurs drops below the level of interest. See Ref. 16.

Data Skew. If the data are presented to the DAC with some skew in time, this causes the various switches in the DAC to turn on at slightly different instances. This causes a transition period that has a signal that is dependent on the data being supplied. The effect leads to results that are identical to those mentioned in the previous paragraph and can be examined in the same way. A reduction in the data skew, by reclocking the data in a high-speed register, for example, can contribute significantly to alleviate this effect.

Implementation Effects. This primarily relates to spur-generating mechanisms that are not inherent in the arithmetic or conversion process, but show up on the output signal nevertheless. One such mechanism is clock and clock subharmonic leakage and mixing in the DAC. Thus, if a strong $F_{clock}/2$ component were present in the system, it could give rise to a spur at $F_{clock}/2 - F_{signal}$. In general, this can lead to spurs at frequencies given by $(M/K)F_{clock} + -NF_{signal}$, where $K = 1, 2, 4, 8$, because it is likely to encounter power of 2 subharmonics of the clock in the digital processing system.

Two Tone Intermodulation Distortion. This is a measure of the dynamic linearity of the signal-generating process. Two equilevel tones at frequencies F_1 and F_2 give rise to spectral lines at $MF_1 - NF_2$ that are undesired. The ratio of the undesired tones to the desired ones in dBc is a figure of merit for the signal generator. Note that quantization gives rise to intermodulation distortion, but this is usually at a level well below that caused by the DAC or the output amplifier for a 12-bit system.

Output Transfer Function Characteristics. Any departure from flat amplitude and linear phase as a function of frequency introduces an error in the signal being created. This is especially apparent when the carrier is being modulated. Thus anti-aliasing filters must be especially precise in these characteristics to reduce the signal distortion arising from this effect. It is possible to compensate for nonideal effects by predistorting the desired signal with just the right amount of amplitude/phase offset to undo the ill effects of the LPF nonlinearities. See Ref. 17 for further details.

Phase Noise. This is caused by the fact that there is jitter in the phase of the signal that is being created. The zero crossings for a sine wave are not distributed uniformly in time; they exhibit some randomness. This leads to a broadening of the spectrum of the otherwise infinitesimally wide spectral line of a sine wave with no phase noise. A figure of merit for phase noise is the ratio of the power or voltage in spectral range 1 Hz wide to that of the carrier itself. A detailed treatment of phase noise can be found in Ref. 20.

DIRECT DIGITAL SYNTHESIS TECHNIQUES 169

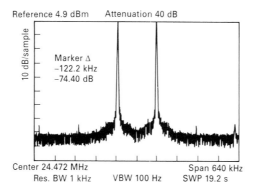

Figure 3-15 Two-tone intermodulation in the HP8770A undesired tones 74 dB down with respect to the two tones at +5 dBm.

There is only one source of phase noise in direct digital synthesis as opposed to multiple sources for phase-locked loop synthesis. That source is the phase noise on the clock signal. In general, any phase modulation on the clock signal is transmitted to the output signal by the ratio F_{signal}/F_{clock}. This means that there is a 20-dB reduction in the phase modulation of a signal that is at 1/10 the clock frequency compared to the phase modulation that exists on the clock signal.

The problem of cleaning up the phase noise in direct digital synthesis is therefore simply that of cleaning up the clock signal. It should be noted, however, that when discreet spurs become very numerous, their effect becomes indistinguishable from that of phase noise for practical applications. The following two pictures show the obtainable quality. Figure 3-15 shows the two-tone intermodulation in the HP8770A undesired tones 74 dB down with respect to the two tones at +5 dBm. Figure 3-16 shows the HP8770A signal generator with a carrier frequency of 37 MHz. The spurious signals are 72 dB down. The internal clock rate for this is 125 MHz.

3-1-4 Future Prospects

The future of digital synthesis is very bright due to the continued high rate of investment occurring in high-speed digital logic technology. While progress in suitable conversion technology is taking place at a much slower pace, the activity is sufficient to provide an appropriate impetus to this method for signal generation.

Present efforts appear to be centered on creating building blocks for digital synthesizers to be used in various receiver applications. A number of companies are actively involved, the following constitutes a partial list of the participants.

Triquint is a GaAs foundry making a high-speed DAC said to provide 8-bit resolution at a conversion rate of several hundred MHz.

Digital RF Solutions (Mountainview, CA) specializes in low-frequency high signal quality board level solutions for digital signal synthesis.

170 SPECIAL LOOPS

Sciteq (San Diego, CA) offers state-of-the-art high-speed components such as digital accumulators and sine lookup chips, as well as board level solutions and complete direct synthesizers. Emphasis is on speed and bandwidth.

Stanford Telecom (Sunnyvale, CA) manufactures high-speed accumulators and sine lookup chips as well as board level products.

LeCroy Instruments (New York, NY) *Analogic*, and *Wavetek* are among several companies who make waveform generators.

Hewlett-Packard has a complete line of signal simulation equipment including waveform synthesizers and accumulator-based digital synthesizers with up-converters.

3-2 MULTIPLE SAMPLER LOOPS

In Chapter 2 we discussed the question of noise in great detail. We concluded that the output of the VCO also contains noise, and it is very clear that it is highly desirable to minimize this noise contribution.

In Chapter 1 we have seen the effect of the loop filter and that higher-order loop filters have better reference suppression.

The tri-state phase/frequency comparator allows us first to obtain frequency and then phase lock, and if it were an ideal circuit, it would provide total suppression of the reference. We go into the details of this circuit in Chapter 4, but it should already be clear that the ideal tri-state phase/frequency comparator, because of its complementary output stage under locked condition, should have no reference output.

It is obvious that such an ideal circuit does not exist. A better approximation is the sample/hold comparator. However, this circuit has a limited pull-in range, as it is a pure phase and not a phase/frequency comparator. The following shows a fairly simple circuit arrangement in which the use of several samplers will increase the pull-in range over the conventional circuit while maintaining low noise and high reference suppression. Figure 3-17 shows the block diagram of a multisampler loop. As can be seen, three samplers are used in a row. The first sampler is operated at 100 kHz, the second sampler at 10 kHz, and the final sampler at 100 Hz. This leads to a one-loop synthesizer arrangement with 100-Hz step size, where the capture range is largely extended.

Previously, such a circuit would have been built by using a frequency detector first. This frequency detector would coarse steer the VCO to a window in which phase lock is possible, and then the loop gain and all other parameters would have to be optimized for smooth phase lock and undisturbed VCO frequency when leaving the desired window.

Conventionally, this coarse tuning could have also been accomplished by tying digital-to-analog converters to the input of the frequency dividers to generate a dc voltage for coarse steering. The drawback here is the large amount of circuitry involved, the possible noise contribution of the coarse steering circuit, and the temperature-compensating requirement for the voltage-controlled oscillator because over the entire operating temperature, the VCO has to track a predetermined voltage/frequency curve. Cascading several sample/hold comparators has the

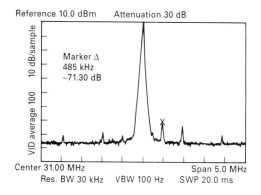

Figure 3-16 HP8770A signal generator with a carrier frequency of 37 MHz. The spurious signals are 72 dB down. The internal clock rate for this is 125 MHz.

advantage that the time constants or the low-pass filters for subsequent sections can be set to a lower freauency, and therefore the acquisition time for each portion can be optimized. To help acquisition further, a digital-to-analog converter can be used to superimpose a dc voltage on the dc output of the sample/hold comparator.

The advantage of this circuit again is that filtering can be applied at the output of the sample/hold comparator, whereby the voltage generated by the D/A converter is applied in a very smooth form, avoiding degradation of noise performance and, specifically, reference frequency suppression. This circuit has been used in the AEG Telefunken E1500 and E1700 shortwave receiver family with great success. The cost/performance ratio of such a loop is extremely attractive. The traditional difficulty of microphonic effects is minimized because of the update of the coarse loop, since the first two sample/hold comparators would detect a major offset of phase or frequency, consequently controlling at a higher speed than the final 100-Hz time constant would permit.

In this configuration, the step size is still limited by the reference frequency, and the following method of applying a sequential phase detector allows increase of resolution. In our example, this will be by a factor of 10. The final system described in this chapter, called the fractional division N synthesizer, will have, theoretically, infinite resolution in a single-loop frequency synthesizer that uses a clever combination of analog and digital circuitry.

3-3 LOOPS WITH DELAY LINE AS PHASE COMPARATORS

The circuit we have just studied, if absolutely ideal components without drift and tolerances were available, would theoretically enable us to come up with infinite resolution. However, the introduction of this circuit will result in some switching noise, probably spikes, and the low-pass filter action of the phase/frequency comparator will cure part of it. However, it is a discontinuous arrangement, and we will now take a look at a system that allows continuous adjustment based on a principle that we have used earlier.

When measuring the phase noise of a frequency synthesizer (Figure 2-28), for

172 SPECIAL LOOPS

all practical purposes we can assume that the delay time τ_d is a constant over time and change of environment, which, of course, is an assumption. We can, by making the phase shifter variable, measure a certain offset from the carrier.

The voltage that is obtained from the mixer was connected to the oscilloscope or the wave analyzer. By integrating this voltage and feeding it through an amplifier back to the VCO, we can close the loop. This arrangement is a noise feedback system. Figure 3-18 shows a circuit in which a PAL delay line and a variable phase shifter are used to stabilize an oscillator if the proper bandwidth and phase offset are chosen. According to Eqs. (2-56) to (2-58), the close-in noise sideband performance of the oscillator can drastically be improved. This circuit has actually been used in some German ham equipment and in the Fluke synthesized signal generator Models 6070A and 6071A. In the case of the Fluke model, a SAW delay line is used, and, as shown in Figure 3-19, by using a subsynthesizer system, a 240- to 520-MHz oscillator is improved by this technique, and the subsynthesizer from 0.8 to 1.2 MHz provides increased resolution down to 1 Hz. Further details on this new Fluke synthesizer can be found in Ref. 8.

The ultimate resolution of this system is determined by the subsynthesizer, which, as discussed earlier, might be a direct digital synthesizer.

The special loops we have seen so far showed an increase in resolution with the trade-off of losing accuracy.

The fractional division N synthesizer offers, theoretically, unlimited resolution and extremely fast settling time. We will learn more about this in the next section.

3-4 FRACTIONAL DIVISION *N* SYNTHESIZERS

Conventional single-loop synthesizers use frequency dividers where the division ratio N is an integer value between 1 and several hundred thousand, and the step size is equal to the reference frequency. Because of the loop filter requirements, the decrease of reference frequency automatically means an increase of settling time. It would be unrealistic to assume that a synthesizer with lower than 100-Hz reference can be built, because the large division ratio in the loop would reduce the loop gain so much that tracking would be very poor, and the settling time would be several seconds.

If it were possible to build a frequency synthesizer with a 100-kHz reference and fine resolution, this would be ideal because the VCO noise from 2 or 3 kHz off the carrier could determine the noise sideband, while the phase noise of frequencies from basically no offset from the carrier to 3 kHz off the carrier would be determined by the loop gain, the division ratio, and the reference. Because of the higher reference frequency, the division ratio would be kept smaller. Traditionally, this conflicting requirement resulted in multiloop synthesizers.

An alternative would be for N to take on fractional values. The output frequency could then be changed in fractional increments of the reference frequency. Although a digital divider cannot provide a fractional division ratio, ways can be found to accomplish the same task effectively. The most frequently used method is to divide the output frequency by $N + 1$ every M cycles and to divide by N the

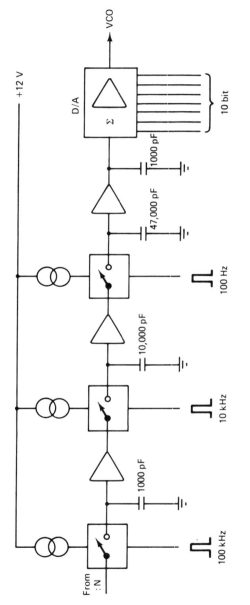

Figure 3-17 Block diagram for a multisampler loop.

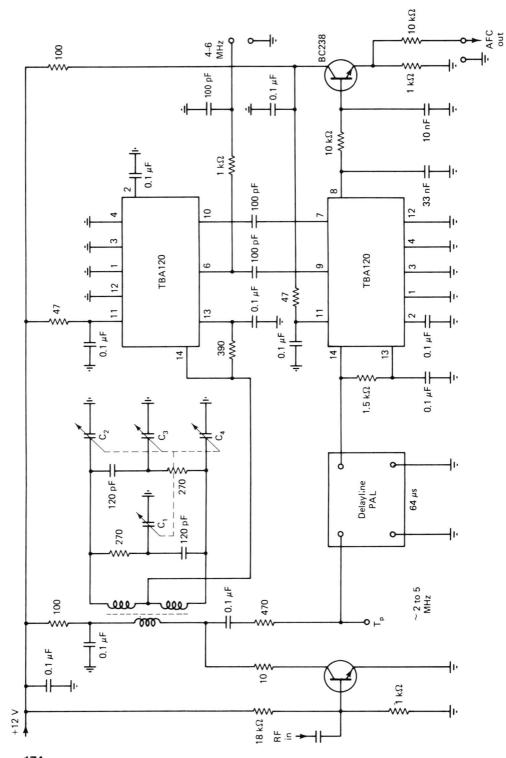

Figure 3-18 Schematic of a delay-line-stabilized oscillator system.

rest of the time. The effective division ratio is then $N + 1/M$, and the average output frequency is given by

$$f_o = \left(N + \frac{1}{M}\right) f_r \qquad (3\text{-}28)$$

This expression shows that f_o can be varied in fractional increments of the reference frequency by varying M. The technique is equivalent to constructing a fractional divider, but the fractional part of the division is actually implemented using a phase accumulator. The phase accumulator approach is illustrated by the following example.

Example 1 Consider the problem of generating 455 kHz using a fractional N loop with a 100-kHz reference frequency. The integral part of the division $N = 4$ and the fractional part $1/M = 0.55$ or $M = 1.8$ (M is not an integer); the VCO output is to be divided by $5(N + 1)$ every 1.8 cycles, or 55 times every 100 cycles. This can easily be implemented by adding the number 0.55 to the contents of an accumulator every cycle. Each time the accumulator overflows (the contents exceed 1) the divider divides by 5 rather than 4. Only the fractional value of the addition is retained in the phase accumulator.

Arbitrarily fine frequency resolution can be obtained by increasing the length of the phase accumulator. For example, with a 100-kHz reference a resolution of $10^5/10^5 = 1$ Hz can be obtained using a five-digit BCD accumulator.

This new method,[*] which we will analyze now in greater detail, is currently used in a number of instruments, such as Hewlett-Packard generators and spectrum analyzers, where it is called fractional N synthesizer, whereas the earlier version, called the digiphase system, is used in the Dana series 7000 synthesizers. A modification of the digiphase system that reduces the low-frequency content of the phase detector output is used in the Racal receiver RA6790. Racal has applied for a patent for this method.

We will now discuss the advantages and drawbacks of this system. At first it may appear that the fractional N loop has unlimited advantages. However, in reality, it is a compromise between resolution, spurious response, and lockup time. Reference 2 may help to clarify the applications further. In reality, the expression fractional N is not quite correct. The loop does not supply a fractional division ratio but rather changes the division ratio periodically over a certain period by the help of an adder driven by the fraction register.

The fractional N phase-locked loop (*NF loop*) is a modified divide-by-N loop. Its unique feature is that it can operate at fractional multiples of the reference signal instead of steps. In the NF loop, N refers to the integer part and F to the fractional part of the divide-by-N number. This number multiplied by the reference signal represents the loop frequency. The integer part is that of a divide-by-N loop. The fractional part represents the offset frequency of the VCO with respect to the integer component of frequency.

---

[*]Part of this description is based on the Hewlett-Packard 3335A signal generator and reproduced with Hewlett-Packard's permission.

176 SPECIAL LOOPS

The description of the NF loop is divided into two parts. The first is a general discussion of the NF loop concept using example frequencies. The second describes the NF loop using simplified block diagrams.

Consider the divide-by-N loop phase detector output under open-loop condition. Assume a reference frequency of 100 kHz, $N = 10$, and VCO frequency of 1.01 MHz ($N = 1.0$ MHz; $F = 0.01$ MHz). The VCO operates at a fractional multiple (10.1) of the reference signal (10.1×0.1 MHz = 1.01 MHz). This configuration is shown in the block diagram of Figure 3-20. The phase detector compares the low-to-high transitions of the reference and divide-by-N signals. Since the VCO is not operating at N times the reference but with a fractional component ($F = 0.01$ MHz), the signal from the divide-by-N block advances on the reference signal. Each time the divide-by-N signal makes a low-to-high transition, the phase detector compares it with the reference and generates an output proportional to the period between the two low-to-high transitions. In the phase-locked condition of a divide-by-N loop, this period remains constant. In the open-loop example, where the VCO contains a fractional component, the period between low-to-high transitions continuously increases, resulting in an increasing phase detector output voltage.

When analyzing the open-loop divide-by-N loop, it is of interest to view the operation in terms of reference periods. A reference period is defined as the time required for the reference signal to complete one cycle. Each reference period, the reference signal goes through one cycle while the VCO, which is operating 10.1 times as fast, goes through 10.1 cycles. We can say the VCO has advanced one-tenth of a cycle of phase on the integer part $N \times f_{ref}$ (f_{ref} = reference frequency) in one reference period. In two reference periods, the VCO has gone 20.2 cycles or advanced two-tenths of a cycle of phase on $N \times f_{ref}$. When the VCO operates with a fractional offset (F), it continually advances phase on $N \times f_{ref}$ each reference period. From the example of Figure 3-18, in 10 reference periods, the VCO signal will have gone 101 cycles, or advanced one cycle of phase (360°) with respect to $N \times f_{ref}$. Table 3-1 illustrates the phase relationship of $N \times f_{ref}$ and NF.

Table 3-1 Phase relationship of $N \times f_{ref}$ and NF

Number of Reference Periods ($f_{ref} = 100$ kHz = 0.1 MHz)	Number of Completed Cycles of:		Phase Advancement of NF on $N \times f_{ref}$
	$N \times f_{ref} = 1$ MHz ($N = 10$)	NF = 1.01 MHz	
1	10	10.1	0.1 cycle of phase
2	20	20.2	0.2 cycle of phase
3	30	30.3	0.3 cycle of phase
4	40	40.4	0.4 cycle of phase
.	.	.	.
.	.	.	.
.	.	.	.
9	90	90.9	0.9 cycle of phase
10	100	101.0	1 full cycle of phase (360°)

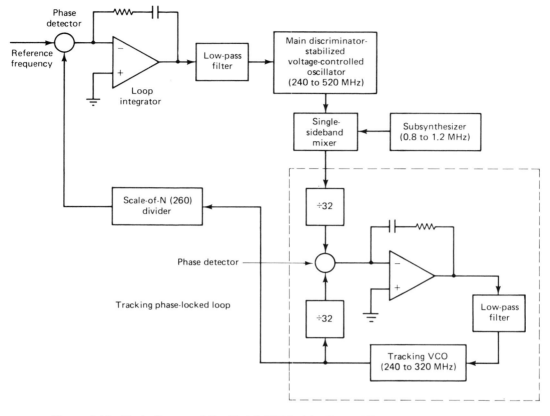

Figure 3-19 Block diagram of the Model 6070A delay-line-stabilized Fluke synthesizer.

While the VCO signal advances phase on $N \times f_{\text{ref}}$, the divide-by-N VCO signal applied to the phase detector advances phase on the reference frequency.

In a divide-by-N loop, the VCO is phase locked to a reference signal and operates at a multiple N of the reference frequency ($N \times f_{\text{ref}}$). In an NF loop, the VCO operates at an integer-plus-fractional multiple of the reference frequency ($N \times f_{\text{ref}} + F = \text{NF}$). As previously illustrated in Figure 3-18, assume again that the VCO operates at 1.01 MHz, the reference is 0.1 MHz, and N equals 10. Each time the reference signal goes through one cycle, the VCO goes through 10.1 cycles. After 10 reference cycles (10 reference periods) the VCO has gone 101 cycles. The VCO has advanced one full cycle of phase (360°) on $N \times f_{\text{ref}}$. If a VCO cycle is removed from the VCO pulse train applied to the divide-by-N block at the point a full VCO cycle has advanced, the phase advancement on the average is canceled and the average frequency applied to the divide-by-N block is $N \times f_{\text{ref}}$ or, in this example, 1 MHz.

Because of the continual removal of a VCO cycle (removal of one cycle of phase) at each point the VCO advances one cycle on $N \times f_{\text{ref}}$, the phase detector output becomes a sawtooth waveform (see Figure 3-21). The waveform increases

178 SPECIAL LOOPS

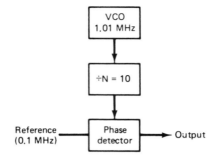

Figure 3-20 Basic diagram of an open-loop divide-by-N loop.

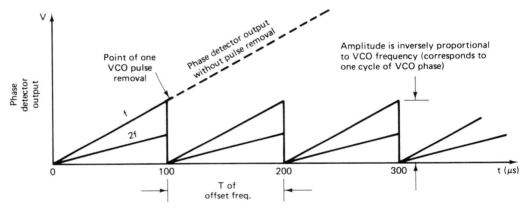

Figure 3-21 Phase detector sawtooth output.

linearly due to the advancing phase of the VCO until the VCO has advanced one cycle of VCO phase (360°). At this point a cycle is removed from the VCO pulse train, canceling the previous advancement of a cycle of phase. The phase detector responds to this sudden one-cycle (360°) phase loss by returning to its initial output. The sequence is repetitive, generating the sawtooth waveform. The maximum amplitude reached represents one cycle of VCO phase. As the VCO frequency is increased, the time interval for the VCO to go through one cycle of phase is less. Therefore, the maximum phase detector amplitude is decreased. The phase detector maximum amplitude is inversely proportional to the VCO frequency.

The necessity to remove one VCO cycle from the VCO output each time the output advances one cycle of phase on $N \times f_{\text{ref}}$ requires that we use a pulse remover block in the divide-by-N loop block diagram (see Figure 3-22). If a VCO pulse is removed each time the VCO advances one cycle of phase, the average frequency applied to the divide-by-N block is $N \times f_{\text{ref}}$ and the average frequency applied to the phase detector is f_{ref}. The relationship of the phase detector sawtooth output and the pulse trains shown in Figure 3-20 is illustrated in Figure 3-23. A method of determining when the VCO has advanced one cycle of phase is required. Such

FRACTIONAL DIVISION *N* SYNTHESIZERS 179

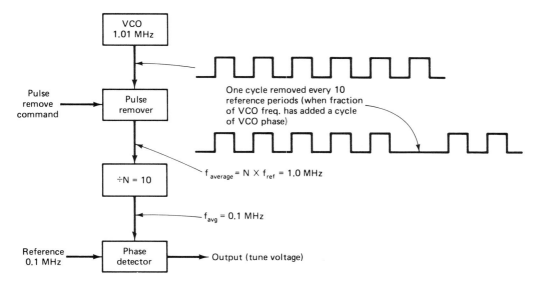

Figure 3-22 Divide-by-*N* loop with pulse remover block.

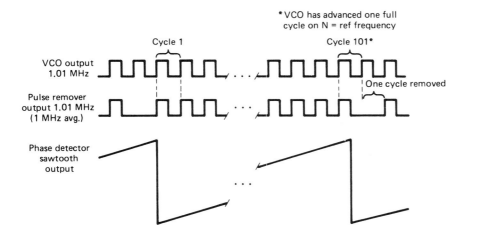

Figure 3-23 Phase detector sawtooth output with respect to pulse remover output.

information can then be used to trigger the pulse remover block and a VCO cycle removed at the appropriate time.

The fractional part of the VCO frequency determines the time required for the VCO to advance one cycle of phase on $N \times f_{\text{ref}}$. The time required is the period of the fractional offset frequency and corresponds to a certain number of reference periods. If the fractional part of the VCO is stored in a register added to a second register each reference period, the second register will contain a running total that represents the VCO phase advancement at any point in time. For this reason the second register is called a *phase register* and the entire configuration is called an *accumulator* (see Figure 3-24). The phase register will reach unity after the same

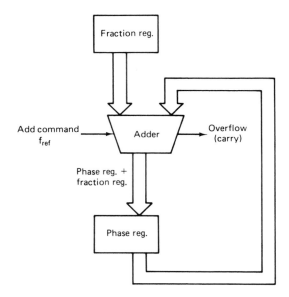

Figure 3-24 Accumulator.

reference period during which the VCO has advanced one full cycle of phase. (Recall the preceding example: in one reference period the VCO has gone 10.1 cycles, in two reference periods the VCO has gone 20.2 cycles, and so on. The summing register will contain 0.1 after one reference period, 0.2 after the second, and so on.) When unity is reached, the phase register overflows and transmits an overflow signal. This signal occurs at the time the VCO has advanced one cycle of phase on $N \times f_{\text{ref}}$ and is applied to the pulse remover block as a pulse remove signal.

If the VCO operates with an offset frequency not evenly divisible into 1 (such as 0.03), a fractional overflow can result when the phase register reaches unity. For example, if the VCO operates at 1.03 MHz instead of 1.01 MHz, after one reference period it has gone 10.3 cycles, 20.6 after two, 30.9 after three, and 41.2 after the fourth reference period. Prior to the fourth reference period, the phase register has accumulated 0.9. The fourth reference period 0.3 is added to the 0.9 from the phase register and results in 1.2. This causes an overflow as the pulse remove signal and the fractional overflow of 0.2 is loaded into the phase register and the next sequence phase begins to accumulate from 0.2 instead of zero.

Up to this point, the discussion has developed the NF loop to include the pulse remove command section. Figure 3-25 is a block diagram of the NF loop with the pulse remove command section. This structure provides a means of automatically removing a VCO cycle whenever the VCO advances one full cycle of phase on the frequency $N \times f_{\text{ref}}$.

The open-loop phase detector output of Figure 3-23 is a sawtooth waveform superimposed on a dc voltage. Only the dc voltage of this output is of interest. A VCO requires a dc tune voltage to maintain a stable output signal. A sawtooth ac signal superimposed on the dc VCO tune voltage would cause VCO frequency modulation.

FRACTIONAL DIVISION N SYNTHESIZERS   181

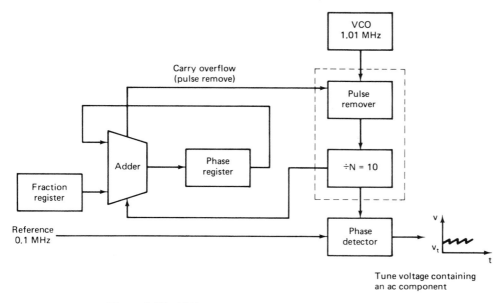

Figure 3-25 NF loop with pulse remove command section.

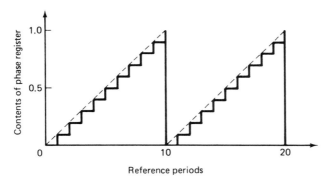

Figure 3-26 Phase register contents.

The ac component must be canceled or removed, leaving the dc component to tune the VCO to the proper frequency.

We know that the VCO output advances a fraction of a cycle of phase on $N \times f_{\text{ref}}$ each reference period. The fraction of a cycle of phase that the VCO is advanced at any one reference period is represented by the fractional sum in the phase register. (Recall that the phase register is incremented by the fractional VCO output each reference period.) For the example of Figure 3-23, the contents of the phase register when viewed with respect to time can be represented as a staircase resetting to zero once unity is reached (see Figure 3-26). The staircase approximates a sawtooth waveform (see dashed lines). The "front edge" of each step represents the phase detector output for that reference period. (Recall that

182 SPECIAL LOOPS

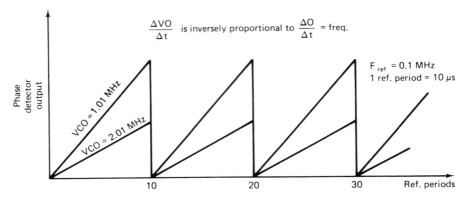

Figure 3-27 Phase detector output for two VCO frequencies with the same offset.

the phase detector does not generate a ramp but samples the VCO with respect to the reference each reference period.)

If the contents of the summing register are applied to a digital-to-analog (D/A) converter, the D/A converter output will follow the steps of the summing register and approximate a sawtooth output. Inverting the D/A converter output and summing it with the phase detector output essentially cancels the ac component (sawtooth) of the phase detector output. This leaves the dc component required as a VCO control signal.

Two requirements exist for the waveform generated by the D/A converter to approximate the phase detector sawtooth output.

1. It must have a variable amplitude.
2. It must have a variable period.

The amplitude is inversely proportional to the frequency of the VCO and changes whenever the VCO frequency is changed. To demonstrate the amplitude dependency on the VCO frequency, refer to Figure 3-27.

In the figure a reference of 0.1 MHz is used (horizontal axis plotted in reference periods) and plots of the phase detector output for VCO frequencies of 1.01 MHz and 2.01 MHz are shown. Note that each VCO frequency example contains a 0.01-MHz offset or fractional frequency. In terms of reference periods (10 μs), the period of the 0.01-MHz offset is 10 reference periods. At this point the offset frequency has completed one cycle and added a cycle of phase to the VCO signal. Since the period of 2 MHz is half the interval of 1 MHz, the phase detector output representing one cycle of phase at 2 MHz is half the amplitude of the output, representing one cycle of 1 MHz phase. When the VCO cycle is removed, a 360° phase loss is detected by the phase detector and it responds by returning to its initial output, causing the high-to-low transition of the sawtooth. If the offset or fractional part of the VCO frequency is changed, the period of the sawtooth changes for these two periods are the same. The sawtooth generated by the D/A converter must change amplitude and period as the phase detector output changes and must be superimposed on zero volts dc. It can then be inverted and summed

FRACTIONAL DIVISION N SYNTHESIZERS

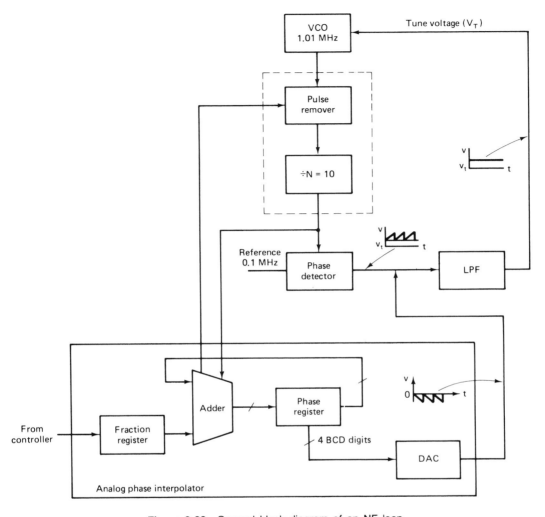

Figure 3-28 General block diagram of an NF loop.

with the phase detector output to remove the sawtooth from the tune voltage applied to the VCO.

A general block diagram of an NF loop is shown in Figure 3-28. The basic elements of a divide-by-N loop are present: the VCO, divide-by-N counter, phase detector, and low-pass filter. In addition to these, a fraction register, adder, and phase register provide the "bookkeeping system" recording the phase advancement from reference period to reference period. This system is known as a *phase interpolator*, and in conjunction with a digital-to-analog converter the system is referred to as an *analog phase interpolator* (API). During each reference period it generates an analog voltage equal and opposite in polarity to the phase advancement voltage generated by the phase detector. The voltage applied to the LPF is then the net VCO tune voltage. Since the "bookkeeping system" must update each reference period (the phase detector output changes each reference

184 SPECIAL LOOPS

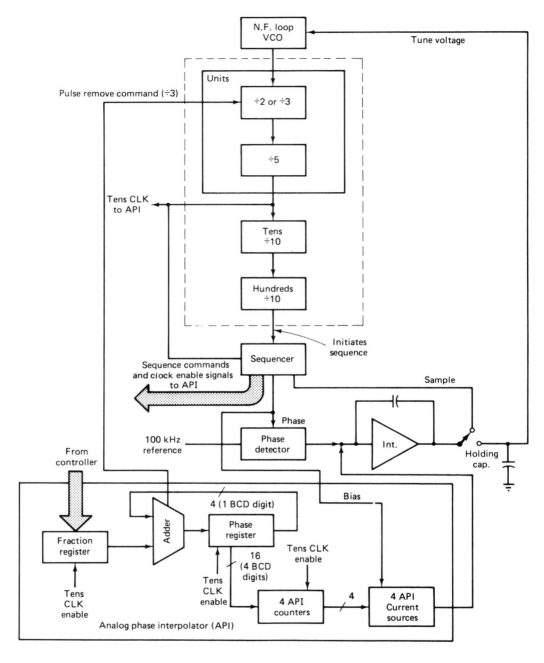

Figure 3-29 Basic block diagram of an NF loop.

period after the VCO/N and reference signal comparison), the system receives its add command (update command) at a VCO/N rate.

The NF loop is a modified divide-by-N loop. It contains all the basic elements of the divide-by-N loop with the addition to several other sections. Figure 3-29 illustrates the additions made to the "loop." These additions are a sequencer, an N counter that can be changed to an $(N + 1)$ counter, and an integrator and sample/hold, which are used to develop the tune voltage. Compare this diagram to Figure 3-26.

The NF loop operates according to an established sequence of events that occur once each reference period. The sequence is initiated at a rate equal to the VCO/N signal. This is accomplished by initiating each sequence of events with the N counter output. The sequencer generates a number of enable and command signals, which are summarized here.

1. *API Tens Clock Enable Signals*. These signals enable the Tens Clock to update the data in the API registers (bookkeeping system).
2. *API Counter Tens Clock Enable*. This signal enables the Tens Clock to clock the four API counters, which are preset each reference period by the four most significant digits of the phase register, which keeps a running total of the phase advancement.
3. *Bias Command*. This signal turns on the four API current sources to establish a current reference point.
4. *Phase Command*. This signal is the Bias command reclocked to the Tens Clock and again reclocked to the NF loop VCO. It is compared with the reference each reference period by the phase detector.
5. *Sample Command*. This signal initiates the sampling of the integrator output each reference period. Once the integrator has settled following the summation of the phase detector and API signals, the integrator voltage is transferred to the holding capacitor.

The rate of events is determined by the Tens Clock, which is the NF loop VCO divided by 10. The sequence of events is initiated once each reference period, but once initiated, the events occur at a rate determined by the VCO frequency.

Figure 3-30 illustrates the loop by a heavy line and separates between digital and analog halves of the loop. The basic structure is shown in Figure 3-31. The major sections of the loop structure are input, decode and data registers, divide-by-N with pulse remove, sequencer, phase detector, API, integrator, sample-and-hold, and VCO.

The input decode section interfaces the loop with the data transmitted by the controller. These data include the loop frequency and instructions that set up the operating modes of the data registers in the phase interpolator. Data register operation is controlled by a steering section.

The data registers comprise the bookkeeping scheme of the phase interpolator. There are three data registers:

1. f_1 frequency register.
2. f_2 frequency register.
3. Phase register.

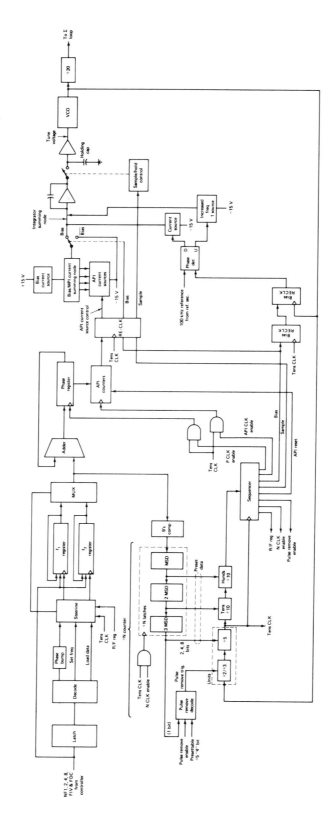

Figure 3-30 Block diagram of a fractional N loop.

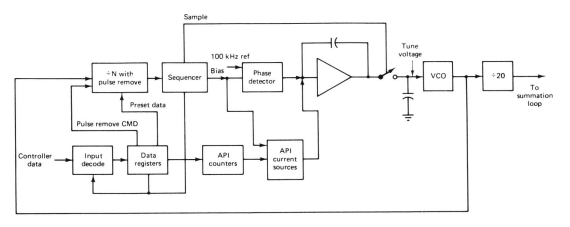

Figure 3-31 Basic structure of an NF loop.

Only one of the frequency registers is active at a time. The frequency register will always contain the current frequency of operation and these data will be circulated (output connected to input and the data shifted until starting state is reached) once each reference period. The other frequency register contains the previous frequency of operation and rests idle but enabled to accept new data when a new output is programmed.

The data steering logic controls the operating modes of the f_1 and f_2 frequency registers. The Load Data command enables the idle frequency register to be clocked by the controller line LDC (Load Data Clock) to enter a new frequency. During this time the operation of the loop is not interrupted because the circulating frequency register continues operation while data are being loaded. Once the data are entered, the Set Freq command interchanges the functions of the f_1 and f_2 registers and the new data now circulate to operate the loop at the new frequency.

Frequency data in the f_1 or f_2 register consist of 16 BCD digits, which are loaded least significant digit first. The 12 least significant digits represent the fractional portion of the frequency, and the next three digits contain the integer or N portion of the frequency. This accounts for 15 of the 16 digits in the f_1 or f_2 register. The sixteenth digit, which is the last digit loaded, is not required and therefore is always loaded as a zero. During circulation of the data in the f_1 or f_2 register, this digit is truncated and does not affect the operation of the loop.

During each reference period the divide-by-N counter initiates a sequence of events by triggering the sequencer. Part of the sequence is the enabling of the f_1 or f_2 register clock, the phase register clock, and the N register clock. The phase register is clocked for the first 12 digits circulated by the f_1 or f_2 register, the N register for the next three. When the sixteenth digit is circulated by the f_1 or f_2 register, neither phase register nor N register is clocked; therefore, this digit has no effect on the loop operation. As a result of the sequence of clocking the registers and N register, the phase register quantity has been increased by the fractional component of the f_1 or f_2 register. The N register contains the three N number digits used to preset the divide-by-N counter.

188 SPECIAL LOOPS

The phase register serves two purposes:

1. Records the total phase advancement of the VCO with respect to each reference period.
2. Causes the adder to overflow in the reference period during which the VCO has advanced a full cycle of phase.

The record of total phase advancement is used each reference period to drive the API section. The four most significant digits of the 12 digits in the phase register are used to preset four API counters. When these counters are clocked by the API clock, they generate an output pulse inversely proportional to the preset number and drive the API section, which develops a signal that counteracts the changing phase detector signal, resulting in an unchanging tune voltage. The overflow of the adder indicates the reference period in which the NF loop VCO has advanced a full cycle of VCO phase. The overflow decode triggers the units counter during the pulse remove enable interval of the loop sequence to divide by 3 for one output pulse of the first stage. Since this stage has been providing an output for every two input pulses (divide by 2), it effectively has removed a VCO cycle by dividing by 3 for one output pulse. The cycle of phase the NF loop VCO has advanced has been removed and the phase relationship of the NF loop VCO and N times the reference is reset.

The API section consists of two parts:

1. The API counters.
2. The API current sources.

All API current sources are turned on by the Bias command each reference period. The four most significant digits of the phase register preset the API counters, which control when each of the four API current sources turn off. The smaller the phase register digits, the longer the API current sources are on.

The phase detector compares the sequencer output "Bias" with a 100-kHz reference signal. The Bias signal is first reclocked to Tens Clock (VCO/10) and then to the NF loop VCO signal itself. If the NF loop VCO is operating with a fractional component, the reclocked Bias signal applied to the phase detector gains phase each reference period with respect to the reference signal. The output applied to the integrator is an increasing voltage. The purpose of the API section is to negate the effects of the increase in the phase detector output.

The method used to generate the NF loop VCO tune voltage is similar to that used in the divide-by-N loop. Currents are integrated and the integrated voltage is transferred to a holding capacitor.

A block diagram of the currents integrated by the NF loop in a phase-locked condition is shown in Figure 3-32. Figure 3-33 illustrates the integrator waveform, showing the contributions of the different currents. A constant-current source, I (Bias), supplies current at all times to the Bias/API summing node. The Bias command from the sequencer goes high each reference period to connect this node to the integrator summing node. Following the Bias command, the phase register data cause the API current source to draw current from the Bias/API summing

FRACTIONAL DIVISION N SYNTHESIZERS   189

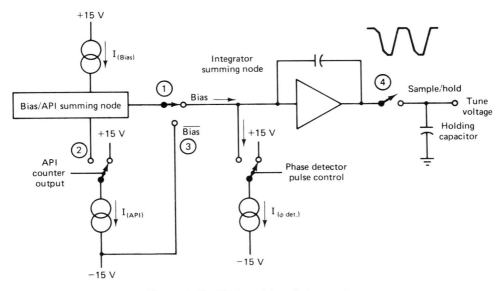

Figure 3-32 NF loop integrated currents.

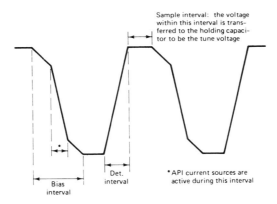

Figure 3-33 Integrator waveform showing the contributions of the different currents.

node and therefore keep this current from being integrated. The amount of API current is determined by the magnitude of the phase register number. Once the Bias event has occurred and the Bias/API summing node is disconnected from the integrator summing node, the phase detector pulse occurs and draws current out of the integrator summing node. When the loop is phase locked, the current entering the integrator node from the Bias/API current sources is equal to the current drawn out by the phase detector current source and the integrated voltage remains constant. After integrating the two currents, the voltage is transferred to a holding capacitor and becomes the tune voltage.

190 SPECIAL LOOPS

The sequence of events is as follows:

1. Bias/API summing node is connected to the integrator and Bias current is integrated.
2. API current source is connected to Bias/API summing node, decreasing the amount of Bias current integrated.
3. Bias/API summing node is disconnected and phase detector is connected to integrator. Phase detector current is integrated.
4. After the phase detector current has been integrated and the voltage has settled, the voltage is transferred to a holding capacitor. This voltage is the NF loop VCO tune voltage.

Note: When phase locked, the Bias/API current is equal to the phase detector current. The API current tracks the increasing phase detector current, canceling the fractional component of VCO phase.

Refer again to Figure 3-28. Assume that the loop operates without a fractional component (data in the phase register remain constant). The loop can be viewed as just a divide-by-N loop with an elaborate method of developing the tune voltage. The data in the phase register are constant; therefore, the API current sources are sinking the same amount of current from the Bias/API summing node each reference period. Since the current entering the integrator from the Bias/API summing node is always a constant value, the phase detector changes the tune voltage just as it does in the divide-by-N loop. A change in the phase relationship causes the phase detector pulse to change in duration, which changes the amount of current the phase detector source draws from the integrator. The result is a change in the integrated voltage after this reference period, and therefore the tune voltage has been changed. The direction of change is such that the NF loop VCO is pulled back into a phase-locked condition.

The Sample command from the sequencer transfers the integrated voltage to the holding capacitor at the appropriate period of the integrator output. This period occurs after the Bias/API summing interval and the phase detector interval have occurred and the integrator output has returned to an unchanging value. This value is the tune voltage.

The increase-frequency current source is shown on the simplified block diagram of Figure 3-28. This current source is also connected to the integrator summing node and is responsible for rapidly changing the tune voltage if a large increase in frequency is programmed; the phase detector connects this current source to the integrator in place of the phase detector current source. Instead of sinking current from the integrator, the current source drives current into the integrating node to add to the current already supplied by the Bias/API interval. This causes the tune voltage to change the NF loop VCO frequency rapidly. Once the newly programmed frequency has been reached, the phase detector again begins operation using the phase detector current source.

The NF loop VCO signal is divided by 20 to aid spur attenuation and reduce phase noise. The division by 20 results in an improvement of 26 dB in the noise sideband and phase noise.

While this system allows extremely high resolution, the synthesizer, so to speak,

/consists of two loops, one being a 100-kHz loop with the lockup time probably 8 to 20 cycles of reference or 800 μs to 2 ms, depending on the loop filter.

The fractional portion of the loop theoretically would lock up within one cycle of reference or 10 μs. However, because of active low-pass filters and speed requirements for the D/A converter, the actual lockup time is somewhat of a compromise between these values and should be in the vicinity of 1 to 2 ms and, therefore, about the same as the 100-kHz loop.

Modern integrated circuits having frequency synthesizers on one chip can be used to build such systems.

Philips Semiconductors has recently introduced the SA 7025 low-voltage 1-GHz Fractional-N Synthesizer. This chip works on the principle just outlined and is ideal for use in cellular telephones.

The noise sideband performance of this synthesizer depends highly on the accuracy of the D/A converter and its ability to remove the reference noise sideband. Ref. 22 discusses a method used in the Racal receiver for which Racal has applied for a patent, and this phase detector output has a zero-running average area. This is effective in reducing the low frequencies produced when the output is near a multiple of the reference and when many corrections are made during each period of the phase detector output [17–37].

3-4-1 Special Patents for Fractional Division N Synthesizers

1. Latched accumulator fractional N synthesis with residual error reduction
   Alexander W. Hietala, Cary, IL; Duane C. Rabe, Rolling Meadows, IL
   Motorola, Inc., Shaumburg, IL
   United States Patent, Patent No. 5,093,632, March 3, 1992
2. Frequency synthesizers having dividing ratio controlled sigma-delta modulator
   Thomas A. D. Riley, Osgoode, Canada
   Carleton University, Ottawa, Canada
   United States Patent, Patent No. 4,965,531, October 23, 1990
3. Phase locked loop variable frequency generator
   Nigel J. R. King, Wokingham, England
   Racal Communications Equipment Limited, England
   United States Patent, Patent No. 4,204,174, May 20, 1980
4. Frequency synthesizers
   John Norman Wells, St. Albans, Hertfordshire (GB)
   Marconi Instruments, St. Albans, Hertfordshire (GB)
   European Patent, Patent No. 0125790B2, July 5, 1995
5. Improvement in or relating to synthesizers
   Thomas Jackson, Twickenham, Middlesex (GB)
   Plessey Overseas Limited, Ilford, Essex (GB)
   European Patent, Patent No. 0214217B1, June 6, 1996
6. Improvement in or relating to synthesizers
   Thomas Jackson, Twickenham, Middlesex (GB)
   Plessey Overseas Limited, Ilford, Essex (GB)
   European Patent, Patent No. WO86/05046, August 28, 1996
7. PLL including an arithmetic unit
   Robert J. Bosselaers
   United States Patent, Patent No. 3913928, October 1975

8. Frequency synthesizer
   R. G. Cox, Hewlett Packard
   United States Patent, Patent No. 2976945
9. Device for synthesizing frequencies with fractional multiplier of a fundamental frequency
   C. A. Kingford-Smith, Hewlett Packard
   United States Patent, Patent No. 3928813
10. PLL frequency synthesizer including fractional digital frequency divider
    A. T. Crowley, RCA
    United States Patent, Patent No. 4468632
12. Enhanced analog phase interpolation for fractional-N frequency synthesizer
    J. K. Crowford, Hughes Aircraft Company
    United States Patent, Patent No. 4586005
13. Frequency synthesizer having jitter compensation
    Y. D. McCann, U.S. Phillips
    United States Patent, Patent No. 4599579
14. Frequency synthesizer with spur compensation
    F. L. Martin, Motorola
    United States Patent, Patent No. 4816774
15. Frequency synthesizer with spur compensation
    F. L. Martin, Motorola
    United States Patent, Patent No. 4918403
16. Fractional-N synthesizer having modulation spur compensation
    W. P. Sheperd et al., Motorola
    United States Patent, Patent No. 5021754
17. Multiple modulation fractional-N divider
    B. M. Miller, Hewlett Packard
    United States Patent, Patent No. 5038117
18. Frequency-modulated PLL with fractional-N divider and jitter compensation
    M. A. Wheatley et al., Racal Dana
    United States Patent, Patent No. 5038120
19. Digital frequency synthesizer
    W. G. Greken, General Dynamics
    United States Patent, Patent No. 3882403
20. Frequency synthesizer having fractional-N frequency divider in PLL
    W. J. Tanis, Engelman Microwave
    United States Patent, Patent No. 3959737
21. Frequency synthesizer with fractional-N division ratio and jitter compensation
    N. G. Kingsbury, Marconi
    United States Patent, Patent No. 4179670
22. Frequency synthesizer including a fractional-N multiplier
    J. Remy, Adret
    United States Patent, Patent No. 4458329
23. Frequency synthesizer of the fractional-N type
    T. Jackson, Plessey
    United States Patent, Patent No. 4800342
24. Fractional-N frequency synthesizer with modulation compensation
    C. Attenborough, Plessey
    United States Patent, Patent No. 4686488
25. Fractional-N division FS for digital angle modulation
    A. Albarello, Thomson-CSF et al.
    United States Patent, Patent No. 4492936

26. Fractional-N frequency divider
    R. O. Yeager, RCA
    United States Patent, Patent No. 4573176
27. Low phase noise radio frequency synthesizer
    A. P. Edwards, Hewlett Packard
    United States Patent, Patent No. 4763083
29. B. G. Goldberg
    United States Patent, Patent No. 5224132, June 1993
30. Digital FS having multiple processing paths
    B. Goldberg
    United States Patent, Patent No. 4898310, September 1990
31. Digital FS
    B. Goldberg
    United States Patent. Patent No. 47587310, June 1988
32. Digital FS
    E. J. Nossen
    United States Patent, Patent No. 4206425, June 1980
33. Digital FS with random jittering for reducing discrete spectral spurs
    C. E. Wheatley
    United States Patent, Patent No. 4410954, October 1983
34. Digital FS
    Leland Jackson
    United States Patent, Patent No. 3735269, May 1973
35. Spurless fractional divider direct digital synthesizer and method
    V. S. Reinhardt, Hughes Aircraft
    United States Patent, Patent No. 4815018

REFERENCES

1. Ulrich Rohde, *Digital PLL Frequency Synthesis Theory & Design*, Prentice-Hall, Englewood Cliffs, NJ, 1983.
2. Roland Hassun, "The Common Denominators of Fractional N," *Microwaves & RF*, June 1984.
3. B. E. Bjerede and G. D. Fischer, "A New Phase Accumulator Approach to Frequency Synthesis," *Proceedings of IEEE NAECON*, May 1976, pp. 928–932.
4. W. D. Stanley, *Digital Signal Processing*, Reston Publishing, Reston, VA, 1975, pp. 51–54.
5. R. Hassun and A. Kovalick, "An Arbitrary Waveform Synthesizer for DC to 50 MHz," Hewlett-Packard Journal, April 1988, Palo Alto, California.
6. L. R. Rabiner and B. Gold, *Theory and Application of Digital Signal Processing*, Prentice-Hall, Englewood Cliffs, NJ, 1975, Chap. 2.
7. H. T. Nicholas and H. Samueli, "An Analysis of the Output Spectrum of Direct Digital Frequency Synthesizers in the Presence of Phase-Accumulator Truncation," *41st Annual Frequency Control Symposium*, IEEE Press, New York, 1987.
8. L. B. Jackson, "Roundoff Noise for Fixed Point Digital Filters Realized in Cascade or Parallel Form," *IEEE Transactions on Audio and Electroacoustics*, AU-18, pp. 107–122, June 1970.
9. Technical Staff of Bell Laboratories, "Transmission Systems for Communication," Bell Labs, Inc., 1970, Chap. 25.

10. A. Kovalick, "Apparatus and Method of Phase to Amplitude Conversion in a SIN Function Generator," U.S. Patent 4,482,974.
11. C. J. Paull and W. A. Evans, "Waveform Shaping Techniques for the Design of Signal Sources," *The Radio and Electronic Engineer*, Vol. 44, No. 10, October 1974.
12. L. Barnes, "Linear-Segment Approximations to a Sinewave," *Electronic Engineering*, Vol. 40, September 1968.
13. R. Hassun and A. Kovalick, "Waveform Synthesis Using Multiplexed Parallel Synthesizers," U.S. Patent 4,454,486.
14. D. K. Kikuchi, R. F. Miranda, and P. A. Thysell, "A Waveform Generation Language for Arbitrary Waveform Synthesis," Hewlett-Packard Journal, April 1988, Palo Alto, California.
15. H. M. Stark, *An Introduction to Number Theory*, MIT Press, Cambridge, MA, 1978, Chap. 7.
16. W. Sagun, "Generate Complex Waveforms at Very High Frequencies," *Electronic Design*, January 26, 1989.
17. G. Lowitz and R. Armitano, "Predistortion Improves Digital Synthesizer Accuracy," *Electronics Design*, March 31, 1988.
18. A. Kovalick, "Digital Synthesizer Aids in Testing of Complex Analog, Digital Circuits," *EDN Magazine*, September 1, 1988.
19. G. Lowitz and C. Pederson, "RF Testing with Complex Waveforms," *RF Design*, November 1988.
20. Catharine M. Merigold, *Telecommunications Measurement Analysis and Instrumentation*, Kamilo Fehrer, Editor, Prentice-Hall, Englewood Cliffs, NJ, 1987.
21. G. C. Gillette, "Digiphase Principle," *Frequency Technology*, August 1969.
22. Jerzy Gorski-Popiel, *Frequency Synthesis: Techniques and Applications*, IEEE Press, New York, 1975.
23. Hewlett-Packard, "Synthesized/Level Generator 3335A," *Instruction Manual*, pp. 816–836.
24. Racal, "RA6790 Receiver," *Instruction Repair Manual*, 1978.
25. Vadim Manassewitsch, *Frequency Synthesis*, 2nd ed., Wiley, New York, 1980.
26. U. L. Rohde, "Modern Design of Frequency Synthesizers," *Ham Radio*, July 1976.
27. William F. Egan, *Frequency Synthesis by Phase Lock*, Wiley, New York, 1981.
28. F. Telewski, K. Craft, E. Drucker, and J. Martins, "Delay Lines Give RF Generator Spectral Purity, Programmability," *Electronics*, August 28, 1980, pp. 133–142.
29. U. L. Rohde, "Low-Noise Frequency Synthesizers Using Fractional N Phase Locked Loops," *Proceedings of Modern Solid-State Devices, Techniques, and Applications for High-Performance RF Communications Equipment*, 1981 Southcon Professional Program, Georgia World Congress Center, January 13–15, 1981, pp. 15/1/1–15/1/10.
30. J. Tierney, C. M. Rader, and B. Gold, "A Digital Frequency Synthesizer," *IEEE Transactions on Audio and Electroacoustics*, Vol. AU-19, No. 1, March 1971, pp. 48–57.
31. AEG Telefunken Instruction and Repair Manual for E1500, E1700 Receiver.
32. U.S. Patent 3,959,737, "Frequency Synthesizer Having Fractional Frequency Divider in Phase Locked Loop," William J. Tanis, Wayne, NJ, and Engelmann Microwave, Montville, NJ, May 25, 1976.
33. Wing S. Djen and Daniel J. Linebarger, "Fractional-N PLL Provides Fast Low Noise Synthesis," *Microwaves & RF*, May 1994, pp. 95–101.

34. Terrence F. Hock, "Synthesizer Design with Detailed Noise Analysis," *RF Design*, July 1993.
35. Jonathan Stilwell, "A Flexible Fractional-*N* Frequency Synthesizer for Digital RF Communications," *RF Design*, February 1993.
36. Ken Mason, "Design Guide to Frequency Synthesis Using the UMA1005," Application Note, Philips Semiconductors, Rept. No: SCO/AN92002, November 1992.
37. *SA 8025 Fractional-N Synthesizer for 2 GHz Applications*, Application Note AN1891, Philips Semiconductors RF Communication Products, September 1994, pp. 1–13.
38. D. A. Sunderland, R. A. Strauch, S. S. Wharfield, H. T. Peterson, and C. R. Cole, "CMOS/SOS Frequency Synthesizer LSI Circuit for Spread Spectrum Communications," *IEEE Journal of Solid State Circuits*, August 1984, pp. 497–505.
39. C. E. Wheatley and D. E. Phillips, "Spurious Suppression in DDS," *Proceedings of the 35th Annual Frequency Control Symposium*, May 1981.
40. C. E. Wheatley, "Spurious Suppression in DDS Using Accumulator Dither," Rockwell International, Document WP81-3055.
41. F. Williams, "A Digital FS," *QST*, April 1984, pp. 24–30.
42. E. M. Mattison and L. M. Coyle, "Phase Noise in DDS," *42nd Annual Frequency Control Symposium*, 1988.
43. Robert P. Gilmore, *DDS-Driven PLL Frequency Synthesizer with Hard Limiter*, U.S. Patent 5028887, July 1991.
44. I. Fobbester, "Spur Reduction in DDS," *Electronic Product Design*, June 1992, pp. 23–24.
45. M. P. Wilson, "Spurious Reduction Techniques for DDS," Colloquium on DDFS, University of Bradford, November 1991, *IEE Digest*, No. 199/172.
46. M. Bozie, "Spurious Redistributing DDS," *IEE Digest*, 1991, p. 172.
47. COMSAT Labs, *Evaluation of 16 KBPS Voice Processor*, February 1991.
48. Y. Matsuya et al., "A 16-Bit Oversampling A/D Conversion Technology Using Triple Integration Noise Shaping," *IEEE Journal of Solid State Circuits*, December 1987, pp. 921–929.
49. Sciteq Electronics, San Diego, DDS-1 application notes.
50. Plessey, U.K., "High Speed DDS: SP2002," data sheet.
51. Analog Devices, "Direct Digital Synthesizer: AD9955," data sheet.
52. Stanford Telecom, "STEL-1 177, STEL 2373," data sheet.
53. L. J. Kushner, "The Composite DDSS: A New DDS Architecture," *Proceedings of the 1993 IEEE International Frequency Control Symposium*, June 2–4, 1993, pp. 255–260.
54. RF Design Staff, "Development of a New Direct VHF Synthesizer," December 1987, pp. 22–26.

Fractional Division *N* Readings

Dana Series 7000 Digiphase, Publication 980428 (Manual), Dana Laboratories, 2401 Campus Drive, Irvine, CA 92664, 1973.
Egan, W. F., *Frequency Synthesis by Phase Lock*, Wiley, New York, 1981.
Gillete, Garry C., "The Digiphase Synthesizer,"*Frequency Technology*, August 1969, pp. 15–29.
Reinhardt, V. et al., "A Short Survey of Frequency Synthesizer Techniques," 40th Annual Frequency Control Symposium, 1986, pp. 355–365.
Hassun, R., "A High-Purity, Fast Switching Synthesized Signal Generator," *Hewlett-Packard J.*, February 1981.

Bjerede, B., "A New Phase Accumulator Approach to Frequency Synthesis," Proc. IEEE NAECON, 1976.

Messerschmitt, D. G., "A New PLL Frequency Synthesis Structure," *IEEE Trans. Comm.*, Vol. COM-26 No. 8, August 1978, pp. 1195–1200.

Rohde, U. L., "Low-Noise Frequency Synthesizers Using Fractional-*N* Phase-Locked Loops," *RF Design*, January/February 1981, pp. 20–34.

Faulkner, T. R. et al., "Signal Generator Frequency Synthesizer Design," *Hewlett-Packard J.*, December 1985, pp. 24–31.

Aken, M. B. and W. M. Spaulding, "Development of a Two-Channel Frequency Synthesizer," *Hewlett-Packard J.*, August 1985, pp. II–18.

Danielson, D. D. and S. E. Froseth, "A Synthesized Signal Source with Function Generator Capabilities," *Hewlett-Packard J.*, Vol. 30, No. I, January 1979.

Browne, J., "Miniature RF Synthesizer Generates Giant Performance," *Microwaves & RF*, December 1984, pp. 135–136.

Hassun, R., "The Common Denominators in Fractional *N*," *Microwaves & RF*, June 1984, pp. 107–110.

Fountain, E., Hughes Ground Systems Group, Fullerton California (personal communication).

Frey, G., Hughes Ground Systems Group, Fullerton, California (personal communication).

O'Leary, P. and F. Maloberti, "A Direct Digital Synthesizer with Improved Spectral Performance," *IEEE Trans. Comm.*, Vol. 39, No. 7, July 1991, pp. 1046–1048.

Rohde, U. L., *Digital PLL Frequency Synthesizers Theory and Design*, Prentice-Hall, Englewood Cliffs, NJ, 1983.

4

LOOP COMPONENTS

4-1 OSCILLATOR DESIGN

The phase-locked loop generally has two oscillators, the oscillator at the output frequency and the reference oscillator. The reference oscillator at times can be another loop that is being mixed in, and the voltage-controlled oscillator is controlled by either the reference or the oscillator loop. The voltage-controlled oscillator is one of the most important parts of the phase-locked loop system because its performance is determined inside the loop bandwidth by the loop and outside the loop bandwidth by its design. To some designers, the design of the voltage-controlled oscillator (VCO) appears to be magic. Shortly, we will go through the mathematics of the oscillator and some of its design criteria, but the results have only limited meanings. This is due to component tolerances, stray effects, and, most of all, nonlinear performance of the device, which is modeled with only a certain degree of accuracy. However, after building oscillators for awhile, a certain feeling will be acquired for how to do this, and certain performance behavior will be predicted on a rule-of-thumb basis rather than on precise mathematical effort. For reasons of understanding, we will deal with the necessary mathematical equations, but I consider it essential to explain that these are only approximations.

4-1-1 Basics of Oscillators

An electronic oscillator is a device that converts dc power to a periodic output signal (ac power). If the output waveform is approximately sinusoidal, the oscillator is referred to as *sinusoidal*. There are many other oscillator types normally referred to as *relaxation* oscillators. For applications in frequency synthesizers, we will try only to build sinusoidal oscillators for reasons of purity and noise sideband performance, and we will deal only with those.

198 LOOP COMPONENTS

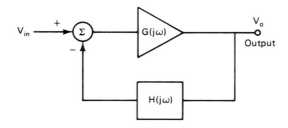

Figure 4-1 Block diagram of an oscillator showing forward and feedback loop components.

All oscillators are inherently nonlinear. Although the nonlinearity results in some distortion of the signal, linear analysis techniques can normally be used for the analysis and design of oscillators. Figure 4-1 shows, in block diagram form, the necessary components of an oscillator. It contains an amplifier with frequency-dependent forward loop gain $G(j\omega)$ and a frequency-dependent feedback network $H(j\omega)$. The output voltage is given by

$$V_o = \frac{V_{in} G(j\omega)}{1 + G(j\omega) H(j\omega)} \tag{4-1}$$

For an oscillator, the output V_o is nonzero even if the input signal $V_i = 0$. This can only be possible if the forward loop gain is infinite (which is not practical), or if the denominator

$$1 + G(j\omega) H(j\omega) = 0 \tag{4-2}$$

at some frequency ω_o. This leads to the well-known condition for oscillation (the *Nyquist criterion*), where at some frequency ω_o

$$G(j\omega_o) H(j\omega_o) = -1 \tag{4-3}$$

That is, the magnitude of the open-loop transfer function is equal to 1:

$$|G(j\omega_o) H(j\omega_o)| = 1 \tag{4-4}$$

and the phase shift is 180°:

$$\arg[G(j\omega_o) H(j\omega_o)] = 180° \tag{4-5}$$

This can be more simply expressed as follows. If in a negative feedback system, the open-loop gain has a total phase shift of 180° at some frequency ω_o, the system will oscillate at that frequency provided that the open-loop gain is unity. If the gain is less than unity at the frequency where the phase shift is 180°, the system will be stable, whereas if the gain is greater than unity, the system will be unstable.

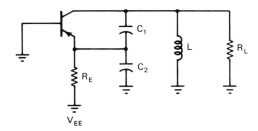

Figure 4-2 Oscillator with capacitive voltage divider.

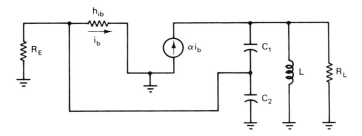

Figure 4-3 Linearized and simplified equivalent circuit of Figure 4-2.

This statement is not correct for some complicated systems, but it is correct for those transfer functions normally encountered in oscillator design. The condition for stability is also known as the *Barkhausen criterion*, which states that if the closed-loop transfer function is

$$\frac{V_o}{V_i} = \frac{\mu}{1 - \mu\beta} \qquad (4\text{-}6)$$

the system will oscillate provided that $\mu\beta = 1$. This is equivalent to the Nyquist criterion, the difference being that the transfer function is written for a loop with positive feedback. Both versions state that the total phase shift around the loop must be 360° at the frequency of oscillation and the magnitude of the open-loop gain must be unity at that frequency.

The following analysis of the relatively simple oscillator shown in Figure 4-2 illustrates the design method. The linearized (and simplified) equivalent circuit of Figure 4-2 is given in Figure 4-3. h_{rb} has been neglected, and $1/h_{ob}$ has been assumed to be much greater than the load resistance R_L and is also ignored. Note that the transistor is connected in the common base configuration, which has no voltage phase inversion (the feedback is positive), so the conditions for oscillation are

$$|G(j\omega_o)H(j\omega_o)| = 1 \qquad (4\text{-}7)$$

and

$$\arg[G(j\omega_o)H(j\omega_o)] = 0° \qquad (4\text{-}8)$$

200 LOOP COMPONENTS

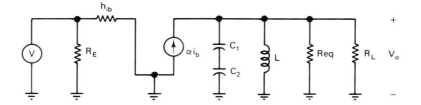

Figure 4-4 Further simplification of Figure 4-2, assuming high-impedance loads.

The circuit analysis can be greatly simplified by assuming that

$$\frac{1}{\omega(C_2 + C_1)} \ll \frac{h_{ib}R_E}{h_{ib} + R_E} \tag{4-9}$$

and also that the Q of the load impedance is high. In this case the circuit reduces to that of Figure 4-4, where

$$V = \frac{V_o C_1}{C_1 + C_2} \tag{4-10}$$

and

$$R_{eq} = \frac{h_{ib}R_E}{h_{ib} + R_E}\left(\frac{C_1 + C_2}{C_1}\right)^2 \tag{4-11}$$

Then the forward loop gain is

$$G(j\omega) = \frac{h_{fb}}{h_{ib}} Z_L = \frac{\alpha}{h_{ib}} Z_L \tag{4-12}$$

and

$$H(j\omega) = \frac{C_1}{C_1 + C_2} \tag{4-13}$$

where

$$Y_L = \frac{1}{Z_L} = \frac{1}{j\omega L} + \frac{1}{R_{eq}} + \frac{1}{R_L} + \frac{1}{j\omega C} \tag{4-14}$$

A necessary condition for oscillation is that

$$\arg[G(j\omega)H(j\omega)] = 0° \tag{4-15}$$

Since H does not depend on frequency in this example, if $\arg GH$ is to be zero, the phase shift of the load impedance Z_L must be zero. This occurs only at the resonant frequency of the circuit,

$$\omega_o = \frac{1}{\sqrt{L[C_1 C_2/(C_1 + C_2)]}} \tag{4-16}$$

At this frequency

$$Z_L = \frac{R_e R_L}{R_{eq} + R_L} \tag{4-17}$$

and

$$GH = \frac{h_{fb}}{h_{ib}} \left(\frac{R_{eq} R_L}{R_{eq} + R_L} \right) \frac{C_1}{C_1 + C_2} \tag{4-18}$$

The other condition for oscillation is the magnitude constraint that

$$G(j\omega) H(j\omega) = \frac{\alpha}{h_{ib}} \left(\frac{R_{eq} R_L}{R_{eq} + R_L} \right) \frac{C_1}{C_1 + C_2} = 1 \tag{4-19}$$

Although the block diagram formulation of the stability criteria is the easiest to express mathematically, it is frequently not the easiest to apply since it is often difficult to identify the forward loop gain $G(j\omega)$ and feedback ratio $H(j\omega)$ in electronic systems. A direct analysis of the circuit equations is frequently simpler than the block diagram interpretation (particularly for single-stage amplifiers). Figure 4-5 shows a generalized circuit for an electronic amplifier. The small-signal equivalent circuit is given in Figure 4-6 (where h_{re} has been neglected). Normally, h_{oe} can also be assumed sufficiently small and can be neglected. The loop equations are then

$$V_{in} = I_1(Z_3 + Z_1 + Z_2) - I_b Z_1 + \beta I_b Z_2 \tag{4-20}$$

$$0 = -I_1 Z_1 + I_b(h_{ie} + Z_1) \tag{4-21}$$

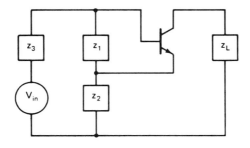

Figure 4-5 Generalized circuit for an oscillator using an amplifier model.

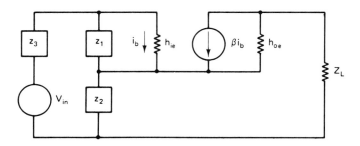

Figure 4-6 Small-signal equivalent circuit of Figure 4-5.

For the amplifier to oscillate, the currents I_b and I_1 must be nonzero even when $V_1 = 0$. This is possible only if the system determinant

$$\Delta = \begin{vmatrix} Z_3 + Z_1 + Z_2 & \beta Z_2 - Z_1 \\ -Z_1 & h_{ie} + Z_1 \end{vmatrix} \tag{4-22}$$

is equal to 0. That is,

$$(Z_3 + Z_1 + Z_2)(h_{ie} + Z_1) - Z_1^2 + \beta Z_1 Z_2 = 0 \tag{4-23}$$

which reduces to

$$(Z_1 + Z_2 + Z_3)h_{ie} + Z_1 Z_2 \beta + Z_1(Z_2 + Z_3) = 0 \tag{4-24}$$

Only the case where the transistor input impedance h_{ie} is real will be considered here (a valid approximation for oscillators operating below 50 MHz). The more complicated case, in which h_{ie} is complex, can be analyzed in the same manner. Assume for the moment that Z_1, Z_2, and Z_3 are purely reactive impedances. [It is easily seen that Eq. (4-24) does not have a solution if all three impedances are real.] Since both the real and imaginary parts must be zero, Eq. (4-24) is equivalent to the following equations if

$$h_{ie}(Z_1 + Z_2 + Z_3) = 0 \tag{4-25}$$

and

$$Z_1[(1 + \beta)z_2 + Z_3] = 0 \tag{4-26}$$

Since β is real and positive, Z_2 and Z_3 must be of opposite sign for Eq. (4-26) to hold. That is,

$$(1 + \beta)Z_2 = -Z_3 \tag{4-27}$$

Therefore, since h_{ie} is nonzero, Eq. (4-25) reduces to

$$Z_1 + Z_2 - (1 + \beta)Z_2 = 0 \tag{4-28}$$

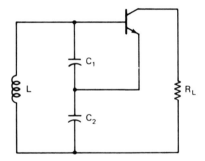

Figure 4-7 Colpitts oscillator.

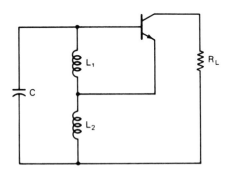

Figure 4-8 Hartley oscillator.

or

$$Z_1 = \beta Z_2$$

Thus, since β is positive, Z_1 and Z_2 will be reactances of the same kind. If Z_1 and Z_2 are capacitors, Z_3 is an inductor and the circuit is as shown in Figure 4-7. It is referred to as a *Colpitts oscillator*, named after the person who first described it. If Z_1 and Z_2 are inductors and Z_3 is a capacitor as illustrated in Figure 4-8, the circuit is called a *Hartley oscillator*.

Example 1 Design a Colpitts circuit to oscillate at 71 MHz, using a transistor that has an input impedance

$$h_{ie} = 1.3 \; k\Omega \quad \text{and} \quad \beta = 100$$

Solution. For the Colpitts circuit Z_1 and Z_2 are negative reactances and Z_3 is an inductive reactance. Let $Z_2 = -j1.33 \; \Omega$ then [from Eq. (4-27)]

$$-Z_3 = (1 + \beta) Z_2 = j133 \; \Omega$$

204 LOOP COMPONENTS

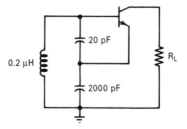

Figure 4-9a Design example of a Colpitts oscillator.

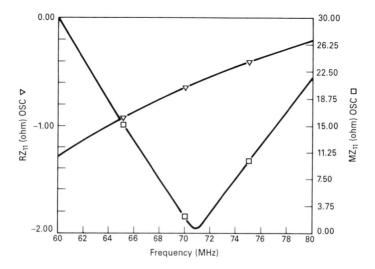

Figure 4-9b Simulation of a Colpitts oscillator using a bipolar model.

and [from Eq. (4-25)]

$$Z_1 = -Z_2 = -Z_3 = -j131.7 \, \Omega$$

At 71 MHz these impedances correspond to component values of

$$C_1 = 20 \text{ pF}$$
$$C_2 = 2000 \text{ pF}$$
$$L = 0.2 \, \mu\text{H}$$

The completed circuit, except for biasing, is shown in Figure 4-9a. Figure 4-9b shows the simulation curve of this oscillator using a bipolar model. The curve was generated by looking into the impedance of the 0.2-μH inductor relative to ground. It is obvious that the value RZ_{11} remains negative from 60 to 80 MHz, which is the condition for oscillation, and MZ_{11} reaches its minimum value at 71 MHz. This is the resonant frequency. As long as RZ_{11} is negative, oscillation starts.

If Z_1, Z_2, Z_3, or h_{ie} is complex, the preceding analysis is more complicated, but the conditions for oscillation can still be obtained from Eq. (4-24). For example, if in the Colpitts circuit, there is a resistor R in series with L ($Z_3 = R + j\omega L$), Eq. (4-24) reduces to the two equations

$$h_{ie}\left(\omega L - \frac{1}{\omega C_1} - \frac{1}{\omega C_2}\right) - \frac{R}{\omega C_1} = 0 \quad (4\text{-}29)$$

and

$$h_{ie}R - \frac{1+\beta}{\omega^2 C_1 C_2} + \frac{1}{\omega C_1}\omega L = 0 \quad (4\text{-}30)$$

Define

$$C_1' = \frac{C_1}{1 + R/h_{ie}} \quad (4\text{-}31)$$

The resonant frequency at which oscillations will occur is found from Eq. (4-29) to be

$$\omega_o = \frac{1}{\sqrt{L[C_1'C_2/(C_1' + C_2)]}} \quad (4\text{-}32)$$

and for oscillations to occur

$$R_e(h_{ie}) \leq \frac{1+\beta}{\omega_0^2 C_1 C_2} - \frac{L}{C_1} \quad (4\text{-}33)$$

If R becomes too large, Eq. (4-31) cannot be satisfied and oscillations will stop. In general, it is advantageous to have

$$X_{C_1} X_{C_2} = \frac{1}{\omega^2 C_1 C_2} \quad (4\text{-}34)$$

with $X \triangleq 1/\omega c$ as large as possible since then R can be large. However, if C_1 and C_2 are too small (large X_{C_1} and X_{C_2}), the input and output capacitors of the transistor, which shunt C_1 and C_2, respectively, become important. A good, stable design will always have C_1 and C_2 much larger than the transistor capacitances they shunt.

Example 2 In Example 1 will the circuit still oscillate if the inductor has a $Q_L = 120$ down from 900 as an ideal value? Since the transistor input capacitance is 5 pF, what effect will this have on the system?

206 LOOP COMPONENTS

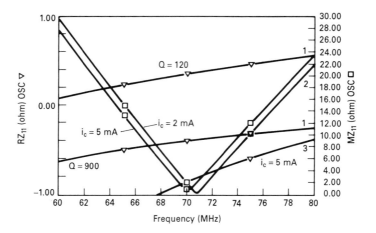

Figure 4-9c Simulation of three conditions.

Solution. In Example 1,

$$X_L = 133, \quad C_1 = 20\,\text{pF}, \quad \text{and} \quad C_2 = 2300\,\text{pF} \quad (\text{value increased})$$

Since $C_1 = 20\,\text{pF}$, adding 5 pF in parallel will change the equivalent C_1 to 25 pF. As the inductor $Q_L = 120$, the equivalent resistance R in series with the lossless inductor is

$$R = \frac{130}{120} = 1.1\,\Omega$$

The new resonant frequency has been determined to be 71 MHz down from 79.6 MHz without the junction input capacitance.

Figure 4-9c shows the simulation of three conditions. The first condition is $Q = 900$. We get a negative resistance of approximately 0.6 Ω and a resonant frequency of 70 MHz. If we reduce the Q_L to a more realistic $Q = 120$, $R_E(Z_{11})$ becomes positive and the oscillator will cease to oscillate. In order to compensate, one has to increase the loop gain and this is done by increasing the collector current. We have chosen to increase the value to 5 mA and now we get a negative resistance of approximately 0.9 Ω. In order to have some safety margin, one has to either increase the loop gain further (done by using more current) or select an inductor with higher Q. The increased dc operating point also changed the resonant frequency by about 1 MHz.

The simulation of this fairly simple circuit was accomplished by changing the value for Q_1 in the circuit and changing the value R_e of the bipolar model for the transistor. R_e is calculated as 26 mV/I_C. For 2 mA, R_e assumes the value of 13 Ω and for 5 mA approximately 5 Ω.

Although Eqs. (4-25) and (4-26) can be used to determine the exact expressions

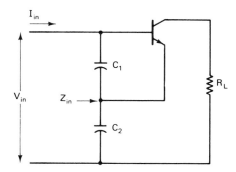

Figure 4-10 Calculation of input impedance of the negative resistance oscillator.

for oscillation, they are often difficult to use and add little insight into the design process. An alternative interpretation, although not as accurate, will now be presented. It is based on the fact that an ideal tuned circuit (infinite Q), once excited, will oscillate infinitely because there is no resistance element present to dissipate the energy. In the actual case where the inductor Q is finite, the oscillations die out because energy is dissipated in the resistance. It is the function of the amplifier to maintain oscillations by supplying an amount of energy equal to that dissipated. This source of energy can be interpreted as a negative resistor in series with the tuned circuit. If the total resistance is positive, the oscillations will die out, while the oscillation amplitude will increase if the total resistance is negative. To maintain oscillations, the two resistors must be of equal magnitude. To see how a negative resistance is realized, the input impedance of the circuit in Figure 4-10 will be derived.

If h_{oe} is sufficiently small ($h_{oe} \ll 1/R_L$), the equivalent circuit is as shown in Figure 4-10. The steady-state loop equations are

$$V_{in} = I_{in}(X_{C_1} + X_{C_2}) - I_b(X_{C_1} - \beta X_{C_2}) \tag{4-35}$$

$$0 = -I_{in}(X_{C_1}) + I_b(X_{C_1} + h_{ie}) \tag{4-36}$$

After I_b is eliminated from these two equations, Z_{in} is obtained as

$$Z_{in} = \frac{V_{in}}{I_{in}} = \frac{(1+\beta)X_{C_1}X_{C_2} + h_{ie}(X_{C_1} + X_{C_2})}{X_{C_1} + h_{ie}} \tag{4-37}$$

If $X_{C_1} \ll h_{ie}$, the input impedance is

$$Z_{in} \approx \frac{1+\beta}{h_{ie}} X_{C_1} X_{C_2} + (X_{C_1} + X_{C_2}) \tag{4-38}$$

$$Z_{in} \approx \frac{-g_m}{\omega^2 C_1 C_2} + \frac{1}{j\omega[C_1 C_2/(C_1 + C_2)]} \tag{4-39}$$

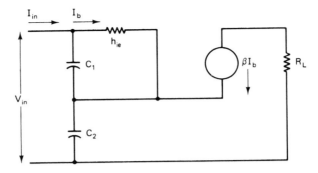

Figure 4-11 Equivalent small-signal circuit of Figure 4-10.

That is, the input impedance of the circuit shown in Figure 4-11 is a negative resistor,

$$R = \frac{-g_m}{\omega^2 C_1 C_2} \tag{4-40}$$

in series with a capacitor,

$$C_{in} = \frac{C_1 C_2}{C_1 + C_2} \tag{4-41}$$

which is the series combination of the two capacitors. With an inductor L (with the series resistance R_s) connected across the input, it is clear that the condition for sustained oscillation is

$$R_s = \frac{g_m}{\omega^2 C_1 C_2} \tag{4-42}$$

and the frequency of oscillation is

$$f_o = \frac{1}{2\pi\sqrt{L[C_1 C_2/(C_1 + C_2)]}} \tag{4-43}$$

This interpretation of the oscillator readily provides several guidelines that can be used in the design. First, C_1 should be as large as possible so that

$$X_{C_1} \ll h_{ie}$$

and C_2 is to be large so that

$$X_{C_2} \ll \frac{1}{h_{oe}}$$

When these two capacitors are large, the transistor base-to-emitter and collector-to-emitter capacitances will have a negligible effect on the circuit's performance. However, Eq. (4-42) limits the maximum value of the capacitances since

$$r \leq \frac{g_m}{\omega^2 C_1 C_2} \leq \frac{G}{\omega^2 C_1 C_2} \tag{4-44}$$

where G is the maximum value of g_m. For a given product of C_1 and C_2, the series capacitance is a maximum when $C_1 = C_2 = C_m$. Thus, Eq. (4-44) can be written

$$\frac{1}{\omega C_m} > \sqrt{\frac{r}{G}} \tag{4-45}$$

This equation is important in that it shows that for oscillations to be maintained, the minimum permissible reactance $1/\omega C_m$ is a function of the resistance of the inductor and the transistor's mutual conductance g_m.

An oscillator circuit known as the *Clapp circuit* or *Clapp–Gouriet circuit* is shown in Figure 4-12a. This oscillator is equivalent to the one just discussed, but it has the practical advantage of being able to provide another degree of design freedom by making C_o much smaller than C_1 and C_2. It is possible to use C_1 and C_2 to satisfy the condition of Eq. (4-44) and then adjust C_o for the desired frequency of oscillation ω_o, which is determined from

$$\omega_o L - \frac{1}{\omega_o C_o} - \frac{1}{\omega_o C_1} - \frac{1}{\omega_o C_2} = 0 \tag{4-46}$$

Example 3 Consider a Clapp–Gouriet oscillator as shown in Figure 4-12a. The transistor used has $G_{\max} = Y_{\max} = 13$ mA/V and it is operated at $g_m = 12$ mA/V. The coil used has an unloaded $Q_u = 120$ at 78 MHz and a reactive impedance of 98 Ω ($R_s = 0.82$ Ω). What are the required conditions for the circuit to oscillate?

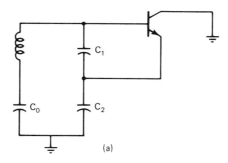

Figure 4-12a Circuit of a Clapp oscillator.

Solution. To satisfy Eq. (4-45), we must have

$$\frac{1}{\omega C_m} \geq \sqrt{\frac{r}{g_m}} = \sqrt{\frac{0.82}{0.012}} = \sqrt{68} = 8.3$$

Therefore, at 78.4 MHz,

$$C_m \leq 150 \text{ pF}$$

C_m corresponds to the case of maximum series capacitance of the combination of C_1 and C_2 and occurs for C_1 and $C_2 = C_m$. If both C_1 and C_2 are 300 pF, the reactance of the series combination of C_1 and C_2 is 27 Ω. Other combinations of C_1 and C_2 can be selected that may provide more gain, provided that Eq. (4-44) is satisfied. C_o must be selected so that $X_L = X_c$ at 78 MHz.

Figure 4-12b shows the impedance plot for RZ_{11} and MZ_{11}. The magnitude MZ_{11} for resonance is less than approximately 2 Ω and the negative resistance at this point is also around 2 Ω. Therefore, oscillation will start up. If the start-up condition is not met, either the capacitive voltage divider is changed (reduced in value) or the dc current increased.

The following is a listing of the Microwave Harmonica/Scope circuit description of the Clapp–Gouriet oscillator. This circuit is essentially the basis of the previous two examples.

```
Compact Software - MICROWAVE SCOPE PC V6.5
File: c:\ap\co13ckt
* TEST OF CLAPP-GOURIET OSCILLATOR
*
*
BLK
  BIP 1 2 3 A=.99 RE=13 F=41GHz T=1E-12 CE=5PF RB1=2
  BASEM: 3POR 1 2 3
END
blk
  basem 1 2 3
  ind 1 12 l=.2µH q1=120 f=100MHz
  cap 12 11 c=20pF
  cap 1 3 c=300pF
  cap 3 0 c=300pF
  res 2 0 r=1000
osc:1por 11
end
*
FREQ
  step 80MHz 90MHz .1MHz
END
*
OUT
  PRI osc z
END
*
```

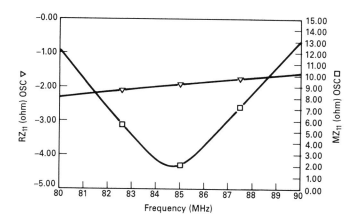

Figure 4-12b Impedance plot for RZ_{11} and MZ_{11}.

Amplitude Stability. Linearized analysis of the oscillator is convenient for determining the frequency but not the amplitude of the oscillation. The Nyquist stability criterion defines the frequency of oscillation as the frequency at which the loop phase shift is 360°, but it says nothing about the oscillation amplitude. If no provisions are taken to control the amplitude, it is susceptible to appreciable drift. Two frequently used methods for controlling the amplitude are operating the transistor in the nonlinear region or using a second stage for amplitude limiting. For the single-stage oscillator, amplitude limiting is accomplished by designing an unstable oscillator; that is, the loop gain is made greater than 1 at the frequency where the phase shift is 180°. As the amplitude increases, the β of the transistor decreases, causing the loop gain to decrease until the amplitude stabilizes. This is a self-limiting oscillator. There are nonlinear analysis techniques predicting the amplitude of oscillation, but their results are approximate except in idealized cases, forcing the designer to resort to an empirical approach.

An example of a two-stage emitter-coupled oscillator is shown in Figure 4-13. In this circuit, amplitude stabilization occurs as a result of current limiting in the

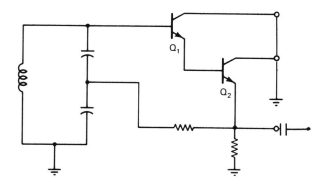

Figure 4-13 Two-stages emitter-coupled oscillator.

second stage. This circuit has the additional advantage that it has output terminals that are isolated from the feedback path. The emitter signal of Q_2, having a rich harmonics content, is normally used as output. Harmonies of the fundamental frequency can be extracted at the emitter of Q_2 by using an appropriately tuned circuit. Note that the collector of Q_2 is isolated from the feedback path.

Phase Stability. An oscillator has a frequency or phase stability that can be considered in two separate parts. First, there is the long-term stability in which the frequency changes over a period of minutes, hours, days, weeks, or even years. This frequency stability is normally limited by the circuit component's temperature coefficients and aging rates. The other part, the short-term frequency stability, is measured in terms of seconds. One form of short-term instability is due to changes in phase of the system; here the term "phase stability" is used synonymously with frequency stability. It refers to how the frequency of oscillation reacts to small changes in phase shift of the open-loop system. It can be assumed that the system with the largest rate of change of phase versus frequency $d\phi/df$ will be the most stable in terms of frequency stability. Figure 4-14 shows the phase plots of two open-loop systems used in oscillators. At the system crossover frequency, the phase shift is $-180°$. If some external influence causes a change in phase, say, it adds $10°$ of phase lag, the frequency will change until the total phase shift is again $0°$. In this case the frequency will decrease to the point where the open-loop phase shift is $170°$. Figure 4-14 shows that Δf_2, the change in frequency associated with the $10°$ change in phase of GH_2, is greater than the change in frequency Δf_1, associated with open-loop system GH_1, whose phase is changing more rapidly near the open-loop crossover frequency.

The qualitative discussion illustrates that $d\phi/df$ at $f = f_0$ is a measure of an oscillator's phase stability. It provides a good means of quantitatively comparing the phase stability of two oscillators. Consider the simple parallel tuned circuit shown in Figure 4-15. For the circuit, the two-port is

$$\frac{V_0(j\omega)}{I(j\omega)} = \frac{R}{1 + jQ[(\omega/\omega_0) - (\omega_0/\omega)]} \tag{4-47}$$

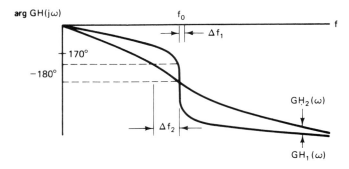

Figure 4-14 Phase plot of two open-loop systems with different Q of the resonator.

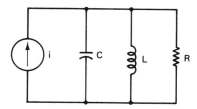

Figure 4-15 Parallel tuned circuit for phase shift analysis.

where

$$\omega_0 = \frac{1}{\sqrt{LC}} \quad \text{and} \quad Q = \frac{R}{\omega_0 L} \tag{4-48}$$

The circuit phase shift is

$$\arg \frac{V_0}{I} = \theta = \tan^{-1} Q\left(\frac{\omega}{\omega_0} - \frac{\omega_0}{\omega}\right) \tag{4-49}$$

and

$$\frac{d\theta}{d\omega} = \frac{1/Q}{1/Q^2 + [\omega^2 + [(\omega^2 - \omega_0^2)/\omega_0\omega]^2} \frac{\omega^2 + \omega_0^2}{(\omega_0\omega)^2} \tag{4-50}$$

at the resonant frequency ω_0,

$$\left.\frac{dQ}{d\omega}\right|_{\omega=\omega_0} = \frac{2Q}{\omega_0} \tag{4-51}$$

The frequency stability factor is S_F, defined as the change in phase $d\phi/d\omega$ divided by the normalized change in frequency $\Delta\omega/\omega_0$: that is,

$$S_F = 2Q \tag{4-52}$$

S_F is a measure of the short-term stability of an oscillator. Equation (4-51) indicates that the higher the circuit Q, the higher the stability factor. This is one reason for using high-Q circuits in oscillator circuits. Another reason is the ability of the tuned circuit to filter out undesired harmonics and noise.

4-1-2 Low-Noise *LC* Oscillators

In Chapter 2 we derived a formula that allows an estimate of the noise performance of an oscillator (Figure 4-16). Under the assumption that the output energy is taken

214   LOOP COMPONENTS

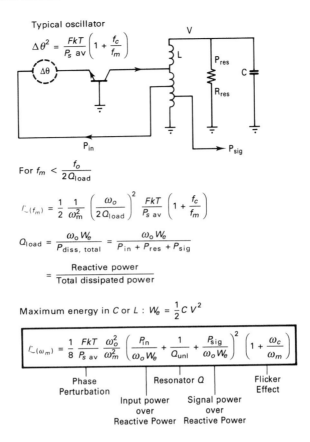

Figure 4-16 Diagram for a feedback oscillator illustrating the principles involved and showing the key components considered in the phase noise calculation and its contribution.

off the resonator rather than from an isolation amplifier, Eq. (2-29) can be rewritten in the form

$$\mathcal{L}(\omega_m) = \frac{1}{8} \frac{FkT}{P_{s\,\mathrm{av}}} \frac{\omega_0^2}{\omega_m^2} \left(\frac{P_{\mathrm{in}}}{\omega_0 W_e} + \frac{1}{Q_{\mathrm{unl}}} + \frac{P_{\mathrm{sig}}}{\omega_0 W_e} \right)^2 \left(1 + \frac{\omega_c}{\omega_m} \right) \quad (4\text{-}53)$$

- Phase perturbation
- Input power/Reactive power ratio
- Resonator Q
- Signal power/Reactive power ratio
- Flicker effect

or a more complete expression for a resonator oscillator's phase noise spectrum

$$s_\phi(f_m) = [\alpha_R F_0^4 + \alpha_E(F_0/(2Q_L))^2]/f_m^3$$
$$+ [(2GFKT/P_0)(F_0/(2Q_L))^2]/f_m^2$$
$$+ (2\alpha_R Q_L F_0^3)/f_m^2$$
$$+ \alpha_E/f_m + 2GFKT/P_0$$

where

G = compressed power gain of the loop amplifier
F = noise factor of the loop amplifier
K = Boltzmann's constant
T = temperature in °K
P_0 = carrier power level (in watts) at the output of the loop amplifier
F_0 = carrier frequency in Hz
f_m = carrier offset frequency in Hz
$Q_L(= \pi F_0 \tau_g)$ = loaded Q of the resonator in the feedback loop
α_R and α_E = flicker noise constants for the resonator and loop amplifier, respectively

In frequency synthesizers, we have no use for LC oscillators without a tuning diode, but it may still be of interest to analyze the low-noise fixed-tuned LC oscillator first and later make both elements, inductor and capacitor, variable.

Later, I will show the performance changes if we utilize the two possible ways of getting coarse and fine tuning in oscillators:

1. Use of tuning diodes.
2. Use of switching diodes.

We will spend some time looking at the effects that switching and tuning diodes have in a circuit because they will ultimately influence the noise performance more strongly than the transistor itself.

The reason is that the noise generated in tuning diodes will be superimposed on the noise generated in the circuit while switching diodes have losses that cause a reduction of circuit Q. The selection of the proper tuning and switching diodes is important, as is the proper way of connecting them. As both types are modifications of the basic LC oscillator, we start with the LC oscillator itself.

Signal generators as they are offered by several companies (e.g., Rohde & Schwarz, Hewlett-Packard, Boonton Electronics, or Marconi), if they are not synthesized, use an air-variable capacitor or, as in the case of one particular Hewlett-Packard generator, the Model 8640, a tuned cavity.

Tuning here is accomplished by changing the value of an air-variable capacitor or changing the mechanical lengths of a quarter-wave resonator.

Using the equations shown previously, it is fairly easy to calculate oscillators and understand how they work, but this does not necessarily optimize their design.

216 LOOP COMPONENTS

For crucial noise application, the oscillator shown in Figure 4-17, used in the Rohde & Schwarz SMDU, is currently the state of the art. Its noise performance is equivalent to the noise found in the cavity tuned oscillator made by Hewlett-Packard, and because of the unique way a tuning diode is coupled to the circuit, its modulation capabilities are substantially superior to any of the signal generators currently offered. To develop such a circuit from design equations is not possible. This circuit is a result of many years of experience and research and looks fairly simple. The grounded gate field-effect transistor circuit provides the best performance because it fulfills the important requirements of Eq. (4-53).

The tuned circuit is not connected directly to the drain, but the drain is put on a tap of the oscillator section. Therefore, the actual voltage across the tuning capacitor is higher than the supply voltage, and thus the energy stored in the capacitor is much higher than in a circuit connected between the gate electrode and ground, the normal Colpitts-type oscillator. In addition, the high output impedance of the field-effect transistor does not load the circuit, which also provides a reduced noise contribution. Since this oscillator is optimized for best frequency modulation performance in the FM frequency range, it becomes apparent that it fits the requirements of low-distortion stereo modulation.

For extremely critical locking measurements in the 2-m band ranging from 140 to 160 MHz, the noise specifications 20 kHz off the carrier are of highest importance, while the peak modulation typically does not exceed 5 kHz. Figure 4-17 shows the schematic of the oscillator section optimized for this frequency range.

It has been found experimentally that these *LC* oscillators should not be used above ~500 MHz; rather, a doubler stage should be employed. Analyzing the signal generators currently on the market, their highest base band ranges typically from 200 to 500 MHz using frequency doublers to 1000 MHz. As can be seen in Figure 4-18, the mechanical layout of such an oscillator is extremely compact.

The Hewlett-Packard equivalent to this approach has been the HP8640 cavity-based low phase noise signal generator. While it has been replaced by more modern synthesized approaches, it was an interesting mechanical approach to low phase noise. A coaxial resonant cavity oscillator was the heart of the famous HP8640 signal generator, which for many years dominated the market. As an active device, a low-noise transistor is operating in common-base configuration, biased by a current source (Figure 4-19). The cavity is less than 1/4 wavelength long so that the short at the bottom end causes the opposite end near the tuning plunger to appear inductive. The capacitance between the tapered tuning plunger and the center post resonates with resulting inductance at selected frequency. The varactor diodes are in series with the capacitance between the varactor end cap and plunger; therefore, the total capacitance is formed in parallel with the plunger capacitance. To avoid microphonics, the emitter and collector loops are embedded in plastic blocks. Therefore, the main possible source of microphonic effects is the cavity center.

4-1-3 Switchable/Tunable *LC* Oscillators

The VCO for a frequency synthesizer, in which the division ratio is larger than 100, is responsible for the noise performance outside the loop bandwidth of 1 to

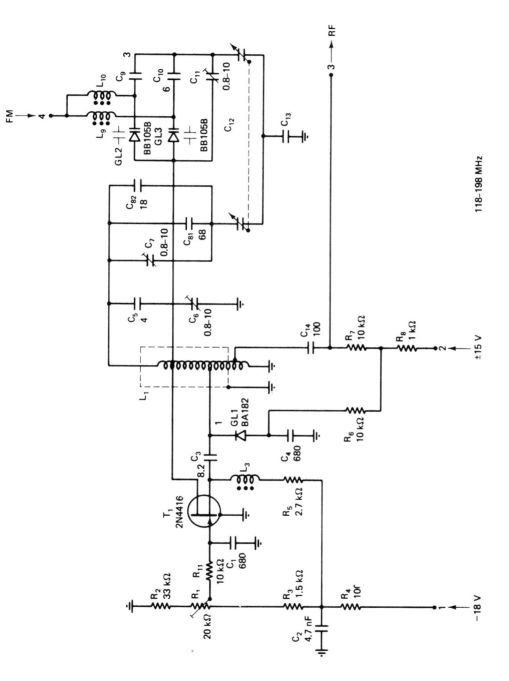

Figure 4-17 A 118–198 MHz oscillator from the Rohde & Schwarz SMDU signal generator.

Figure 4-18 Photograph of the helical resonator system from the Rohde & Schwarz SMDU signal generator.

10 kHz, while multiple-loop synthesizers using division ratios of less than 100 typically operate at extremely large bandwidths to increase the switching speed and frequency agility of the oscillator for critical applications such as frequency hopping and spread-spectrum techniques. Because of the wide loop bandwidth (>100 kHz) and the relatively small division ratio, the loop is able to clean up oscillator phase noise, making the oscillator noise performance be of second order. In a one-loop synthesizer using a wide-range diode tuned oscillator, the oscillator sensitivity K_0 at the lower end is much higher than at the higher end. This ratio can be as high as 1:10 and cause loop instabilities and pickup problems on the control line. It is therefore desirable to reduce the diode tuned range and coarse tune the oscillator by switching inductors or capacitors in parallel. It is somewhat questionable which approach is better. I have generally found that higher Q values are obtained if capacitors are switched in; as the capacitance is increased toward the low-frequency end the division ratio decreases, the higher voltage gain of the VCO is offset to a degree. One might call this "linearization of the tuning range."

Figure 4-20 shows a VCO using switchable capacitors of the Rohde & Schwarz EK070 shortwave receiver. This oscillator operates from 40 to 70 MHz and has a noise power of ∼145 dB/Hz at 25 to 30 kHz off the carrier. The loop bandwidth of the PLL using this oscillator is set to ∼3 kHz, which takes care of microphonics. To improve noise performance, AGC is used.

A different route was chosen by Hewlett-Packard in their 8662A synthesized oscillator. Figure 4-21 shows the schematic of this oscillator. The inductors shown in this circuit are in reality transmission lines (see Section A-5). According to

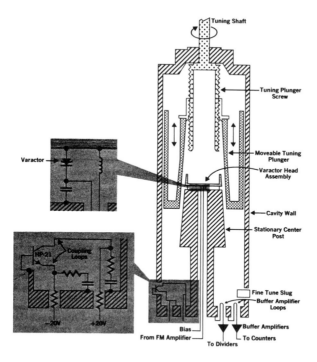

Figure 4-19 Mechanical drawing of a high-Q UHF cavity oscillator HP8640. (Reproduced with permission.)

Hewlett-Packard, this loop is also operated at several kilohertz bandwidth, and therefore the noise performance outside the loop bandwidth is determined by the losses of the oscillator.

Another method of coarse steering is the use of tuning diodes. This oscillator, while avoiding diode switching current, is somewhat noisier because of tuning diode noise.

Figure 4-22 shows the oscillator used in the Rohde & Schwarz ESH2/ESH3 receiver. The three oscillators cover a frequency range from 75 to 105 MHz in 10-MHz increments. A digital-to-analog converter combined with the synthesizer decoder generates the voltage to coarse set the frequency, while fine tune input is used for actual frequency locking.

A clever technique is used to select the different ranges, and an isolation power amplifier stage decouples the oscillator from the loop circuitry. Figure 4-23 shows the noise sideband performance of this oscillator section. It is interesting to compare this with the SSB noise for the Rohde & Schwarz SMDU oscillator, where different types of tuning diodes are applied. While the one line refers to a tuning diode of 10 kHz/V tuning sensitivity, the other curves refer to the noise performance for tuning diodes of higher sensitivity. Apparently, the noise performance already is degraded from 200 Hz off the carrier to as much as 10 MHz off the carrier. Above 10 MHz all curves meet.

Let us look next at some oscillator circuits used for less demanding applications. Figure 4-24 shows three oscillators that cover the range 225 to 480 MHz using the

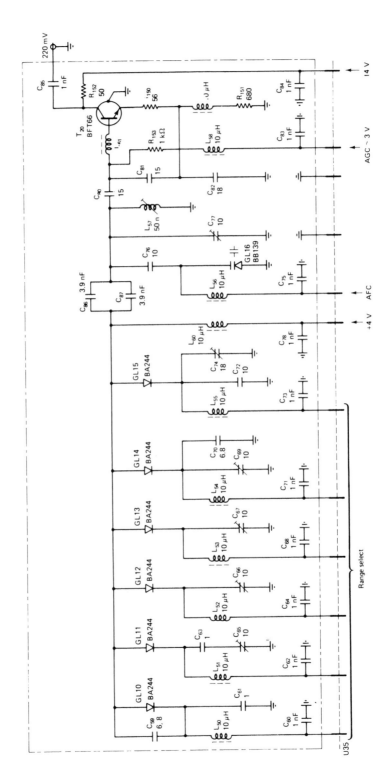

Figure 4-20 Schematic of the VCO from the Rohde & Schwarz EK070 receiver.

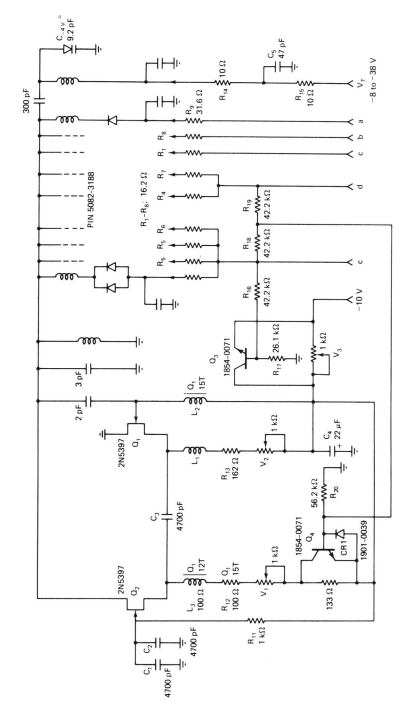

Figure 4-21 Schematic of the HP8662A VCO operating from 260 to 520 MHz.

222 LOOP COMPONENTS

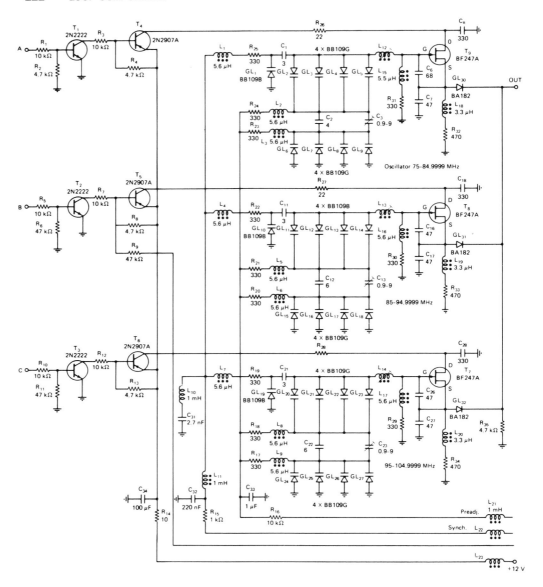

Figure 4-22 Oscillator and switching section of the Rohde & Schwarz ESH2/ESH3 test receiver.

low-noise Siemens transistor BFT66. The tuned circuit is loosely coupled to the oscillator transistor and output is taken from the collector in order to increase the isolation. In this particular case, no attempts were taken to reduce any harmonic contents, as these outputs have to drive digital dividers.

Finally, let us take a look at Figure 4-25, which shows a wideband oscillator operating in the 500-MHz range taken from the Rohde & Schwarz SMS signal generator and a wideband oscillator taken from the Rohde & Schwarz ESM500

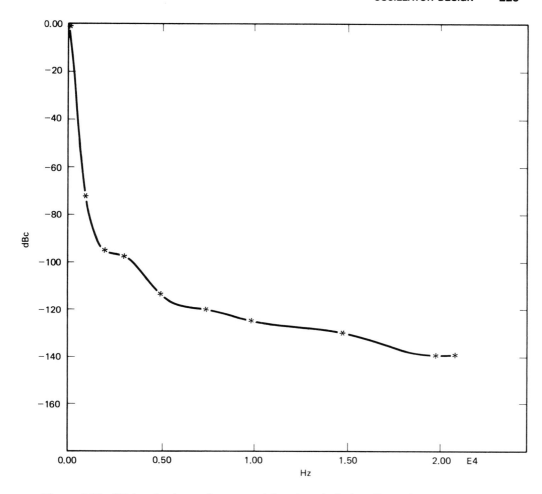

Figure 4-23 Sideband noise performance of the phase-locked oscillator shown in Figure 4-22.

receiver (Figure 4-26). The oscillator noise of the ESM500 is extremely small and is used with a loop bandwidth of about 10 Hz. Outside the 10-Hz bandwidth, this circuit by itself is responsible for the noise performance of the synthesizer, and data taken from blocking measurements indicate that the noise floor of 135 dB/Hz 25 kHz off the carrier is at least 20 dB better than previous circuit designs. The oscillator section shown in Figure 4-27 from the Rohde & Schwarz ESV receiver covering the same frequency range is based on the same principle and has a coarse and fine tuning input. This oscillator operates together with the HEF4750 and 4751 LOCMOS frequency divider and synthesizer ICs.

The selection and use of tuning diodes are extremely crucial, and in order to understand and distinguish between the various diodes offered on the market, the following section is devoted to understanding these principles.

224 LOOP COMPONENTS

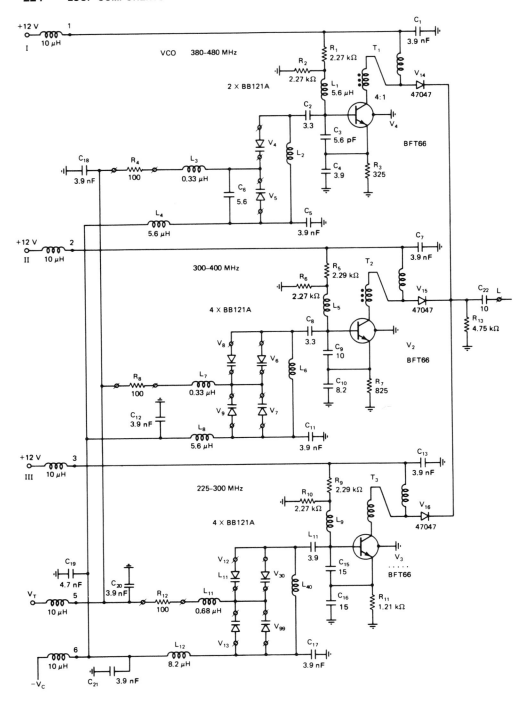

Figure 4-24 VCO schematic of a set of three oscillators covering the range from 225 to 480 MHz with bipolar transistors.

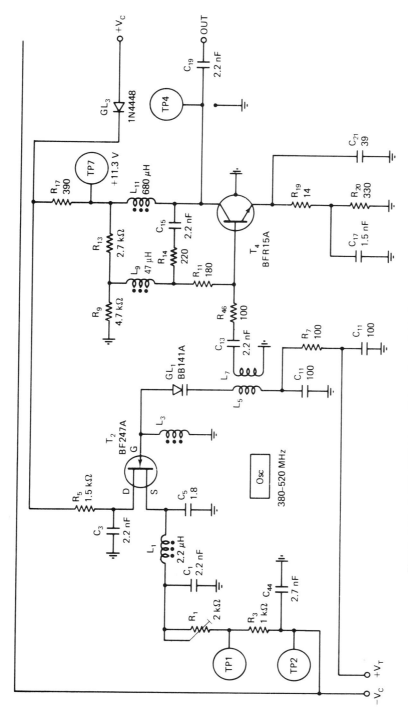

Figure 4-25 A 380–520 MHz oscillator using a field-effect transistor with decoupling stage.

226 LOOP COMPONENTS

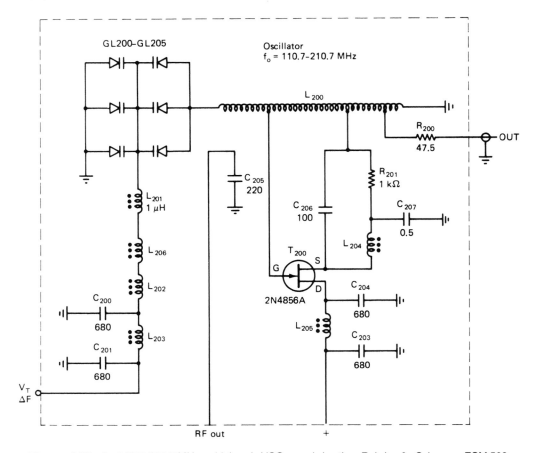

Figure 4-26 A 110.7–210.7 MHz wideband VCO used in the Rohde & Schwarz ESM 500 receiver.

4-1-4 Use of Tuning Diodes

In order to tune the oscillator within the required range, so-called tuning diodes are used. These diodes are often called varactors or voltage sensitive diodes. By way of approximation, we can use the equation

$$C = \frac{K}{(V_R + V_D)^n} \qquad (4\text{-}54)$$

wherein all constants and all parameters determined by the manufacturing process are contained in K. The exponent is a measure of the slope of the capacitance/voltage characteristic and is 0.5 for alloyed diodes, 0.33 for single diffused diodes, and (on average) 0.75 for tuner diodes with a hyperabrupt PN

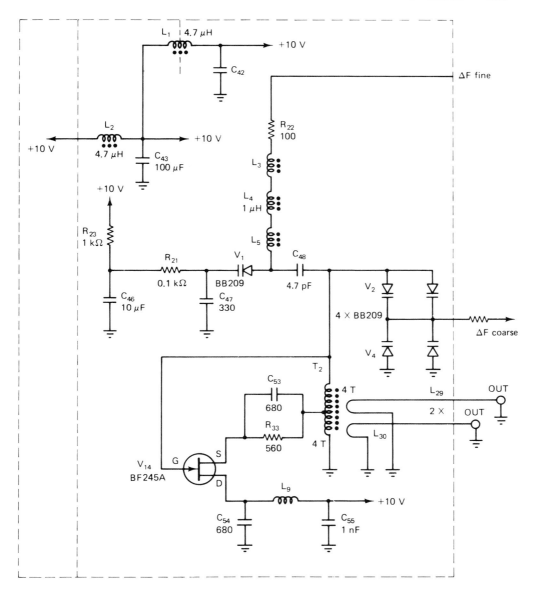

Figure 4-27 VCO operating from 160 to 260 MHz in the Rohde & Schwarz ESV receiver.

junction [7, 8]. Figure 4-28 shows the capacitance/voltage characteristics of an alloyed, a diffused, and a tuner diode.

Recently, an equation was developed that, although purely formal, describes the practical characteristic better than Eq. (4-54):

$$C = C_0 \left(\frac{A}{A + V_R} \right)^m \qquad (4\text{-}55)$$

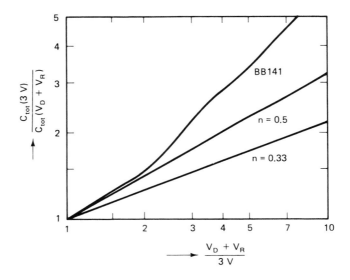

Figure 4-28 Capacitance/voltage characteristic for an alloyed capacitance diode ($n = 0.33$), a diffused capacitance diode ($n = 0.5$), and a wide-range tuner diode (BB141).

wherein C_0 is the capacitance at $V_R = 0$, and A is a constant whose dimension is a volt. The exponent m is much less dependent on voltage than the exponent n in Eq. (4-54).

The operating range of a capacitance diode or its useful capacitance ratio,

$$\frac{C_{\max}}{C_{\min}} = \frac{C_{\text{tot}}(V_{R\min})}{C_{\text{tot}}(V_{R\max})} \qquad (4\text{-}56)$$

is limited by the fact that the diode must not be driven by the alternating voltage superimposed on the tuning voltage either into the forward mode or the breakdown mode. Otherwise, rectification would take place, which would shift the bias of the diode and considerably affect its figure of merit.

There are several manufacturers of tuning diodes. Motorola is a typical supplier in this country; Siemens of Germany and Philips also provide good diodes. The following table contains information for three typical tuning diodes as they might be considered useful for our applications.

Capacitance	MV209	BB105	MVAM125
At $V_R = 1$ V, C_{tot}	40 pF	18 pF	500 pF
At $V_R = 25$ V, C_{tot}	6 pF	2 pF	33 pF
Useful capacitance ratio, $C_{\text{tot}}(1\text{ V})/C_{\text{tot}}(25\text{ V})$	6	9	15

Diode Tuned Resonant Circuits

Tuner Diode in Parallel Resonant Circuit. Figures 4-29, 4-30, and 4-31 illustrate three basic circuits for the tuning of parallel resonant circuits by means of

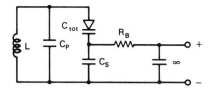

Figure 4-29 Parallel resonant circuit with tuner diode, and bias resistor parallel to the series capacitor.

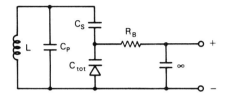

Figure 4-30 Parallel resonant circuit with tuner diode, and bias resistor parallel to the diode.

capacitance diodes. In the circuit diagram of Figure 4-29, the tuning voltage is applied to the tuner diode via the tank coil and the bias resistor R_B. Series-connected to the tuner diode is the series capacitor C_S, which completes the circuit for the alternating current but isolates the cathode of the tuner diode from the coil and thus from the negative terminal of the tuning voltage. Moreover, a fixed parallel capacitance C_P is provided. The decoupling capacitor preceding the bias resistor is large enough to be disregarded in the following discussion. Since for high-frequency purposes the biasing resistor is connected in parallel with the series capacitor, it is transformed into the circuit as an additional equivalent shunt resistance R_c. We have the equation

$$R_c = R_B \left(1 + \frac{C_S}{C_{\text{tot}}}\right)^2 \tag{4-57}$$

If in this equation the diode capacitance is replaced by the resonant circuit frequency ω, we obtain

$$R_c = R_B \left(\frac{\omega^2 L C_S}{1 - \omega^2 L C_P}\right)^2 \tag{4-58}$$

The resistive loss R_c, caused by the bias resistor R_B, is seen to be highly frequency dependent, and this may result in the bandwidth of the tuned circuit being dependent on frequency if the capacitance of the series capacitor C_S is not chosen sufficiently high.

Figure 4-30 shows that the tuning voltage can also be applied directly and in parallel to the tuner diode. For the parallel loss resistance transformed into the circuit, we have the expression

$$R_c = R_B \left(1 + \frac{C_{\text{tot}}}{C_S}\right)^2 \tag{4-59}$$

230 LOOP COMPONENTS

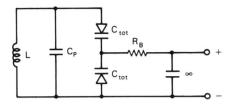

Figure 4-31 Parallel resonant circuit with two tuner diodes.

and

$$R_c = R_B \left[\frac{\omega^2 L C_S}{\omega^2 L (C_S + C_P) - 1} \right]^2 \tag{4-60}$$

The influence of the bias resistor R_B in this case is larger than in the circuit of Figure 4-29, provided that

$$C_S^2 > C_S(C_{\text{tot}} + C_P) + C_{\text{tot}} C_P$$

This is usually the case because the largest possible capacitance will be preferred for the series capacitor C_S, and the smallest for the shunt capacitance C_P. The circuit of Figure 4-29 is therefore normally preferred to that of Figure 4-30. An exception would be the case in which the resonant circuit is meant to be additionally damped by means of the bias resistor at higher frequencies.

In the circuit of Figure 4-31, the resonant circuit is tuned by two tuner diodes, which are connected in parallel via the coil for tuning purposes, but series-connected in opposition for high-frequency signals. This arrangement has the advantage that the capacitance shift caused by the ac modulation takes effect in opposite directions in these diodes and therefore cancels itself. The bias resistor R_B, which applies the tuning voltage to the tuner diodes, is transformed into the circuit at a constant ratio throughout the whole tuning range. Given two identical, loss-free tuner diodes, we obtain the expression

$$R_c = 4R_B \tag{4-61}$$

Capacitances Connected in Parallel or in Series with the Tuner Diode. Figures 4-29 and 4-30 show that a capacitor is usually in series with the tuner diode, in order to close the circuit for alternating current and, at the same time, to isolate one terminal of the tuner diode from the rest of the circuit with respect to direct current, so as to enable the tuning voltage to be applied to the diode. If possible, the value of the series capacitor C_S will be chosen such that the effective capacitance variation is not restricted. However, in some cases, for example, in the oscillator circuit of receivers whose intermediate frequency is of the order of magnitude of the reception frequency, this is not possible and the influence of the series capacitance will then have to be taken into account. By connecting the capacitor C_S, assumed to be loss-free, in series with the diode capacitance C_{tot}, the tuning capacitance is reduced to the value

$$C^* = C_{tot} \frac{1}{1 + C_{tot}/C_S} \qquad (4\text{-}62)$$

The Q-factor of the effective tuning capacitance, taking into account the Q-factor of the tuner diode, increases to

$$Q^* = Q\left(1 + \frac{C_{tot}}{C_S}\right) \qquad (4\text{-}63)$$

The useful capacitance ratio is reduced to the value

$$\frac{C^*_{max}}{C^*_{min}} = \frac{C_{max}}{C_{min}} \frac{1 + C_{min}/C_S}{1 + C_{max}/C_S} \qquad (4\text{-}64)$$

wherein C_{max} and C_{min} are the maximum and minimum capacitances of the tuner diode.

On the other hand, the advantage is gained that, due to capacitive potential division, the amplitude of the alternating voltage applied to the tuner diode is reduced to

$$\hat{v}^* = \hat{v} \frac{1}{1 + C_{tot}/C} \qquad (4\text{-}65)$$

so that the lower value of the tuning voltage can be smaller, and this results in a higher maximum capacitance C_{max} of the tuner diode and a higher useful capacitance ratio. The influence exerted by the series capacitor, then, can actually be kept lower than Eq. (4-63) would suggest.

The parallel capacitance C_P that appears in Figures 4-29 to 4-31 is always present, since wiring capacitances are inevitable and every coil has its self-capacitance. By treating the capacitance C_P, assumed to be loss-free, as a shunt capacitance, the total tuning capacitance rises in value and, if C_S is assumed to be large enough to be disregarded, we obtain

$$C^* = C_{tot}\left(1 + \frac{C_P}{C_{tot}}\right) \qquad (4\text{-}66)$$

The Q-factor of the effective tuning capacitance, derived from the Q-factor of the tuner diode, is

$$Q^* = Q\left(1 + \frac{C_P}{C_{tot}}\right) \qquad (4\text{-}67)$$

232 LOOP COMPONENTS

or, in other words, it rises with the magnitude of the parallel capacitance. The useful capacitance ratio is reduced:

$$\frac{C_{max}^*}{C_{min}^*} = \frac{C_{max}}{C_{min}} \frac{1 + C_P/C_{max}}{1 + C_P/C_{min}} \qquad (4\text{-}68)$$

In view of the fact that even a comparatively small shunt capacitance reduces the capacitance ratio considerably, it is necessary to ensure low wiring and coil capacitances in the layout stage.

Tuning Range. The frequency range over which a parallel resonant circuit (according to Figure 4-29) can be tuned by means of the tuner diode depends on the useful capacitance ratio of the diode and on the parallel and series capacitances present in the circuit. The ratio is

$$\frac{f_{max}}{f_{min}} = \sqrt{\frac{1 + \dfrac{C_{max}}{C_P(1 + C_{max}/C_S)}}{1 + \dfrac{C_{max}}{C_P(C_{max}/C_{min} + C_{max}/C_S)}}} \qquad (4\text{-}69)$$

In many cases the series capacitor can be chosen large enough for its effect to be negligible. In that case, Eq. (4-69) is simplified as follows:

$$\frac{f_{max}}{f_{min}} = \sqrt{\frac{1 + C_{max}/C_P}{1 + C_{min}/C_P}} \qquad (4\text{-}70)$$

From this equation, the diagram shown in Figure 4-32 is computed. With the aid of this diagram the tuning diode parameters required for tuning a resonant circuit over a stipulated frequency range (i.e., the maximum capacitance and the capacitance ratio) can be determined. Whenever the series capacitance C_S cannot be disregarded, the effective capacitance ratio is reduced according to Eq. (4-64).

Tracking. When several tuned circuits are used on the same frequency, diodes have to be selected for perfect tracking.

Practical Circuits. After so much theory, it may be nice to take a look at some practical circuits, such as the one shown in Figure 4-22. This oscillator is being used in the Rohde & Schwarz ESN/ESVN40 field strength meter and in the HF1030 receiver produced by Cubic Communications, San Diego. This circuit combines all the various techniques shown previously. A single diode is being used for fine-tuning a narrow range of less than 1 MHz; coarse tuning is achieved with the antiparallel diodes.

Several unusual properties of this circuit are apparent:

1. The fine tuning is achieved with a tuning diode that has a much larger capacitance than that of the coupling capacitor to the circuit. The advantage of this technique is that the fixed capacitor and the tuning diode form a voltage divider whereby the voltage across the tuning diode decreases as the

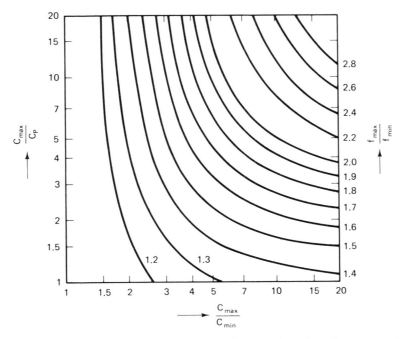

Figure 4-32 Diagram for determining the capacitance ratio and maximum capacitance.

capacitance increases. For larger values of the capacitance of the tuning diode, the Q changes and the gain K_0 increases. Because of the voltage division, the noise contribution and loading effect of the diode are reduced.

2. In the coarse-tuning circuit, several tuning diodes are used in parallel. The advantage of this circuit is a change in LC ratio by using a higher C and storing more energy in the tuned circuit. There are no high-Q diodes available with such large capacitance values, and therefore preference is given to using several diodes in parallel rather than one tuning diode with a large capacitance, normally used only for AM tuner circuits.

I have mentioned previously that, despire this, the coarse-tuning circuit will introduce noise outside the loop bandwidth, where it cannot be corrected. It is therefore preferable to incorporate switching diodes for segmenting ranges at the expense of switching current drain.

Figure 4-33 shows a circuit using a combined technique of tuning diodes for fine- and medium-resolution tuning and coarse tuning with switching diodes. The physics and technique of using switching diodes are explained in the next section.

4-1-5 Use of Diode Switches

The diode switches described here differ somewhat from the switching diodes used in computer and pulse technology. In normal diodes, the signal itself triggers the

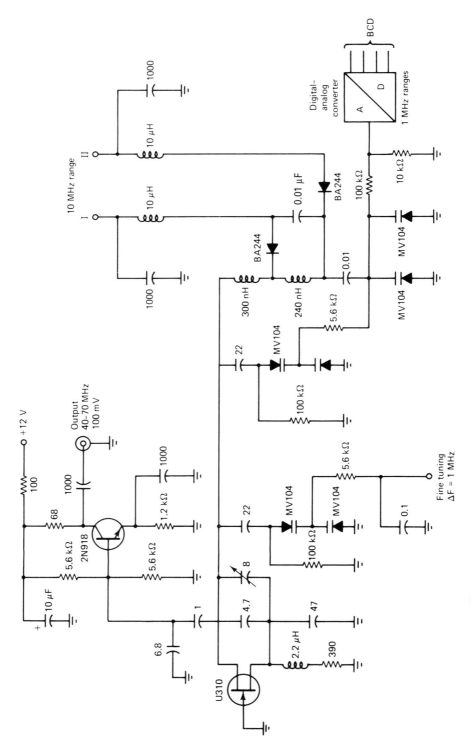

Figure 4-33 A 40–70 MHz VCO with two coarse-steering ranges and fine-tuning range of 1 MHz.

switching operation—current does or does not pass through the switching diode in dependence on the signal level. Diode switches allow an alternating current to be switched on or off by application of a direct voltage or a direct current. The diode switches BA243, BA244, and the later version BA238 by ITT or the Motorola MPN 3401 series were developed especially for such a purpose and constitute the present state of the art. However, diodes can also be employed to advantage for switching audio signals (i.e., in tape recorders or amplifiers).

Diode Switches for Electronic Band Selection. The advantages of the electronic tuning of VHF–UHF circuits become fully effective only when band selection also takes place electronically and no longer by means of mechanically operated switching contacts subject to wear and contamination. Figure 4-34 shows an example of the use of diode switches.

Diode switches are preferable to mechanical switches because of their higher reliability and virtually unlimited life. Since the diode switches BA243 and BA244 permit range switching without mechanical contacts, and since they can be controlled in a similar way as capacitance diodes by the application of a direct current, there should be many new applications for these devices in remote control receivers. Their use obviates the need for mechanical links between the front-panel control and the tuned circuit to be switched, allowing a VHF–UHF circuit to be located in the most favorable place with regard to electrical or thermal influence, giving the designer more freedom in front-panel styling. Moreover, because the

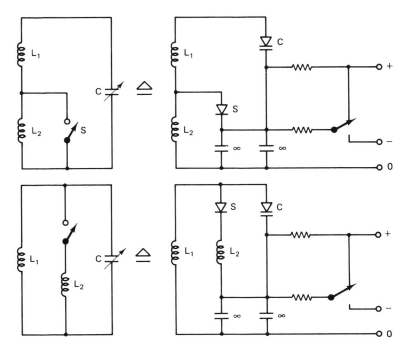

Figure 4-34 Comparison of mechanically and electronically tuned and switched resonant circuits.

tuner is no longer subject to mechanical stress, its chassis may be injection-molded from a plastic material, which can be plated for screening purposes. All this makes for small, more compact tuners and results in considerable savings in production.

Let us take a look at three oscillators that are designed around switching diodes. Figure 4-20 shows an oscillator that is used in the Rohde & Schwarz EK070 shortwave receiver and that presets the value of the frequency within a few hundred kHz off the final frequency. The fine tuning then is accomplished by the use of varactor diodes or tuning diodes.

The advantage of using one oscillator for the entire frequency range lies in the fact that the switching speed is not slowed down by the settling time of an oscillator circuit being activated and showing the familar initial drift phenomena.

The gain of the oscillator K_o now changes due to the parallel capacitance switched into the tuning diodes, and the loop therefore requires some gain adjustments. This is further discussed in Section 4-7.

While the above circuit uses external AGC, which is frequently used with bipolar transistors, Figure 4-35 shows a similar circuit operating from 42 to 72 MHz using field-effect transistors and switching diodes.

Those previously shown circuits switch capacitors rather than inductors. Figure 4-21 shows a circuit that is being used in the HP signal generator type 8962. High-Q inductors are being switched in and out rather than capacitors, thereby avoiding the gain variation of the oscillator to a large degree.

This oscillator also uses a differential amplifier feedback circuit, and the advantage of this circuit is that the signal-to-noise ratio is further improved. Details on differential limiter low-noise design can be found in the literature [35].

4-1-6 Use of Diodes for Frequency Multiplication

So far we have been dealing with free-running oscillators, which are being locked to a reference with the PLL. I mentioned earlier that some synthesizer simplification is possible when using a heterodyne technique. The auxiliary frequency for this heterodyne action can be obtained from the frequency standard by multiplication. There are a number of ways in which to obtain harmonic outputs, and probably the best one is the highest frequency of operation and the use of special diodes, such as step recovery diodes or snap-off diodes. This application would lead us into microwave techniques, which are beyond the scope of this book. Figure 4-36 shows a schematic of a 100- to 1700-MHz frequency multiplier. More information about frequency multiplication is found in the references at the end of this section.

4-2 REFERENCE FREQUENCY STANDARDS

4-2-1 Requirements

Frequency standards are the heart of the synthesizer, as they control the accuracy of the frequency (if we are dealing with a coherent synthesizer) and, within the loop bandwidth, the noise sideband performance of the synthesizer.

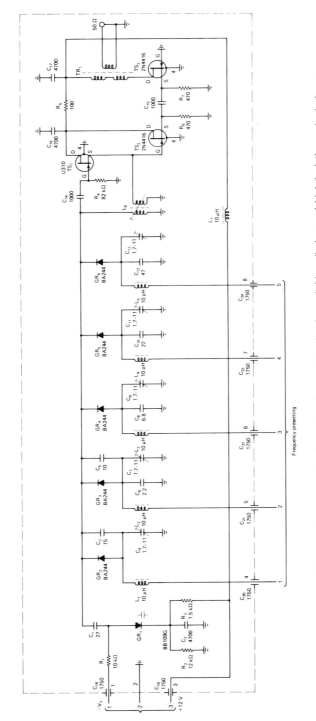

Figure 4-35 VCO operating from 42 to 72 MHz with coarse tuning by switching diodes and high-isolation output stage.

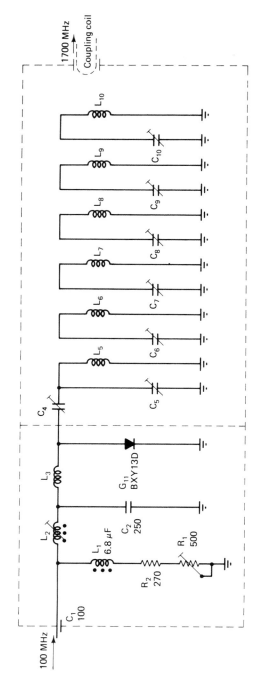

Figure 4-36 A 100–170 MHz frequency multiplier.

REFERENCE FREQUENCY STANDARDS

There are several frequency standards available, and basically they can be put into three categories:

1. Cesium frequency standards.
2. Rubidium frequency standards.
3. Crystal oscillators.

While it must be noted that at least one manufacturer in the United States is selling a commercial hydrogen maser [1], the maser has not achieved widespread usage. Cesium frequency standards are used as primary standards where extremely accurate stability is needed for long periods of time. Crystal oscillators are currently the most popular choice for reference standards in frequency synthesizers. The short-term stability and noise floor of the crystal oscillator are typically equal to or better than the rubidium standard, and considerably better than the cesium standard. Depending on a number of factors, such as price, performance, size, and power, we find frequency standards ranging from simple crystal oscillators to very high-stability double-oven crystal oscillators to ultrastable cesium beam standards.

4-2-2 Specifying Oscillators

Although it is fairly simple to design a crystal oscillator that has moderate stability, it is not a trivial task to design a high-stability oscillator with low noise. Consequently, it is generally more economical to purchase oscillators for use in synthesizers. This section provides guidelines for specifying crystal oscillators. Specifying the proper parameters (and, perhaps as important from a cost standpoint, *not* specifying unnecessary parameters) is an important matter. Developing a proper specification is most important when specifying custom oscillators. The following can be considered as a guideline when specifying oscillators. Generally, not every parameter listed below needs to be, or should be, specified. Rather, the parameters can be used as a checklist when developing a specification.

1.0 *Scope.* This is a general description of the type of oscillator, whether it be a TCXO, OCXO, DCXO, rubidium, or cesium.
2.0 *Reference Documents.* Any reference documents such as test methods or military specifications are listed here. Such examples might be MIL-0-55310 [2] (which is an excellent reference for specifying crystal oscillators).
3.0 *Electrical Requirements.*
3.1 *Nominal Frequency.*
3.2 *Frequency Stability.* This is considered over operating temperature range.
3.3 *Aging.* This should be specified on a per day, per month, or per year basis. Depending on the test method, it is possible to obtain greatly different aging results for the same oscillator.
3.4 *Power Supply Voltages and Tolerances.*
3.5 *Frequency Change.* This results from power supply variations.
3.6 *Power Supply Currents or Power.* Note that the peak heater current as well as steady-state currents should be specified.

240 LOOP COMPONENTS

3.7 *Output Power and Signal Type.* If a sine wave is desired, the power into 50 Ω is normally specified. If digital output is desired, the type (TTL, CMOS, etc.), duty cycle, and fanout should be specified.

3.8 *Load Stability.* This is normally specified as the amount of frequency change allowed when the load VSWR is varied over some range, say, 2:1.

3.9 *Phase Noise and/or Allan Variance.* If the application requires operation during vibration, the phase noise in vibration should be specified as well.

3.10 *Electrical or Mechanical Tuning Range.* Sufficient tuning must be allowed to accommodate for anticipated frequency aging.

3.11 *Acceleration Sensitivity.* This is a specification of the frequency change due to acceleration. It is often specified in terms of a two-G tipover test, where the oscillator is physically turned over and the resulting frequency change measured. The units are most commonly specified in terms of parts per G. Acceleration sensitivity is discussed in more detail in Section 4-2-4.

3.12 *Magnetic Field Sensitivity.* The sensitivity to magnetic fields can be specified in terms of spurious signals induced by a time-varying field.

3.13 *Radiation Hardness.* Since the frequency of an oscillator may change significantly upon exposure to radiation, it is important to specify this parameter for applications where exposure to radiation may be encountered, such as a space environment.

4.0 *Environmental Requirements.* This section includes factors such as operating temperature range, shock, vibration, and any other factors as necessary.

5.0 *Quality Assurance Requirements.* Any special screening, parts, or other quality assurance requirements should be specified. Since these requirements can add considerable cost to the unit, proper specification of these requirements is essential to maintaining reasonable costs with good performance.

4-2-3 Typical Examples of Crystal Oscillator Specifications

As in all design work, the optimal crystal oscillator design depends on the particular application. For instance, there are trade-offs that can be made between long-term aging and short-term stability, and between the noise floor and close-in noise.

Table 4-1 shows the phase noise for various frequency sources, while Table 4-2 shows typical specifications for an ovenized crystal oscillator of moderately high stability. This oscillator does not represent the state of the art but is representative for good-quality commercial oscillators. It employs a third overtone AT crystal in a Colpitts oscillator. The entire assembly is hermetically sealed. Table 4-3 shows typical specifications for a precision analog TCXO. Note that the floor of the TCXO is much better, but close-in noise performance is not as good. In general, the floor of TCXOs can be almost as good as high-stability ovenized oscillators, but the close-in phase noise and short-term stability are very poor relative to an oven. This is because the need to pull the oscillator frequency of a TCXO mandates a low resonator Q relative to an ovenized oscillator.

REFERENCE FREQUENCY STANDARDS

Table 4-1 Frequency stability of various frequency standards (Rohde & Schwarz)

Offset from Signal f (Hz)	Cesium XSC	Rubidium XSRM	Quartz XSD2
10^{-2}	−30	−62	
10^{-1}	−50	−80	
10^{0}	−85	−105	−90
10^{1}	−125	−132	−120
10^{2}	−140	−140	−140
10^{3}	−144	−145	−157
10^{4}	−150	−150	−160

Table 4-2 Partial specification for a moderately high-performance commercial-grade oscillator (Ovenaire 49-5-2)

Center frequency	10 MHz
Aging	$\pm(5\cdot 10)^{-10}$/day
Allan variance (1 s)	$3\cdot 10^{-11}$
Temperature range	0 to +60 °C
Ambient stability	$\pm 3\cdot 10^{-9}$
Supply voltage	+20 Vdc, ±5%
Warm-up power	9 W
Warm-up time	20 min
Steady-state power at 25 °C	2.2 W
Voltage stability	$\pm(5\cdot 10)^{-10}$
Phase noise at offset of	
10 Hz	−120 dBc
100 Hz	−140 dBc
1000 Hz	−145 dBc
10000 Hz	−145 dBc

Table 4-3 Typical specifications for a moderately high-performance commercial-grade TCXO (McCoy Electronics)

Center frequency	10 MHz
Temperature range	−45 to +85 °C
Ambient stability	±0.8 ppm
Supply voltage	+12 Vdc
Supply power	200 mW max
Output	+7 dBm into 50
Aging	0.3 ppm/year max
Phase noise at offset of	
10 Hz	−90 dBc
100 Hz	−120 dBc
1000 Hz	−145 dBc
10000 Hz	−160 dBc

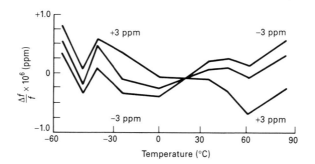

Figure 4-37 Effect of varying frequency adjustment on ambient stability of a typical TCXO. The frequency scale is normalized so that all curves intersect at +25 °C.

The TCXO uses a 3.333-MHz fundamental crystal that is multiplied to the output frequency. This approach produces better aging and compensation. A 10-MHz monolithic crystal filter is used to reduce subharmonics to -70 dBc and lower the noise floor to less than -160 dBc. This is achieved with a crystal power dissipation of approximately 50 μW.

By using digital compensation rather than analog compensation, the ambient stability can be reduced to less than ± 0.1 ppm [3, 4]. If the frequency is later adjusted for aging, however, the compensation may be altered [5, 6]. Considering these and other subtle effects such as hysteresis in the frequency/temperature characteristics of the crystal, it is very difficult to make a digitally compensated oscillator that is stable and repeatable to better than ± 0.05 ppm under all conditions. CXOs also often have a noise spur in the output spectrum at the clock rate of the temperature correction circuitry, although the latest techniques of digital compensation currently being developed may eliminate this problem [7].

If it is expected that the frequency will be adjusted to correct for aging, the customer would be wise to measure the frequency stability over temperature at extremes of the frequency adjust trimmer, and while varying temperature from cold to hot and hot to cold. It is generally found that the compensation varies considerably [6]. Figure 4-37 shows the frequency stability of a typical TCXO as the frequency is varied ± 3 ppm to compensate for aging. It must be noted though that the new compensation techniques mentioned above may allow DCXOs consistently to maintain a stability of around ± 0.05 ppm [7], while maintaining good short-term stability.

4-2-4 Crystal Resonators

The crystal is the most critical element in determining oscillator performance and the proper selection of crystal parameters is essential. Most oscillators at this time employ resonators cut from AT quartz, but double rotated cuts such as the SC and IT are becoming increasingly popular. These cuts are considerably more expensive but offer significant advantages when employed in ovenized oscillators. The significant advantages of crystal resonators are that the equivalent electrical circuit can have a Q of over 1 million, and that the electrical characteristics are very stable with respect to variations in temperature and time.

REFERENCE FREQUENCY STANDARDS

Equivalent Circuit. Figure 4-38 shows the equivalent circuit of a crystal resonator and enclosure that is valid around a resonance [8, 9]. There are also capacitances between the resonator and the package, and lead inductance of the package. However, they can generally be neglected for frequencies up to VHF.

This circuit has a low impedance series resonant frequency given by

$$f_s = \frac{1}{2\pi\sqrt{L_1 C_1}} \qquad (4\text{-}71)$$

and an antiresonant or parallel resonant frequency given by

$$f_p = \frac{1}{2\pi\sqrt{L_1\left(\dfrac{C_1 C_0}{C_1 + C_0}\right)}} \qquad (4\text{-}72)$$

Note that if additional capacitance is placed across the crystal, only the parallel resonant frequency is affected.

The difference between series and parallel resonance can be expressed as

$$\frac{\Delta f}{f} = \frac{C_1}{2(C_0 + C_L)} \qquad (4\text{-}73)$$

where C_L is the additional capacitance placed in parallel with the crystal. If we multiply the result of Eq. (4-73) by 10^6, the result will be expressed in parts per million (ppm). Alternatively, we could simply enter f in MHz.

By differentiating Eq. (4-73) with respect to C_L, another important relationship can be obtained:

$$\frac{\partial(\Delta f/f) \times 10^6}{\partial C_L} = \frac{-C_1 \times 10^6}{2(C_0 + C_L)} \quad \text{ppm/pF} \qquad (4\text{-}74)$$

Adding the 10^6 factor expresses the result in ppm/pF. Expressing the result in this manner is generally more convenient when using the crystal in an oscillator, because we can quickly obtain the frequency change resulting from a capacitance change by multiplying the results of Eq. (4-74).

It is important to note that when a capacitor is placed in series with a crystal, the new series resonant frequency is displaced slightly to a higher frequency than

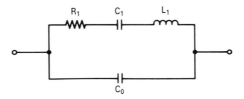

Figure 4-38 Equivalent circuit of a crystal resonator.

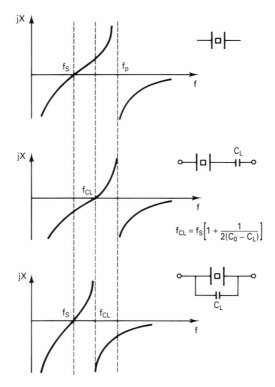

Figure 4-39 Crystals with load capacitances.

the original series resonance. Conversely, when the capacitor is placed in parallel with the resonator, the new parallel resonant frequency is displaced slightly to a lower frequency. Indeed, the new series resonant frequency is displaced to almost exactly the same frequency as the parallel resonant frequency would be, had the capacitor been placed in parallel with the crystal.

Unfortunately, this simple phenomenon has caused a great deal of confusion between the terms series and parallel resonance. Since the equations express either the change in the series resonance or the change in the parallel resonance, we can see that eqs. (4-73) and (4-74) may be used to calculate either the series resonant frequency or parallel resonant frequency, depending on the circuit configuration. This is expressed graphically in Figure 4-39.

Note that since it is generally easier to measure the low impedance series resonant frequency, C_1 can easily be determined from the change in series frequency resulting from a change in the series capacitance. By measuring the series resonance of the crystal and then measuring the new series resonance with a load capacitor in series with the crystal, C_1 can be calculated. By solving Eq. (4-73) for C_1 and inserting the frequency change, we obtain

$$C_1 = \frac{2\Delta f(C_0 + C_L)}{f \cdot 10^6} \tag{4-75}$$

An example will clarify the use of this equation. Suppose that we had measured the plate capacitance, C_0, of a 10-MHz resonator to be 3.0 pF. The crystal series resonant frequency was found to be 9.999900 MHz, and the series resonant frequency of the crystal and series capacitor of 32 pF was measured at 10.000002 MHz. Solving for C_1 gives

$$C_1 = \frac{2(10.000002 - 9.999900)(3 + 32)}{9.9999} = 0.0007 \text{ pF}$$

If the capacitor had been placed in parallel with the crystal, the series resonant frequency would still be 9.9999 MHz, but the parallel resonance would occur at 10.000002 MHz. The advantage of measuring at series resonance is that the low impedance at series, relative to parallel, makes the measurement much less dependent on stray capacitance.

The sensitivity of this circuit to changes in load capacitance can be calculated from Eq. (4-74) to be

$$-\frac{0.0007 \times 10^6}{2 \cdot 35^2} = -0.29 \text{ ppm/pF}$$

Since 2 Hz is 0.2 ppm at 10 MHz (ppm × MHz = Hz), the capacitance needed to set this crystal to exactly 10 MHz can be found by approximating

$$\frac{\partial(\Delta f/f) \times 10^6}{\partial C_L} \approx \frac{\Delta f \text{ (in ppm)}}{\Delta C_L} = -0.29 \text{ ppm/pF}$$

to get

$$\Delta C_L \approx \frac{\text{pF}}{-0.29 \text{ ppm}} \times 0.2 \text{ ppm} = -0.69 \text{ pF}$$

This small change can easily be accommodated by using a trimmer capacitor or varactor diode as part of the load capacitance. Since manufacturers cannot set the frequency of a crystal exactly, it is common to specify either the range of capacitance needed to set the crystal to a certain frequency, or the tolerance on frequency with a fixed value of load capacitance. If the crystal is specified as series resonant, the frequency tolerance should, of course, be specified.

Another important equation allows us to compute the real portion of the impedance as a function of load capacitance. The real portion is given by

$$\text{Re} = R_1 \left(\frac{C_0 + C_L}{C_L} \right) \qquad (4\text{-}76)$$

Note that Re increases as C_L decreases. For the crystal considered above, if the series resistance of the crystal were 50 Ω, the effective resistance in circuit would be approximately 59.8 Ω.

The increased loss in the circuit limits the upper tuning range in parallel resonant

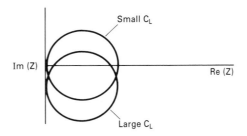

Figure 4-40 Locus of crystal impedance with increasing parallel capacitance.

oscillators. Eventually, the resistance increases so much that the sustaining amplifier no longer has enough gain to maintain oscillations. The lower frequency tuning limit is determined by how close one can bring the pole to the circuit zero and still realize a resonance. As the shunt load capacitance is increased beyond a critical value, the circuit never reaches resonance and the total circuit impedance stays capacitive. This is illustrated in Figure 4-40.

As the capacitance across the crystal is increased, the locus moves downward until at some point the circle is tangent to the real axis. The value of load capacitance at which this happens is given by

$$C_L = CQ/2 \qquad (4\text{-}77)$$

At high frequencies, the plate capacitance C_0 may be large enough to cause problems, in that the reactance of C_0 may be so small that oscillations will run "through" C_0. In this instance, an inductor may be placed across the crystal to tune out the shunt capacitance. This technique is recommended only for series resonant oscillators, however, as other spurious oscillations can occur "through" the shunt inductor in parallel resonant oscillators. Oscillations can occur in these circuits when either the crystal or some other path appears inductive. Pierce and Colpitts oscillators are particularly susceptible to spurious oscillations when one attempts to tune out C_0 with a parallel inductance. Tuning out C_0 is not recommended for these circuits.

Overtone Crystals. Due to the acoustic wavelength, crystals have a response at odd integer multiples of the fundamental resonant frequency. These responses are termed "overtones". By including a resonant branch for each overtone, the equivalent circuit can be expanded as shown in Figure 4-41.

The motional C_1 of each response will decrease by the square of the overtone. Thus, the C_1 of the third overtone will be 1/9 that of the fundamental C_1, and the fifth overtone C_1 will be 1/25 of the fundamental. Since C_1 is directly proportional to the "pullability" of the resonator, operation on the overtone greatly reduces the circuit sensitivity to reactance changes. Depending on the application, this may or may not be desirable.

Because of subtle differences in the response characteristics, if the resonator is to be operated on the overtone, the crystal should be specified as such. Crystal

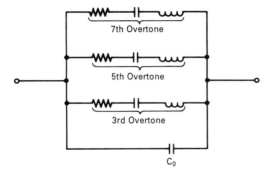

Figure 4-41 Equivalent circuit of a quarter resonator including overtone responses. Each resonant branch corresponds to a particular overtone.

manufacturers can then design and test for the critical characteristics on the desired overtone. Neglecting to do this can lead to unanticipated but serious consequences. For example, the frequency versus temperature characteristics will be different for different overtones. Other parameters, such as resistance and spurious responses, may also be considerably different from what may have been anticipated. By specifying the crystal correctly, one can avert these problems.

Temperature Behavior of Quartz Resonators. The resonant frequency of crystal resonators can be modeled by a third-order polynomial of the form

$$\frac{\Delta f}{f} = a(T - T_0) + b(T - T_0)^2 + c(T - T_0)^3 \tag{4-78}$$

where T is the inflection temperature [10]. Note that the inflection temperature is the temperature at which the frequency varies linearly with temperature. The constants a, b, and c as well as the inflection temperature vary greatly with the crystalline orientation of the resonator.

Figure 4-42 graphically shows the crystal upper and lower turning points along with the inflection point. To minimize frequency changes resulting from small changes in the operating temperature of the oven in ovenized oscillators, the temperature of the oven is adjusted to the turning point of the crystal.

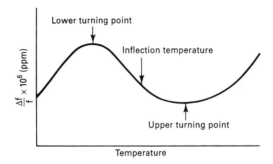

Figure 4-42 Crystal upper and lower turn points (UTP, LTP) and inflection temperature.

248 LOOP COMPONENTS

Crystal Cuts. Small changes in the crystalline orientation angle can cause large changes in temperature behavior. Large changes in crystal orientation can yield entirely different resonator properties. One speaks rather loosely of a family of crystals as a certain cut. This specifies the general crystal orientation. Small changes from this general orientation then alter the temperature performance significantly. By properly cutting the resonator from a bar of quartz, crystal manufacturers can vary the temperature behavior considerably for a particular cut or family.

Figures 4-43, 4-44, and 4-45 show typical frequency versus temperature plots for several cuts. The various plots are obtained by varying the basic angle of cut by only a few minutes. Because of the sensitivity to angular changes, extreme process control must be maintained during resonator processing, and some variation in frequency versus temperature performance must be expected. One must allow for this in the crystal specification.

Note that the inflection temperature is at approximately 27 °C for the AT cut but is much higher for the other cuts. This is why AT cuts are typically used for TCXOs. The frequency is very flat around the turnover temperatures for the IT and SC cuts. This is one of the reasons why they are being increasingly used for oven oscillators.

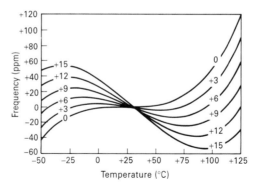

Figure 4-43 Frequency versus temperature for AT cut resonators.

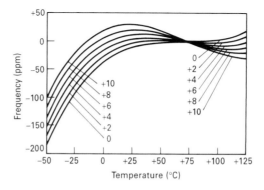

Figure 4-44 Frequency versus temperature for IT cut resonators.

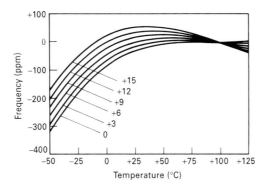

Figure 4-45 Frequency versus temperature for SC cut resonators.

Dynamic Thermal Characteristics. Another significant benefit of the SC cut is associated with dynamic or transient thermal effects of SC cut versus AT cut resonators. It has been found that the frequency of AT cut resonators not only varies with temperature but is also dependent on thermal gradients across the surface of the resonator caused by dynamic temperature variations [11]. The SC cut and, to a lesser extent, the IT cut are nearly immune to frequency changes caused by thermal gradients across the resonator.

Since the SC cut is nearly insensitive to the thermal gradients that invariably occur during oven warm-up, one would expect that SC cut resonators can give much faster warm-up in ovenized oscillators. Figure 4-46 shows typical warm-up for ovenized oscillators using AT and SC cut resonators. The packaging and control circuitry play a critical role in warm-up time as well, and most oscillator manufacturers have proprietary packaging schemes.

It can also be seen that due to the very high inflection temperature in the SC cut, it is usually desirable to operate the crystal at the lower turning point. Since this is typically around 80 to 85 °C, the maximum operating temperature for ovenized oscillators incorporating SC cut crystals is generally 70 to 75 °C. To maintain control of the oven, a margin of 5 to 15 °C between the oven temperature and the ambient temperature must be maintained.

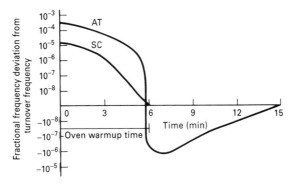

Figure 4-46 Typical warm-up for AT and SC cut oven oscillators.

If the oscillator is to be used at higher temperatures, the crystal can be operated either at the inflection temperature or at the upper turn. Both options have drawbacks, however. These are explained below.

If an SC cut crystal oscillator is operated at the upper turn, the resulting high temperature may greatly reduce the reliability of the electronics unless special precautions are taken in the design and selection of components. One may use a double oven with the crystal set at the UTP and the electronics at 5 or 10 °C above the maximum operating temperature, but the added complexity of the double oven increases cost and size. In addition, aging performance is generally thought to decrease as the operating temperature of the crystal increases, although some evidence to the contrary exists [12].

Unless the oven control is extremely good, operation at the inflection temperature can be used only with those crystals that have a very flat frequency versus temperature characteristic. Because of the difficulty in maintaining angular control in resonator manufacturing, not all resonators will have a near-zero temperature coefficient at the inflection temperature. The resulting yield loss generally prohibits operation at the inflection temperature for oscillators that must maintain very low drift over temperature.

b-Mode. The SC and IT cuts have an additional resonance termed the b-mode, which is about 10% higher in frequency than the desired or c-mode [13]. The b-mode has a linear temperature characteristic and can be used as a sensitive thermometer of the resonator temperature. This effect has been used to design extremely stable microcomputer-compensated oscillators, where the frequency of the b-mode is used to compensate the frequency of the main mode and oscillations are present on both modes at all times [7].

More typically, however, the crystal is operated on a single mode. A resonant circuit must then be employed to trap out the b-mode and force the oscillation to occur on the desired mode. Depending on the circuit configuration, several different methods can be employed. Trapping the b-mode is discussed in more detail later.

It can be very convenient to measure thermal performance of an oven using the b-mode. By retuning the resonant circuit to force oscillations to occur on the b-mode, the linear frequency versus temperature characteristics can provide a means of precisely measuring the performance of the thermal enclosure. To accomplish this, the frequency variations caused by temperature changes of the b-mode are first measured with the oven off. This step provides a means of characterizing the frequency–temperature characteristics of the b-mode. Next, the frequency change caused by external temperature variations with the oven on is then measured, and temperature variations are computed from this. Of course, the trap can then be retuned to set the oscillation back to the desired mode.

Maximum Obtainable Q. Another important factor to consider in selecting crystal resonators is that the product of maximum obtainable Q times frequency is a constant. For AT cut resonators, it has been empirically found that [10]

$$Q_{\max} = 16 \times 10^6/f \tag{4-79}$$

where f is in MHz. For example, the maximum Q for a 10-MHz crystal would be 1.6 million. Although this relationship is independent of overtone, in practice, third and fifth overtones generally yield a higher Q. A similar relationship holds for other cuts as well.

Crystal Aging. Crystal aging is dependent on many factors, including the overtone. This can easily be explained by the fact that aging is strongly affected by the cleanliness of the resonator and enclosure. A thin resonator will be affected by a thin film of contamination more than a thick resonator. Since a third overtone is approximately three times thicker than a fundamental, the overtone will have better aging. In practice, many factors influence aging, but in general a thicker resonator will give better aging performance. Factors other than cleanliness that may affect aging include stress relaxation of the mount structure and/or the crystal electrode, excessive crystal drive power, and outgassing of materials in the crystal support structure or enclosure. The subject of crystal aging is very complex and remains an area of intense study.

The choice of oscillator frequency is very important and depending on whether long-term aging, floor noise, or some other parameter is to be optimized, it may be advantageous to start at some lower frequency and multiply up. For instance, to obtain the best aging, a 5-MHz fifth overtone would be an excellent choice. But if a low phase noise floor is desired at, say, 100 MHz, it would be preferable to utilize a 100-MHz fifth overtone oscillator since the noise floor of the signal multiplied up from 5 MHz would be increased by a minimum of $20\log(20)$, or 26 dBc. Wide pull voltage-controlled oscillators (VCXOs) typically utilize fundamental crystals with the largest possible motional C_1. Overtone operation is generally precluded if a wide tuning range is desired.

Acceleration Sensitivity. Acceleration forces affect the resonant frequency of a crystal and should be considered when designing oscillators that must perform in mobile platforms or other applications, where vibration is likely to be encountered. This is critical since mechanical vibration can induce a severe degradation in phase stability. The effect is so strong that even the slight vibration from cooling fans may induce large spurious signals.

It has been found empirically that the acceleration sensitivity can vary by more than an order of magnitude from crystal to crystal. This variation has even been found to occur within a production lot of crystals that have the same design and have been processed in a supposedly identical manner. We now quantify this effect.

The change in resonant frequency has been found to be dependent on the direction of acceleration, as well as the magnitude, so that a vector representation is required.

Representing acceleration by $\vec{A}$ (in Gs), the series resonant frequency of a resonator can be described by [14]

$$f_s = f_{s(\text{nominal})}[1 + \vec{A} \times \vec{\Gamma}] \qquad (4\text{-}80)$$

Table 4-4 Typical specifications for a high-stability 10-MHz third overtone resonator

Factor	Cut				Units
	AT	FC	IT	SC	
C_i	.009	.007	.0023	.0018	pF
C_o	3.2	3.2	3.2	3.2	pF
R_l	3.5	50	80	80	ohms
Q	700	800	800	800	10^3
Aging/day	10	10	5	5	10^{-10}
Tin reflection	25	45	74	95	c
G-sensitivity (magnitude)	20	10	5	2	10^{-9}/G

Source: Piezo Crystal Company.

where the vector $\vec{\Gamma}$ is referred to as the acceleration sensitivity of the resonator. A typical value for $\vec{\Gamma}$ might be

$$[3\hat{i} + 5\hat{j} - 9\hat{k}] \times 10^{-10}/G$$

Thus an acceleration of 10 Gs in the $\hat{i}$ direction will cause a frequency shift of 3 parts in 109. Often, one may refer to the acceleration or "G" sensitivity of the resonator when speaking of the magnitude of the vector. This is simply a convenience, since the magnitude would represent the worst-case condition.

4-2-5 Crystal Specifications

The general format given in Section 4-2-1 can also be used for crystal resonators. Key specifications for typical 10-MHz resonators are given in Table 4-4.

4-2-6 Crystal Oscillators

Before proceeding into the design of crystal oscillators, a few facts must be kept in mind. Although not a few engineers consider the design of crystal oscillators to be largely an art, one can design for good performance using established principles. However, it must still be noted that although formulas and equations can be used to obtain the initial design, some trimming is usually needed to obtain the best performance. The equations should always be kept nearby when "tweaking," as they point out the direction in which to change the values of components. Even if a formula does not give exact results, because of component tolerances or assumptions made in the derivation, the derivative of the equation with respect to any component value is fairly accurate.

Seen in this light, trimming an oscillator design in the laboratory is really, in essence, a form of gradient optimization. Although this may seem obvious, it is often helpful to recall this as one attempts to tune a circuit to obtain the best performance. Just as one can obtain local minima when performing numerical gradient optimization, one can tune a circuit for what appears to be good performance, but in reality the circuit is capable of much better performance.

A realization of these concepts is essential to working with oscillators, since, by their very nature, oscillators are nonlinear devices that are extremely difficult to model. Approximations can be made based on small signals models, but the fundamental limitations of the small signal model must be kept in mind. Obtaining the best results is generally the result of a thorough design, followed by careful "optimization" on the laboratory bench.

Methods of Analysis. To obtain a true analytical solution to the steady-state oscillator circuit would require the solution of nonlinear simultaneous equations. This is an extremely difficult task and certainly not within the reach of most practicing engineers. Perhaps the intense work now being done in the field of chaos and nonlinear dynamics will eventually yield new insight into the workings of crystal oscillators [15]. For the present, though, an exact analytical solution is not practical.

Another alternative would be to obtain a numerical solution. Because of the very slow buildup of oscillations, a numerical solution in the time domain is very time consuming at best. It has been the author's experience that these methods generally do not converge for crystal oscillators having even a moderately high Q. This is not surprising since it may take tens of thousands to millions of cycles to reach the final steady-state limit cycle.

The principle of harmonic balance is a numerical technique that currently is being applied to analyze microwave oscillators [15]. Since the solution is obtained directly in the time domain, rather than the frequency domain, this may prove to be an excellent method of solving for the steady-state oscillator conditions.

At the current time, however, small signal models are the most practical way to analyze the oscillator. Even though the models do not truly describe the real oscillator, considerable insight can still be obtained from these models. The method described in this section is based on performing the following steps.

1. Choose frequency based on desired performance attributes. This involves choosing the frequency of operation based on the requirement of frequency multiplication of the oscillator signal or other special considerations. In general, 5- to 15-MHz third or fifth overtone oscillators give the best aging and short-term stability. Up to about 100 MHz, the lowest phase noise floor can generally be obtained by operating with an overtone directly at the desired frequency and eliminating the multiplication. At higher frequencies, SAW oscillators become attractive for the lowest noise floor.

2. Determine the desired crystal power dissipation. Since many of the oscillator performance attributes are dependent on crystal power, this should be determined early in the design cycle. Too often, the crystal power is measured as an afterthought, when it may be difficult to correct without changing other parameters. This is particularly true when both low phase noise and low crystal power are desired.

3. Select oscillator configuration. Certain configurations give superior performance in one area at the expense of another area. Choice of the circuit configuration is generally predicated by the required performance. Typical

characteristics of particular configurations are discussed in more detail later in this section.

4. Analyze the circuit based on small signal models. These models can be based on either negative resistance/conductance or open-loop gain phase requirements. Either method will work, but one method is sometimes more convenient to apply to the actual circuit. Designers often become more proficient in one method or the other, and one method becomes the preferred "tool."
5. Determine limiting conditions. Several methods are available to limit the growth of oscillations, ranging from simple cutoff or saturation to automatic level control (ALC). Since proper limiting is often essential to maximizing oscillator performance, this is a key step.
6. Experimentally optimize for best performance on the real oscillator. Generally, more care taken in the previous steps will mean less time required at the bench "tuning up" the circuit.

Crystal Power. One key to obtaining good performance is the crystal power dissipation. It is here that one must make an assumption based on past experience. To obtain the best aging performance, the crystal power dissipation should be around 10 to 50 μW for AT cut crystals and around 30 to 100 μW for SC and IT cut crystals. This can be increased somewhat for larger-diameter resonators. It must be noted though that some evidence exists that SC cut resonators may be driven considerably harder without significant increases in aging [12]. To obtain the best phase noise, the power dissipation into the sustaining amplifier should be as large as possible. AT cut crystals of 5 to 20 MHz can typically be driven with 2 mW and SC cut crystals can be used at 5 mW without damage.

It is sometimes useful to consider the current density per surface area of resonator electrode diameter. Due to smaller size, the drive should be reduced somewhat for higher frequency/smaller resonators with the rule of thumb being 50 μA/mm^2 for the AT cut [10] and about twice that for the SC cut. The advantage of the SC cut for phase noise becomes obvious when considered in this light.

Oscillator Configuration. Once the crystal power is assumed, the choice of oscillator configuration should be made. The guidelines in this section can be used as an aid in selecting the oscillator type. Note that there are many different types of oscillators from those listed here, some of which offer excellent performance in certain areas. The author has tried to include those configurations that are fairly simple to design and yet offer good performance. As with most circuits, one can usually obtain better oscillator performance with a simple configuration that is carefully designed than with a complex circuit that is hastily designed. A few simple modifications to the common Colpitts–Pierce–Butler configurations, which can yield excellent performance will be shown.

Series–Parallel Resonance. Crystal oscillators can be considered in a rather loose sense to be parallel or series resonant. This division is the most fundamental means of categorizing the oscillator.

For purposes of this chapter, a series resonant oscillator is one in which the crystal and any load capacitance or inductance in series with the crystal operates

REFERENCE FREQUENCY STANDARDS   255

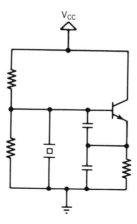

Figure 4-47 Colpitts crystal oscillator operating in parallel resonant configuration. The crystal appears inductive in the parallel resonant configuration, neglecting the small effect of the crystal C_o.

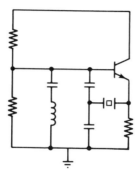

Figure 4-48 Colpitts crystal oscillator operating in series resonant configuration. The crystal appears inductive in the parallel resonant configuration and resistive in the series resonant configuration, neglecting the small effect of the crystal C_o.

at the transmission zero through the network. This is usually at or near the crystal series resonant frequency.

A parallel resonant oscillator is one in which the crystal and any load capacitance across the crystal operate at a pole. This can be somewhat confusing, though, since several components in series can function as a net capacitor or inductor at the frequency of operation. Thus, a capacitor and inductor may be placed in series with the crystal in a Colpitts oscillator, yet it will still be a parallel resonant oscillator.

Note that one could take a Colpitts LC oscillator and place the crystal between the transistor emitter and capacitor junctions as shown in Figures 4-47 and 4-48. This oscillator would be considered series resonant since the crystal could be replaced with a small resistor. However, the LC split tank would still be operating at its high impedance or parallel resonant state. This illustrates the confusion that can be generated by the use of the term series or parallel resonant. One must note that the only reason to specify an oscillator as series or parallel resonant is so that one can specify the crystal properly. The proper crystal load designation is usually obvious from the circuit configuration.

Crystal Oscillator Configurations. Historically, crystal oscillators have been called by certain names [3, 10]. The most familiar types are the Pierce, Colpitts, Clapp, and Butler. The principal difference between the Pierce and Colpitts configurations is the choice of signal ground, as shown in Figure 4-49 [3, 10]. The Butler configuration is shown in Figure 4-50. One must note that since the only change is the choice of signal ground, the negative resistance equation that governs the Colpitts oscillator for small signals [Eq. (4-40)] will also be valid for the Pierce oscillator. This means that the small signal gain requirements are the same and the circuit will oscillate if

$$g_m X_{C_1} X_{C_2} > \text{Re}$$

where Re is real part of the series impendance across the crystal. This would normally be given by Eq. (4-76) for Pierce or Colpitts oscillators. Even though

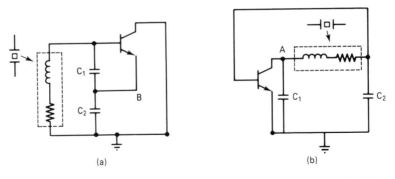

Figure 4-49 (a) Pierce oscillator configuration for fundamental crystals. Note that if the emitter is grounded for the Pierce, the configuration reverts to the Colpitts. (b) Colpitts oscillator configuration for fundamental crystals.

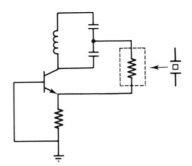

Figure 4-50 Butler oscillator configuration.

the small signal gain requirements for the Pierce and Colpitts oscillators are the same, the real oscillators will perform quite differently. This is principally due to different limiting action, occurring under large signal conditions, and different characteristics of the sustaining amplifier. The large signal characteristics and limiting conditions should always be considered for the sustaining amplifier, regardless of the configuration. With this in mind, a few general comments are in order regarding the different oscillator configurations.

Configurations. It is beyond the scope of this chapter to make a complete analysis of the various oscillator configurations. A good comparison of several oscillator circuits can be found in Ref. 17. Some general comments are in order, however.

The Colpitts configuration works very well for low aging/good performance oscillators up to about 50 MHz. It is fairly simple to design and the performance is moderate to good. Since there is no collector resistor in the typical Colpitts oscillator, the limiting is generally by cutoff in the transistor. This method of limiting is preferable to saturation of the transistor, as will be discussed later.

The Pierce oscillator is capable of very good performance, but the impedance at the collector may cause transistor saturation if the oscillator gain is too high. Crystal power may also become excessive for low aging unless ALC or low bias current is used in the Pierce configuration. The Pierce oscillator can be used at frequencies up to about 50 MHz, although there is some tendency for spurious oscillations to develop at the higher frequencies.

The Butler oscillator gives good performance for frequencies from 10 MHz to VHF. This is because the collector to base capacitance is effectively grounded in the Butler configuration and the series mode operation. Because of the true series mode configuration, the Butler oscillator is also recommended when C_0 must be tuned out with a shunt inductor. Butler oscillators are quite susceptible to spurious resonances, however. One common method of setting up a Butler oscillator is to short out the crystal and adjust the free-run frequency of the oscillator to the crystal. When the crystal is reinstalled, the oscillator will be controlled by the crystal. Typical circuits and corresponding values are shown for the Pierce and Colpitts oscillators in Figures 4-51 and 4-52. The Colpitts oscillator works well up to about 20 MHz, while the inductor in the Pierce oscillator allows the use of overtone crystals up to about 50 MHz. Note that in the Pierce oscillator the signal is taken from the input of the oscillator sustaining amplifier. Since the signal-to-noise ratio is always better at the input to an amplifier, the noise floor of this circuit can be quite good, with typical values in the range of −155 to −165 dBc/Hz. Figure 4-53 shows a Butler oscillator with typical values up to 100 MHz. While crystal oscillators above 100 MHz can easily be built, it is often preferable to use a lower frequency oscillator followed by a frequency multiplier. This arrangement is often more practical in a production environment.

An assortment of modifications to the Pierce and Colpitts oscillators is given in Figures 4-54 to 4-59. The design of these oscillators is very similar to the standard configuration, but the modified versions offer some advantages (and disadvantages) in certain aspects. Several of the designs are given for 10 MHz, but the oscillator frequencies may easily be modified by a factor of 2 or 3 by simply scaling the reactive components. Some allowance must also be given for stray reactance

258 LOOP COMPONENTS

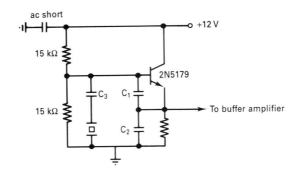

	3 MHz	6 MHz	10 MHz	20 MHz	30 MHz
C_1 (pF)	330	270	180	82	43
C_2 (pF)	430	360	220	120	68
C_3 (pF)	39	43	43	36	32
R_E (kΩ)	4.7	4.7	4.3	3.6	2.7
Crystal C_L (pF)	32	32	30	20	15

C_3 is the approximate value needed to set oscillator on frequency using indicated crystal C_L.

Figure 4-51 A 3–30 MHz Colpitts oscillator.

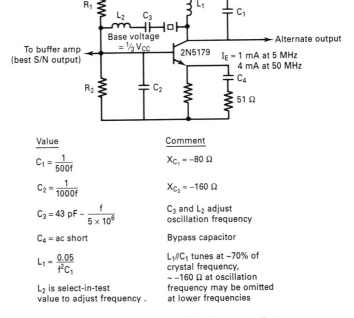

Value	Comment
$C_1 = \dfrac{1}{500f}$	$X_{C_1} \approx -80\ \Omega$
$C_2 = \dfrac{1}{1000f}$	$X_{C_2} \approx -160\ \Omega$
$C_3 = 43\ \text{pF} - \dfrac{f}{5 \times 10^6}$	C_3 and L_2 adjust oscillation frequency
$C_4 =$ ac short	Bypass capacitor
$L_1 \approx \dfrac{0.05}{f^2 C_1}$	$L_1 // C_1$ tunes at ~70% of crystal frequency, ~ –160 Ω at oscillation frequency may be omitted at lower frequencies

L_2 is select-in-test value to adjust frequency.

Figure 4-52 A 5–50 MHz Pierce oscillator.

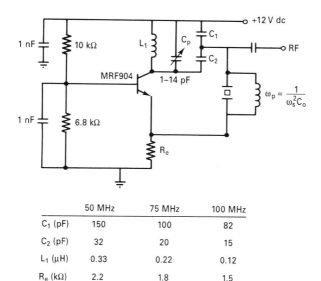

Figure 4-53 A 50–100 MHz Butler oscillator.

	50 MHz	75 MHz	100 MHz
C_1 (pF)	150	100	82
C_2 (pF)	32	20	15
L_1 (μH)	0.33	0.22	0.12
R_e (kΩ)	2.2	1.8	1.5

at higher frequencies. The exception is the semi-isolated Colpitts oscillator of Figure 4-54. Due to the coupling of load through C_{cb}, this circuit is not recommended for frequencies above 10 MHz, unless the output is tuned to a harmonic. The cascode arrangement of Figure 4-56 is a superior circuit for higher frequencies.

Advantages and disadvantages of the semi-isolated Colpitts oscillator with power taken from the collector are as follows:

Advantages	Disadvantages
High ratio of P_{out} to crystal power Simple design Output can be tuned to harmonic, giving frequency multiplication Greater isolation to load than a normal Colpitts oscillator	Limited frequency range due to Miller capacitance; can be used at higher frequencies if output is tuned to harmonic Coupling of collector load into oscillator through C_{cb} reduces loaded Q Transistor may saturate if collector load is too high

The principal advantages of the semi-isolated Colpitts oscillator are that the power extracted from the oscillator can be larger than the crystal power and that isolation to the load is better than the normal Colpitts oscillator. In the normal Colpitts oscillator, the power extracted from the load is much less than the crystal power. More power can be extracted only at the expense of loaded Q.

A similar effect occurs with the semi-isolated circuit in that if excessive power

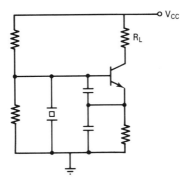

Figure 4-54 Semi-isolated Colpitts oscillator with power taken from collector [10].

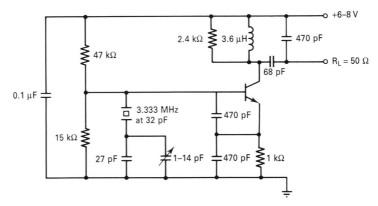

Figure 4-55 Semi-isolated Colpitts with 10-MHz output.

is extracted from the collector, the output load couples into the loop and reduces loaded Q. Because of this property, this circuit operates best at frequencies only up to about 10 MHz. Beyond this frequency, the collector output can be fed into a common base amplifier in a cascode arrangement, which generally gives better results.

By tuning the collector to a harmonic of the oscillator frequency, the circuit also acts as a frequency multiplier. Isolation from load to the oscillator circuit is also much better when the output is a harmonic of the oscillator frequency. A 3.333-MHz circuit based on this oscillator is shown in Figure 4-55. The output is tuned to 10 MHz.

As mentioned previously, the output of the semi-isolated Colpitts oscillator can be dc coupled into a common base amplifier in a cascode arrangement. This modification gives excellent isolation to the load and allows for a much larger collector load in the second transistor. By changing the configuration of the crystal, the output current can be forced to flow through the resonator. This modification reduces the noise floor considerably. Figure 4-56 shows the form of this circuit, while Figure 4-57 is a circuit that uses a 10-MHz fundamental resonator. The

REFERENCE FREQUENCY STANDARDS 261

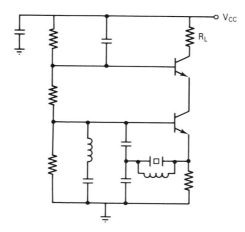

Figure 4-56 Modified semi-isolated Colpitts oscillator with power extracted through cascode stage amplifier [18].

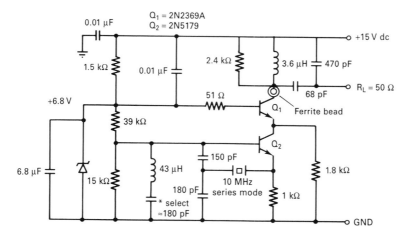

Figure 4-57 Modified semi-isolated Colpitts in cascode configuration.

additional bleed resistor from the collector of Q_2 allows the current in the second stage to be much larger than the Q_1 current. This is very desirable since Q_2 operates in the linear mode. If the current in Q_2 is several times that of Q_1, the oscillator stage will limit by a light cutoff of Q_1 but Q_2 will remain active. This is the preferred mode of operation. An ALC can also be conveniently added to this circuit by controlling the current in the lower transistor with a current source. Since the high impedance of the current source will force a larger amount of the out-of-band noise to flow through the resonator, the crystal can function more effectively as a filter on the noise floor.

262 LOOP COMPONENTS

Note that since the crystal operates at series resonance, we can tune out C_0. This allows the circuit also to operate as a VCXO by inserting a series tuned circuit in series with the crystal. By tuning out C_0, excellent linearity can be obtained.

Advantages and disadvantages of the semi-isolated Colpitts oscillator with power extracted through a cascode stage amplifier are as follows:

Advantages	Disadvantages
Moderate to high loaded Q	Slightly more complex to design and manufacture
Higher useful frequency range than the Colpitts oscillator in Figure 4-51	Requires higher supply voltage
Good isolation to load	Some Q degradation caused by emitter impedance of lower transistor
Crystal acts as a filter on output signal, reducing noise floor	

The next oscillator to consider is the modified Colpitts oscillator with the power extracted through the resonator (Figure 4-58). By extracting the resonator power through the crystal, the crystal acts as a very narrowband filter on noise generated by the oscillator stage. Since a common base stage has a very low input impedance, the Q degradation will be very slight. The principal disadvantage of this circuit is that the output power will be a small percentage of the crystal power. The implication of this on phase noise is discussed later in this section.

Advantages and disadvantages of the Colpitts oscillator with power extracted through the resonator are as follows:

Advantages	Disadvantages
Crystal filters output signal	Low power into common base stage may cause poor floor noise *or* high crystal power dissipation; ratio of crystal power to oscillator output power is equal to ratio of crystal resistance to common base input resistance
Good isolation	
Very low harmonic output	

A typical realization of this circuit is shown in Figure 4-59. This circuit is capable of producing very low noise levels at 5 MHz. Note that an ALC is also included to stabilize the oscillator amplitude and maintain circuit linearity. The 4.3-μH inductor and 510-pF shunt capacitor allow the use of third overtone precision crystals.

In a similar manner to what was done with the Colpitts, we can also modify the Pierce oscillator to extract power through the resonator (Figure 4-60). The crystal operates at series resonance in this circuit, so that we may tune out the crystal C_0 to suppress more effectively out-of-band noise. Series mode operation

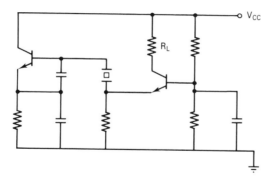

Figure 4-58 Modified Colpitts oscillator with power extracted through resonator.

also means that the crystal current will be less than the circulating power in the oscillator reactive loop components.

It is important to note that the collector of the oscillator transistor circuit is not tuned to the frequency of operation. Since the effective collector load must be capacitive, parallel resonance must be set lower than the frequency of operation. If the collector tank is tuned to a frequency between the fundamental and the overtone, the circuit will look inductive at the fundamental frequency and capacitive at the third overtone (Figure 4-61). Hence, the circuit can operate only at the overtone.

Note that if the base were placed at the ac ground, this circuit would have the same ac configuration as the Butler oscillator.

Advantages and disadvantages of the Pierce oscillator with power extracted through the resonator are as follows:

Advantages	Disadvantages
Crystal filters output signal but oscillator transistor is operating in a potentially lower noise configuration	Low power into common base stage may cause poor floor noise *or* high crystal power dissipation; ratio of crystal power to oscillator output power is equal to ratio of crystal resistance to common base input resistance
Crystal current is lower than circulating current in oscillator loop	

Once the basic configuration is selected, some means of mode selection is necessary for oscillators that employ overtone, IT cut, or SC cut resonators.

Mode Selection. A crystal resonator is essentially a plate that can readily convert electrical signals into mechanical displacements and vice versa. We should then expect that, like any mechanical resonator, several different modes of vibration will exist. In the case of the crystal resonator, modes will exist at frequencies that are odd integer multiples of the fundamental mode: that is, a 100-MHz fifth

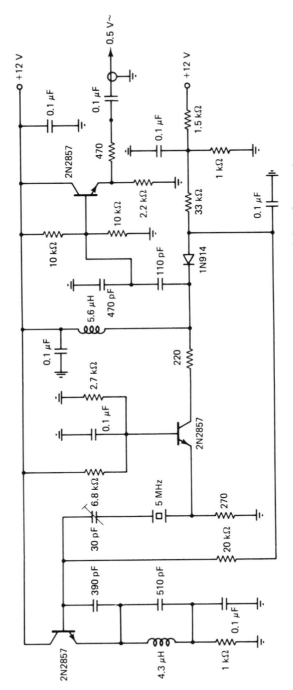

Figure 4-59 A 5-MHz Colpitts oscillator with power extracted through resonator.

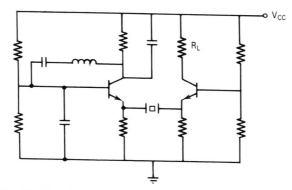

Figure 4-60 Modified Pierce oscillator with power extracted through resonator.

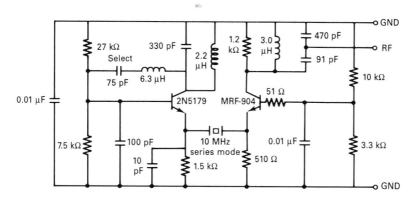

Figure 4-61 Third overtone 10-MHz Pierce oscillator.

overtone resonator will have responses at the third overtone frequency of 60 MHz and the fundamental frequency of 20 MHz. Unless special precautions are taken, the oscillator may not oscillate in the desired mode. Depending on the particular circuit, the final frequency of operation may be either the response that grows the fastest or the response that has the highest open-loop gain. Although it might at first seem that the response with the highest open-loop gain will be the one that grows most rapidly, this is not necessarily the case. Since Q is defined as the ratio of energy stored to energy lost per cycle, a high-Q response may grow more slowly than a low-Q response, even though the loop gain is greater for the former. This can often be seen in time-domain simulations of the oscillator during starting conditions. In any case, oscillator circuits often must utilize mode traps.

As mentioned previously, a resonator should be used only at the overtone for which it was designed. In fact, it is quite possible to produce a resonator that has a good response at the third overtone and poor response at the fundamental frequency. We have previously seen that SC cut and IT cut crystals also have a strong response called the b-mode, which is approximately 10% above the desired response. To use overtone, SC cut, or IT cut resonators, some means of forcing the oscillations to occur in the desired mode must be provided. This can be accomplished in the Colpitts or Pierce oscillator by forcing one of the capacitive splitting elements to be inductive at the frequency of the unwanted mode.

An example of this is shown in Figure 4-62. In this case we have a Darlington Colpitts oscillator operating on the third overtone. The crystal is an SC cut, so a b-mode trap is also required.

Inductor L_2 and capacitor C_2 are series resonant at a frequency just *below* the fundamental. This means that they appear inductive at the fundamental frequency. A parallel resonance occurs at a frequency above the fundamental, but below the desired frequency. At the desired frequency, the net impedance between the emitter of Q_2 and ground is capacitive and slightly greater than the impedance of capacitor C_3. This circuit has the same equivalent circuit as a crystal resonator, so one would expect a similar response but with much greater pole zero spacing, of course.

Inductor L_3 and capacitor C_4 function in a similar manner to the fundamental trap, but the series resonance is now close to the b-mode. They serve as the b-mode trap. The net impedance of this circuit is shown in Figure 4-63. Due to component

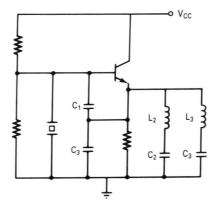

Figure 4-62 Colpitts oscillator with b-mode and fundamental trap.

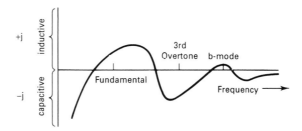

Figure 4-63 Reactance of fundamental and b-mode trap.

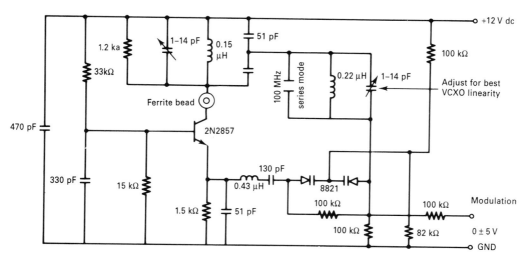

Figure 4-64 Fifth overtone Butler oscillator.

tolerances and the close spacing between the desired mode and the b-mode, it is usually necessary to make C_4 a select-in-test capacitor. The b-mode trap can also be placed between the emitter and base. This configuration generally gives better selectivity since the fairly low impedance at the emitter node reduces the resonance Q of the trap.

Figure 4-64 shows a Butler oscillator that uses a fifth overtone 100-MHz crystal. In this case, the collector impedance for the other responses is very low, and oscillations can be sustained only at the fifth mode.

The crystal C_0 is tuned out by L_2 to prevent spurious oscillations through the static capacitance. This particular circuit is still fairly susceptible to spurious oscillations, so the collector tank must be tuned to force the oscillations to occur at the crystal response.

Another benefit of removing C_0 is that the voltage linearity is greatly improved. Since the parallel resonance between C_0 and L_2 is a high impedance, the net circuit between the junction of C_1–C_2 and the emitter can be considered to be a series circuit for frequencies close to the crystal series resonant frequency. A more detailed analysis of the circuit, which includes losses, transistor junction capacitances, and other effects, shows that this is only approximately true, but the linearity is greatly improved when C_0 is tuned out.

For the best linearity, a trimmer capacitor should be placed across the crystal and adjusted to obtain the best linearity. Note the use of the varactor configuration of Figure 4-31. Figure 4-65 shows a computer analysis of the circuit after optimizing for linearity. The computed linearity was less than 1%. Actual measurements correlated very closely.

Tuning the collector can also be used for mode selection in the Pierce oscillator. Figure 4-66 shows a Pierce oscillator that uses a third overtone. Recall that the collector impedance must be capacitive at the desired frequency in the Pierce oscillator, so that the tank must be tuned at a frequency somewhat lower than

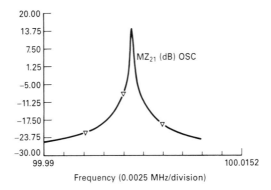

Figure 4-65a Simulated frequency versus voltage and open-loop power gain for a Butler oscillator.

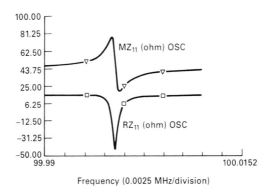

Figure 4-65b Simulated frequency versus voltage and open-loop power gain for a Butler oscillator.

the crystal frequency. The effective capacitance of the tank at the operating frequency should be used in series with other loop capacitances when calculating the total load capacitance of the crystal.

Once the desired crystal mode and power are chosen, and an oscillator configuration is selected, the design can be finalized. One final assumption regarding limiting must be made, however. The assumption is that limiting is achieved by lightly cutting off the transistor during a small portion of the limit cycle. Because of the low resistance between transistor junctions in saturation and the resultant severe degradation in operating Q, saturation of the oscillator transistor is to be avoided at all costs in oscillators designed for good phase noise.

Although the best noise performance is usually obtained with an ALC, very good performance can be obtained by designing the oscillator transistor to limit by light cutoff. Even if an ALC is used, the initial design should proceed in this

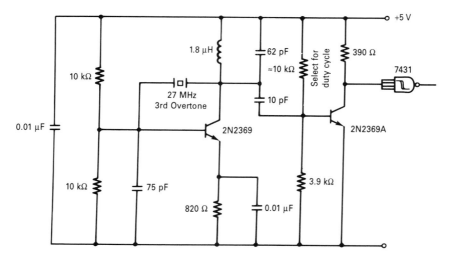

Figure 4-66 Third overtone Pierce oscillator with TTL output.

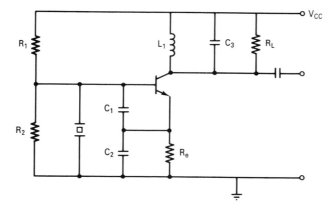

Figure 4-67 Basic Colpitts oscillator circuit with tuned collector.

manner, assuming light cutoff. Then the effect of the ALC will need only to be moderate. This is important because if the AGC control of the sustaining amplifier is excessive, the noise induced by ALC components can be greater than that of the oscillator.

Perhaps an example of this design approach will clarify the procedure. Let us design a 5-MHz Colpitts oscillator. We must first begin with a suitable model.

Figure 4-67 shows the oscillator configuration. Using the techniques developed earlier in this chapter, and assuming that the impedance of the bias resistors is much larger than the ac impedances, the oscillator portion of the circuit (excluding the collector load) can be simplified to a negative resistance model, as shown in

270 LOOP COMPONENTS

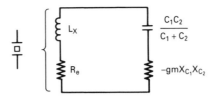

Figure 4-68 Negative resistance model of oscillator in Figure 4-67.

Figure 4-68. For sustained oscillations, the positive resistance must be exactly equal to the negative resistance. To ensure good starting, the negative resistor should be made about three to five times larger than the positive resistor.

For example, with an effective crystal resistance of 70 Ω, the bias current can be found as follows:

$$R_e = \frac{g_m X_{C_1} X_{C_2}}{5}$$

$$5 \times 70 = -38 I_c X_{C_1} X_{C_2}$$

Good performance is generally obtained with X_{C_1} and X_{C_2} in the range of −75 Ω. Using these values, I_c can be obtained:

$$I_c = \frac{70 \times 5}{75 \times 75 \times 38} = 0.00164 \text{ or } 1.6 \text{ mA}$$

This approach tells us that the circuit will oscillate, but it tells us nothing about the steady-state voltages, crystal power, and so on. However, with the assumption that oscillations will grow until the transistor just begins to cut off, we can develop approximations for these parameters. In essence, the oscillations will grow until at some instant the peak ac current out of the base is equal to the base dc bias current into the base. At that intant, the transistor will cut off. If the gain is not too excessive, the oscillations will not grow very much beyond this point.

By replacing the transistor with a large signal model [19] and assuming that the crystal is replaced with an equivalent inductance (X_1) and series resistance (R_e), a large signal model of the oscillator can be developed. Using the large signal model of Figure 4-69, and assuming that the transistor is just beginning to cut off, we can begin to write the circuit equations.

Note that I_b is flowing into the base from the bias circuitry. We now calculate the ac voltage V_1, which produces an ac current out of the base equal to the bias current I_b. This will cut off the transistor. It is also assumed that the resistances of the bias resistors R_1, R_2, and R_3 are much larger than the circuit ac impedances. With this in mind, we have

$$-i_{\text{in}} = \frac{V_1 + V_2}{X_1 + R_e} = \frac{(i_{\text{in}} - i_b) X_{C_1} + (i_{\text{in}} + \beta i_b) X_{C_2}}{X_1 + R_e} \qquad (4\text{-}81)$$

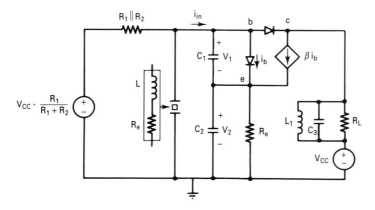

Figure 4-69 Large-signal model of oscillator circuit.

Rewriting gives

$$i_{in} = \frac{i_b(X_{C_1} - \beta X_{C_2})}{X_{C_1} + X_{C_2} + X_1 + R_e} \qquad (4\text{-}82)$$

But it has previously been shown that

$$X_{C_1} + X_{C_2} + X_1 \approx 0 \qquad (4\text{-}83)$$

Since βX_{C_2} is normally much greater than X_{C_1},

$$i_{in} \approx \frac{-i_e X_{C_2}}{R_e} \qquad (4\text{-}84)$$

If I_e (bias) + i_e (instantaneous ac) = 0, then the transistor is just cutting off at that instant. This implies that the *peak* current i_{in} is equal to the bias current. There will, of course, be some phase shift but at the present we are interested in peak current only. Thus,

$$i_{in}\ (\text{peak}) = \frac{I_e |X_{C_2}|}{R_e} \qquad (4\text{-}85)$$

The justification for using the magnitude of X_{C_2} is that we are relating the peak value of a signal to a dc value, and the phase relationship really has no meaning in this context. This equation is very significant in that it allows one to relate the ac signal to the dc bias current. Although this technique is not as elegant, perhaps, as true large signal oscillator models, the author has found this approach to be very useful. Crystal power can now be solved from

$$P_{\text{crystal}} = \frac{[i_{in}(\text{peak})]^2 R}{2} = \frac{[I_e |X_{C_2}|]^2}{2R_e} \qquad (4\text{-}86)$$

If $X_{C_1} \ll r_\pi$, then $V_1 \approx i_{in} X_{C_1}$. This gives

$$V_1(\text{peak}) = \frac{I_e |X_{C_1} X_{C_2}|}{R} \tag{4-87}$$

and

$$V_2 = \left(\frac{V_1}{r_\pi}(\beta + 1) + \frac{V_1}{X_{C_1}} \right) X_{C_2} \tag{4-88}$$

or

$$\frac{V_2}{V_1} = \frac{\beta + 1}{r_\pi} + \frac{X_{C_2}}{X_{C_1}} \approx \frac{X_{C_2}}{X_{C_1}} \tag{4-89}$$

Returning to the design of the 5-MHz oscillator, the bias circuitry is designed for 9.1-V operation. For low aging, the crystal power is chosen as 100 μW initially. AGC will be used to reduce this to approximately 50 μW. For now, we will assume 100. Using the original estimate of $-j75 \, \Omega$ for X_{C_2} and re-solving for the collector current, we find

$$I_e = \frac{\sqrt{2 P_{\text{crystal}} R}}{|X_{C_2}|} = \frac{\sqrt{2(0.0001)(75)}}{75} = 1.63 \text{ mA} \tag{4-90}$$

It would be more correct to assume the crystal power and then calculate a value of X_{C_1} to obtain the proper gain, since gain corresponds directly to X_{C_1}, but the steady-state values are affected more by X_{C_2}. Since our initial estimates were close, we will proceed.

Since emitter current is nearly equal to collector current, the output voltage can be found from $I_c R_L$. To avoid saturation, the collector load should be chosen so that 2 to 3 V is always present from collector to base. A collector load of about 2 K is chosen, and matching is provided to 50 Ω. We now consider limiting.

Limiting in Crystal Oscillators. To obtain the best performance, one must pay close attention to the means of limiting amplitude in a crystal oscillator. There are five basic means of limiting: (1) saturation of the sustaining amplifier, (2) cutoff of the sustaining amplifier, (3) gain reduction caused by large signal reduction in transistor g_m without hard cutoff or saturation, (4) voltage clamping by some nonlinear device in the loop such as a Shottky diode, and (5) AGC using a detector with some type of feedback to the loop.

Saturation should be avoided at all costs in oscillators designed for low phase noise. Limiting by gain reduction is possible, but component tolerances make it difficult to achieve in a production setting. Although very good performance can be achieved by lightly cutting off the transistor, the best performance is obtained with voltage clamping or AGC. A well-designed oscillator using cutoff will give much better performance than a poorly designed circuit with AGC, however. In

REFERENCE FREQUENCY STANDARDS 273

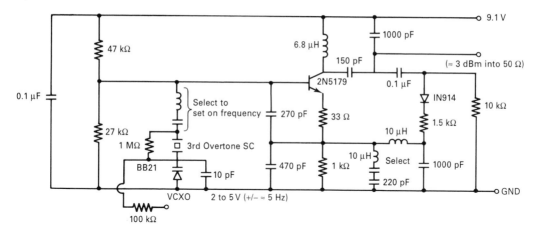

Figure 4-70 Oscillator with AGC and traps.

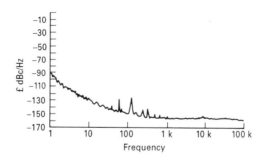

Figure 4-71 Measured phase noise of circuit in Figure 4-70.

general, the signals in the oscillator should be fairly sinusoidal since linear operation reduces the mixing of low-frequency noise with the oscillator signal.

We will add AGC and mode traps to the previous oscillator. One very simple mechanism is shown in Figure 4-70. This configuration is very simple and provides moderate level control over the output power. The effect of the AGC is fairly small, and the author has found that no additional noise is induced by the AGC. A small emitter degeneration resistor has been added to stabilize the gain and reduce flicker noise [20]. When the peak ac collector voltage exceeds the emitter bias voltage, the diode begins to conduct and reduce the bias current. Thus, the collector signal is limited to approximately the emitter voltage, -0.65 V.

This circuit was constructed and the phase noise was measured as shown in Figure 4-71. Phase noise at 100 Hz is about -150 dBc/Hz with a measured crystal power dissipation of approximately 60 μW. However, the floor is only -157 dBc/Hz. Although this is acceptable for some applications, other applications may require a lower noise floor. We can lower the noise floor by coupling the

274 LOOP COMPONENTS

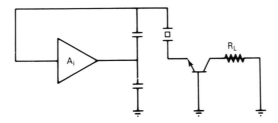

Figure 4-72 Modified Colpitts with power extracted through resonator.

output power through the resonator. Since the output current must flow through the crystal, the crystal will function as a narrowband filter on the output signal. Figure 4-72 shows a common method of accomplishing this.

The crystal acts as a filter on noise generated by the oscillator transistor. It is important to recognize a fundamental trade-off in this circuit. Assuming that about one-half of the thermal noise will be phase noise and one-half will be amplitude noise, the phase noise floor of the output amplifier is given by

$$\mathcal{L}(f) = -177 + \text{noise figure (dB)} - \text{input power (dBm)} \qquad (4\text{-}91)$$

Since the current into the common base will be the same as the resonator current, the ratio of resonator power to input power will simply be the ratio of the resistances. For example, if the crystal resistance is $50\,\Omega$ and the input impedance to the output amplifier is $5\,\Omega$ (a typical value for a common base amplifier), the input power will be only $-20\,\text{dBm}$ for a crystal power dissipation of $100\,\mu\text{W}$. Thus, the floor will be approximately $-154\,\text{dBc}$ if the output amplifier has a noise figure of $0\,\text{dB}$! But, if the input resistance to the output amplifier increases (or impedance transformation is used), P_{in} increases and the floor decreases. This is at the expense of the loaded Q, however.

One immediately notices the trade-off between resonator power dissipation, loaded Q (which affects close-in noise), and the noise floor. In general, trade-offs exist among these three parameters. One can decrease floor noise at the expense of resonator power dissipation; or floor noise may be decreased by coupling more power out of the loop at the expense of close-in noise. The real "trick" of designing low-noise oscillators is how to extract substantial power from the oscillator loop without causing a significant degradation in the operating Q or open-loop bandwidth.

Recognition of this fundamental trade-off brings us to the next oscillator configuration. This oscillator provides a fairly good floor, but, more importantly, the loaded Q can actually be higher than the resonator Q. This important benefit can yield extremely good close-in phase noise and short-term stability. These oscillators can yield phase noise levels of $-145\,\text{dBc}$ at 10 Hz. This represents an extremely high level of performance.

Bridge Oscillators. The phase slope across an impedance bridge can become very high as the bridge approaches balance. Bridge oscillators exploit this characteristic,

REFERENCE FREQUENCY STANDARDS   275

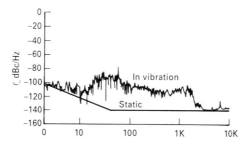

Figure 4-73 Measured phase noise during random vibration.

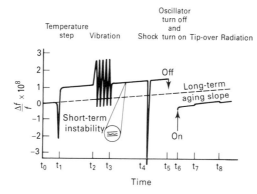

Figure 4-74 Effect of various conditions on oscillator frequency stability.

in effect, to increase the inherent phase slope, or Q, of the resonator [21]. Figure 4-73 shows the basic configuration of a bridge oscillator. Since the crystal operates at series resonance, it can be considered to be a resistor. Solving for the voltage difference gives the equation

$$\frac{R_2}{R_1+R_2} - \frac{R_x}{R_x+R_3} = \frac{1}{A_d} \qquad (4\text{-}92)$$

where A_d is the differential mode voltage gain. Although bridge oscillators are commonly built with operational amplifiers at lower frequencies, precision crystal oscillators have typically not used the bridge configuration. Benjaminson [22, 23] has proposed several bridge oscillator configurations that are suitable for use as frequency standards and has developed general oscillator models that clearly show the Q-enhancing benefits of negative feedback.

One implementation of a bridge oscillator that can produce excellent short-term stability is shown in Figure 4-74. The differential mode gain of the differential pair amplifies the difference signal between points A and A' of the impedance bridge. note that the maximum phase slope occurs when the bridge is balanced, but that the differential signal will be zero. To maintain oscillation, the positive feedback must exceed the negative feedback. Since the closed-loop gain must equal unity

276 LOOP COMPONENTS

for sustained oscillations, a high differential mode gain will yield a high phase slope.

The AGC causes the bridge to be initially unbalanced so that a rapid buildup of oscillation occurs. When the desired level is reached, the resistance of the FET increases and the bridge becomes nearly balanced. To maintain the highest phase slope, the differential gain must be set as high as possible. Again, we see a fundamental trade-off between close-in noise and floor noise. As the differential mode gain is increased, the phase slope increases and the loaded Q increases. Thus, the close-in noise is improved. However, as the bridge approaches balance, the power into the differential amplifier decreases, so the noise floor increases. Measured results on this type of oscillator indicate that the close-in noise and short-term stability can be much better than oscillators of the Pierce family.

Note that in this circuit, even though the difference signal from the bridge is small due to the balanced configuration, a fairly large common mode current will be flowing through the resonator. We may then place the output load in series with the crystal, so that the crystal filters the load current. Due to the Q multiplication, the load resistance can be higher than in the other configurations. Consequently, the power into the load can be made higher without the attendant large decrease of loaded Q, which occurs in most oscillator configurations. This effect is extremely beneficial because the real problem of designing low-noise oscillators is how to extract sufficient power from the loop without degrading Q.

One method of achieving this benefit is shown in Figure 4-75. Integrated circuit differential video amplifiers can be used to provide a high differential mode gain.

To illustrate the advantage of this circuit, suppose that the crystal power is 100 μW, or -10 dBm. Since the crystal current will be equal to the current through the transformer, the net power into the load will be the ratio of crystal resistance to the transformed load resistance. If this value is 1000 Ω, the output power will be 0 dBm. With a noise figure of 2 dB for the buffer amplifier, the noise floor will be -175 dBc. This is obtained without the usual reduction in loaded Q and attendant increase in close-in noise that occur with parallel resonant oscillators. Of course, this also assumes that the load resistor is free from noise and is stable. In practice, the load resistor would be the input impedance of a low-noise buffer amplifier.

4-2-7 Effect of External Influences on Oscillator Stability

Power supply noise, vibration, and other effects can greatly influence the short-term stability of an oscillator. The fact that acceleration forces can affect the crystal frequency was discussed previously. When the crystal is used to stabilize the frequency of an oscillator, sinusoidal vibrations will induce sidebands with an amplitude given approximately by

$$\mathscr{L}(f) = 20 \log \left(\frac{F_o \vec{A} \cdot \vec{\Gamma}}{2 f_v} \right) \qquad (4\text{-}93)$$

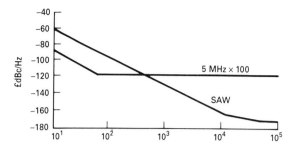

Figure 4-75 Typical phase noise of SAW oscillator at 500 MHz versus 5 MHz signal multiplied by 100.

where f_v is the frequency of vibration. Reference 14 provides an excellent review of this effect, as well as the effect of random vibration.

Figure 4-73 shows the effect of random vibration on a precision ovenized oscillator. The phase noise is increased by as much as 30 dB over static conditions.

In practice, many factors can influence frequency stability. Figure 4-74 shows an idealized plot of frequency under various conditions. If a crystal oscillator is to be subjected to various influences, it is essential that the sensitivity to these parameters be specified.

4-2-8 High-Performance Oscillator Capabilities

We conclude our discussion by examining what the author believes to be the current state of the art in crystal oscillator performance. As this is somewhat subjective, the author apologizes in advance for any omissions or errors in Table 4-5. Several manufacturers are currently selling oscillators that approach or meet these specifications, including Austron, Cinox, CTS, Efratom, McCoy, Piezo Crystal, Vectron, and Wenzel Associates. This list of companies is by no means exhaustive and is included for reference only.

Certainly, one cannot expect to obtain an oscillator that meets all of these performance levels simultaneously, since many factors of oscillator performance require trade-offs. Yet, the excellent performance shown in Table 4-5 indicates that the modern crystal oscillator is capable of very high levels of performance. The progress made in oscillator technology has been impressive, and, while the field is relatively old in comparison with some areas of electronics, the stability of oscillators continues to improve. We can only look forward to additional advances in oscillator performance with anticipation. The field of surface acoustic wave oscillators also shows great promise. We now consider these devices.

4-2-9 Surface Acoustic Wave (SAW) Oscillators

Due to their inherently high operating frequencies, SAW oscillators can generate signals with a very low floor noise in the VHF and UHF ranges. The use of a SAW oscillator should be considered when generating low-noise signals in these

Table 4-5 Representative specifications for state-of-the-art oscillators

	Phase Noise	
Offset Frequency (Hz)	10-MHz Oscillator	100-MHz Oscillator
1	−100	−70
10	−130	−100
100	−155	−130
1000	−165	−155
10000	−170	−170

	Size and Power for Ovenized Oscillators	
Stability over −55 to +71 °C	Size (in.^{3})	Power (W)
±0.1 ppm	<1	<1
±0.01 ppm	<2	<1

Aging: $<10^{-10}$/day in production basis
Warm-up time from −55 °C to a stability of 3×10^{-8}: 3 minutes

---

frequency ranges. Since the noise due to multiplication will increase by $20 \log N$, where N is the multiplication factor, it may be advantageous to select the highest possible frequency for the oscillator and reduce or eliminate multiplier chains.

In addition, SAW resonators and delay lines are capable of handling power levels greater than +10 dBm without damage. Note that the Q of SAW resonators is typically much lower, however, than bulk wave devices. This means that the close-in noise of a signal, which has been multiplied up from, say, a precision 10-MHz oven, will typically be lower than that generated by a SAW oscillator. Figure 4-75 shows the typical noise spectrum of signals derived by multiplication of a 5-MHz bulk acoustic wave oscillator (BAW) versus direct generation with a SAW oscillator.

The long-term stability of bulk wave resonators is also somewhat better than SAW devices, although significant progress has recently been made in this area. Precision SAW resonators are capable of aging rates in the range of a few parts in 10^{-9} per day after a stabilization period [24], while more typical devices age a few ppm per year.

It is important to note that there are two fundamentally different types of SAW devices that can be used to stabilize the frequency of an oscillator. The delay line is inherently a two-port device that typically has a fairly high loss, on the order of 10 to 25 dB in a matched environment, although the loss can be reduced somewhat through special design techniques. The resonator is a fairly low-loss device, on the order of 1 to 3 dB, and is preferred for low-noise applications.

Due to the low Q, delay line SAW oscillators have poor phase noise performance relative to oscillators stabilized with resonators. SAW oscillators can also, of course, be ovenized (OCSO), voltage controlled (VCSO), or temperature

REFERENCE FREQUENCY STANDARDS

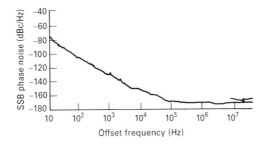

Figure 4-76 Phase noise for ovenized SAW oscillator (OCSO) with moderate voltage tuning capabilities. (Courtesy of SAWTEK.)

Table 4-6 Typical SAW oscillator specifications

Parameter	SO	TCSO	OCSO
Frequency range (MHz)	100–1250	100–1250	100–1250
Temperature stability			
0 to +70 °C	80 ppm	±20 ppm	5 ppm
−40 to 85 °C	200 ppm	±40 ppm	10 ppm
Set tolerance	±50 ppm	(Included in temperature stability)	(Included in temperature stability)
Output power	\multicolumn{3}{c}{0 to +10 dBm for all types (user specified)}		
Power consumption (+15-V operation)	50 mA	60 mA	550 mA

Source: Courtesy of SAWTEK.

compensated (TCSO). The delay line oscillator is preferable for wide pull VCSO. Table 4-6 shows a partial listing of parameters for various SAW oscillators.

The VCSO has performance similar to the SO, but the frequency can be varied with a control voltage by 100 to 10,000 ppm, depending on whether a resonator or delay line is used.

Phase noise for an oven-controlled SAW oscillator with moderate voltage control is shown in Figure 4-76. Being voltage controlled, the ovenized oscillator can also be phase locked to an external reference standard. Thus, the long-term stability and close-in noise would be controlled by the reference oscillator, while the short-term stability and noise floor would essentially be that of the SAW oscillator.

SAW Oscillator Design Techniques. The design of SAW oscillators proceeds in the same general fashion as bulk wave oscillators, with the exception that, due to the higher frequency, impedance matching becomes more critical. We still have the basic requirements of a sustaining amplifier and some type of feedback network that provides the necessary phase shift. Although it can be shown that the

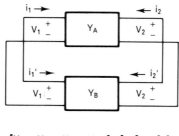

$$\begin{bmatrix} Y_{A11}+Y_{B11} & Y_{A12}+Y_{B12} \\ Y_{A21}+Y_{B21} & Y_{A22}+Y_{B22} \end{bmatrix} \begin{bmatrix} V_1 \\ V_2 \end{bmatrix} = \begin{bmatrix} 0 \\ 0 \end{bmatrix}$$

Figure 4-77 Y-parameter characterization of oscillator.

requirements are in fact identical, the oscillator can be designed using negative resistance techniques or open-loop gain phase methods.

Since most SAW oscillators are designed in the range of 100 MHz to about 1.2 GHz, feedback models and networks used in HF bulk wave oscillators may not be appropriate. Conversely, techniques used for microwave oscillators, where the feedback path is intimately connected with the amplifier, may be complex enough to hinder an intuitive understanding of the feedback mechanism. A method of analysis that is appropriate to the frequency range of SAW oscillators is desirable.

Several different options are available to analyze SAW oscillators, including the use of Y parameters, $ABCD$ parameters, and T parameters. As there are advantages with each method of design, the best choice depends on the particular circuit configuration and the individual preference of the designer. Alternatively, the designer may simply choose experimentally to modify an existing design, although this method often does not result in optimal performance if the new specifications are considerably different. It must be emphasized that since the network parameters are based on linear circuits, an analytical solution can only be an approximation of the real oscillator.

The Y parameters of two networks that are in parallel can be added as shown in Figure 4-77, where network A represents the sustaining amplifier and network B the feedback network. Since the port voltages and currents can be equated, we can write

$$[i] + [i'] = [0] \qquad (4\text{-}94)$$

giving

$$[Y_A + Y_B][V] = [0] \qquad (4\text{-}95)$$

The only nontrivial solution to this equation occurs when

$$\Delta[Y_A - Y_B] = 0 \qquad (4\text{-}96)$$

If a current gain factor, G_i, is included in the equation, the result

$$(Y_{11A} + G_i Y_{11B})(Y_{22A} + G_i Y_{22B})$$
$$-(Y_{21A} + G_i Y_{21B})(Y_{12A} + G_i Y_{12B}) \quad (4\text{-}97)$$

is obtained. This equation provides a very general method of analyzing oscillators but is most useful for computer-generated solutions.

The chain matrix using $ABCD$ parameters provides another method of analyzing an oscillator. By defining the input and output variables as shown in Figure 4-78, it can be shown that the resulting chain matrix must be equal to the identity matrix for sustained oscillations. Note that the source impedance driving the open-loop network must be equal to the output impedance of the network. Similarly, the open-loop network must be terminated with the input impedance. This is necessary to ensure oscillations when the loop is terminated onto itself (i.e., when the loop is closed).

Since V_2 and i_2 must be precisely equal to V_1 and i_1, respectively, in both phase and magnitude when the loop is closed, the open-loop chain matrix must be exactly zero for sustained oscillations. Since the gain margin is simply the ratio of V_2 to V_1, the gain margin can also be included in the model, resulting in the equation

$$\begin{bmatrix} A & B \\ C & D \end{bmatrix} = G \begin{bmatrix} 1 & 0 \\ 0 & 1 \end{bmatrix} \quad (4\text{-}98)$$

where G is the gain margin. The advantage of using chain parameters is that the resulting chain matrix of cascaded networks is simply the product of the matrices of the individual networks. Thus, we can simply cascade different transistors, matching elements or other networks, and multiply the additional chain parameters together. Note that although we are principally discussing SAW oscillators, chain parameters can easily be used to analyze BAW oscillators as well. A simple analysis

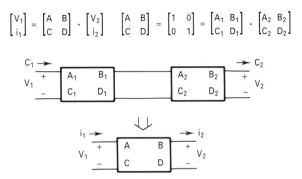

Figure 4-78 Open-loop oscillator model using chain parameters.

program can be written, which, by sweeping the input frequency, calculates open-loop gain, phase shift, and group delay. Since both Q and group delay are directly proportional, by optimizing for maximum group delay, the maximum Q will also be realized.

Chain scattering parameters provide a convenient method of characterizing cascaded microwave circuits and can also be used to characterize oscillators. Chain scattering parameters can be combined to describe the behavior of cascaded networks in the same manner as is done with the normal chain matrix at lower frequencies. Chain scattering parameters are defined by

$$\begin{bmatrix} a_1 \\ b_1 \end{bmatrix} = \begin{bmatrix} T_{11} & T_{12} \\ T_{21} & T_{22} \end{bmatrix} \begin{bmatrix} b_2 \\ a_2 \end{bmatrix} \quad (4\text{-}99)$$

where a and b denote incident and reflected waves [25].

The relationship between T parameters and S parameters is given by

$$\begin{bmatrix} T_{11} & T_{12} \\ T_{21} & T_{22} \end{bmatrix} = \begin{bmatrix} \dfrac{1}{S_{21}} & -\dfrac{S_{22}}{S_{21}} \\ \dfrac{S_{11}}{S_{21}} & S_{12} - \dfrac{S_{11}S_{22}}{S_{21}} \end{bmatrix} \quad (4\text{-}100)$$

T parameters may also be converted to S parameters through the use of

$$\begin{bmatrix} S_{11} & S_{12} \\ S_{21} & S_{22} \end{bmatrix} = \begin{bmatrix} \dfrac{T_{21}}{T_{11}} & T_{22} - \dfrac{T_{21}T_{12}}{T_{11}} \\ \dfrac{1}{T_{11}} & -\dfrac{T_{12}}{T_{11}} \end{bmatrix} \quad (4\text{-}101)$$

If two networks are cascaded as shown in Figure 4-79, we may combine the matrices in the following manner:

$$\begin{bmatrix} a_1 \\ b_1 \end{bmatrix} = \begin{bmatrix} T_{11}^a & T_{12}^a \\ T_{21}^a & T_{22}^a \end{bmatrix} \begin{bmatrix} T_{11}^b & T_{12}^b \\ T_{21}^b & T_{12}^b \end{bmatrix} \begin{bmatrix} b_2' \\ a_2' \end{bmatrix} \quad (4\text{-}102)$$

If the input and output ports are properly matched to the system impedance, when the oscillator loop is closed (terminated onto itself), then

$$\begin{bmatrix} a_1 \\ b_1 \end{bmatrix} = \begin{bmatrix} b_2' \\ a_2' \end{bmatrix} \quad (4\text{-}103)$$

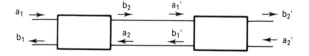

Figure 4-79 Cascaded networks using T parameters.

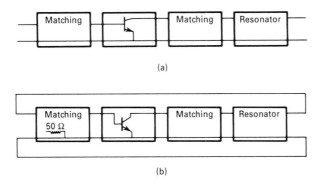

Figure 4-80 (a) Open-loop oscillator model using *T* parameters. (b) Closed-loop model terminated onto itself with 50-Ω load.

In the same manner that we had found with $ABCD$ parameters, Eq. (4-100) implies that the product of the two chain scattering matrices must be the identity matrix. By extending this concept to several cascaded networks, we can write

$$[T^1][T^2][T^3]\cdots = G\begin{bmatrix} 1 & 0 \\ 0 & 1 \end{bmatrix} \tag{4-104}$$

where T^n represents the T matrix for the nth network and G is again the open-loop gain margin.

In essence, what this equation states is that the open loop output power will be equal to the input power in both phase and magnitude. In fact, it is not necessary for the input and output ports to be matched. However, if they are not matched, then they must be calculated with Z_L set to Z_{in} and Z_S set to Z_0 of the network. This can easily be accounted for in computations but is not convenient for measurements. If open-loop measurements are to be performed, it may be simpler to match input and output ports to the system impedance. The load resistor can be incorporated into one of the networks.

A typical application of this method of design can be outlined in the following manner. Suppose that we wish to build a SAW oscillator using a two-port SAW oscillator with power delivered to a 50-Ω load. The general circuit configuration can be divided into networks, as shown in Figure 4-80a for the open-loop model and in Figure 4-80b for the closed-loop model. The output impedance of the transistor can easily be matched to the desired impedance level of the resonator. Similarly, the resonator can be matched to the output impedance of the system. With the exception of the first stage, this allows the chain scattering parameters for the network to be computed. Finally, the $[T]$ matrix of the first matching network simply becomes the reciprocal of the product of the other matrices. Since b_1 and a'_2 will be zero in a matched system, the problem can be simplified considerably. Thus, it is advantageous to design for a matched condition. This is particularly desirable in that the open-loop parameters, including loaded Q, can then easily be measured in a network analyzer.

A power splitter, which provides an impedance match among three ports, is

284 LOOP COMPONENTS

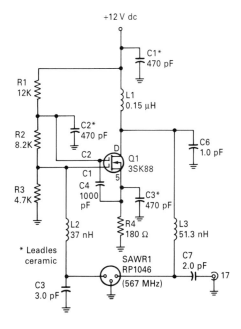

Figure 4-81 SAW oscillator schematic diagram.

an alternative method of extracting power from the oscillator. The advantage of the power splitter is that open-loop measurements can conveniently be performed using a network analyzer. The use of 50-Ω modular hybrid amplifiers should also be considered, since open-loop measurements can easily be performed. This approach has been used with BAW oscillators to provide signal sources with exceptionally low phase noise [18].

An additional benefit of this particular configuration is that the SAW resonator will act as a filter on the noisy signal from the output of the transistor. Out-of-band noise will be reduced by the same mechanism as was described previously for BAW devices.

Lastly, we consider a general-purpose SAW oscillator that can be used up to about 1000 MHz. This design is based on Application Note No. 3 by RF Monolithics [25]. Since it incorporates a MOSFET, the noise performance is not quite as good as can be obtained with a BJT. The design is fairly simple, however, and has been well characterized. Figure 4-81 shows the schematic for the oscillator, and Figure 4-82 shows a representative circuit layout and coil winding details. As with all RF circuits, careful layout to minimize parasitic reactances and coupling is essential. Typical component values are shown in Table 4-7.

4-3 MIXER APPLICATIONS

In multiloop synthesizers, the heterodyne principle is used, and various frequencies are combined with mixers. For frequency synthesizer applications, only double-

MIXER APPLICATIONS   285

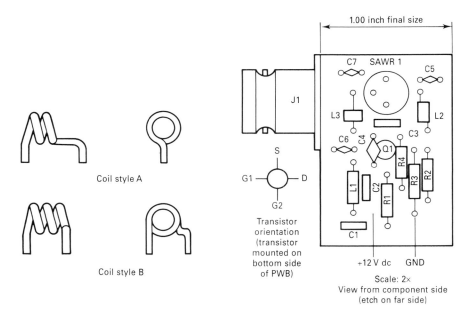

Figure 4-82 PCB layout and coil winding details.

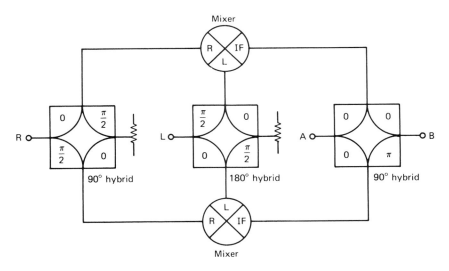

Figure 4-83 Single-sideband mixer.

balanced mixers should be used. They fall into the following categories and are considered a component. Active mixers typically do not show enough suppression of unwanted frequencies as can be obtained in passive double-balanced mixers.

1. *Single-Sideband Mixer.* A single-sideband mixer is capable of delivering an IF output composed of one sideband only. Figure 4-83 shows a combination that provides the upper sideband at port A and the lower sideband at port B.

Table 4-7 SAW oscillator component values

		335.5 RP1048 −30 to +70	567.0 RP1046 −130 to +70	674.0 RP1032 −130 to +70	840.0 RP1069 −130 to +70	1000.0 RP1000 −200 to +40
Oscillator frequency	MHz					
RFM resonator P/N						
Oscillator pull range	kHz					
Transistor parameters						
y_i	mS	$0.7 + j5.0$	$1.5 + j8.0$	$2.1 + j10$	$3.6 + j10$	$6.0 + j14$
y_f	mS	$17 - j6.0$	$16 - j10$	$15 - j12$	$12 - j17$	$10 - j22$
y_r	mS	$0.004 + j0$	$0.045 + j0$	$0.07 + j0.025$	$0.13 + j0.08$	$0.17 + j0.15$
y_o	mS	$0.4 + j1.5$	$0.75 + j3.3$	$0.9 + j4.0$	$1.0 + j4.5$	$1.0 + j5.0$
SAW resonator values						
C_o	pF	2.30	1.90	1.50	1.40	1.35
L_m	mH	0.900	0.267	0.186	0.170	0.150
C_m	pF	0.00025 0042	0.00029 5096	0.00029 9783	0.00021 1170	0.00016 8869
R_m	Ω	150	150	150	150	200
Circuit values						
L_1	nH	150	150	150	150	150
L_2	nH	100.30	36.90	26.91	20.27	14.26
L_3	nH	90.75	53.33	44.50	34.65	24.28
C_5	pF	2.0	2.0	2.0	1.0	1.0
C_6	pF	4.0	1.0	0.6	NA	NA
C_7	pF	4.0	2.0	1.0	1.0	1.0
Coil dimensions						
L_2 form		10-32	10-32	8-32	8-32	4-40
L_2 turns		$6\frac{1}{2}$, #22	$2\frac{1}{2}$, #22	$2\frac{1}{2}$, #22	$1\frac{1}{2}$, #22	$1\frac{1}{2}$, #22
L_2 styles		B	B	B	B	B
L_3 form		10-32	10-32	10-32	8-32	6-32
L_3 turns		5, #22	3, #22	3, #22	3, #22	2, #22
L_3 style		A	A	A	A	A
Network voltage gains						
Drain matching		0.55 ∠−151.2°	0.47 ∠−158.7°	0.61 ∠−162.2°	0.49 ∠−165.3°	0.51 ∠−162.6°
SAW resonator		0.76 ∠180°	0.72 ∠180°	0.73 ∠180°	0.77 ∠180°	0.59 ∠180°
Gate 1 matching		1.43 ∠−162.2°	1.34 ∠−165.6°	1.20 ∠−166.5°	0.94 ∠−166.9°	0.94 ∠−160.5°
Transistor		9.0 ∠−226.6°	11.45 ∠−215.8°	10.09 ∠−211.3°	11.05 ∠−207.7°	10.62 ∠−216.8°
Total loop		5.39 ∠0°	5.19 ∠0°	5.41 ∠0°	3.91 ∠0°	2.99 ∠0°

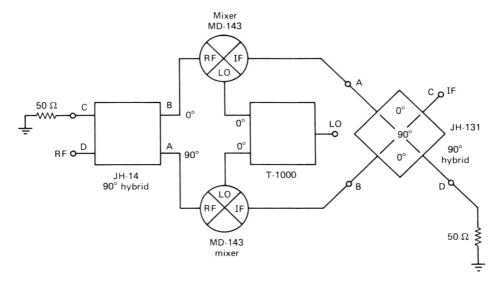

Figure 4-84 Image-rejection mixer.

2. *Image-Rejection Mixers.* An LO frequency of 75 MHz and a desired RF frequency of 25 MHz would produce an IF difference frequency of 50 MHz. Similarly, an image frequency of 125 MHz at the mixer RF port would produce the same 50-MHz difference frequency. The image-rejection mixer shown in Figure 4-84 produces a desired IF difference frequency at port C while rejecting difference frequencies from RF signals, which are greater than the LO frequency.

3. *Termination-Insensitive Mixers.* While the phrase "termination insensitive" is somewhat misleading, a combination as shown in Figure 4-85 results in a mixer design that allows a fairly high VSWR at the output without third-order intermodulation distortion being much affected by port mismatches.

Double-balanced mixers are manufactured by several companies; the best known are probably Anzac, Lorch, Mini-Circuit Laboratories, and Watkin–Johnson. Some of them specialize in high-performance mixers at very high costs, others in large-volume inexpensive devices. It is advisable to contact the manufacturer before deciding on a particular mixer, as technology changes and new mixer combinations are being introduced. Most manufacturers supply detailed application reports and information about their mixers. Therefore, these details do not have to be covered here.

Table 4-8 is of interest as it shows the typical spurious response of a high-level double-balanced mixer and information about unwanted products to be gathered.

Important: Make absolutely sure that the mixer sees a 50-Ω resistive load; this is achieved either by diplexer, resistive padding, or the use of feedback amplifiers that have 50-Ω input impedance or grounded gate FET amplifiers using devices such as the CP643.

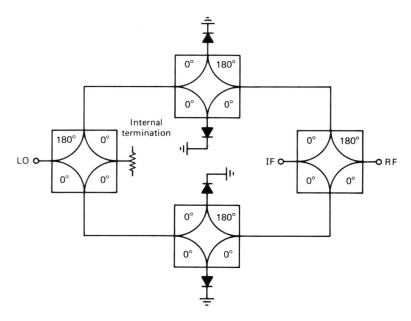

Figure 4-85 Termination-insensitive mixer.

Table 4-8 Typical spurious response: high-level double-balanced mixer[a]

Harmonics of f_{RF}									
$8f_{RF}$	100	100	100	100	100	100	100	100	100
$7f_{RF}$	100	97	102	95	100	100	100	90	100
$6f_{RF}$	100	92	97	95	100	100	95	100	100
$5f_{RF}$	90	84	86	72	92	70	95	70	92
$4f_{RF}$	90	84	97	86	97	90	100	90	92
$3f_{RF}$	75	63	66	72	72	58	86	58	80
$2f_{RF}$	70	72	72	70	82	62	75	75	100
$1f_{RF}$	60	0	35	15	37	37	45	40	50
		60	60	70	72	72	62	70	70
	$1f_{LO}$	$2f_{LO}$	$3f_{LO}$	$4f_{LO}$	$5f_{LO}$	$6f_{LO}$	$7f_{LO}$	$8f_{LO}$

[a] RF harmonic referenced to RF input signal; LO harmonic referenced to LO input signal. Spurious responses caused by internal harmonic generation and mixing of the input signals are shown. The mixing products are referenced in dB below the desired $f_{LO} \pm f_{RF}$ output or 0 level at f_{IF}. This performance can typically be attained with f_{LO} and f_{RP} at approximately 100 MHz, f_{LO} at +17 dBm, and f_{RF} at −0 dBm using broadband resistive terminations at all ports.

4-4 PHASE/FREQUENCY COMPARATORS

In Chapter 1 we looked at the phase-locked loop as the fundamental building block of any synthesizer that uses its principle, and we decided to use two classifications—analog and digital loops. The main criterion was the phase/frequency comparator.

The phase/frequency comparator can be divided into two types:

1. Phase detectors.
2. Phase/frequency comparators.

This means that the phase comparator has limited means to compare two signals and accepts only phase, not frequency, information. In this case, particular measures have to be taken to pull the VCO into the locking range. The phase comparators require special locking help, and we dealt with this in Section 1-10. Here, we are analyzing only the performance.

The phase detectors we will treat are the diode ring, the exclusive-OR gate, and the sample/hold comparator. The digital phase/frequency comparator (the exclusive-OR gate, because of the waveforms, is a digital device, and the sample/hold comparator, because of its special signal processing, can also be considered in this category) comes in several versions. Here our main interest will be in the tri-state or sequential phase/frequency comparator.

4-4-1 Diode Rings

The diode ring is normally driven with two signals with sinusoidal waveform and also is some sort of a mixer. Here it will suffice to derive the gain characteristic K_θ of the device. If the input signal is $\theta_i = A_i \sin \omega_0 t$, and the reference signal is $\theta_r = A_r \sin(\omega_0 t + \phi)$, where ϕ is the phase difference between the two signals, the output signal θ_e is

$$\theta_e = \theta_i \theta_r = \frac{A_i A_r}{2} K \cos \phi - \frac{A_i A_r}{2} K \cos(2\omega_0 t + \phi) \tag{4-105}$$

where K is the mixer gain. One of the primary functions of the low-pass filter is to eliminate the second harmonic term before it reaches the VCO. The second harmonic will be assumed to be filtered out and only the first term will be considered, so

$$\theta_e = \frac{A_i A_r}{2} K \cos \phi \tag{4-106}$$

When the error signal is zero, $\phi = \pi/2$. Thus, the error signal is proportional to phase differences from 90°. For small changes in phase $\Delta\phi$,

$$\begin{aligned}\theta_e &\simeq \frac{\pi}{2} + \Delta\phi = \frac{A_i A_r}{2} K \left[\cos\left(\frac{\pi}{2} + \Delta\phi\right) \right] \\ &= \frac{A_i A_r}{2} K \sin \Delta\phi \end{aligned} \tag{4-107}$$

For a small phase perturbation $\Delta\phi$,

$$\theta_e \simeq \frac{A_i A_r K}{2} \Delta\phi \tag{4-108}$$

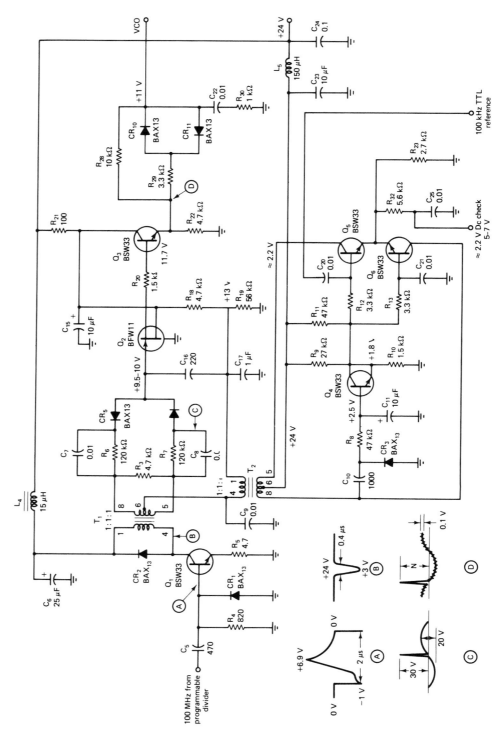

Figure 4-86 Phase detector circuit and loop filter for the Rohde & Schwarz EK47 receiver.

Since the phase detector output was assumed to be

$$\theta_e = K_\theta(\theta_i - \theta_o) \qquad (4\text{-}109)$$

the phase detector scale factor K_θ is given by

$$K_\theta = \frac{A_i A_r K}{2} \qquad (4\text{-}110)$$

The phase detector scale factor K_θ depends on the input signal amplitudes; the device can be considered linear only for constant-amplitude input signals and for small deviations in phase. For larger deviations in phase,

$$\theta_e = K_\theta \sin \Delta\phi \qquad (4\text{-}111)$$

which describes a nonlinear relation between θ_e and ϕ.

In frequency synthesizers, the reference is typically generated from a reference oscillator and is lower than the VCO frequency, which is divided by a programmable divider. Both signals are therefore square waves rather than sine waves, and theoretically, a diode ring can be driven from those two signals.

A drawback is that the output voltage of the diode ring is very small, about several hundred millivolts at most, and a postamplifier is required, which is bound to generate noise. Some modification of this analog circuit is possible to increase the voltage.

Figure 4-86 shows a phase detector circuit used in the frequency synthesizer of the Rohde & Schwarz EK47 shortwave receiver. This balanced mixer arrangement has a limited capture range but supplies enough output voltage and therefore does not require an additional amplifier. There are several possible combinations of this circuit, and because it is not a double-balanced mixer, some harmonics may be at the output, and care has to be taken to avoid having any unwanted spikes on the control line.

4-4-2 Exclusive ORs

The exclusive-OR gate is, to a certain degree, the equivalent of a balanced mixer. However, there are certain restrictions. Let us take a look at several waveform combinations. Figure 4-87 shows the case where two waveforms of equal frequency and different phase are applied, and the resulting output from the exclusive-OR gate. If the reference and the VCO waveform have the same duty cycle, the output of the phase/frequency discriminator is clearly defined. In the case of a phase shift of $\pi/2$, the output results in a square wave of twice the frequency, with a duty cycle of 50%. If the waveforms do not have the same duty cycle, things become more complicated. If the VCO frequency is divided by a programmable counter, the pulses from the programmable divider are fairly narrow and thus the duty cycle becomes very small. It is therefore possible, because of the unsymmetrical form of the waveform, that the output voltage is the same for two different phase errors, depending on the duty cycle. To avoid difficulties, it is necessary to add an additional stage that acts as a pulse stretcher, and this pulse stretcher will make the waveform approximately symmetrical.

292 LOOP COMPONENTS

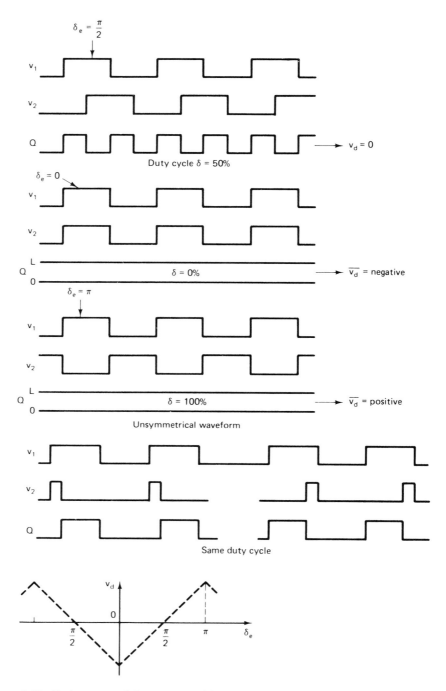

Figure 4-87 Performance of the exclusive-OR phase detector relative to different waveforms at the input.

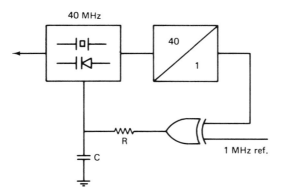

Figure 4-88 Block diagram of a 40-MHz single-loop synthesizer using an exclusive-OR gate as phase detector.

The exclusive-OR gate phase/frequency discriminator should really be used only in cases where the reference and VCO frequency are fairly high and very close together. Figure 4-88 shows a typical application where the exclusive-OR gate is recommended. A 40-MHz crystal oscillator is divided down by 10 and by 4 and provides 1 MHz of output. The reference oscillator is divided down to 1 MHz. Since the crystal oscillator at most will be 1 or 2 kHz off the reference frequency, it is well within the capture range of the exclusive-OR gate, and if both duty cycles are equal and 50%, the circuit can be made stable with a fairly simple RC low-pass filter. This is a type 1 second-order phase-locked loop and will follow the equations in Chapter 1. As the loop bandwidth can be made as narrow as 10 Hz, it will probably compensate only for temperature drift and aging of the crystal relative to the frequency standard.

Example 4 Let us assume that we have a crystal that can be pulled 2 kHz with 10 V, and that our phase detector operates from 0 to 5 V. The product $K = K_0 K_\theta / N$ then equals 100 Hz. The 3-dB bandwidth of this simple synthesizer without any loop filter added would result in a loop bandwidth of 100 Hz. An additional filter would be required to reduce the lop bandwidth down to 10 Hz. In Section 1-10 we learned that there is a beat frequency generated at the output of the phase detector, and the 10-Hz low-pass filter will attenuate the output voltage at the beat frequency.

Let us assume that our initial condition is that the 40-MHz crystal is aligned to be within 1 Hz of final frequency and that the loop is closed. For most receivers, a proportionally controlled crystal oscillator is used that has a warm-up time of 1 or 2 min for the internal standard to reach final frequency. As a result, the initial offset at the output can be 2000 Hz or 200 Hz at the phase/frequency comparator. In this case, our formula for the pull-in range applies,

$$T_p = \frac{\Delta \omega_0^2}{2\zeta \omega_n^3} \qquad (4\text{-}112)$$

294 LOOP COMPONENTS

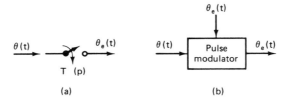

Figure 4-89 Switch shown as a pulse modulator.

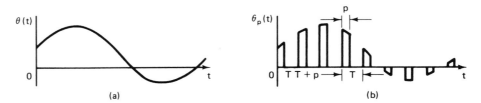

Figure 4-90 Input and output waveform of a uniform sampling device.

and we will insert the values

$$\Delta\omega = 2000 \times 2\pi$$
$$\zeta = 0.7$$
$$\omega_n = 10 \times 2\pi$$

or

$$T_p = \frac{2000}{1.4\pi} = 454 \text{ s}$$

However, as the final error after 1 min or 90 s approaches 1 Hz, the pull-in time T_p becomes 4.49 ms.

Most oven-controlled frequency standards have a type 2 second-order servo control system, and therefore the frequency of the frequency standard will go through the desired value and become higher, and then it will settle at the final frequency. Because of this, the frequency standard, so to speak, sweeps the 40-MHz crystal, and therefore the locking is made much faster than 454 s, even under the assumption that the initial frequency error was 2000 Hz because of aging or drift.

4-4-3 Sample/Hold Detectors

Phase detection can also be accomplished with a linear time-varying switch that is closed periodically. Mathematically, the switch can be described as a pulse modulator, as shown in Figure 4-89. If the operation of the sampling switch is periodic, that is, if the sampler closes for a short interval P at instants $T = 0$, T, $2T, \ldots, nT$, the sampling is uniform. The wave shapes of the input and output signals of a uniform rate sampling device are shown in Figure 4.90. The output can be considered to be

$$\theta_e(t) = \theta_i(t)\, \theta r(t) \tag{4-113}$$

where $\theta_r(t)$ can be assumed to be a periodic train of constant-amplitude pulses of amplitude A_r, width p, and period T. Since $\theta_r(t)$ is periodic, it can be expanded in a Fourier series as [plot $\theta_r(t)$ here]

$$\theta_r(t) = \sum_{n=0}^{\infty} C_n \cos n\omega_0 t \tag{4-114}$$

where

$$C_n = \frac{2}{T}\int_{-P/2}^{P/2} \frac{A_r}{2} \cos n\omega_0 t\, dt$$

$$= 2\frac{A_r}{T}\left[\left(\sin \frac{n\omega_0 P}{2}\right) - \frac{1}{n\omega_0}\right]\quad (n \neq 0) \to \frac{A_r}{T} P \quad (n \approx 0) \tag{4-115}$$

Thus,

$$\theta_r(t) = \frac{A_r}{T}P + \sum_{n=1}^{\infty} 2\frac{A_r}{T} \sin \frac{n\omega_0 P/2}{n\omega_0} \cos n\omega_0 t \tag{4-116}$$

If the input signal is a sine wave,

$$\theta_i(t) = A_i[\sin(\omega_i t + \theta_i)] \tag{4-117}$$

then

$$\theta_e(t) = \theta_r(t)\,\theta_i(t)$$
$$= \frac{A_i A_r}{T}\left\{P \sin(\omega_i t + \theta) + \sum_{n=1}^{\infty} \sin \frac{n\omega_0 P/2}{n\omega_0}\right. \tag{4-118}$$
$$\left. \times [\sin(n\omega_0 + \omega_i)t + \theta_i] + \sin(\omega_i t + \theta_i - n\omega_0 t)\right\}$$

when the loop is in lock ($\omega_i = \omega_0$). The dc term is

$$\theta_e(t)_{\text{dc}} = \frac{A_i A_r}{T} \sin \frac{\omega_0 P/2}{\omega_0} (-\sin \theta_i) \tag{4-119}$$

For small θ_i, the error signal is proportional to the phase difference θ_i. Therefore, the linear time-varying switch is able to serve as a phase detector. It differs from

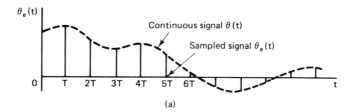

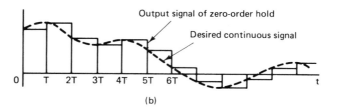

Figure 4-91 Zero-order data hold filter.

the mixer in that the dc output is zero when $\sin\theta_i = 0$, that is, when the oscillator and reference signal are in phase. The mixer type of phase detector is nulled when the two signals are in phase quadrature. Also, when the loop is in lock ($\omega_i = \omega_0$), the mixer output contains a dc term and the second harmonic, whereas the sampled output contains a dc term plus all harmonics of the input frequency. Therefore, the low-pass filter requirements for the sampling type of phase detector are more stringent than those for the sinusoidal mixer.

Fortunately, there are filters that can easily be implemented for the sampling PD. The most commonly used is the *zero-order data hold* (ZODH) or "boxcar generator." The zero-order data hold is a device that converts the pulse of width P to constant-amplitude pulses of width T, as shown in Figure 4-91. The output of the zero-order data hold $\theta_0(t)$ between the sampling instants t_i and t_{i+1} is

$$\theta_0(t) = \theta_e(t_i)[u(t) - u(t_i)] \qquad (4\text{-}120)$$

where $\theta_e(t)$ is the value of $\theta_e(t)$ at the sampling time t_i. Although the exact analysis of the finite-pulse-width sampler and ZODH combination is complex, the frequency response can be approximated closely if the sampling process is replaced by an "ideal sampler" whose output is a train of impulses. That is, the sampled signal $\theta^*(t)$ is a train of amplitude-modulated impulses

$$\theta^*(t) = \theta_i(t)\delta_T(t) \qquad (4\text{-}121)$$

where $\delta_T(t)$ is a unit impulse train of period T:

$$\delta_T(t) = \sum_{n=-\infty}^{\infty} \delta(t - nT) \qquad (4\text{-}122)$$

where $\delta(t - nT)$ represents an impulse of unit area occurring at time $t = nT$. Since $\delta_T(t)$ is periodic, it can be expressed by the Fourier series

$$\delta_T(t) = \sum_{n=-\infty}^{\infty} C_n e^{-jn\omega_0 t} \tag{4-123}$$

where

$$\omega_0 = \frac{2\pi}{T} \tag{4-124}$$

The constants C_n are determined from

$$C_n = \frac{1}{T}\int_{-T/2}^{T/2} \delta_T(t) e^{-jn\omega_0 t} dt = \frac{1}{T} \tag{4-125}$$

That is, the frequency spectrum of impulse train of period T contains a dc term plus the fundamental frequency and all harmonics, all with an amplitude of $1/T$.

$$\delta_T(t) = \frac{1}{T}\sum_{n=-\infty}^{\infty} e^{jn\omega_0 t} \qquad \omega_0 = \frac{2\pi}{T} \tag{4-126}$$

and since

$$e^{jn\omega_0 t} + e^{-jn\omega_0 t} = 2\cos n\omega_0 t \tag{4-127}$$

$$\delta_T(t) = \frac{1}{T} + \frac{2}{T}\sum_{n=1}^{\infty} \cos n\omega_0 t \tag{4-128}$$

Therefore, Eq. (4-121) can be written

$$\theta^*(t) = \theta_i(t)\frac{1}{T} + \frac{2}{T}\sum_{n=1}^{\infty} \cos n\omega_0 t \tag{4-129}$$

If the input $\theta_i(t)$ is a sine wave $\theta_i(t)_i = A_i \sin(\omega_i t + \theta_i)$,

$$\theta^*(t) = \frac{A_i}{T}\left[\sin(\omega_i t + \theta_i) + 2\sum_{n=1}^{\infty} \cos n\omega_0 t \sin(\omega_i t + \theta_i)\right] \tag{4-130}$$

This equation is similar to the result obtained in Eq. (4-118) for the more realistic finite-pulse-width model of the sampler. The difference is that for the finite-pulse-width model, the harmonics are attenuated by the factor

$$\frac{\sin(n\omega_0 P/2)}{n\omega_0}$$

With the impulse sampler, all harmonics are of amplitude $2/T$.

LOOP COMPONENTS

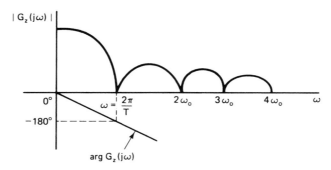

Figure 4-92 Transfer characteristic of the ZODH filter.

The impulse response of the ZODH is

$$u(t) - u(t - T) = \frac{1 - e^{-sT}}{s} \qquad (4\text{-}131)$$

and the ZODH frequency response is

$$\frac{1 - e^{-j\omega T/2}}{j\omega} = \frac{2}{T} e^{-j\omega T/2} \frac{e^{j\omega T/2} - e^{-j\omega T/2}}{j\omega T/2} \qquad (4\text{-}132)$$

$$G_z(j\omega) = \frac{2}{T} e^{-j\omega T/2} \frac{\sin(\omega T/2)}{\omega T/2} \qquad (4\text{-}133)$$

which is a digital low-pass filter with a linear phase response, as illustrated in Figure 4-92. An important feature of this filter is that it has zero gain at the sampling frequency and at all harmonics thereof. As Eq. (4-118) or (4-130) shows, when the input and sampling frequencies are equal, the output of the sampler contains a dc term and all harmonics of the sampling frequency. Since the ZODH has zero gain at these frequencies, the unwanted harmonics are completely removed by the filter. This is one of the primary reasons for the widespread application of samplers in phase-locked loops. The ZODH has a phase lag that increases linearly with frequency. This negative phase shift can seriously degrade loop stability.

Whenever the only frequency-sensitive components in the PLL are the VCO and the sampler plus zero-order data hold, the loop stability and frequency response are readily analyzed. When the loop is in frequency lock, the system can be represented as shown in the block diagram of Figure 4-93. The open-loop transfer function is

$$G(j\omega) = \frac{2}{T} \frac{K_v e^{-j\omega T/2}}{j\omega} \frac{\sin(\omega T/2)}{\omega T/2} \qquad (4\text{-}134)$$

At the crossover frequency ω_c, the open-loop phase shift is

$$\phi = -\frac{\pi}{2} - \frac{\omega_c T}{2} \qquad (4\text{-}135)$$

and the phase margin is

$$\phi_m = \pi + \phi = \frac{\pi}{2} - \frac{\omega_c T}{2} \quad (4\text{-}136)$$

Since the magnitude of the open-loop gain at ω_c is unity,

$$K_v \frac{\sin(\omega_c T/2)}{(\omega_c T/2)^2} = 1 \quad (4\text{-}137)$$

or

$$K_v = \frac{(\omega_c T/2)^2}{\sin(\omega_c T/2)}$$
$$= \frac{(\pi/2 - \phi_m)^2}{\sin(\pi/2 - \phi_m)} \quad (4\text{-}138)$$

Equation (4-138) describes the relation between phase margin ϕ_m and loop gain K_v. The plot of K_v as a function of ϕ_m given in Figure 4-94 shows that for each value of ϕ_m, there is a single value of K_v. For a $K_v = (\pi/2)^2$, the phase margin is 0°. As K_v is decreased, the phase margin increases and reaches 90° for $K_v = 0$.

The effect on loop performance of changes in the sampling rate T can be determined in the same manner. Since at the crossover frequency ω_c the magnitude of the open-loop gain is unity,

$$K_v \frac{\sin(\omega_c T/2)}{(\omega_c T/2)^2} = 1 \quad (4\text{-}139)$$

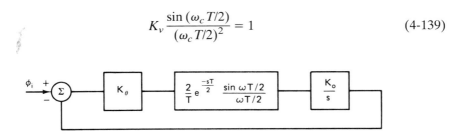

Figure 4-93 Block diagram of a PLL with a sample/hold comparator.

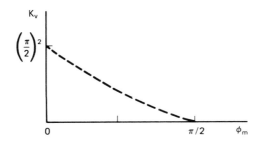

Figure 4-94 Plot of K_v as a function of ϕ_m.

If K_v remains constant and the sampling rate T is changed, ω_c must change such that

$$\omega_c T = \text{constant}$$

That is, if the sampling rate is decreased (T increases), the crossover frequency must decrease so that $\omega_c T$ remains constant. Therefore, changing the sampling rate of the system has no effect on the system phase margin or system stability; it affects only the loop bandwidth.

Example 5 Calculate the value of K_v required for a 45° phase margin in a PLL whose open-loop transfer function is given by Eq. (4-134).

Solution. In order to have a 45° phase margin, the phase lag of the sample/hold comparator must be 45° at the crossover frequency. Therefore,

$$\frac{\omega_c T}{2} = \frac{\pi}{4} \tag{4-140}$$

and the crossover frequency must be

$$\omega_c = \frac{\pi}{2T} \tag{4-141}$$

Since the magnitude of the open-loop gain is unity at the crossover frequency, K_v is determined from

$$K_v \frac{\sin(\pi/4)}{(\pi/4)^2} = 1 \tag{4-142}$$

or

$$K_v = \left(\frac{\pi}{4}\right)^2 \sqrt{2} \tag{4-143}$$

The sample/hold comparator, however, has a somewhat limited frequency range. In frequency synthesizer applications, it is not very likely that it will be used below a few hundred hertz, and the upper limit is determined by the speed of the following circuit and the crosstalk. Crosstalk depends on the isolation of the CMOS switch.

Let us take a look at a practical circuit using the sample/hold comparator. It is typically used in a cascaded form, which means that there are two samplers. The reason for this is the fact that one gets better reference suppression. Figure 4-95 shows a dual sample/hold comparator with the additional filtering circuits. The 10-kHz output from the divider, depending on the division ratio, is 200 to 500 ns wide, and a special circuit acts as a pulse stretcher to increase the width of these pulses to 2 μs.

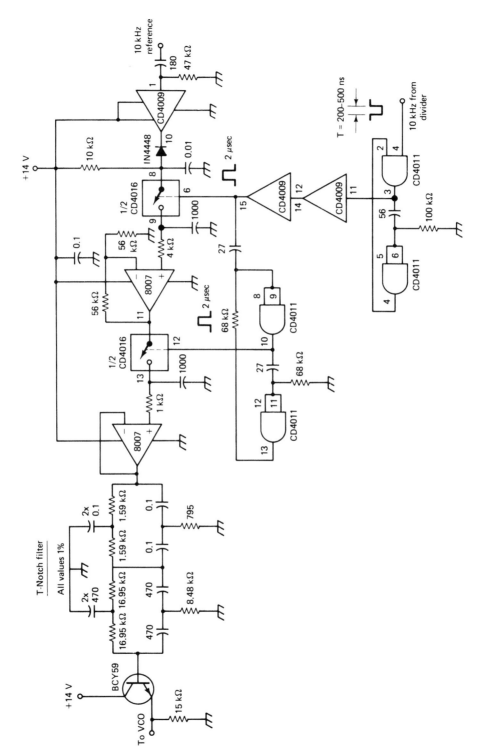

Figure 4-95 Dual sample/hold comparator with additional filtering.

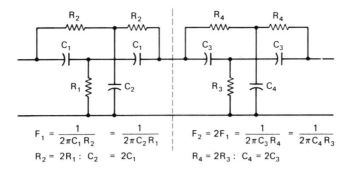

Figure 4-96 T-notch filter with design equations.

The two CD4009s decouple the circuit. Two CD4016s are being used as CMOS switches. The first CD4009 generates the ramp, and the two 8007 operational amplifiers again provide the necessary decoupling and high input impedances.

In the output of the second 8007, we have a *T-notch filter*, which is a minimal phase shift filter with about 40 dB of suppression of the reference. In this particular circuit two T-notch filters are being used to suppress the fundamental and first harmonic frequency. In Section 4-7 it will be shown that at times the active second-order low-pass filter is a good replacement for the T-notch filter because it attenuates all harmonics rather than two discrete frequencies and requires fewer low-tolerance components.

Figure 4-96 shows the T-notch filter with the design formulas.

The highest recommended frequency of operation for the phase/frequency comparator is approximately 5 kHz. At frequencies above this, the CMOS switches are not fast enough. The high impedance and parasitic capacitance produce crosstalk, and only 60 dB of attenuation is possible.

While sample/hold comparators will typically operate from 12 to 24 V, the minimum crosstalk voltage that determines the resolution and the reference suppression is about $10\,\mu V$. The attenuation possible with the sample/hold comparator is therefore about 110 dB. This is a highly theoretical value and in practice depends on the relationship of reference frequency and desired cutoff frequency. In many cases, the loop bandwidth is set to one-half of the reference frequency, and then delays and stray effects become very critical.

4-4-4 Edge-Triggered JK Master/Slave Flip-Flops

The fundamental idea of the sequential phase comparator we will be dealing with is that there are two outputs available, one to charge and one to discharge a capacitor. Output 1 then is high if the signal 1 frequency is greater than the signal 2 frequency; or if the two frequencies are equal, if signal 1 leads signal 2 in phase.

Output 2 is high if the frequency of signal 2 is greater than that of signal 1, or if the signal frequencies are the same and signal 2 leads signal 1 in phase.

Figure 4-97 shows the minimum configuration to build such a phase comparator. It can be operated from -2π to $+2\pi$, and an active amplifier is recommended

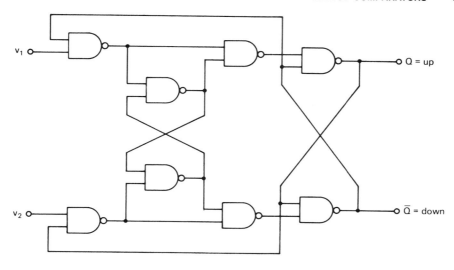

Figure 4-97 Edge-triggered JK master/slave flip-flop.

as a charge pump. The Q output of the JK master/slave flip-flop is set to 1 by the negative edge of the signal 1, while the negative edge of the signal 2 resets it to zero. Therefore, the output $\bar{Q}$ is the complement of Q. The output voltage $\bar{V}$ is defined as the weighted duty cycle of Q and $\bar{Q}$. This means that a positive contribution is made when $Q = 1$ and a negative contribution (discharge) is made when $Q = 0$. The averaging and filtering of the unwanted ac component are done by the following integrator. The integrator circuit then is called a *charge pump*, as the loop capacitor is being charged and discharged depending on whether Q is high or low.

If the system using the JK flip-flop is not in lock and there is a large difference between frequencies F1 and F2 at the input, the output is not going to be zero but will be positive or negative relative to one-half the supply voltage. This is an advantage and indicates that this system is frequency sensitive. We therefore call it a *phase/frequency comparator* because of its capability to detect both phase and frequency offsets. In its locking performance and pull-in performance, it is similar to the exclusive-OR gate.

For better understanding, let us look at a few cases where the system is in lock. It should be noted that whereas the exclusive-OR gate was sensitive to the duty cycle of the input signals, the JK flip-flop responds only to the edges, and therefore the phase/frequency comparator can be used for unsymmetrical waveforms. Let us assume first that the input signals 1 and 2 have the same frequency. Figure 4-98 shows what happens if the phase error is about 0, π, and 2π. In those cases, the duty cycle at the output is about 0, 50%, or 100%, respectively. The narrow output pulses may cause spikes on the power supply line and lines in the vicinity, and certain precautions have to be taken to filter them.

The output voltage $\bar{V}$ is the average of the signal Q and is a linear function of the phase error.

Now let us take a look at several cases where the system is not in lock. Figure 4-99 shows the case where frequency 1 is substantially higher than frequency 2.

304 LOOP COMPONENTS

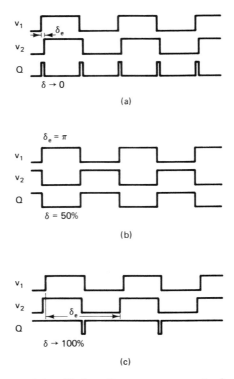

Figure 4-98 Performance of the JK phase/frequency comparator for different input signals.

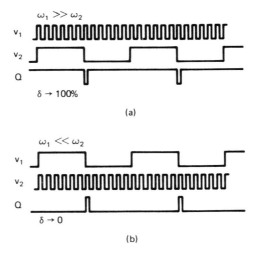

Figure 4-99 Phase detector output for two input frequencies that are substantially different.

As a result, the output duty cycle is close to 100% and the VCO frequency is being pulled up to higher frequencies. If the frequency at input 2 is much higher than that at input 1, the opposite is true. This proves that this device is sensitive to frequency changes.

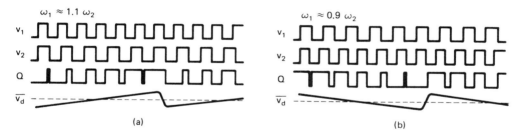

Figure 4-100 Performance of the phase detector for small frequency errors.

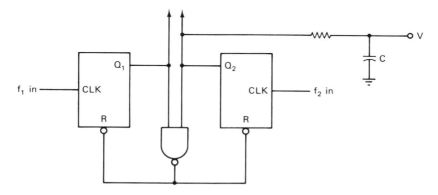

Figure 4-101 Phase detector with two D flip-flops and a NAND gate. This type of phase detector will be called a tri-state comparator.

In cases where both frequencies are about the same, as shown in Figure 4-100, the crossover area is not clearly defined. The first picture shows the case where frequency 2 is 10% higher than frequency 1 and the duty cycle is changing periodically between 0 and 100%. Therefore, the ac voltages look like a sawtooth, with a rate equal to the difference of both frequencies. The same holds true if the two inputs are reversed. In the case where both frequencies are identical, the JK flip-flop behaves the same way as an exclusive-OR gate. From this discussion it can be concluded that, while this phase/frequency comparator was included to explain how it works, it is not a very desirable device because of the uncertainty of its behavior close to lock.

4-4-5 Digital Tri-state Comparators

The digital tri-state phase/frequency comparator is probably the most universally used and most important next to the sample/hold comparator. Although the ring and exclusive-OR gate have some applications, the tri-state phase/frequency comparator can be used widely. Even in cases where a sample/hold comparator theoretically could be used, it may be inferior as far as reference attenuation or noise is concerned, but it is generally well behaved. Unfortunately, the tri-state system is very complex and shows a number of unusual phenomena. Such a digital tri-state comparator is shown in Figure 4-101 using two D flip-flops and a NAND gate. The Q_2 output signal is filtered with the low-pass filter. The operation of

306 LOOP COMPONENTS

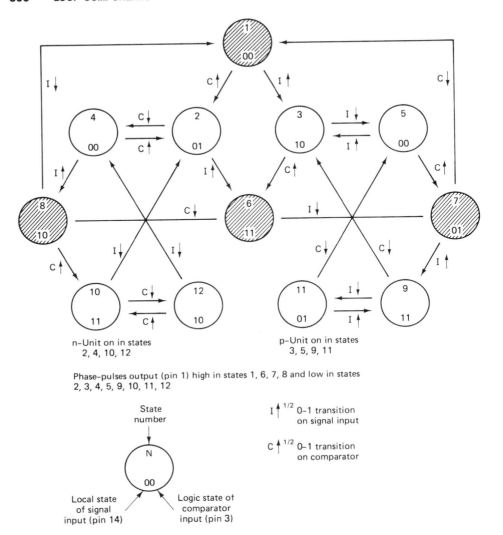

Figure 4-102 Logic diagram of the tri-state detector.

this logic circuit is readily analyzed using the state transition diagram shown in Figure 4-102. The D flip-flop outputs go high on the leading edge of their respective clock inputs and remain high until they are reset. The rest signal occurs when both inputs are high. When both signals are in phase and of the same frequency, both outputs will remain low, and no signal will be applied to the operational amplifier. When the two signal frequencies are the same, the dc output voltage transfer characteristic will be as shown in Figure 4-103. If the two signal frequencies are not the same, the output voltage will depend on both the relative frequency difference and the phase difference. The timing diagram of Figure 4-104 illustrates the case in which $f_2 = 3f_1$. In part (a) of the figure, the leading edge of f_1 occurs just after that of f_2, so that Q_2 is high 50% of the time, and the average value

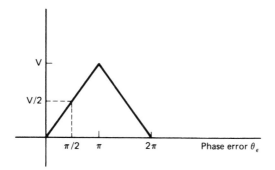

Figure 4-103 Transfer characteristic of the tri-state phase/frequency comparator.

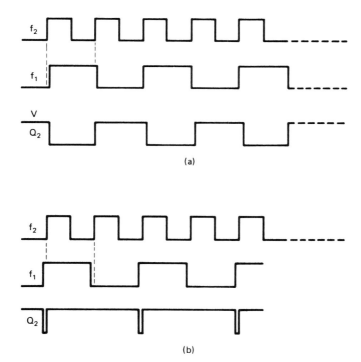

Figure 4-104 Output waveforms of the tri-state frequency comparators for different input frequencies.

of the PD output is 50%. In part (b) of the figure, the leading edge of f_1 occurs just before that of f_2, so Q_2 is high almost all the time and the average output voltage is approximately V. The output voltage averaged over all phase differences is then 67% for $f_2 = 3f_1$. In general, it can be said that the average output (averaged over all phase differences) is given by

$$V_{\text{ave}} = 1 - \frac{f_1}{f_2} V \qquad (4\text{-}144)$$

308 LOOP COMPONENTS

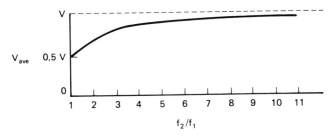

Figure 4-105 Average output voltage as a function of frequency ratio.

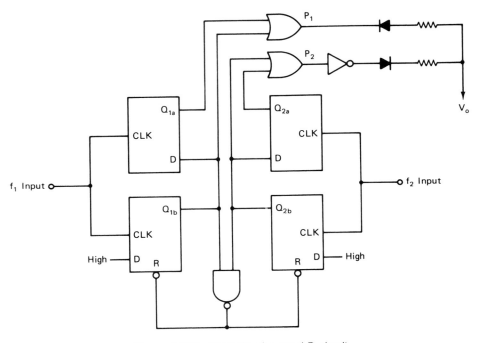

Figure 4-106 Example of a quad-D circuit.

provided that f_2 is greater than f_1. This expression is plotted in Figure 4-105 together with the cases in which f_1 is greater than f_2.

The digital network used in this realization is only one of a large number of logic circuits that could be used. Many IC manufacturers now produce a *quad-D circuit*, which functions much like the dual D flip-flop; the main difference is that when the frequency of one signal is more than twice that of the other signal, the corresponding output will be high all of the time. Therefore, a larger voltage is applied to the VCO and the loop response is faster. An example of a quad-D circuit is shown in Figure 4-106.

The most popular digital tri-state phase/frequency comparator on the market is the one used in the CD4046 PLL IC, shown in Figure 4-107. It contains an

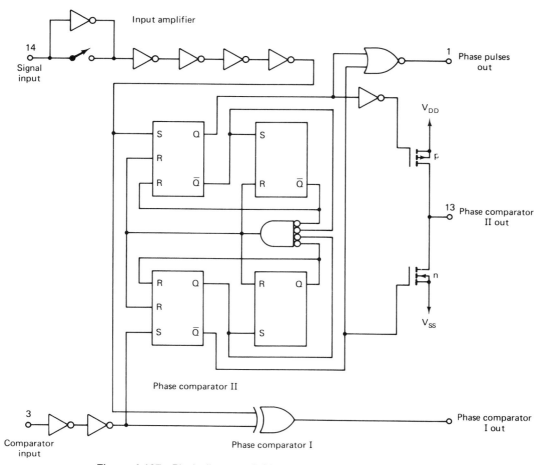

Figure 4-107 Block diagram of CD4046 phase/frequency comparator.

additional phase comparator, an exclusive-OR gate that can be used as a lock indicator. In addition, two field-effect transistors are used to sum the two outputs. A slightly faster version in TTL techniques is the Motorola MC4044.

The fastest version in ECL is the MC12040, also made by Motorola, shown in Figure 4-108. Sometimes it is convenient to build the phase/frequency comparator in discrete technique to add additional features. Figure 4-109 shows an example.

This particular tri-state phase/frequency comparator has a peculiarity that was first mentioned by Egan and Clark [1]. When actually building a phase-locked loop with this phase/frequency comparator or the CD4046 type by going through the normal mathematical design routine, it will become apparent that the expected performance and the actual results are not the same.

1. The reference suppression will be better than expected.
2. The phase error or tracking will be worse than expected.

310 LOOP COMPONENTS

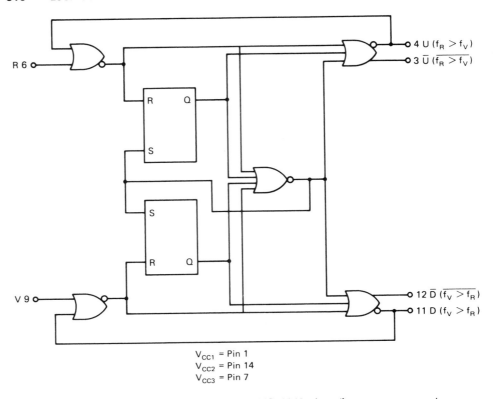

Figure 4-108 Block diagram of Motorola MC12040 phase/frequency comparator.

3. The phase margin will differ and the system may not lock despite the fact that the calculation is correct.

The reason for this is due to two effects:

1. The flip-flops are not absolutely alike, and as a result of this, the output in the crossover region is not zero.
2. If there is very little or no correction voltage required, the gain of the phase detector will drop substantially.

Let us assume the ideal situation where the output of the phase/frequency comparator feeding the charge pump does not have to correct any error, the system is drift free, and there are no leakage currents. The holding capacitor of the charge pump would maintain constant voltage, and, as there is no drift, no correction voltage is necessary.

The flip-flops, however, introduce a certain amount of jitter, and a certain amount of jitter is also introduced by the frequency dividers, both the reference divider and the programmable divider. This jitter results in an uncertainty regarding the zero crossings, and extremely narrow pulses will appear at the output of this summation amplifier used in the CD4046.

PHASE/FREQUENCY COMPARATORS

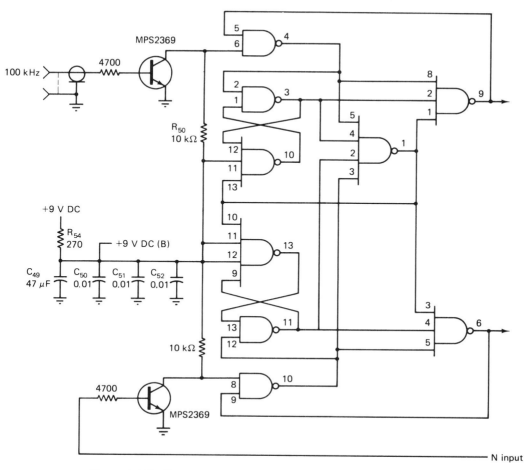

Figure 4-109 Possible version of tri-state phase/frequency comparator.

Under the ideal assumption that there are no corrections required and those pulses would not exist, the reference suppression would be infinite, as there is no output and, therefore, the reference suppression, disregarding the effect of the loop filter, depends only on how well this condition is met.

The change of gain seems somewhat surprising, but as we think of it, if there is no correction and no update, there is also no gain. It is impossible to meet this condition, which is fortunate, but with regard to the temperature stability and aging effect of some devices, we may have some difficulties as far as predicting the actual performance.

There are several remedies to this problem. A simple version is to introduce a controlled amount of leakage. While the electrolytic capacitor required in the charge pump will have a leakage current, it is better to use a leakage current that is independent of temperature and aging. This can be accomplished by putting a 1-MΩ resistor from the output of the CD4046 to ground. The phase/frequency comparator then has to deliver an output current, and this output current is determined by a resistor that can be independent of temperature and other effects. As a result of this, the duty cycle of the output pulses of the phase/frequency comparator will change and the pulses will become wider. As these pulses contain more energy, the reference suppression will suffer.

It is theoretically possible to put one side of the 1-MΩ resistor, instead of to ground, to the wiper of a potentiometer and set the voltage in such a manner that this offset is compensated, but again, because the phase will shift theoretically, one has to adjust the potentiometer according to the actual phase error. This is not a very convenient arrangement.

A somewhat better method was proposed by Fairchild several years ago, but the hardware was never realized. It was proposed to insert a gate in one of the output arms of the phase/frequency comparator, before the signal is fed to the summation amplifier, and a periodic current disturbance is introduced. This disturbance has the same rate as the reference frequency and is of extremely small duration, so that the output contains only fairly high harmonics of reference, which is easily filtered as it contains very little energy. This periodic disturbance offsets the output of the phase/frequency comparator and therefore has an effect similar to that of a leakage resistor. The advantage of this method, however, is that this is done at a fairly high frequency and does not introduce low-frequency noise generated by the 1-MΩ resistor.

Figure 4-110 shows the circuit that accomplishes this, and Figure 4-111 shows the effect on the output pulses. The charge pump output exhibits a short negative-going pulse followed immediately by a short positive-going pulse. This can also be called an *antibacklash* feature, and it prevents operating in the dead zone. This zone is not really a dead zone because of the leakage currents in the tuning diode. The duration and proximity of these pulses are such that they cause no net change to the charge of the integrator. Figure 4-112 shows the response of a phase/frequency detector near loop lock, including the dead zone; this may not be true for ECL.

Another method, developed by Mr. Fritze of Rohde & Schwarz, Munich, is used in the Rohde & Schwarz ESM 500 receiver and described below.

Tri-state Phase/Frequency Comparator with Pulse Compensation. Because of the leakage in capacitors and the tuning diodes, the phase/frequency comparator has to supply current pulses until the phase difference becomes zero, and as mentioned, under ideal conditions (i.e., without drift), the output would be zero. Since the ideal configuration cannot be realized and because of the finite switching times of the current sources, there are some pulses required all the time.

Using the 1-MΩ resistor, the effect that is occurring is that the charge, even under extremely narrow conditions contained in those pulses, may be more than is required to compensate for the drift caused by the leakage. While the resistor now has effectively prevented the dead zone, the 1 MΩ may have overcompensated this effect. As a result, we have avoided the instabilities of the loop caused by the dead zone and traded them for another type of oscillation that appears like residual FM very close to the carrier as a result of the overcompensation.

The sources that can be practically built have residual currents that do not fully cancel, and the two residual currents are not identical. This, together with any charging phenomena, generates a disruptive effect and causes a permanent charge

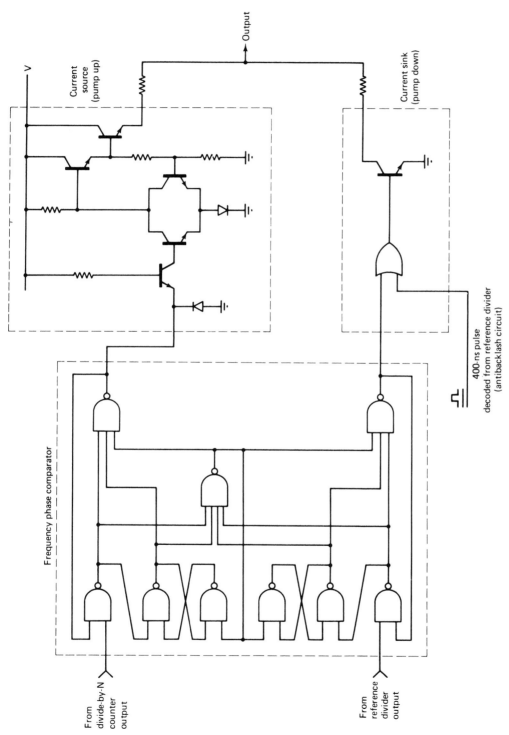

Figure 4-110 Tri-state detector with antibacklash circuit included.

314 LOOP COMPONENTS

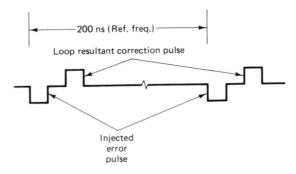

Figure 4-111 Output of frequency/phase detector with antibacklash circuit.

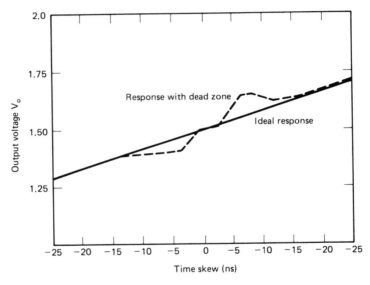

Figure 4-112 Response of frequency/phase detector near loop lock resulting in a dead zone.

and discharge of the holding capacitor, and a frequency modulation occurs. The invention by Mr. Fritze avoids the disadvantages in the steady-state condition described, and most important, fluctuations due to residual currents from the current sources or reverse currents from the tuning diodes are avoided without negatively influencing the noise characteristic of the oscillator. Let us take a look at the schematic, Figure 4-113. It contains the familiar phase/frequency comparator and the two charging pumps. An auxiliary circuit has been added to overcome the difficulties described. Capacitor 6 is charged, and capacitor 7, capacitor 6, and the parallel resistor form the familiar time constant for the lag network. An additional auxiliary circuit, consisting of an operational amplifier and several capacitors and diodes, generates a correction current that is proportional to the integral of the difference of the charge pulses supplied by the current sources; that is, the current is directly proportional to the steady-state condition of the two

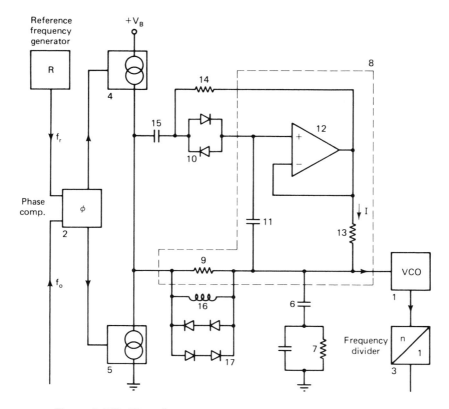

Figure 4-113 Phase/frequency comparator with spike compensation.

mutually canceling charge amounts in the ideal state. By the use of such an auxiliary circuit, the constant recharging current pulses across the phase comparsion circuit, which are required in conventional circuits for the compensation of such errors, are avoided, and the charging errors that may occur as a result of residual currents in the current sources or as a result of reverse currents of the varactor diodes are compensated by the correction current. No additional circuit arrangements are necessary, which may unfavorably influence the noise characteristic of the oscillator, and therefore the noise performance is not changed.

A phase-controlled high-frequency oscillator is shown in Figure 4-114, having oscillator 1, which is frequency controlled by the application of a dc voltage. Oscillator 1 has an output frequency f_o that is compared in phase comparator 2 with a reference frequency f_r, which is generated by reference frequency generator R. The output frequency of oscillator 1 may, for example, be changed in a known manner by means of interconnected frequency divider 3, which may have an adjustable division ratio. Depending on the phase relation of the oscillator frequency f_o with respect to the reference frequency f_r, phase comparison circuit 2 supplies appropriate control pulses to adjustable positive direct current source 4 and to a correspondingly adjustable negative direct current source 5. Current source 4 is connected to a positive voltage U_B, while the negative current source is connected to ground. Control pulses from phase comparison circuit 2 are

316 LOOP COMPONENTS

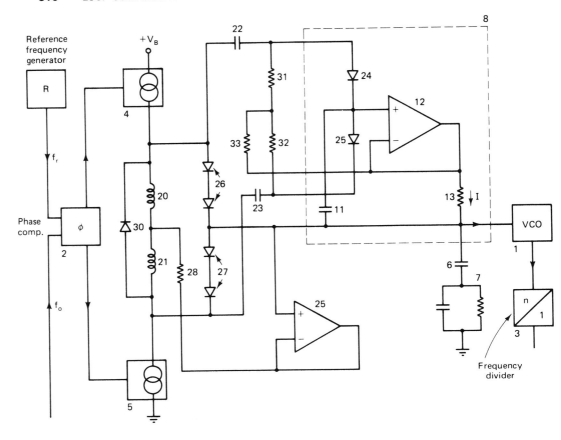

Figure 4-114 Schematic of the phase/frequency comparator with spike suppression and linearized performance.

converted in switchable current sources 4 and 5 into current pulses of corresponding pulse duration, which are then supplied to charging capacitor 6, where the pulses are integrated to form the control voltage for oscillator 1. Oscillator 1 may be tuned, for example, by varactor diodes. To avoid drift, charging capacitor 6 is connected in series to RC circuit 7.

In accord with the invention, a correction dc I is generated by auxiliary circuit 8 enclosed in the dashed lines of Figure 4-113.

It is approximately proportional to the integral of the difference between the charge pulses supplied from the respective current sources 4 and 5. Additional circuit elements, described in greater detail later, ensure that only relatively short charge pulses, which in the ideal case would mutually cancel, generate a corresponding correction current in auxiliary circuit 8. The term "relatively short charge pulses" designates pulses attributed only to potential disruptions or to the overlap in the steadystate and that are short in comparison to the charge pulses occurring as a result of deviation or fluctuation between f_o and f_r.

Such pulses, which result from center frequency deviation, have a duration, depending on the particular design of phase comparison circuit 2, which fluctuates between the period of the reference frequency f_r (given a very large center frequency deviation) and substantially zero (for the steady-state condition in the ideal state). In contrast, the "relatively short charge pulses" have a width of approximately 0.1% of the period of the reference frequency f_r and depend on the dimensioning of the remainder of the components in the oscillator circuit.

In the circuit shown in Figure 4-113, in the simplest case auxiliary circuit 8 may consist of resistor 9 connected between the mutually connected current sources 4 and 5 and charging capacitor 6, at which a voltage drop occurs that is proportional to the difference between the current pulses supplied by the respective current sources 4 and 5. After rectification by rectifiers 10, the voltage is integrated by a second capacitor, 11, into a plurality of successive periods so that finally a voltage occurs at auxiliary capacitor 11, which under ideal conditions is proportional to the difference between the current pulses. By "ideal conditions" is meant the conditions wherein the two current sources, 4 and 5, respectively, supply current pulses of identical size and length in the quiescent state. As stated earlier, the current pulses would only be identical in the steady-state condition when the current sources have no residual currents and when no reverse currents resulting from the varactor diodes in oscillator 1 are present to contribute to the voltage formed by charging capacitor 6. When such residual currents or reverse currents exist, the charge pulses are no longer identical even in the steady-state condition, and auxiliary circuit 8 generates a correction current I from such differences. The correction current I, which is proportional to the voltage at the second capacitor, 11, is generated across a "times 1" amplifier 12 with a very low input current by means of impedance transformation and is supplied directly to charging capacitor 6 across resistor 13. The charging current circuit for rectifier diodes 10 is terminated by resistor 14, which is directly connected to the output of amplifier 12. This ensures that rectifier diodes 10 are biased with a voltage that corresponds to the rectified output voltage of amplifier 12. Thus, no quiescent voltage appears at rectifiers 10; therefore, no undesired residual current is generated and, moreover, it is not necessary to supply constant pulses to the circuit for compensation of self-generated residual currents.

Auxiliary circuit 8 is separated by capacitor 15 from current sources 4 and 5. Capacitor 15 has a relatively small capacitance value, so that only relatively short charge pulses, such as mutually overlapping pulses in the steady-state condition, are evaluated in auxiliary circuit 8 for generating the residual current. This is expedient so that longer current pulses, such as occur, for example, as a result of a frequency change and that should recharge as fast as possible capacitor 6 to the new frequency value, are not significantly considered in auxiliary circuit 8. That is, given a frequency change by means of relatively long current pulses, auxiliary circuit 8 retains the correction current previously generated in the steady-state condition, which is then directly utilized after termination of the frequency change. This is because one can make the valid assumption that errors due to residual currents or reverse currents are independent of the frequency; that is, such errors present at an initial frequency will also be present at the new frequency setting for oscillator 1. Auxiliary circuit 8 thus does not supply correction currents that would disrupt the tuning operation for a frequency change.

The blocking of auxiliary circuit 8 for relatively long switching pulses can be further improved as shown in Figure 4-113. An additional inductor 16 ensures that only relatively short current pulses generate a corresponding voltage drop evaluated in auxiliary circuit 8 and allows relatively long pulses to be transmitted unimpeded to capacitor 6. In the same manner, this separation effect between short and long pulses can be further improved by means of the interconnection of additional rectifiers 17, which, in comparison to rectifiers 10 of auxiliary circuit 8, have a higher forward-bias voltage. Relatively short current pulses, therefore, first arrive at rectifiers 10, and only when the higher transmission voltage of rectifiers 17 is exceeded do longer current pulses, which occur as a result of a frequency change, arrive at charging capacitor 6 via rectifiers 17.

It was assumed in the circuit of Figure 4-113 that the two current sources, 4 and 5, supply ideal current pulses of identical size as well as of identical length. Usually, this cannot be achieved in practice. A further development of the circuit of Figure 4-113, shown in Figure 4-114, ensures that such potential asymmetries of the current pulses are also compensated.

As shown in Figure 4-114, rectifiers 24 and 25 of auxiliary circuit 8 are driven separately by the current pulses from the two current sources, 4 and 5, respectively. Two inductors, 20 and 21, are connected at the outputs of the two current sources, 4 and 5. Inductors 20 and 21 do not affect the short current pulses, which are supplied via blocking capacitors 22 and 23 to rectifiers 24 and 25. Inductors 20 and 21 serve in the quiescent condition, that is, between the individual current pulses, to avoid voltages at additional rectifiers 26 and 27 connected in parallel, and thus have a higher transmission voltage than rectifiers 24 and 25. Symmetry is produced via a resistor 28 and an additional "times 1" amplifier, 29. An additional rectifier, 30, limits voltage pulses at the turn-off of the current pulses. A voltage that is proportional to the integral of the difference of the current pulses from the two current sources, 4 and 5, is again integrated by the second capacitor, 11. Bias voltage resistors 31, 32, and 33 are allocated to rectifiers 24 and 25. Resistor 33 is connected to the output of amplifier 12 so that for every voltage present at capacitor 11 it is ensured that the voltage at diodes 24 and 25 beyond the pulse time is equal to zero, and therefore no residual currents flow through these circuit elements.

Blocking capacitors 22 and 23 are dimensioned so that capacitor 11 is not too strongly recharged upon the occurrence of a frequency change, and thus longer current pulses from current sources 4 and 5 flow only across additional rectifiers 26 and 27. Inductor 21 may be eliminated if resistor 28 has an appropriate resistance value.

The circuit shown in Figure 4-114 allows conduction of the entire charge of short pulses supplied by the current sources to charging capacitor 6 via the second capacitor, 11. Therefore, for short current pulses, the voltage change at capacitor 11 coincides with the voltage change at charging capacitor 6 not only with respect to polarity but also with respect to proportionality. In the aligned condition, that is, for short current pulses, for example, the charge flows from voltage source 4 via separating capacitor 22 and rectifier 24 to capacitor 11, and from the current source 5 analogously via separating capacitor 23 and rectifier 25 to capacitor 11. From capacitor 11 the combined voltage is supplied to charging capacitor 6. A positive or negative voltage for the duration of the short pulses arises at diodes 26 and 27 during the charge pulses and capacitors 22 and 23, which are first discharged, can thus conduct the charge pulses to diodes 24 and 25. As long as capacitor 11 is discharged, the output voltage of amplifier 12 is identical to the oscillator control voltage at charging capacitor 6, so that no current flows across resistor 13. This is the case as long as the positive and the negative charge pulses produce equal charge amounts. Blocking capacitors 22 and 23 are quickly discharged across the resistor combination comprised of resistors 31, 32, and 33 after each pulse in such a manner that no voltage remains at diodes 24 and 25. Therefore, no undesired residual current can reach capacitor 11 via those diodes.

When the aligned or balanced condition is disrupted, for example, due to residual currents from one of the two current sources or from the varactor diodes in oscillator 1, the current pulses have different lengths. During a frequency change, one pulse may temporarily disappear. Because short current pulses are not conducted via diodes 26 and 27, which have a higher transmission voltage, but rather are conducted primarily across diodes 24 and 25, capacitor 11 is charged and a current flows across resistor 13. The current across resistor 13 increases until the positive and negative current pulses are again of identical size. A false voltage at capacitor 11 after a frequency change is compensated in the same manner, as is a potential oscillator drift or recharing effect of capacitor 6.

Amplifier 12 has a very low input current and may be, for example, a MOSFET amplifier having an input current that is less than the current of the blocked semiconductors by several magnitudes. It is important for capacitor 11 to have good insulating properties. The noise from amplifier 12 is decoupled from the control voltage at capacitor 6 by resistor 13. The time constant represented by capacitor 6 and resistor 13 is chosen to suppress noise interference. Moreover, the value of resistor 13 is made large in comparison to the resistor of *RC* circuit 7.

The circuit described above has the additional advantage that a potential stepwise detuning of the oscillator is made continuous. In phase-controlled high-frequency oscillators of this type, for example, the reference frequency is often supplied from a synthesizer, which is adjustable in decades with crystal frequency precision. Frequency divider 3 may also be adjustable in decades. The frequency of such oscillators is thus variable in very fine steps, and when tuning of the synthesizer or of the frequency divider occurs over a number of decades via a suitable pulse control circuit, a quasicontinuous, that is, stepwise, tuning of oscillator 1 is achieved in a predetermined frequency range. The presence of such relatively small frequency steps results in correspondingly short charge puls03es from the current sources, which, as described earlier, lead to the generation of a correction current via auxiliary circuit 8. Oscillator 1 is therefore no longer adjusted in terms of its frequency by jumps or steps but is instead continuously adjusted via the auxiliary current.

4-5 WIDEBAND HIGH-GAIN AMPLIFIERS

4-5-1 Summation Amplifiers

Depending on the application, we have two types of high-gain amplifiers that require a fairly high bandwidth.

1. Operational amplifiers for the loop filter.

320 LOOP COMPONENTS

2. Wideband amplifiers that either act as isolation amplifiers or that transform sine-wave voltages into square-wave voltages (differential limiters), or power amplifiers that raise the output level of a synthesizer. Although they belong in the category of isolation amplifiers, they also require some different considerations.

Let us start with the operational amplifiers required for the loop filters. Although we will be dealing with loop filters, specifically active loop filters, in great detail, we should touch on the requirement for operational amplifiers. As in some cases the reference frequency will be as high as several megahertz, in order to maintain a wide loop bandwidth, the operational amplifier has to be able to track the frequencies involved. The *slew rate* of the operational amplifier, which determines how many volts per microsecond the operational amplifier can follow, is one figure of merit that has to be taken into consideration, and it is interesting that most operational amplifiers that do not have fixed internal frequency compensation are fairly poor in slew rate. In addition, the more familiar cutoff frequency of the operational amplifier indicates the 3-dB drop in gain in open-loop configuration. The open-loop configuration, however, is not used, as the loop filter will reduce the passband, and there are only a few cases in which the input frequency would come close to the cutoff frequency of the operational amplifier. As operational amplifiers are introduced very frequently, it is somewhat dangerous to indicate a particular type. However, it was found that the RCA operational amplifier CA3160 for low frequencies is a good choice (CA3130, CA3100).

The tri-state phase/frequency comparator has two outputs that are being fed together in a circuit which I call a *summation amplifier*. The high and low pulses generated by the digital portion are being fed simultaneously to two ports, and the amplifier then has to combine the two pulses into a single output. The simplest way of accomplishing this it to take an operational amplifier in which the two inputs are connected to the inverting and noninverting input and the biasing is provided with identical resistors to avoid any offset problems. Many of the application reports, such as the one on the recent Motorola MC145156, recommend this circuit.

The drawback is that, as the pulses become faster in frequency, the operational amplifier will be unable to follow, and a discrete amplifier has to be built. There are several choices for discrete loop amplifiers, whose main purposes are:

1. To be able to follow the input pulses up to several hundred kilohertz or even several megahertz (good choice: CA3100, 9906, 9909).
2. To have very little or no dc drift.
3. To not introduce any significant noise.
4. To be as symmetrical as possible and therefore to avoid any leakage of reference at the output.
5. To avoid a dead zone (the dead zone was mentioned previously in Section 2-6-3).

Figure 4-115 shows a dc amplifier that takes the two outputs from the phase/frequency comparator type 4044, made by Motorola, and then operates into

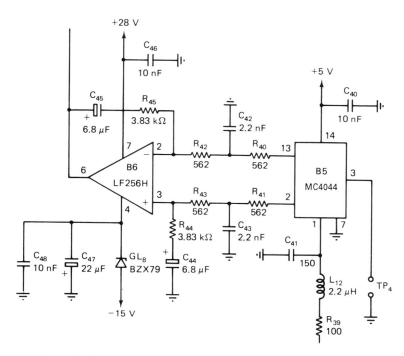

Figure 4-115 Schematic of the phase/frequency comparator MC4344 and an active loop filter using both inverting and noninverting input for stabilization. The first *RC* network is for transient suppression.

the loop filter. This is a fairly complex arrangement, and Figure 4-116 shows a circuit with basically the same performance that operates from an ECL phase/frequency comparator, which by definition has less dc output voltage. Figure 4-117 shows a phase/frequency comparator with the discrete operational amplifier containing several diodes for temperature compensation, as well as voltage shifts.

When analyzing operational amplifiers, it becomes apparent that those having very little output saturation voltage (i.e., CMOS outputs) cannot tolerate a high supply voltage. The discrete amplifier can be operated up to 30 to 40 V.

It was indicated, however, that most tuning diodes become noisy, and the operating voltage should be restricted to less than 30 V. In addition, using such a high voltage, one will find that the voltage gain of the diode becomes so small that the loop gain has to be adjusted to avoid sluggish performance at the top of the band.

Several combinations of these circuits are possible, and depending on the transistors or new integrated circuits, these loop amplifiers may change from time to time. The additional dc gain has to be taken into consideration as the simple amplifier that is used in monolithic devices, such as the CD4046, has unity gain, and the more complex circuits, such as those above, have gain that can be adjusted by selecting component values. In the case of the temperature-compensated amplifier, the two resistors in series with the emitters determine the open-loop gain. In order to provide sufficient frequency response, very fast switching

322 LOOP COMPONENTS

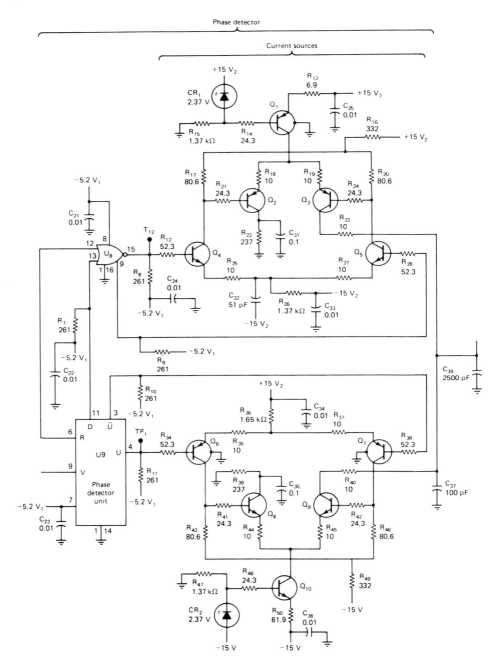

Figure 4-116 Phase/frequency comparator and discrete summation amplifier for low-noise operation. The loop filter is not shown.

WIDEBAND HIGH-GAIN AMPLIFIERS 323

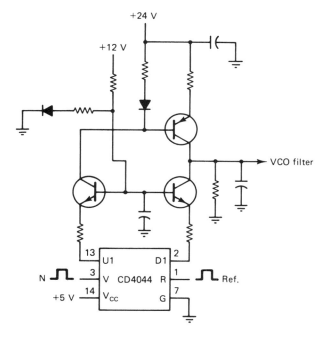

Figure 4-117 Discrete summation amplifier recommended for use with the CD4044 up to 1 MHz.

transistors should be used and matched where possible. So far, none of these amplifiers is available in integrated form, but this may be only a question of time.

We will now take a look at RF applications, the first one being the requirement to drive frequency dividers from a sine-wave source.

4-5-2 Differential Limiters

The easiest way to convert a sine wave into a square wave is to use an integrated circuit called a *line receiver*. A line receiver is an amplifier, practically always a differential limiter, that converts a sine wave into an ECL- to TTL-compatible waveform. Integrated TTL line receivers are relatively slow (about 5 MHz maximum), and it is common practice to build a discrete line receiver using two PNP transistors, as shown in Figure 4-118. The gain of such a device depends on the ratio of collector resistor to emitter differential resistance (RD = $26\,\text{mV}/I_e$). For a current of 5 mA, this RD is about 5 Ω, and if a 500-Ω collector resistor is used, the gain is 100 or 40 dB. Depending on the collector resistor and the collector voltage, these devices can be made to work for ECL, TTL, or 12-V CMOS logic. The differential limiter should be made as fast as possible, which means that the rise and fall time should be extremely short. It is important to minimize hysteresis or zero-crossing errors. It has been shown several times that if a poorly designed differential limiter follows a low-noise frequency standard, the introduction of noise by the slow limiter determines the system's noise rather than the crystal

324 LOOP COMPONENTS

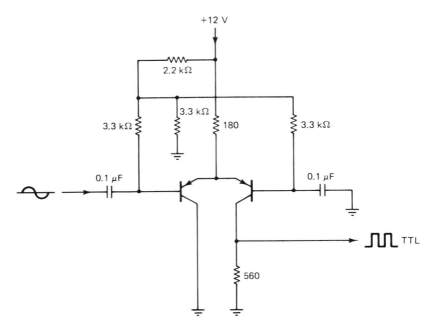

Figure 4-118 Schematic of a differential amplifier operating up to 10 MHz.

oscillator. Modern integrated circuits, such as the Plessey swallow counters, can be driven with sine-wave inputs and have open collector outputs. The open collector outputs allow the use of the necessary resistor and voltage from the power supply to adapt to the following circuits, and therefore no line receivers are required. The line receivers are absolutely necessary for TTL and CMOS logic.

In most cases, however, we do not want to limit the output, but rather isolate output from input without distorting the transferred waveform output of a required output power, so that reverse feedback and isolation have to be considered. These isolation amplifiers are described next.

4-5-3 Isolation Amplifiers

The output from the voltage-controlled oscillator (VCO) is fed into an amplifier that has to drive the frequency divider chain, and an additional output provides the desired frequency $N \times f_{\text{ref}}$.

There are several ways to accomplish this task. Obviously, it is important to use a low-noise amplifier, which suggests the use of a dual-gate MOSFET amplifier, which, driven by the VCO, is excellent for low noise and high reverse isolation. Because of the high-impedance nature of these devices, they will work well only in conjunction with an impedance transformation circuit as shown in Figure 4-119. In the frequency range 30 to 120 MHz, this circuit has enough reverse isolation (at least 60 dB) and is capable of driving stages such as high-level mixers requiring +23 dBm of drive. For wider frequency ranges, Figure 4-120 shows a multistage amplifier that will meet this requirement.

WIDEBAND HIGH-GAIN AMPLIFIERS   325

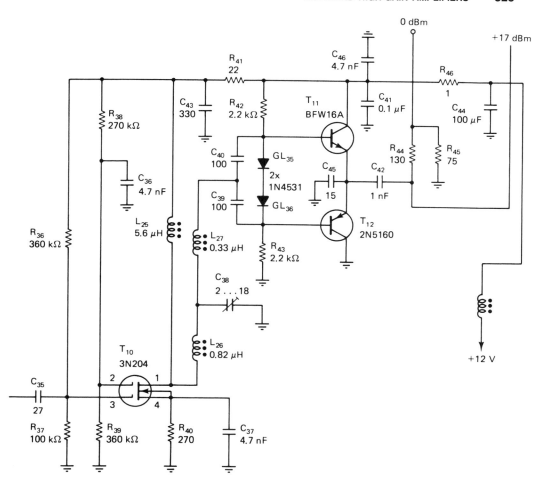

Figure 4-119 High-power output stage with isolation amplifier delivering about +17 dBm. The circuit works well up to 100 MHz.

Another solution is offered in Figure 4-121. It shows the combination of a discrete transistor type 2N5179 driving a power splitter, sometimes called a *hybrid coupler*, and a wideband monolithic amplifier type 733, which is manufactured by several companies, including Motorola and Fairchild.

The 2N5179 transistor has a very low reverse feedback, and the hybrid coupler is capable of up to 40-dB isolation. Although in many cases designers use two different amplifiers, the use of a hybrid coupler should be preferred, as it is a passive device with no power consumption and a long life expectancy.

The 733 amplifier is a unique device that allows programming internal gain. By setting the appropriate bypass capacitor, different gain can be chosen, and this amplifier can even be used in a linear mode exhibiting good intermodulation distortion.

As it is an internal push–pull configuration, the use of balanced output terminals is recommended. Several monolithic amplifiers are available as isolation stages,

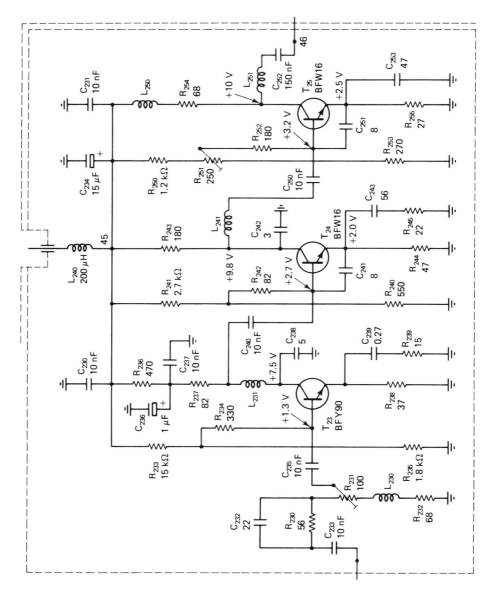

Figure 4-120 Wideband amplifier from 100 kHz to 600 MHz.

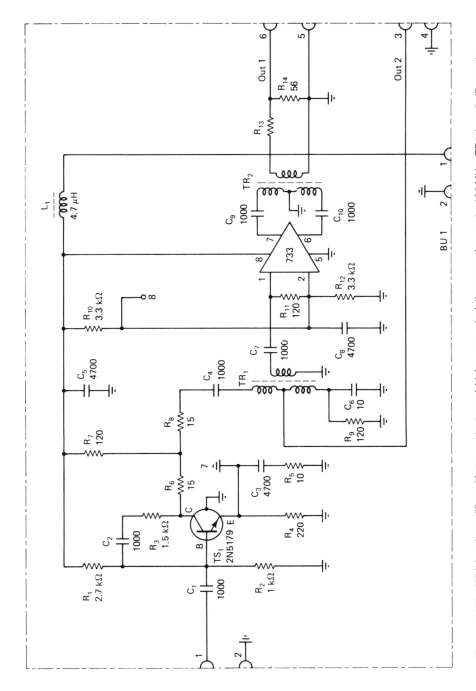

Figure 4-121 Wideband amplifier with two outputs and high reverse isolation operating up to 100 MHz. TR1 is a 3-dB coupler.

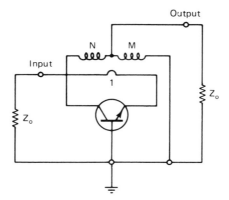

Figure 4-122 Feedback amplifier using the noiseless feedback technique.

and amplifiers such as the RCA CA3028 or similar can be used in a differential mode and AGC can be applied.

In Section 4-3 we learned that the proper termination of the mixer is absolutely essential. The feedback amplifier shown in Figure 4-122 will provide extremely low noise and perfect matching. This is an example of several feedback amplifiers that can be constructed, and this combination seems to be the best. It is based on Patent No. 3891934 issued to David Norton. This type of lossless feedback amplifier consists of a three-winding transformer connected to a common-base transistor in such a manner as to provide gain and impedance matching.

This circuit can be analyzed under the simplifying assumptions that the common-base transistor has a zero input impedance, an infinite output impedance, and unity current gain, while the transformer is considered to be ideal. With these assumptions it can easily be shown that a two-way impedance match to Z_o will be obtained if the transformer turns ratio is chosen such that $n = m^2 - m - 1$.

With this choice, the power gain is m^2, the load impedance presented to the collector is $(n + m)Z_o$, and the source impedance presented to the emitter is $2Z_o$. Turns ratios for m equal to 2, 3, and 4 yield gains of 6, 9.5, and 12 dB and load impedances of 3, 8, and $15Z_o$, respectively.

It is seen that, similar to a conventional common-base amplifier, the gain of the stage is determined by the ratio of the load impedance, Z_1, to the input impedance, Z_{in}. In this case the gain is given by $Z_1/Z_{in} + 1$, whereas it is just Z_1/Z_{in} in the conventional configuration. The significant difference is that the transformer-coupled device provides a two-way impedance match, which is obtained by coupling the load impedance to the input, and the source impedance to the output by transformer action.

The dynamic range considerations for this device are similar to those of the directional coupler circuit, but with some important differences. First, the operation of the circuit depends on the completely mismatched conditions presented by the transistor to the circuit (i.e., the emitter presents a short circuit and the collector an open circuit). Hence, there is no requirement to introduce resistive elements for impedance matching as in the directional coupler circuit. Therefore, a noise figure advantage is obtained with this circuit. Second, the source

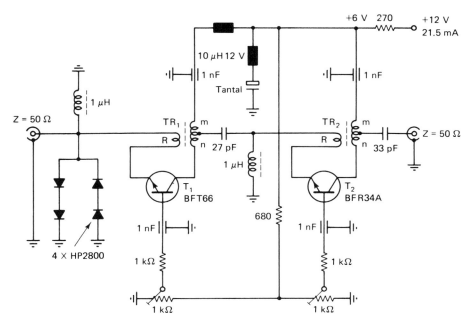

Figure 4-123 Two-stage ultra-low-noise high dynamic amplifier.

impedance of $2Z_o$ presented to the emitter tends to give optimum noise figure performance with low collector currents, which also favors lower noise figures. Finally, in spite of the small currents involved, relatively large output powers can be provided because of the high load impedance, which goes along with the higher-gain versions.

The main disadvantage of the circuit is that the high load impedance tends to limit the bandwidth. Nevertheless, sufficient bandwidth can be achieved to provide broadband IF gain with noise figures competitive with those that could be obtained previously only in very narrowband units.

As this is a very convenient circuit, let us look at the actual working design shown in Figure 4-123. Both transistors are made by Siemens.

The antiparallel diodes at the input can be omitted. The core material used in the transformers depends on the frequency range. Siemens ferrite material type B62152-A8-X17, U17 for frequencies from 100 MHz up and K30 material for frequencies below, should be used.

Figure 4-124 shows the winding arrangement for the transformer. This amplifier has the following characteristics:

1. Power gain 19 dB.
2. Noise figure 1.35, equivalent to 1.3 dB.
3. Third-order intercept point 14 dBm at the input or 33 dBm at the output.
4. 1-dB compression +18 dBm.
5. Input impedance $50 \pm 2\,\Omega$.

330 LOOP COMPONENTS

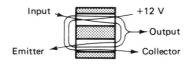

Figure 4-124 Details on the transformer for schematic of Figure 4-123.

6. Bandwidth 70 to 570 MHz (can be made to work at lower frequencies with the higher-permeability core).
7. Dynamic range 102 dB determined from the fact that two signals, separated in frequency of 3.17 mV at the input of the amplifier, result in two intermodulation–distortion products of 25.4 nV. The noise figure of 1.3 dB at 2.4-kHz bandwidth results in the same noise voltage of 25.4 nV. If the noise floor of -138.8 dBm equivalent to the 25.4 nV is subtracted from the -36.96 dBm or the 3.17 mV, which generates the two intermodulation–distortion products, the difference is approximately 102 dB.
8. Power supply 12 V/21 mA.

4-6 PROGRAMMABLE DIVIDERS

4-6-1 Asynchronous Counters

In frequency synthesizers, the VCO is operating at a much higher frequency than the reference. There are two ways to provide information for the phase/frequency comparator.

1. The use of harmonic sampling. Here the reference generates extremely narrow high harmonic contents, pulses that are being fed together with the VCO frequency to the phase detector. In addition, a coarse steering circuit pretunes the oscillator so that the oscillator frequency is very close to the required harmonic of the comb generator.
2. The input frequency is divided to the reference frequency. The same is correct for the reference frequency, as the reference oscillator may be much higher than the desired reference frequency or step size. In most cases, the reference frequency is at 5 or 10 MHz, sometimes even as high as 100 MHz, while the comparison frequency applied to the phase/frequency comparator may lie between 1 MHz and 1 kHz.

The reference divider for most cases requires only a simple asynchronous counter, whereas the programmable divider chain requires synchronous and resettable counters.

While the simplest counter or divider is a flip-flop dividing by 2, modern design uses dividers of high integration.

Originally, the dividers were built in RTL logic, then TTL logic, then Schottky clamped, and then CMOS integrated circuits were developed.

In parallel, the ECL technology was expanded up in frequency, and we now have dividers that operate well above 1000 MHz. Power consumption of these devices naturally is very high, while the CMOS technology was developed to reduce power consumption. TTL devices are probably the least expensive but very noisy, and we will find that most synthesizers now go from ECL to CMOS directly, avoiding TTL as much as possible. The more advanced frequency dividers, such as the one made by Plessey, have the ECL to TTL or CMOS translators built in. TTL circuitry is probably used only to support the dual-modulus counters, as you will see later.

The following types of counters are currently offered in CMOS:

1. Seven-stage ripple counter.
2. Decade counter/divider.
3. Presettable divide-by-N counter.
4. Decade counter (asynchronous clear).
5. Decade counter (synchronous clear).
6. 4-bit Presettable up/down-counter.
7. BCD up/down-counter.
8. Programmable divide-by-N 4-bit counter (BCD).
9. 12-bit Binary counter.
10. 14-bit Binary counter.
11. Octal counter/divider.
12. 4-bit Binary counter (asynchronous clear).
13. 4-bit Binary counter (synchronous clear).
14. Binary up/down-counter.
15. Programmable divide-by-N 4-bit counter (binary).
16. Dual BCD up-counter.
17. Dual binary up-counter.
18. Dual programmable BCD/binary counter.
19. Three-digit BCD counter.
20. Real-time five-decade counter.
21. Industrial time-base generator.

Similar dividers are available in ECL and TTL. Special devices in ECL are bi/quinary counters and swallow counters.

Figure 4-125 shows a speed/power characteristic of major logic lines. It is apparent that CMOS is indicated only in quiescent dissipation. The reason for this is the fact that the power consumption is a function of actual input frequency. Depending on the particular device and the manufacturing process, it is very possible that TTL dividers at 10 MHz require less power than a similar CMOS device. However, the additional interface circuit may justify the use of CMOS rather than involving three different logic families.

The simplest divider is a flip-flop. The flip-flop divides by 2, and input frequency ranges up to 2300 MHz are handled.

332 LOOP COMPONENTS

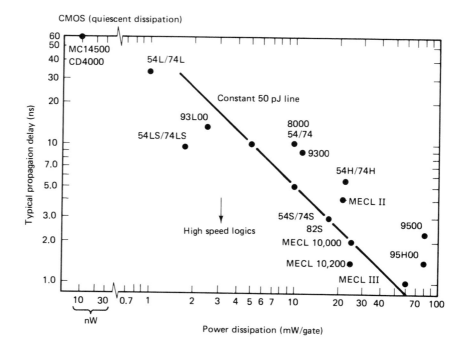

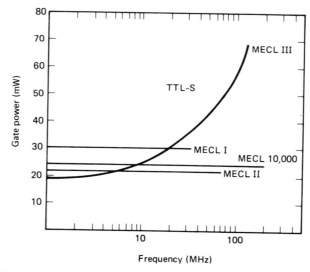

Figure 4-125 Power consumption of different divider technologies as a function of frequency.

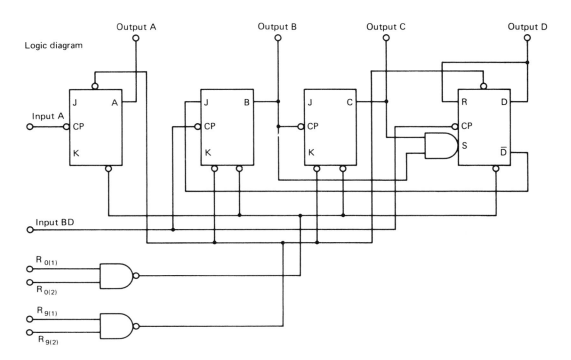

Figure 4-126 Schematic of the 7490 divide-by-10 counter.

Several of these flip-flops can be cascaded, and the drawback of this circuit is the resulting division

$$N = 2^n$$

where n is the number of stages. The figures referred to below show configurations where flip-flops are being cascaded to provide dividers with random division rates. This is accomplished by feedback loops.

A typical application of this technique is used in the familiar 7490 divide-by-10 counter, which is an asynchronous decade counter. Figure 4-126 shows the internal arrangement. It consists of four master/slave flip-flops. The asynchronous dividers are slow, as the input signal triggers the first flip-flop, which then triggers the second flip-flop, and so on. In the synchronous divider, the clock is fed to all the clock inputs simultaneously, and, therefore, the delay is avoided.

4-6-2 Programmable Synchronous Up/Down-Counters

In frequency synthesizers, asynchronous dividers have very little application, while synchronous dividers are of greater importance. In this section we deal with synchronous dividers and a special version of them, presettable dividers, as the synthesizer requires that the division rate be selective rather than hard-wired. Figure 4-127 shows synchronous counters consisting of several flip-flops with the according waveforms.

Several applications require up/down-counters. The reason for this is that for programmed division ratio, the decoding of the circuit becomes easy. The series 74192/74193 counters provide this facility.

Both the 74192 and 74193 are synchronous reversible (up/down) counters with four master/slave flip-flop stages. Inputs include the separate up/down count, load preset, overriding clear, and individual preset data input to each stage. The 74192 BCD counter is capable of counting either up from zero to BCD nine or down from BCD nine to zero. The 74193 is a four-stage binary counter operating in exactly the same manner, except that it can count up to binary fifteen from zero and down from binary fifteen to zero.

The state of the counter outputs depends on the number of count (clock) input pulses received on either the count-up or count-down input. The counter is advanced to its next appropriate state on the positive transitions of the count pulses, while the unused count input is held high. To count in the up direction, the count-up input is pulsed, and to count down, the count-down input is pulsed. The count direction is changed by taking both count inputs high before entering the count signal on the other count input.

In addition to changing the counter state in the normal counting mode, when the counter outputs respond to the incoming pulses, the counter state can easily be taken to any desired state within its range. This is achieved using the fully presettable facilities of the counter. The desired new count state is entered in parallel on the preset data inputs, and the load preset input is taken low, enabling the preset data to be presented to the outputs. The signals on the outputs then agree with those on the preset data inputs independent of any further clock information received while the load preset input is low. This preset state is stored in the counter when the load preset input goes high. Further count input pulses then clock the counter to its next appropriate state.

To reset all outputs to zero, the clear input is taken high. The overriding clear is independent of load and count inputs.

These counters can be cascaded without additional logic. When the counter overflows, the carry output produces a pulse of width equal to that of the count-up input pulse. Similarly, the borrow output pulse width equals that of the count-down input pulse when the counter underflows.

All inputs have input clamping diodes to reduce line termination effects and outputs are of standard 74 series configuration.

Figure 4-128 shows the various signals in the clear, load, and count sequences for the 74192. Figure 4-129 shows the same for the 74193. Figure 4-130 is a logic diagram for the 74192 and Figure 4-131 is a logic diagram for the 74193.

The 7496 is a similar powerful divider that provides:

1. Preset parallel input.
2. Parallel or serial input.
3. Parallel and serial output.
4. Buffered clear, clock, and serial inputs.
5. Serial to parallel/parallel to serial converter capability.

PROGRAMMABLE DIVIDERS   335

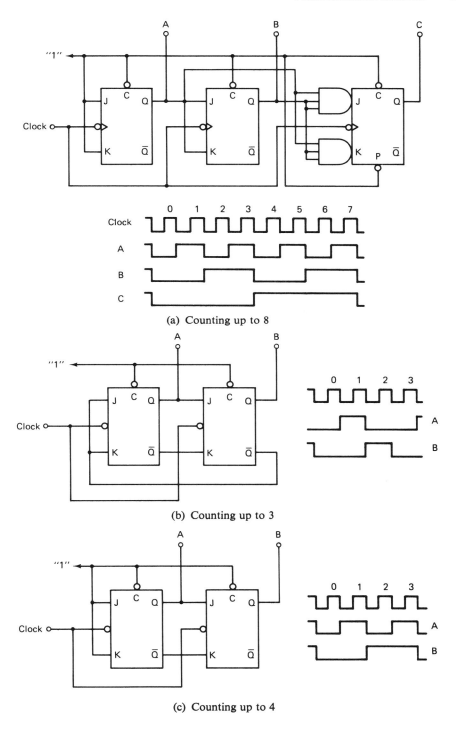

Figure 4-127 Synchronous counters.

336 LOOP COMPONENTS

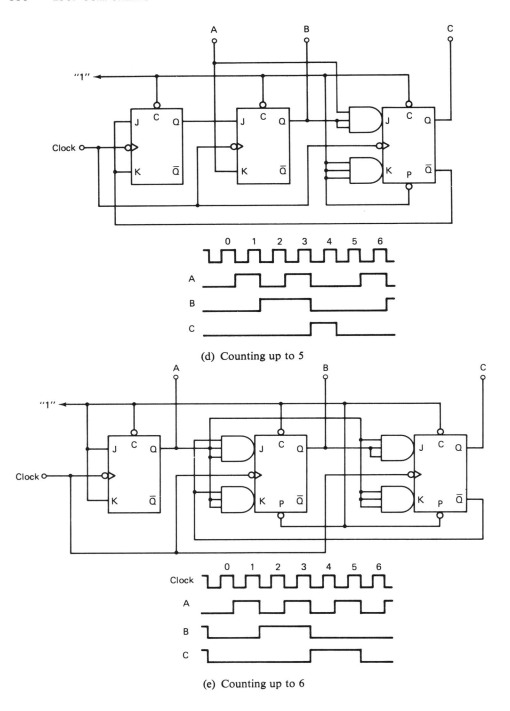

(d) Counting up to 5

(e) Counting up to 6

Figure 4-127 (*Continued*).

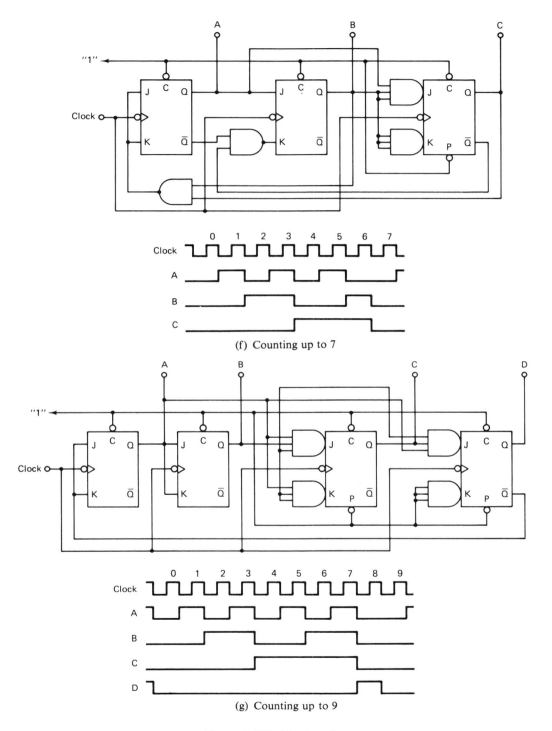

Figure 4-127 (Continued).

338 LOOP COMPONENTS

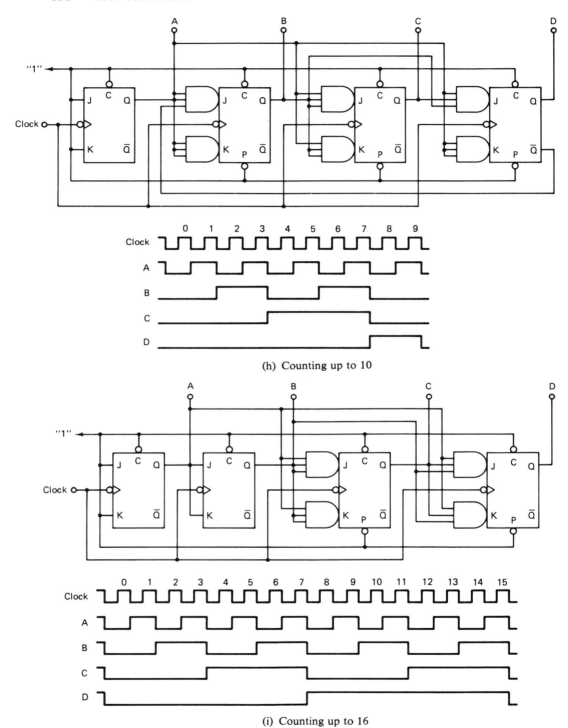

(h) Counting up to 10

(i) Counting up to 16

Figure 4-127 (*Continued*).

8	4	2	1	Decimal value
0	0	0	0	0
0	0	0	L	1
0	0	L	0	2
0	0	L	L	3
0	L	0	0	4
0	L	0	L	5
0	L	L	0	6
0	L	L	L	7
L	0	0	0	8
L	0	0	L	9

(j) Truth table to convert 1248 BCD to decimal

Clock	2 D	4 C	2 B	1 A
0	0	0	0	0
1	0	0	0	L
2	0	0	L	0
3	0	0	L	L
4	0	L	0	0
5	0	L	0	L
6	0	L	L	0
7	0	L	L	L
8	L	0	L	L
9	L	L	L	L

(k) Synchronous 1218 BCD counter

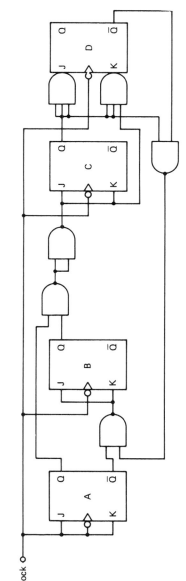

Figure 4-127 (*Continued*).

340 LOOP COMPONENTS

Typical clear, load, and count sequences for the 74192 Decade Counter

Illustrated below is the following sequence:

1. Clear all outputs to zero.
2. Preset to BCD seven.
3. Count up to eight, nine, carry, zero, one, and two.
4. Count down to one, zero, borrow, nine, eight, and seven.

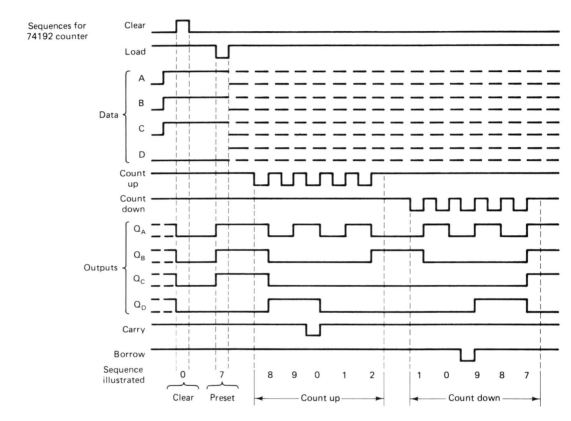

Notes:
1. Clear overrides all other inputs.
2. When counting up, count-down input must be high; when counting down, count-up input must be high.

Figure 4-128 Timing table to the 74192 decade counter.

Figure 4-132 shows the block diagram. The 7496 consists of five RS master/slave flip-flops connected to form a 5-bit shift register. The clock, clear, and serial inputs are buffered by four inverters. The preset inputs A to E are connected to the flip-flop preset inputs via five two-input NAND gates. The second input of each gate is connected to the common preset input, which when at logical 1 enables all preset inputs. A logical 1 on these inputs sets the flip-flops to the logical 1 state. A logical 1 on the clear input sets all flip-flops to the logical 0 state simultaneously. Right shift in the register occurs on the positive edge of the clock pulse. The serial

Typical clear, load and count sequences for the 74193 Binary Counter
Illustrated below is the following sequence:

1. Clear outputs to zero
2. Preset to binary thirteen
3. Count up to fourteen, fifteen, carry, zero, one and two
4. Count down to one, zero, borrow, fifteen, fourteen and thirteen

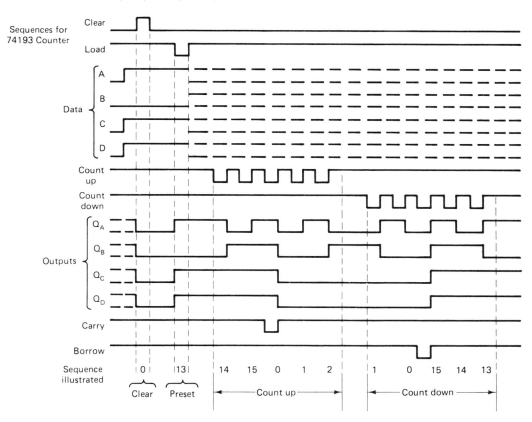

Notes:

1. Clear overrides all other inputs
2. When counting up, count-down input must be high, when counting down, count-up input must be high

Figure 4-129 Timing table for the 74193 binary counter.

input data must be present prior to the clock edge. All inputs have clamping diodes and outputs are standard 74 series configuration. The registers may be cascaded to provide any length of register.

There are basically two forms of parallel input shift registers: the preset parallel input type such as the 7494 and 7496 and the clocked parallel input type such as the 7495.

The inputs of the preset type are connected to the flip-flop preset inputs via appropriate gating. The application of a logical 1 voltage to the input sets the

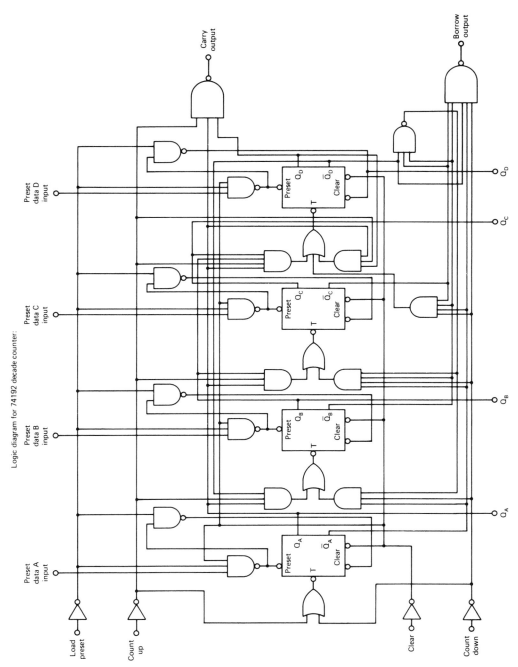

Figure 4-130 Logic diagram for the 74192 decade counter.

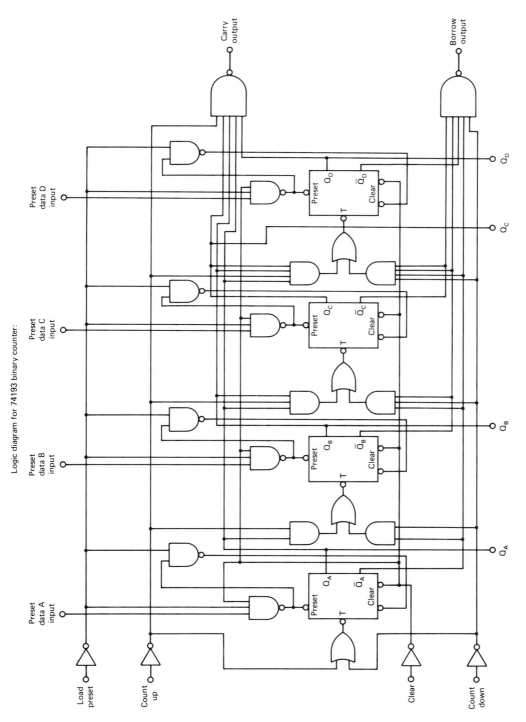

Figure 4-131 Logic diagram for the 74193 binary counter.

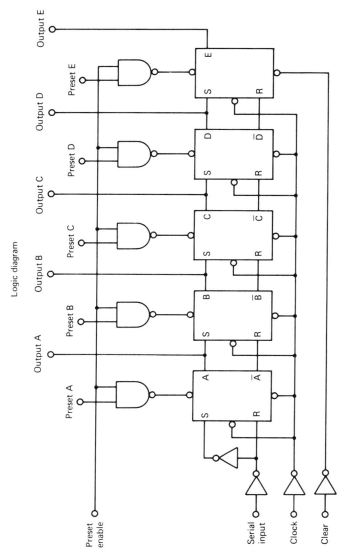

Figure 4-132 Block diagram of the 7496 counter.

flip-flop to logical 1 state, but the application of a logical 0 voltage then has no effect. To set flip-flops to logical 0 state, the clear input must be used.

The inputs of the clocked type are connected to the RS flip-flop inputs via appropriate gating. The application of a logical 1 voltage at the input prior to clocking sets the flip-flops to a logical 1 state. Similarly, the application of a logical 0 voltage at the input sets the flip-flops to a logical 0 state. Hence, the clocked parallel register is most suitable for applications where the parallel data are applied for a predetermined time, as is the case for the accumulator of a multiplier. The preset parallel input register is more suitable for applications where the input data arrive randomly or are present for a very short time, as can be the case in some types of analog-to-digital conversion systems.

The various counters mentioned earlier, available in CMOS and TTL, are spin-offs that may be advantageous for certain applications. It is important to consult the data books of the various manufacturers to determine which divider is the best for a particular application. What this means is that the divider that has the highest flexibility may also be the most expensive, and in some cases, not all capabilities are required simultaneously. In order to select the right device, a price/performance analysis has to be done. As a result of price/performance analysis, as I have mentioned earlier, choosing CMOS devices and avoiding TTL may be the most judicious choice.

Although these divider changes have to remain programmable, the simultaneous availability of parallel and serial loading may not be a requirement. In some designs, however, serial loading may ease the concept; microprocessor applications, for instance, will make the design much easier and require fewer wires if serial loading is used.

Several modern integrated circuits are offered that have all the required dividers on one chip. This may be a convenience for a particular design, but only in cases where the integrated circuit is specifically designed for the application will the integrated circuit have sufficient flexibility. Generally, there is a trade-off between high integration and flexibility. As the number of available pins on the integrated circuit and decoding format limits the number of different tasks, some particular design efforts may require the use of discrete integrated circuits rather than LSI circuits.

A frequency synthesizer for CB application is a typical example where the market requires certain capabilities, and several manufacturers offer basically the entire synthesizer on one chip. AM/FM radios are another typical example. However, for high-performance synthesizers, these devices are probably not suitable. Motorola introduced the MC145156 one-chip frequency synthesizer family, and Philips has introduced the HEF4750 and HEF4751 frequency synthesizer integrated circuits. We will now take a look at those two devices to see what advantages can be obtained from these very latest designs.

Motorola is manufacturing a family of low-power frequency synthesizer chips, and for microprocessor application, the MC145156 is probably the most interesting one. Figure 4-133 shows a block diagram of this device.

The MC145156 is programmed by a clocked, serial input, 19-bit data stream. The device features consist of a reference oscillator, selectable reference divider, digital phase detector, 10-bit programmable divide-by-N counter, 7-bit programmable divide-by-A counter, and the necessary shift register and latch circuitry

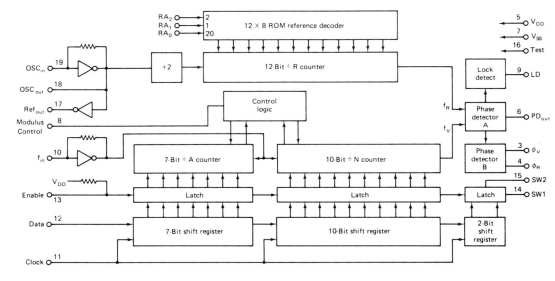

Figure 4-133 Block diagram of the MC145156 synthesizer IC.

for accepting the serial input data. When combined with a loop filter and VCO, the MC145156 can provide all the remaining functions for a PLL frequency synthesizer operating up to the device's frequency limit. For higher VCO frequency operation, a down mixer or a dual-modulus prescaler can be used between the VCO and MC145156.

It can be used for the following applications:

1. General-purpose applications: CATV, AM/FM radios, two-way radios, TV tuning, scanning receivers, amateur radio.
2. Low power drain.
3. 3.0 to 9.0 V dc supply range.
4. Typical input capability of >30 MHz at 5 V dc.
5. Eight user-selectable reference divider values: 8, 64, 128, 256, 640, 1000, 1024, 2048.
6. On- or off-chip reference oscillator operation with buffered output.
7. Lock detect signal.
8. Two open-drain switch outputs.
9. Dual-modulus/serial programming.
10. Divide-by-N range = 3 to 1023.
11. "Linearized" digital phase detector, which enhances transfer function linearity.
12. Two error signal options: single ended (three state) and double ended.

Figure 4-134 shows the phase comparator output waveforms. It was mentioned earlier that this type of integrated circuit fits well together with a microprocessor, and Figure 4-135 shows a block diagram of an Avionics NAV and COM

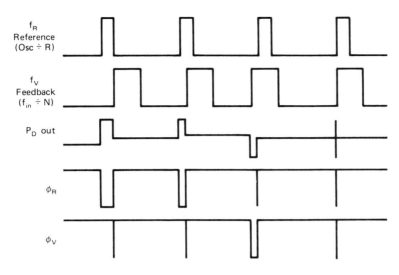

Note:
The P_D output state is equal to either V_{DD} or V_{SS} when active. When not active, the output is high impedance and the voltage at that pin is determined by the low-pass filter capacitor.

Figure 4-134 Phase detector output waveforms of the MC145156 PLL IC.

synthesizer, while Figure 4-136 shows an FM/AM broadcast radio synthesizer. Such devices will be made available by various companies, and their usage extends over several years; improvements will yield to a different design.

The MC145156 is not really a general-purpose frequency synthesizer chip, but the reason why the frequency synthesizer chips are included under programmable dividers is because most integrated circuits contain logic cirsuits either for reference, division, or VCO frequency division.

The general-purpose frequency synthesizer chips HEF4750 and HEF4751 have been mentioned several times already. The two integrated circuits together form a versatile LSI frequency synthesizer system. The HEF4750 integrated circuit contains a reference oscillator, the reference divider, a phase modulator, and two phase detectors on board. Its internal structure is shown on Figure 4-137. The programmable divider section is contained in the HEF4751. This universal divider contains several dual-modulus dividers and other programmable dividers, which results in substantial flexibility. Programmability of $4\frac{1}{2}$ decades, with a guaranteed input frequency of 9 MHz at a 10-V supply voltage, makes this a very universal chip. The additional internal circuitry allows the use of up to three external 10/11 prescalers.

Figure 4-138 shows the internal structure. With the external prescalers, a maximum configuration of $6\frac{1}{2}$ decades and a maximum input of 4.5 GHz can theoretically be obtained. Programming is performed in BCD code in a bit-parallel decade serial format. Details about the programming can be found in the data sheet. One other programmable feature is the fractional channel selection and half-channel offset.

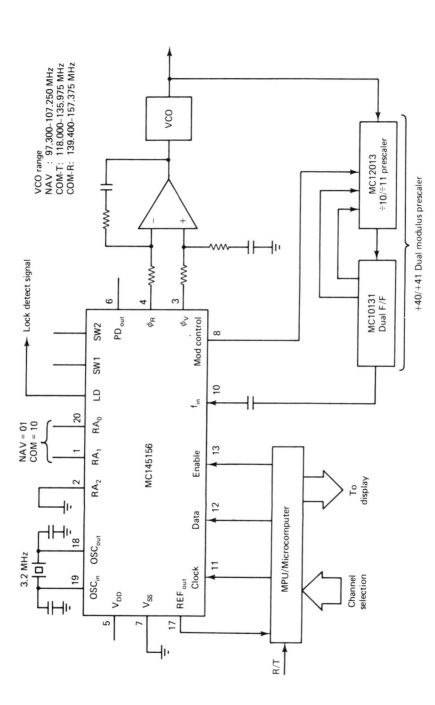

Figure 4-135 Recommended use of the MC145156 PLL synthesizer IC for NAV/COM applications.

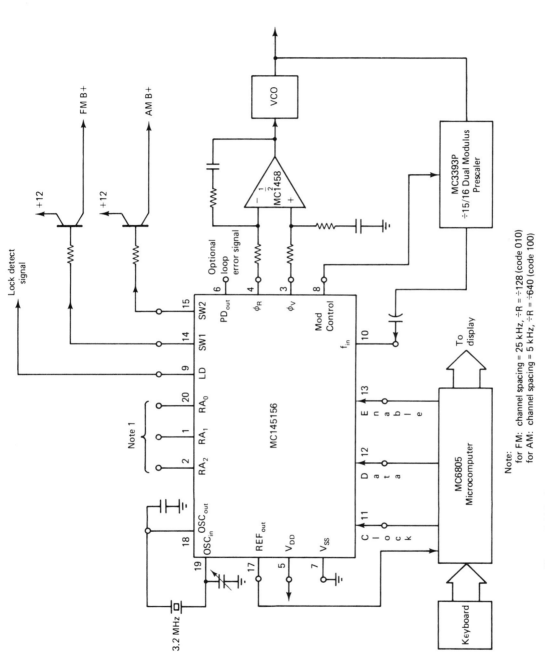

Figure 4-136 Recommended use of the MC145156 in a car radio (AM–FM range).

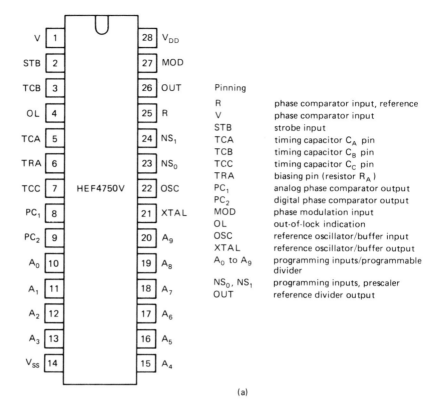

Figure 4-137a Pin layout of the HEF4750 LOCMOS IC.

The fractional channel selection allows a decrease of the step size up to a factor of 16, which means that the step size can be 1/16 of the reference. This may cause some difficulties in programming, and use of this feature is recommended only up to one-tenth of the reference. A typical application would be a one-loop synthesizer with a 10-kHz reference and 1-kHz resolution.

One of the two-phase comparators is a dual cascaded sample/hold detector, which provides a jitter-free output used for fine phase control of the synthesizer. Although not all details of how this is achieved are known, this integrated circuit appears extremely promising for future designs. It is currently being used in the Rohde & Schwarz ESV 20 to 1000 MHz EMI/RFI and field strength receiver.

I have mentioned several times that the TTL and CMOS devices operate up to only 30 MHz at most, and higher input frequencies require faster dividers. A way around the speed requirements is to use ECL swallow counters. Swallow counters are treated in the following section.

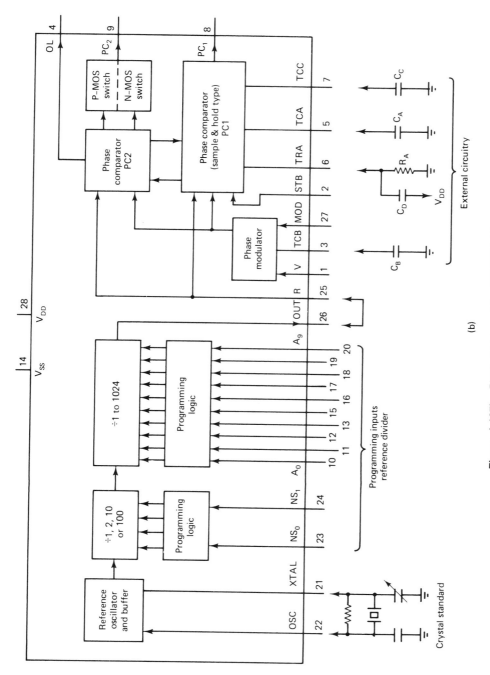

Figure 4-137b Block diagram of the HEF4750 IC.

352 LOOP COMPONENTS

Universal divider

The HEF4751V is a universal divider (U.D.) intended for use in high performance phase lock loop frequency synthesizer systems. It consists of a chain of counters operating in a programmable feedback mode. Programmable feedback signals are generated for up to three external (fast) ÷10/11 prescaler.

The system comprising one HEF4751V U.D. together with prescalers is a fully programmable divider with a maximum configuration of: 5 decimal stages, a programmable mode M stage ($1 \leqslant M \leqslant 16$, non-decimal fraction channel selection), and a mode H stage (H = 1 or 2, stage for half channel offset). Programming is performed in BCD code in a bit-parallel, digit-serial format.
To accommodate fixed or variable frequency offset, two numbers are applied in parallel, one being subtracted from the other to produce the internal program.
The decade selection address is generated by an internal program counter which may run continuously or on demand. Two or more universal dividers can be cascaded, each extra U.D. (in slave mode) adds two decades to the system. The combination retains the full programmability and features of a single U.D. The U.D. provides a fast output signal FF at output OFF, which can have a phase jitter of ± 1 system input period, to allow fast frequency locking. The slow output signal FS at output OFS, which is jitter-free, is used for fine phase control at a lower speed.

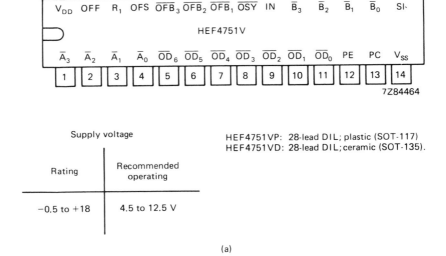

Figure 4-138a Pin layout of the HEF4751 universal divider.

4-6-3 Swallow Counters/Dual-Modulus Counters

To extend the frequency range beyond 30 MHz, swallow counters are being used; they are often referred to as two- or dual-modulus prescalers. Figure 4-139 shows the block diagram of such a device. The three flip-flops are wired in such a manner that by changing the decoding, the division ratio can be changed between a division ratio of 2, 5, 6, 10, 11, or 12. The Motorola MC12012 shown in the figure can be used for this. There are several ways to interface these dual-modulus prescalers, which I prefer to refer to as swallow counters. What actually happens is that out of a chain of pulses, one or two pulses are being swallowed; but let us first take a look at how the system works.

The most popular swallow counters are the 95H90 (350 MHz) and 11C90

PROGRAMMABLE DIVIDERS 353

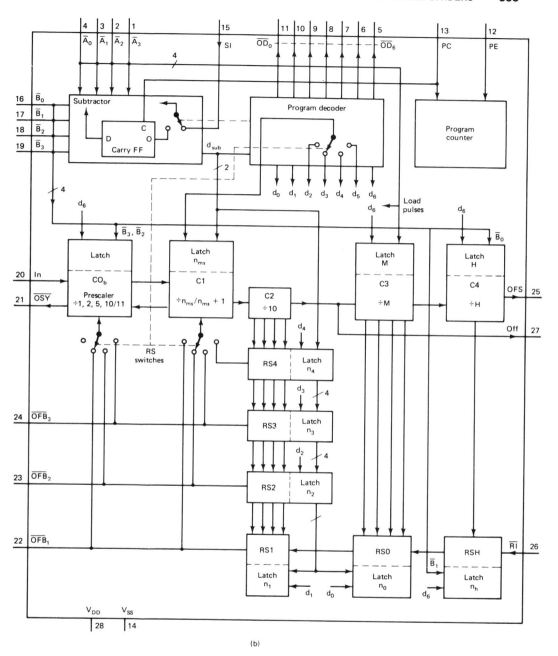

Figure 4-138b Block diagram of the HEF4751 universal divider.

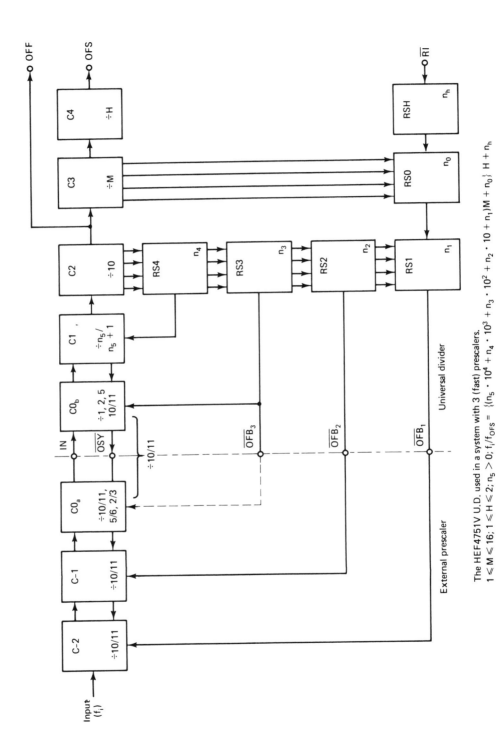

The HEF4751V U.D. used in a system with 3 (fast) prescalers.
$1 \leq M \leq 16; 1 \leq H \leq 2; n_5 > 0; f_i/f_{OFS} = \{(n_5 \cdot 10^4 + n_4 \cdot 10^3 + n_3 \cdot 10^2 + n_2 \cdot 10 + n_1)M + n_0\} H + n_h$

(c)

Figure 4-138c Block diagram of the universal divider type HEF4751 including external prescalers.

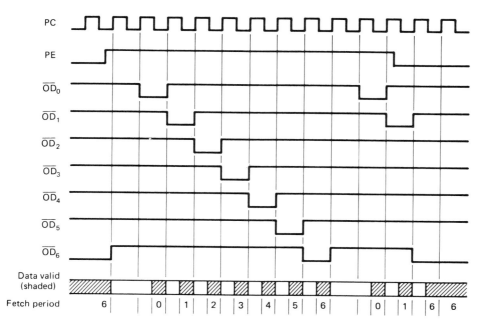

Allocation of data input

Fetch period	Inputs								SI
	$\overline{A}_3$	$\overline{A}_2$	$\overline{A}_1$	$\overline{A}_0$	$\overline{B}_3$	$\overline{B}_2$	$\overline{B}_1$	$\overline{B}_0$	
0		n_{0A}				n_{0B}			b_{in}
1		n_{1A}				n_{1B}			X
2		n_{2A}				n_{2B}			X
3		n_{3A}				n_{3B}			X
4		n_{4A}				n_{4B}			X
5		n_{5A}				n_{5B}			X
6		M			CO_b control		$\frac{1}{2}$ channel control		X

Allocation of data input $\overline{B}_3$ to $\overline{B}_0$ during fetch period 6

$\overline{B}_3$	$\overline{B}_2$	CO_b division ratio
L	L	1
L	H	2
H	L	5
H	H	10/11

$\overline{B}_1$	$\overline{B}_0$	$\frac{1}{2}$ channel configuration
L	L	H = 1
L	H	H = 2; n_h = 0
H	H	H = 2; n_h = 1
H	L	Test state

H = HIGH state (the more positive voltage)
L = LOW state (the less positive voltage)
X = state is immaterial

(d)

Figure 4-138d Timing diagram of the HEF4751 universal divider.

Program data input

The programming process is timed and controlled by input PC and PE. When the program enable (PE) input is HIGH, the positive edges of the program clock (PC) signal step through the internal program counter in a sequence of 8 states. Seven states define fetch periods, each indicated by a LOW signal at one of the corresponding data address outputs ($\overline{OD}_0$ to $\overline{OD}_6$). These data address signals may be used to address the external program source. The data fetched from the program source are applied to inputs $\overline{A}_0$ to $\overline{A}_3$ and $\overline{B}_0$ to $\overline{B}_3$. When PC is LOW in a fetch period and internal load pulse is generated, the data are valid during this time and have to be stable. When PE is LOW, the programming cycle is interrupted on the first positive edge of PC. On the next negative edge at input PC fetch period 6 is entered. Data may enter asynchronously in fetch period 6.

Ten blocks in the U.D. need program input signals. Four of these (CO_b, C3, C4 and RSH) are concerned with the configuration of the U.D. and are programmed in fetch period 6. The remaining blocks (RS_0 to RS4 and C1) are programmed with number P, consisting of six internal digits n_0 to n_5.

$$P = (n_5 \cdot 10^4 + n_4 \cdot 10^3 + n_3 \cdot 10^2 + n_2 \cdot 10 + n_1) \cdot M + n_0$$

These digits are formed by a subtractor from two external numbers A and B and a borrow-in (b_{in}).

$$P = A - B - b_{in} \text{ or if this result is negative; } P = A - B - b_{in} + M \cdot 10^5$$

The numbers A and B each consisting of six 4-bit digits n_{0A} to n_{5A} and n_{0B} to n_{5B}, are applied in fetch period 0 to 5 to the inputs $\overline{A}_0$ to $\overline{A}_3$ (data A) and $\overline{B}_0$ to $\overline{B}_3$ (data B) in binary coded negative logic.

$$A = (n_{5A} \cdot 10^4 + n_{4A} \cdot 10^3 + n_{3A} \cdot 10^2 + n_{2A} \cdot 10 + n_{1A}) \cdot M + n_{0A}$$
$$B = (n_{5B} \cdot 10^4 + n_{4B} \cdot 10^3 + n_{3B} \cdot 10^2 + n_{2B} \cdot 10 + n_{1B}) \cdot M + n_{0B}$$

Borrow-in (b_{in}) is applied via input SI in fetch period 0 (SI = HIGH: borrow, SI = LOW: no borrow).

Counter C1 is automatically programmed with the most significant non-zero digit (n_{ms}) from the internal digits n_5 to n_2 of number P. The counter chain $C-2$ to C1 is fully programmable by the use of pulse rate feedback.

Rate feedback is generated by the rate selectors RS4 to RS0 and RSH, which are programmed with digits n_4 to n_0 and n_h respectively. In fetch period 6 the fractional counter C3, half channel counter C4 and CO_b are programmed and configured via data B inputs. Counter C3 is programmed in fetch period 6 via data A inputs in negative logic (except all HIGH is understood as: M = 16). The counter C0 is a side steppable 10/11 counter composed of an internal part CO_b and an external part CO_a. CO_b is configured via $\overline{B}_3$ and $\overline{B}_2$ to a division ratio of 1 or 2 or 5 or 10/11; CO_a must have the complementary ratio 10/11 or 5/6 or 2/3 or 1 respectively. In the latter case CO_b comprises the whole C0 counter with internal feedback, CO_a is then not required.

The half channel counter C4 is enabled with $\overline{B}_0$ = HIGH and disabled with $\overline{B}_0$ = LOW. With C4 enabled, a half channel offset can be programmed with input $\overline{B}_1$ = HIGH, and no offset with $\overline{B}_1$ = LOW.

(e)

Figure 4-138e Program data input instructions for the HEF4751 universal divider.

(520 MHz) made by Fairchild, and the Plessey SP8692 (200 MHz, 14 mA, 5/6), SP8691 (200 MHz, 14 mA, 8/9), SP8690 (200 MHz, 14 mA, 10/11), and SP8786 (1300 MHz, 85 mA, 20/22). The division ratio of a swallow counter is controlled by two inputs. The counter will divide by 10 when either input is in the high state and by 11 when both inputs are in the low state.

This 10/11 division ratio enables one to build fully programmable dividers to 500 MHz. The switch counting principle means that high-frequency prescaling occurs without any reduction in comparison frequency. The disadvantage of this technique is that a fully programmable divider is required to control the 10/11 division ratio and that a minimum limit is set on the possible division ratio, although this is not a serious problem in practice. Figure 4-140 uses a division ratio of $P/(P+1)$, which is set to 10/11. The A counter counts the units, and the B counter counts the tens.

PROGRAMMABLE DIVIDERS   357

Feedback to prescalers

The counters C1, C0, C – 1 and C – 2 are side-steppable counters, i.e. its division ratio may be increased by one, by applying a pulse to a control terminal for the duration of one division cycle. Counter C2 has 10 states, which are accessible as timing signals for the rate selectors RS1 to RS4. A rate selector, programmed with n(n_1 to n_4 in the U.D.), generates n of 10 basic timing periods as active signal. Since $n \leqslant 9$, 1 of 10 periods is always non-active. In this period RS1 transfers the output of rate selector RS0, which is timed by counter C3 and programmed with n_0. Similarly, RS0 transfers RSH output during one period of C3. Rate selector RSH is timed by C4 and programmed with n_h. In one of the two states of C4, if enabled, or always, if C4 is disabled, RSH transfers the LOW active signal at input $\overline{R1}$ to RS0. If $\overline{R1}$ is not used it must be connected to HIGH. The feedback output signals of RS1, RS2 and RS3 are externally available as active LOW signals at outputs $\overline{OFB}_1$, $\overline{OFB}_2$ and $\overline{OFB}_3$.

Output $\overline{OFB}_1$ is intended for the prescaler at the highest frequency (if present), $\overline{OFB}_2$ for the next (if present) and $\overline{OFB}_3$ for the lowest frequency prescaler (if present). A prescaler needs a feedback signal, which is timed on one of its own division cycles in a basic timing period. The timing signal at $\overline{OSY}$ is LOW during the last U.D. input period of a basic timing period and is suitable for timing of the feedback for the last external prescaler. The synchronization signal for a preceding prescaler is the OR-function of the sync. input and sync. output of the following prescaler (all sync. signals active LOW).

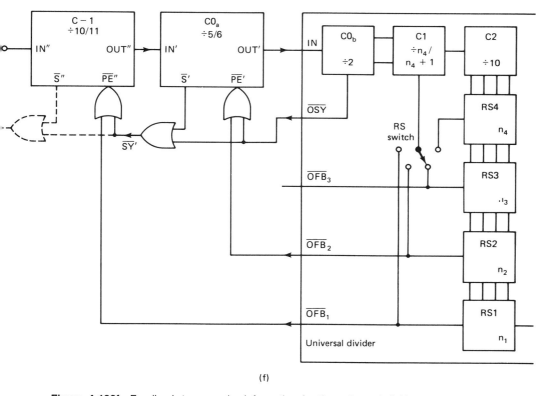

(f)

Figure 4-138f Feedback to prescaler information for the universal divider type HEF4751.

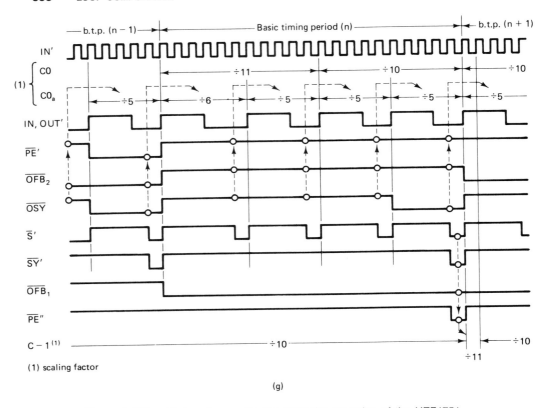

Figure 4-138g Timing diagram for the prescaler operation of the HEF4751.

Consider the system shown in Figure 4-140. If the $P/(P+1)$ is a 10/11 divider, the A counter counts the units and the M counter counts the tens. The mode of operation depends on the type of programmable counter used, but the system might operate as follows. If the number loaded into A is greater than zero, then the $P/(P+1)$ divider is set to divide by $P+1$ at the start of the cycle. The output from the $P/(P+1)$ divider clocks both A and M. When A is full, it ceases counting and sets the $P/(P+1)$ divider into the P mode. Only M is then clocked, and when it is full, it resets both A and M and the cycle repeats.

The divider chain therefore divides by

$$(M-A)P + A(P+1) = MP + A \qquad (4\text{-}145)$$

Therefore,

$$f_{\text{out}} = (MP + A)f_{\text{ref}} \qquad (4\text{-}146)$$

If A is incremented by one, the output frequency changes by f_{ref}. In other words, the channel spacing is equal to f_{ref}. This is the channel spacing that would be obtained with a fully programmable divider operating at the same frequency as the $P/(P+1)$ divider.

Cascading of U.D.s

A U.D. is programmed into the 'slave' mode by the program input data: $n_{2A} = 11$, $n_{2B} = 10$, $n_{3A} = n_{4A} = n_{3B} = n_{4B} = n_{5B} = 0$. A U.D. operating in the slave mode performs the function of two extra programmable stages C2' and C3' to a 'master' (not slave) mode operating U.D. More slave U.D.s may be used, every slave adding two lower significant digits to the system.

Output $\overline{OFB}_3$ is converted to the borrow output of the program data subtractor, which is valid after fetch period 5. Input SI is the borrow input (both in master and in slave mode), which has to be valid in fetch period 0. Input SI has to be connected to output $\overline{OFB}_3$ of a following slave, if not present, to LOW. For proper transfer of the borrow from a lower to a higher significant U.D. subtractor, the U.D.s have to be programmed sequentially in order of significance or synchronously if the program is repeated at least the number of U.D.s in the system.

Rate input $\overline{RI}$ and output OFS must be connected to rate output $\overline{OFB}_1$ and the input IN of the next slave U.D. The combination thus formed retains the full programmability and features of one U.D.

Output

The normal output of the U.D. is the slow output OFS, which consists of evenly spaced LOW pulses. This output is intended for accurate phase comparison. If a better frequency acquisition time is required, the fast output OFF can be used. The output frequency on OFF is a factor M · H higher than the frequency on OFS. However, phase jitter of maximum ±1 system input period occurs at OFF, since the division ratio of the counters preceding OFF are varied by slow feedback pulse trains from rate selectors following OFF.

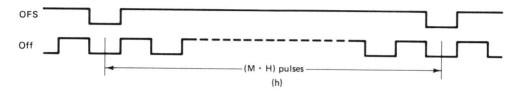

Figure 4-138h Connection information for cascading of several universal dividers.

For this system to work, the A counter must underflow before the M counter does; otherwise, $P/(P + 1)$ will remain permanently in the $P + 1$ mode. Thus, there is a minimum system division ratio, M_{min}, below which the $P/(P + 1)$ system will not function. To find that minimum ratio, consider the following.

The A counter must be capable of counting all numbers up to and including $P - 1$ if every division ratio is to be possible, or

$$A_{max} = P - 1 \qquad (4\text{-}147)$$

$$M_{min} = P \quad \text{since } M > A \qquad (4\text{-}148)$$

The divider chain divides by $MP + A$; therefore, the minimum system division ratio is

$$\begin{aligned} M_{min} &= M_{min}(P + A_{min}) \\ &= P(P + 0) = p^2 \end{aligned} \qquad (4\text{-}149)$$

Using a 10/11 ratio, the minimum practical division ratio of the system is 100.

In the system shown in Figure 4-140, the fully programmable counter, A, must be quite fast. With a 350-MHz clock to the 10/11 divider, only about 23 ns is

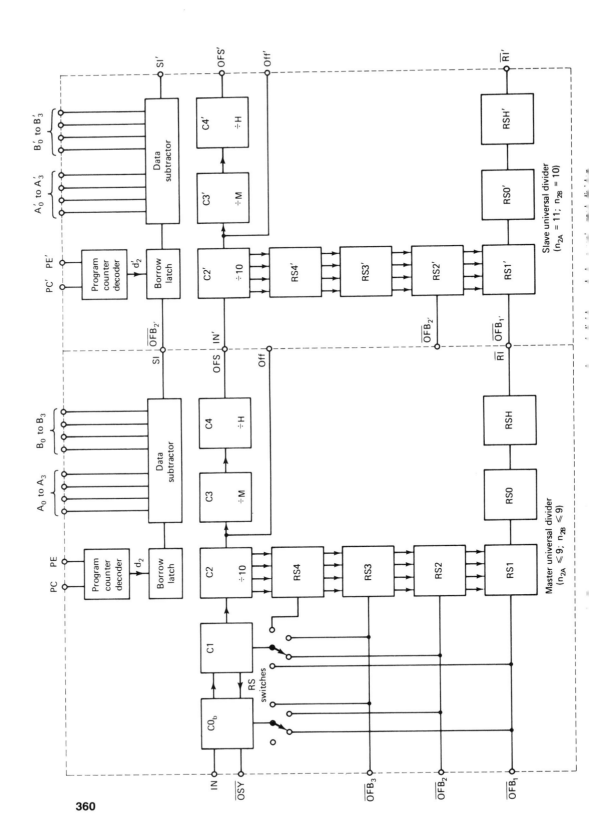

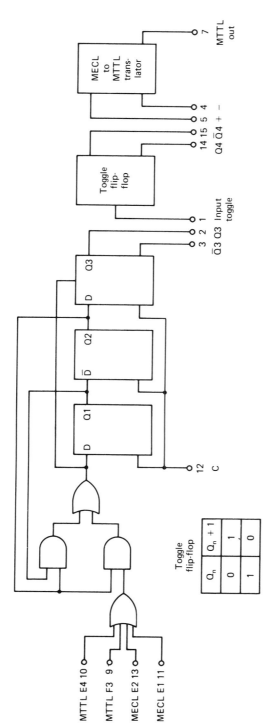

Figure 4-139 Block diagram of the Motorola MC12012 universal dual-modulus counter. (Courtesy of Motorola Semiconductor Products, Inc.)

361

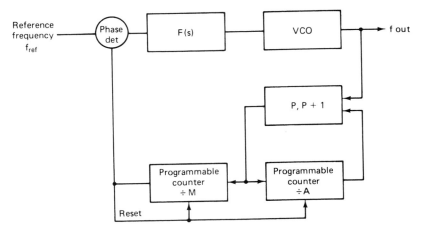

Figure 4-140 System using dual-modulus counter arrangement.

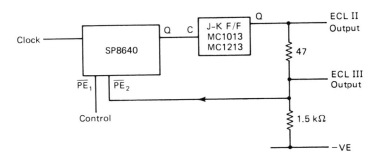

Figure 4-141 Level shifting information for connecting the various ECL2 and ECL3 stages.

available for counter A to control the 10/11 divider. For cost reasons it would be desirable to use a TTL fully programmable counter, but when the delays through the ECL-to-TTL translators have been taken into account, very little time remains for the fully programmable counter. The 10/11 function can be extended easily, however, to give a $\div N(N+1)$ counter with a longer control time for a given input frequency, as shown in Figures 4-141 and 4-142. Using the 20/21 system shown in Figure 4-141, the time available to control 20/21 is typically 87 ns at 200 MHz and 44 ns at 350 MHz. The time available to control the 40/41 (Figure 4-142) is approximately 180 ns at 200 MHz and 95 ns at 350 MHz.

This frequency-division technique can, of course, be extended to give 80/81, which would allow the control to be implemented with CMOS, but which would increase the minimum division ratio to 6400 (80^2). This ratio is too large for many synthesizer applications, but it can be reduced to 3200 by making the counter an 80/81/82. Similarly, a 40/41 can be extended to 40/41/42, as shown in Figure 4-143 to reduce the minimum division ratio from 1600 to 800. The available time to

PROGRAMMABLE DIVIDERS

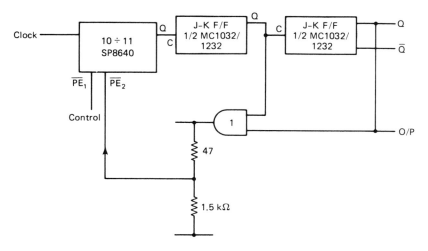

Figure 4-142 Level shifter diagram to drive from ECL2 and ECL3 levels.

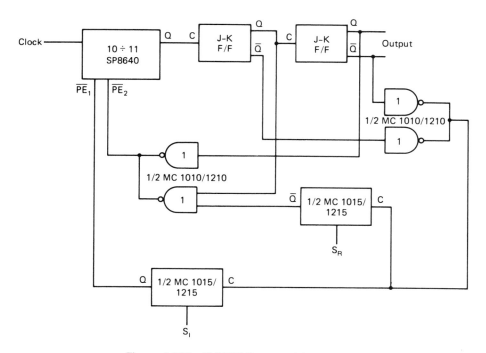

Figure 4-143 40/41/42 three-modulus counter.

control the 40/41/42 is a full 40 clock pulses (i.e., 200 ns with a 200-MHz input clock or 110 ns at 350 MHz). The principle of operation is as follows:

$$\text{Minimum division ratio}$$
$$800 = (20 \times 40) + (0 \times 41) + (0 \times 42)$$
$$801 = (19 \times 40) + (1 \times 41)$$
$$802 = (19 \times 40) + (2 \times 42)$$

More information can be found in Ref. 1.

4-6-4 Look-Ahead and Delay Compensation

The swallow counter can be used, as we have seen, as a synchronous counter, often referred to as a prescaler. The term "prescaler" for a dual-modulus or swallow counter is really not recommended because a prescaler refers to a divider inserted between the VCO and the programmable divider, which is not necessarily resettable at the same clock rate, and therefore, one loses resolution. Let us consider a typical example. Assume that we have a VCO operating from 100 to 200 MHz, and we use a divide-by-10 prescaler followed by a programmable divider that drives a phase comparator with a reference input of 1 kHz. If this phase-locked loop is closed, we will find that, for each 1-kHz step, the programmable divider is changed; in reality, we get a 10-kHz step at the output frequency of the VCO. This indicates that the prescaler is not in synchronous condition with the other counters. The dual-modulus technique, when applied properly, allows the frequency extension of standard dividers without losing resolution.

All the various integrated circuits have a propagation delay, especially the CMOS integrated circuits, as they are much slower than the ECL.

We are now concerned with analyzing a programmable divider using a 95H90 divide-by-10/11 counter.

Let us take look at Figure 4-144. The frequency range of 21.500 to 49.990 from a VCO will be divided down to a 10-kHz reference. Four counters and several flip-flops are required to provide the necessary timing. It seems convenient to divide this module into two portions and analyze them first.

1. 10/11 divider with control counter.
2. Programmable dividers with decoding.

The programmable dividers, as well as decoders, can be built in discrete CMOS technology. A programmable divider is the CD4018. They can be used as divide-by-N counters and will operate at a maximum clock frequency of 5 MHz. Therefore, we will use the 95H90 Fairchild divide-by-10/11 counter. [Note that the device is called a $P/(P+1)$ counter rather than a prescaler.] It was mentioned earlier that there are several different dividers available in ECL technology, and this particular circuit was later used in the design of the Plessey 8940 to reduce power consumption. However, this change does not affect the operating principle. The three programmable dividers permit a division ratio from 002 to 999. The

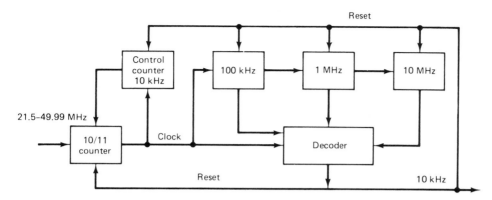

Figure 4-144 Simplified block diagram of the four-stage counter using swallow counter principle and programmable counters.

dividers are being programmed via five lines using the Johnson code, where a 00001 equals 0 in decade count, 00011 is equivalent to 1, and so on. As the counter always returns to the zero position, the input must be arranged so that the 9 is equivalent to a 0 count in the counter, position 8 is equivalent to a 1 count, position 7 to a 2 count, and so on. In this case, the counter gets the command to divide by 10, which means that the following counter gets one-tenth of the input. At the same time, the outputs $\overline{Q_1}$ to $\overline{Q_5}$ of the divider can be decoded, and it can be checked whether the end pulse has reached the divider and one cycle is finished. This information will be used to reset the counter to its original condition. In the case of the three-digit counter, one cycle is finished after all three stages deliver the pulses simultaneously, which are decoded and form the output and reset pulse. Figure 4-145 shows the CMOS divider chain including the truth table. The programmable inputs J_1 through J_5 are tied to Vdd via 100-kΩ resistors. If the switch is closed, a logic 1 applies. The complete divider is being reset via PE after one cycle is complete and the output flip-flop goes to logic 1. Since each of the dividers contains five flip-flops, $\overline{Q_5}$ outputs are available, which, connected with D, allow the determination of the division.

$\overline{Q_5}$-D results in divide by 10, $\overline{Q_4}$, $\overline{Q_5}$-D results in divide by 9, $\overline{Q_4}$-D results in divide by 8, and so on. It is important to understand that the five outputs $\overline{Q_1}$ to $\overline{Q_5}$ provide a pulse sequence that allows a function of the clock pulses to obtain the output pulse. This is shown in the previously indicated truth table. Let us assume that the third input pulse should be detected at the $\overline{Q}$ outputs; the 3-equivalent, 6, has to be programmed. Therefore, the divider requires three steps until $\overline{Q_4}$ is at 1 and $\overline{Q_5}$ is at 0. Now it becomes apparent that, regardless of what combination is chosen, the final result at the $\overline{Q}$ outputs is always the same. That is the reason why $\overline{Q_4}$ and $\overline{Q_5}$ of the 100-kHz and 10-kHz dividers are being used. The 1-MHz divider decodes at $\overline{Q_2}$ and $\overline{Q_3}$, which means that, two pulses prior to the cycle count, the information is extended. The reason for this, as we will see in more detail, is to compensate for the delays in the decoders, gates, and flip-flops.

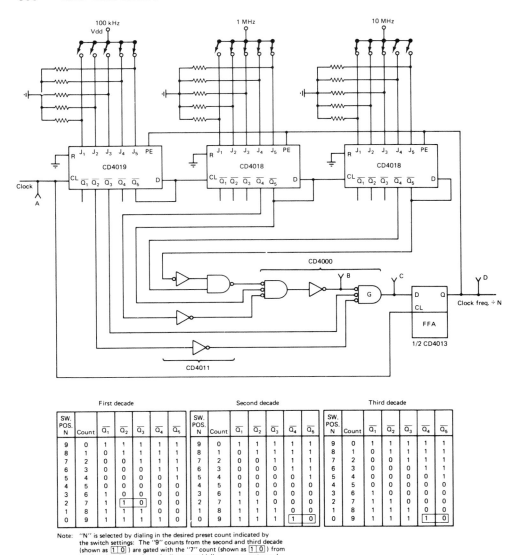

Figure 4-145 Schematic of the CMOS divider section and its truth table.

Division by 584. Let us assume that $\overline{Q_4}$ and $\overline{Q_5}$ of the 1-MHz divider decode t. Also, the 100-kHz and 10-kHz dividers have reached their final count. Point B of the decoder is at zero, and the 1-MHz divider now supplies the pulses as shown in Figure 4-146. It becomes apparent that two pulses of the cycle are lost because of the delay in gate 3 and flip-flop A. If the 1-MHz divider is decoded in such a way that it provides an output two pulse counts earlier, the delay is compensated.

PROGRAMMABLE DIVIDERS 367

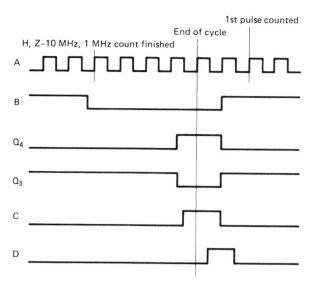

Figure 4-146 Timing diagram of the decoder.

E_1	E_2	÷
0	0	11
0	1	10
1	0	10
1	1	10

Figure 4-147 Truth table of the 10/11 divider.

10/11 Counter (ECL). We have learned that the CMOS divider can operate up to a maximum of 5 MHz. Higher frequencies can be handled by using the 10/11 counter. The combination of the 10/11 counter and the control circuit together with the additional counter is referred to as a "swallow" counter. This term is used because of the fact that one pulse count is "swallowed" from time to time using the divide-by-10 or divide-by-11 principle. Figure 4-147 shows the truth table determining divide by 10 or divide by 11 as a function of the input at E_1 and E_2 of the 10/11 counter shown in Figure 4-148. To control this facility, a control counter that determines how often one divides by 10 or 11 is required. This circuit works on the principle that, as required, either 10 or 11 input pulses produce one output pulse. At times, one pulse gets removed or swallowed.

Example 6 Let us assume that a two-digit counter is constructed using the 95H90 with a control counter following a CMOS divider. If a division ratio of 73 is required, the 10/11 counter will divide three times by 11; then the control counter is full and the 10/11 counter is set to divide by 10. The programmable divider must take four more sequences, during which it divides by 10. As a result, the output

368 LOOP COMPONENTS

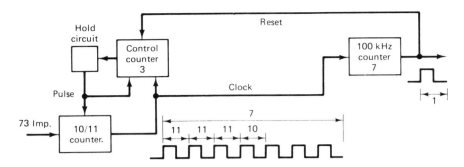

Figure 4-148 Timing of the 10/11 divider.

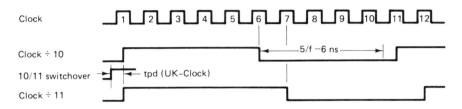

Figure 4-149 Switching and delay times of the 95H90 counter.

pulse occurs after 73 input pulses. Figure 4-148 shows a block diagram with the timing. The control counter is responsible for the 1-MHz digit, and the programmable counter is responsible for the 10-MHz digit.

Delays. A limitation of this system is that there is a clearly defined minimum division ratio at which the system will no longer function. More critical is the fact that there is a time delay in the 10/11 division path. The delay must not be more than $1/(10/f_{in})$, typically 6 ns, determined by the ECL/CMOS level shifters, the CMOS logic, and back through the CMOS/ECL level shifters. This means that 6 ns prior to the critical pulse it has to be known whether division by 10 or 11 is required. Only $5/f_{in}-6$ ns time is available between two pulses. This can be seen from the pulse diagram, Figure 4-149, showing the switching time of the 95H90. Thus, there are limits for the delay.

Figure 4-150 shows the complete schematic of the divider chain. Before we analyze this circuit in further detail, let us note that the 10/11 counter is shown in the last portion of the figure; the 2N2907 is used as a level shifter from ECL to CMOS and drives the counter chain.

The first CD4018 is the control counter. The swallow counter therefore consists of the 10/11 counter, the control counter, the flip-flop (CD4013), the ECL-to-CMOS level shifter transistor, and the CMOS-to-ECL level shifter being clamped by the two diodes. The gates shown in the circuit are used for decoding purposes, and this function will be discussed when we go into the details of the programmable

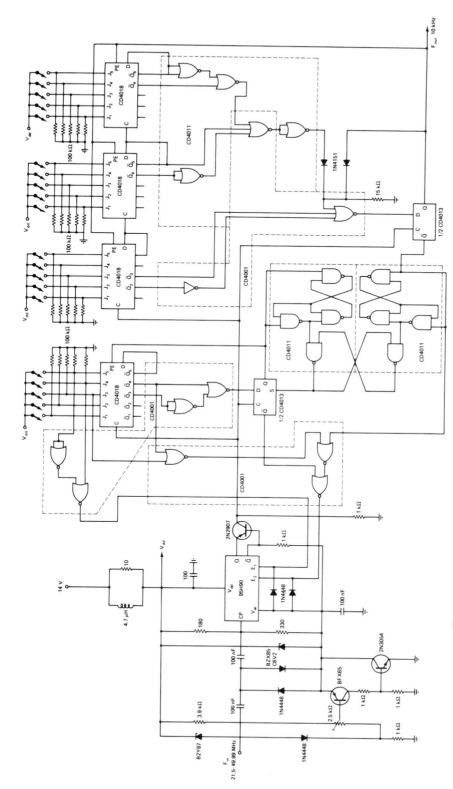

Figure 4-150 Schematic of the 21.5–49.99 MHz divider used for determining the design of delay compensation and look-ahead schematics.

counter that uses the three CD4018s on the right of the figure. Let us go back to our delays.

1. There is a limit for the delay. The control counter has to determine the division ratio, divide 10 or 11. It has to be remembered that the prescaler has to be clamped to 10 and the logic that is doing this attains additional delay.
2. There is a delay from the flip-flop CD4013 via PE and $\overline{Q_4}$ of the control counter to the logic to reset to divide by 11 at the end of the cycle.

According to the data sheet, the CD4018 has a propagation delay for PE–$\overline{Q_4}$ and clock–$\overline{Q_4}$ of 125 ns at 10 V.

To compensate for the delay from clock to $\overline{Q_4}$, it is sufficient to decode the output pulse one count before it actually occurs.

To compensate for the delay in the output flip-flop and PE–$\overline{Q_4}$ a different circuit is required. It is not possible to take the reset pulse for the divide by 10 from the $\overline{Q}$ output of FFN to the 95H90; instead, it has to go through the control counter. In the case where a straight division by 10 is possible, the 10/11 counter must be prevented from dividing by 11. This decision can be made only by the control counter, with the disadvantage that the additional circuit adds additional delay. In a test circuit, it was found that despite the attempted presetting of the control counter, the timing was still not correct. The width of the PE pulse is a function of the clock frequency, and the delays are constant. As a result of this, there are crossovers between the two areas, which prevent proper functioning of the circuit.

Since this is a technological problem, it has to be taken into consideration in the decoding circuit. To understand this better, let us analyze the divider cycle. Let us assume that the divider of the whole system is set at 3754, which means that the output pulse has to occur after 3751 pulses at the input. Regarding the 10/11 counter, the following happens:

$$\begin{array}{rrl} 4x \div 11 & 44 & \text{Control counter full} \\ 296x \div 10 & 2960 & \text{10-MHz counter full} \\ 70x \div 10 & 700 & \text{1-MHz counter full} \\ 5x \div 10 & \underline{50} & \text{100-kHz counter full} \\ & 3754 & \end{array}$$

Figure 4-151 shows the complete cycle. The critical time intervals and positions a, b, c, d, e, f, g, h, and i are indicated.

a. In this moment, the control counter has to have told the 10/11 counter to switch the division ratio.
b. Two pulses before final count, decoding is activated.
c. The D input of the flip-flop A receives information "Logic High."
d. The next clock pulse (e) triggers flip-flop A.
f. This information appears much later at the control counter.
g. It is now too late to be fed into the 10/11; the same applies for the changing for h and i.

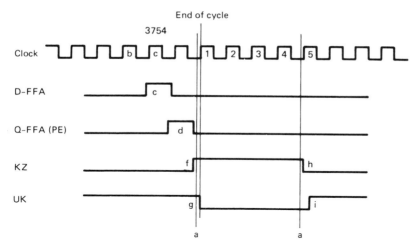

Figure 4-151 Timing diagram for an entire cycle, showing the effects of the various delays.

Analysis. During the entire cycle, we divide N times by 11 and it does not matter when in time this happens. It is important only how often the division by 11 occurs. Let us assume that the 10/11 divider receives the "change to" information only after the second pulse, and the control counter follows. This only means that the entire pulse chain is shifted by one count. The final result remains the same. We then get

$$\begin{array}{rr} 1x \div 10 & 10 \\ 4x \div 11 & 44 \\ 295x \div 10 & 2950 \\ 70x \div 10 & 700 \\ 5x \div 10 & 50 \\ \hline & 3754 \end{array}$$

As the 10/11 counter is still set to divide by 10, it will divide by 10 first and then switch over. To do this, a second flip-flop B is required, as well as a NAND latch and two edge detectors FD1 and FD2, which provide only negative pulses at positive edges.

Function. The rising signal at $\overline{Q}$ is fed to FD1; at the output of FD1 a very narrow pulse is generated that switches the NAND latch and FFB is set to zero via the S input. The $\overline{Q}$ output of FFB changes the divide ratio of the 10/11 counter using some gates in line. These gates are necessary to support the control counter. As the control counter is being decoded to 1 before its final count, it is not possible to determine the division ratio $1x \div 11$. In this case, the output is taken directly from the control counter, connected with $\overline{Q}$ of FFA and applied to the input E_1.

Figure 4-152 shows the minimum configuration, consisting of the swallow

372 LOOP COMPONENTS

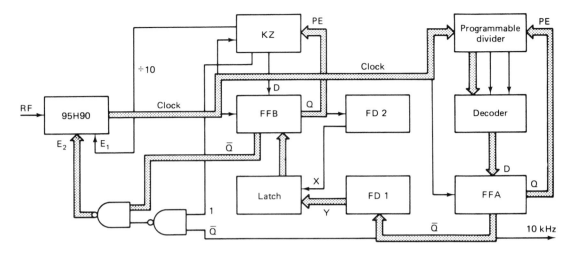

Figure 4-152 Presetting for division-by-11 beginning of the cycle.

counter (10/11 counter, 95H90 control counter, FFB, latch, FD1, FD2) and the programmable divider chain with its decoder and FFA.

In case there is no requirement to divide by 11, we can take the output from the control counter and apply it directly to the second input of the 10/11 counter. This blocks any switching into the other mode. This can be seen from the truth table of the 10/11 counter that was listed earlier. For all other division ratios after the initial division by 11, the 10/11 counter has to be switched to divide by 10. One pulse before the control counter is full, it will supply a logic high to the D input of FFB. With the next clock pulse, this information appears at $\overline{Q}$ via the gates to the input E_2. At the same time, FD2 resets the latch and is holding FFB, and therefore the control counter, via the input S. Now this cycle is repeated. To further help understanding, the following figures will give more insight. Figure 4-153 shows how the divider chain is set to divide by 10 after the control counter is full; Figure 4-154 shows division $1x \div 11$; Figure 4-155 shows constant division by 10. This circuit was built as indicated by the previously shown complete schematic, and the result confirmed the theoretical discussions. The pulse diagram in Figure 4-156 explains this. The 10-kHz digit (swallow counter) is set $3x \div 11$. The clock pulses are numbered, and the delays are indicated (ns). The critical positions are marked a, b, c, d, and e (see Figure 4-156).

a. Because of crosstalk on the printed circuit board, a pulse stretching of the output of FFA became apparent. Two diodes were used to force input D of FFA immediately after the output pulse $\overline{Q}$ appears at logic 0. This guarantees that FFA is reset to the next cycle.
b. Start of the cycle; this can also be seen, as pulse counts start with 1. The clock pulse is shown from -2 to $+5$.
c. Maximum time delay is rest.
d. Maximum time limit for switching divide by 11 to divide by 10.
e. Maximum time safety margin.

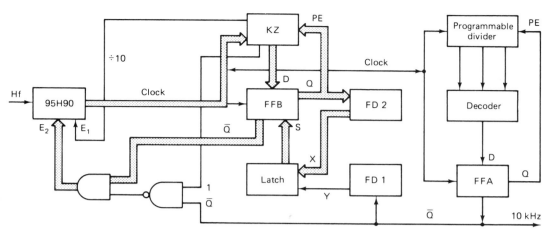

Figure 4-153 Programming of the divider chain to divide by 10 after the control counter is finished.

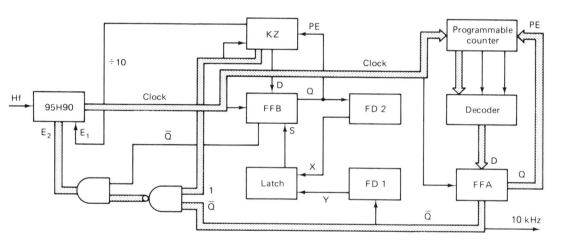

Figure 4-154 $1x \div 11$.

We can deduce from this that the second cycle pulse is being considered as the first pulse to the control counter, while the programmable divider chain reacts to the first clock count of the cycle. Depending on the timing, FFB-$\overline{Q}$ the swallow counter will divide 2 to $9x \div 11$. To divide by 11 only once, a different arrangement is used that sees to it that the changeover command for the divide by 10/11 does not occur at the same time as the other reset pulses occur. This is because the pulse is being fed from FFA-$\overline{Q}$ to E_2 independent of FFB. A several-nanosecond time delay is available, and the actual circuit operated up to 55 MHz without difficulties.

374 LOOP COMPONENTS

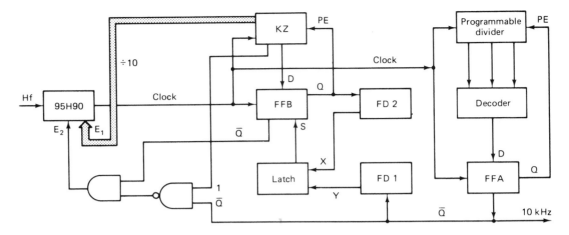

Figure 4-155 Constant division by 10.

Final Comments. The circuit has some peculiarities:

1. Since the entire circuit operates from 14 V and the ECL divider can only be operated from 5.2 V, additional circuitry to provide a voltage drop of 8.8 V, using transistors BFX65 and 2N3054, was used. In a practical battery-operated device, this two-transistor circuit would be replaced by something like the IF and/or RF stages of the receiver to conserve energy, instead of using a voltage regulator, which dissipates the energy.
2. The input of the 95H90 has to be biased with the two resistors. The voltage is set in such a way that 50 mV rms is sufficient to drive the divider at 50 MHz.
3. There is a voltage difference between the ECL and CMOS dividers, as the ECL operates at 5.2 V and the CMOS as 14 V. The transistor 2N2907 performs the necessary level shift. In order to interface the CMOS to the ECL voltage, a configuration was chosen to avoid a transistor circuit for each input E_1 and E_2. Two diodes clamp the high-level signals to the V_{ee} level of the 95H90.
4. Because of the peculiarities of the divider, the 9-equivalent had to be programmed, and nonstandard encoder switches had to be used. This can be avoided by connecting the main contact of the switch to +Vdd.

Finally, the complete circuit was temperature cycled. At 80 °C, the current was 145 mA, and the maximum frequency was 50 MHz. At −20 °C, the current was 128 mA, and the cutoff frequency was 54.5 MHz.

Although this system today could be constructed simply by a few integrated circuits, or even one, the circuit discussed has the advantage that one is forced to go through this detailed analysis, and a better understanding is possible. Some of the other circuits we use later are much easier to deal with. In Chapter 6 a somewhat simpler circuit is analyzed using the same principles.

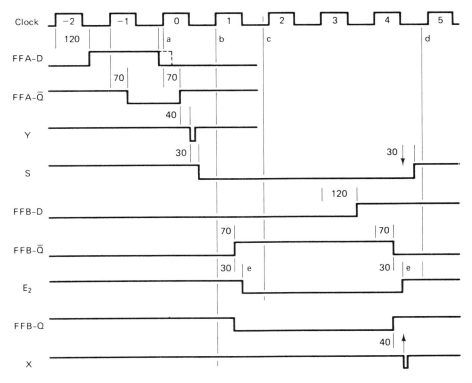

Figure 4-156 Pulse diagram for reset and switchover area for the $3x \div 11$ of the entire system.

4-7 LOOP FILTERS

Earlier, we became acquainted with some loop filters, and therefore this may be a repeat of some of the things already treated. As a convenient reference, the four most important filter types are described below.

4-7-1 Passive *RC* Filters

Figure 4-157 shows the simple *RC* filter with the transfer equation

$$F(s) = \frac{1/sC}{R + 1/sC} \tag{4-150}$$

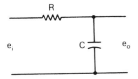

Figure 4-157 *RC* network.

or the frequency response

$$F(j\omega) = \frac{1}{j\omega CR + 1} \qquad (4\text{-}151)$$

The magnitude of the frequency response is

$$|F(j\omega)| = \frac{1}{\sqrt{1 + \omega^2 R^2 C^2}} \qquad (4\text{-}152)$$

and the phase is

$$\phi(\omega) = -\arctan \omega CR \qquad (4\text{-}153)$$

The lag filter, Figure 4-158, has the transfer function

$$F(s) = \frac{sCR_2 + 1}{sC(R_1 + R_2) + 1} \qquad (4\text{-}154)$$

The frequency response is

$$|F(j\omega)| = \sqrt{\frac{1 + \omega^2 R_2^2 C^2}{1 + \omega^2 C^2 (R_1 + R_2)^2}} \qquad (4\text{-}155)$$

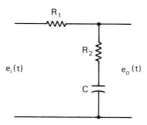

Figure 4-158 Lag filter.

4-7-2 Active *RC* Filters

Any high-performance loop will use an active filter because of the second integrator. Figure 4-115 shows such a circuit. In this particular case, the phase/frequency comparator, B5, having two outputs, drives the inverting and noninverting inputs of an operational amplifier B6. While the same equations hold true as those used with the passive device, note that the same bias as found in the inverting input is provided in the noninverting input to provide a precisely symmetrical load. The additional *RC* network ahead of the loop filter is used for spike suppression. The 560 Ω/2200 pF *RC* combination can be calculated, and its

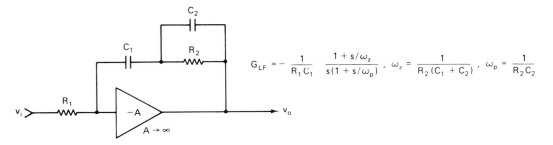

Figure 4-159 Type 2 third-order loop active filter.

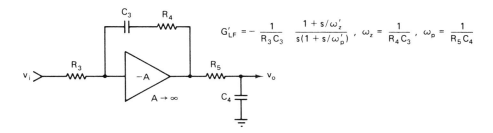

Figure 4-160 The additional RC output filtering converts this into a type 2 third-order loop filter.

effect can be determined by using the last of the computer programs. However, in the practical world, this filter is optimized experimentally.

A type 2 third-order loop, as well as higher-order loop versions, is created by adding additional RC time constants.

The real-life situation always requires some additional filtering, or the parasitic elements in a circuit act as such. Figure 4-159 shows the filter of the type 2 third-order loop that was discussed earlier.

Taking the three time constants as

$$T_1 = C_1 R_1 \qquad (4\text{-}156)$$

$$T_2 = R_2(C_1 + C_2) \qquad (4\text{-}157)$$

$$T_3 = C_2 R_2 \qquad (4\text{-}158)$$

the transfer characteristic is

$$F(s) = -\frac{1}{sT_1} \frac{1 + sT_2}{1 + sT_3} \qquad (4\text{-}159)$$

Sometimes higher orders occur because of some additional low-pass filter requirement. Figure 4-159 can be redrawn in the form of Figure 4-160, which is electrically equivalent provided that V_0 is unloaded. The correlations are

$$G_{LF1} = -\frac{1}{R_1 C_1} \frac{1 + s/\omega_z}{s(1 + s/\omega_p)} \quad (4\text{-}160)$$

$$\omega_z = \frac{1}{R_2(C_1 + C_2)} \quad (4\text{-}161)$$

$$\omega_p = \frac{1}{R_2 C_2} \quad (4\text{-}162)$$

$$G_{LF2} = -\frac{1}{R_3 C_3} \frac{1 + s/\omega_z}{s(1 + s/\omega_p)} \quad (4\text{-}163)$$

$$\omega_z = \frac{1}{R_4 C_3} \quad (4\text{-}164)$$

$$\omega_p = \frac{1}{R_5 C_4} \quad (4\text{-}165)$$

Using these equations, one schematic can be transformed into the other.

In a type 2 fifth-order loop, an active low-pass filter was added. The following sections give some insight into the mathematics of the active low-pass filter, which then becomes part of the loop filter.

4-7-3 Active Second-Order Low-Pass Filters

Active filters, compared to their passive counterpart, have a more rectangular frequency response, allowing for better noise suppression without sacrifice of reference suppression.

In Section 1-8 we saw the transient response of the type 2 fifth-order loop, which was generated from a type 2 third-order loop with the addition of a second-order low-pass filter. Although there are various configurations available with which to build a second-order low-pass filter, the one shown in Figure 4-161 is probably

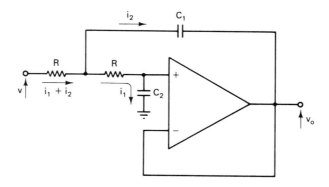

Figure 4-161 Active second-order low-pass filter.

the most useful if noninverting operation is desired. The following is a short derivation of this type of filter and some applications of its use. I would like to state here that with this filter, depending on whether a Butterworth or Chebyshev response or something in between is chosen, different phase shift and amplitude responses are obtained. By trading off skirt slope versus phase shift, it can be adapted for the purpose. The advantage of an active low-pass filter over a passive attenuator such as the T-notch filter is its low-pass filter action; the T-notch filter filters out only the particular frequency to which it is tailored. It seems that generally the second-order low-pass filter is a better choice provided that the operational amplifier used for this circuit does not introduce intolerable noise.

Let us use the abbreviation $j\omega = p$ and assume that the inverting and noninverting inputs have the same potential. We obtain the following equations:

$$V_o = V - i_1 R - i_2 R - i_1 R \tag{4-166}$$

$$V_o = V - i_1 R - i_2 R - i_2 \frac{1}{pC_1} \tag{4-167}$$

$$V_o = i_1 \frac{1}{pC_2} \tag{4-168}$$

or

$$i_1 = V_o p C_2 \tag{4-169}$$

Substituting Eq. (4-169) into Eq. (4-166), we obtain

$$V_o = V - 2RV_o pC_2 - i_2 R \tag{4-170}$$

and from Eq. (4-170),

$$i_2 = \frac{V - V_o - 2RV_o pC_2}{R} \tag{4-171}$$

With Eq. (4-169), and Eq. (4-171) in Eq. (4-167), we obtain

$$V_o = V - RV_o pC_2 - V + V_o + 2RV_o pC_2 - \frac{V - V_o - 2RV_o pC_2}{RpC_1} \tag{4-172}$$

$$R^2 V_o p^2 C_1 C_2 - V + V_o + 2RV_o pC_2 = 0 \tag{4-173}$$

$$V_o(R^2 p^2 C_1 C_2 + 2RpC_2 + 1) = V \tag{4-174}$$

The transfer function is

$$\frac{V_o}{V} = \frac{1}{R^2 p^2 C_1 C_2 + 2RpC_2 + 1} \tag{4-175}$$

380 LOOP COMPONENTS

from

$$R^2 C_1 C_2 = \frac{1}{\omega_n^2} \tag{4-176}$$

We define the frequency as

$$\omega_n = \frac{1}{R\sqrt{C_1 C_2}} \tag{4-177}$$

Normalizing the values p, C_1, and C_2 yields

$$p_o = \frac{p}{\omega_c} \tag{4-178}$$

where ω_c is the cutoff frequency, and

$$C_{1N} = RC_1 \omega_c \tag{4-179}$$
$$C_{2N} = RC_2 \omega_c \tag{4-180}$$

We finally obtain

$$\frac{V_o}{V} = \frac{1}{p_o^2 C_{1N} C_{2N} + 2p_o C_{2N} + 1} \tag{4-181}$$

To solve this equation, we find the binomial expression for the denominator,

$$\frac{1}{p_o^2 C_{1N} C_{2N} + 2p_o C_{2N} + 1} = \frac{1}{(p - P_1)(p - P_2)} \tag{4-182}$$

where P_1 and P_2 are conjugate roots of the form $(a \pm jb)$ and

$$\frac{1}{(p - P_1)(p - P_2)} = \frac{1}{(p + a \pm jb)^2} = \frac{1}{p^2 + 2ap + a^2 + b^2} \tag{4-183}$$

or

$$p_o^2 C_{1N} C_{2N} + 2p_o C_{2N} + 1 = p_o^2 + 2ap_o + a^2 + b^2 \tag{4-184}$$

with

$$p_o^2 C_{1N} C_{2N} = p_o^2 \tag{4-185}$$
$$2p_o C_{2N} = 2ap_o \tag{4-186}$$
$$1 = a^2 + b^2 \tag{4-187}$$

We finally obtain the coefficients C:

$$(a^2 + b^2) C_{2N} 2 p_o = 2 a p_o \tag{4-188}$$

$$C_{2N} = \frac{a}{a^2 + b^2} \tag{4-189}$$

$$C_{1N} = \frac{1}{a} \tag{4-190}$$

The conjugate roots of the Chebyshev approximation are

$$P_v = -\sin\frac{2v-1}{2n}\pi \sinh\phi + j\cos\frac{2v-1}{2n}\pi\cosh\phi \tag{4-191}$$

where

$$\sinh\phi = \frac{1}{2}\left\{\left[\left(1+\frac{1}{\epsilon^2}\right)^{1/2}+\frac{1}{\epsilon}\right]^{1/n} - \left[\left(1+\frac{1}{\epsilon^2}\right)^{1/2}+\frac{1}{\epsilon}\right]^{-1/n}\right\} \tag{4-192}$$

$$\cosh\phi = \frac{1}{2}\left\{\left[\left(1+\frac{1}{\epsilon^2}\right)^{1/2}+\frac{1}{\epsilon}\right]^{1/n} + \left[\left(1+\frac{1}{\epsilon^2}\right)^{1/2}+\frac{1}{\epsilon}\right]^{-1/n}\right\} \tag{4-193}$$

with ϵ^2 the ripple tolerance.

Rather than looking up normalized values from published tables (Chebyshev filters have good stopband attenuation with small passband phase distortion), a ripple factor can be chosen and then the loop should be analyzed for phase margin. If the ripple is too high, it can be changed to obtain the desired response. This is done very easily by defining a damping factor,

$$d^2 = \frac{1}{a^2 + b^2} = \frac{C_2}{C_1} \tag{4-194}$$

The ripple is inversely proportional to the damping factor. With $d = 1/\sqrt{2}$, the ripple is zero (i.e., Butterworth response). Smaller d values render increased ripple and increased stopband attenuation. What remains to be determined is ω_n, of which a good starting value is five times the required loop cutoff frequency. Modifying

$$d = \sqrt{\frac{C_2}{C_1}} \tag{4-195}$$

and ω_n in the overall open-loop transfer function while Bode plotting will render the required phase and gain margins for stable loop performance. A phase margin of $>30°$ to $<70°$ for $|A(p)| \doteq 1$ and a gain margin of -10 dB for a phase value of $180°$ should be sought.

382 LOOP COMPONENTS

The overall open-loop transfer function is followed by including the damping factor d and the natural pole frequency ω_n:

$$A^* = \frac{\omega_n^2}{-\omega^2 + 2dp\omega_n + \omega_n^2} \quad (4\text{-}196)$$

and therefore the open-loop gain of a type 2 fifth-order loop as an example would become

$$A(p) = \frac{\omega_n^2 K_o K_\theta}{NT_1 \omega^2} \frac{-pT_2 - 1}{p[2d\omega_n + T_3(\omega_n^2 - \omega)] + (\omega_n^2 - \omega^2 - 2dT_3\omega_n\omega^2)} \quad (4\text{-}197)$$

4-7-4 Passive *LC* Filters

While the active loop *RC* filter is a convenient way of improving reference suppression, as the realization of *LC* filters at times may be difficult, the use of normalized tables and an analysis program permits optimizing loops with *LC* filters. The *LC* filter in the loop has the distinct advantage that the passive devices do not introduce noise, and filter design is well established. Figure 4-162 shows a low-pass filter with its proper termination at the input and output.

The 741 operational amplifier driving the filter theoretically would have zero impedance, but since this is not permissible, the proper filter source resistor R_{30} has to be added. Similarly, R_{31} serves as output termination since pin 6 of B22 is essentially at ac ground potential.

The particular filter chosen here is of the order of 7. Table 4-9 shows the values that can be taken from Anatol Zverev's *Handbook of Filter Synthesis*, page 287 [1]. Similar elliptical filters are found in Rudolf Saal's *Handbook of Filter Design* (AEG Telefunken, West Germany). A description of how to design filters is given in the book and is not repeated here.

4-8 MICROWAVE OSCILLATOR DESIGN

4-8-1 Introduction

The difference between oscillators and microwave oscillators has to do with the fact that microwave oscillators do not rely on lumped elements but rather on distributed ones. Also, certain types of oscillator circuits just cannot be implemented. A good example is that while we can design and build a wideband transformer for the purpose of phase inversion, there is no equivalent to this for microwave application. This is also true for specialty oscillators such as YIG oscillators, where the actual resonant portion is a three-dimensional (3-D) structure. Other types of millimeter-wave oscillators use SAW devices, ceramic resonators, and dielectric resonators. In certain applications, such as wireless data transmission, the low-cost design requires printed circuit production and a similar approach is true for MMIC oscillators up to millimeter-wave application. Here the resonator is part of the layout and generally the phase noise is determined both by the low Q of the resonator and the low Q of the tuning diodes.

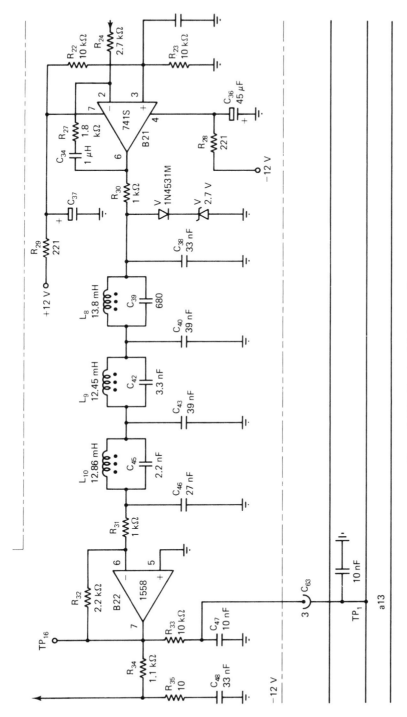

Figure 4-162 Passive LC low-pass filter.

Table 4-9 Elliptical low-pass filter CC300750

θ	30.0
C_1	0.7085
C_2	0.0389
L_2	1.3566
C_3	1.6029
C_4	0.1604
L_4	1.4917
C_5	1.6792
C_6	0.1016
L_6	1.5876
C_7	1.4569
Ω_S	2.0000
A_{min}	105.37
σ_0	0.2036327
σ_1	0.0364992
σ_3	0.1103824
σ_5	0.1756643
Ω_1	0.9936072
Ω_2	4.3544
Ω_3	0.8183516
Ω_4	2.0445
Ω_5	0.4701232
Ω_6	2.4903

In designing microwave oscillators, we continue to look for the negative impedance generated by the transistor, which will overcome the losses and therefore provide oscillation. The two major differences between RF oscillators and microwave oscillators are the following:

1. To discrete the active devices, we will use S-parameters.
2. For higher frequencies, we will move away from the lumped elements to distributed elements.

We will use different types of resonators, however; their Q will be less than those of crystal oscillators.

The conditions for oscillation can be expressed as

$$k < 1 \qquad (4\text{-}198)$$

$$\Gamma_G S'_{11} = 1 \qquad (4\text{-}199)$$

$$\Gamma_L S'_{22} = 1 \qquad (4\text{-}200)$$

The stability factor should be less than unity for any possibility of oscillation. If this condition is not satisfied, either the common terminal should be changed or positive feedback should be added. Next, the passive terminations Γ_G and Γ_L must be added to resonate the input and output ports at the frequency of oscillation.

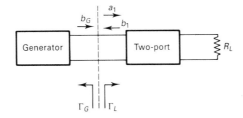

Figure 4-163 Two-port connected to a generator.

This is satisfied by either Eq. (4-198) or (4-200). It will be shown that if Eq. (4-199) is satisfied, Eq. (4-200) must be satisfied, and vice versa. In other words, if the oscillator is oscillating at one port, it must be simultaneously oscillating at the other port. Normally, a major fraction of the power is delivered only to one port, since only one load is connected. Since $|\Gamma_G|$ and $|\Gamma_L|$ are less than unity, Eqs. (4-199) and (4-200) imply that $|S'_{11}| > 1$ and $|S'_{22}| > 1$.

The conditions for oscillation can be seen from Figure 4-163, where an input generator has been connected to a two-port. Using the following representation,

$$a_1 = b_G + \Gamma_L \Gamma_G a_1 \qquad (4\text{-}201)$$

and defining

$$\Gamma_L = S'_{11} \qquad (4\text{-}202)$$

$$S'_{11} = \frac{b_1}{a_1} \qquad (4\text{-}203)$$

we find

$$\begin{aligned} b_G &= a_1(1 - \Gamma_{L1}\Gamma_G) \\ &= \frac{b_1}{S'_{11}}(1 - S'_{11}\Gamma_G) \end{aligned} \qquad (4\text{-}204)$$

$$\frac{b_1}{b_G} = \frac{S'_{11}}{1 - S'_{11}\Gamma_G} \qquad (4\text{-}205)$$

Thus, the wave reflected from the two-port is dependent on Γ_G and Γ_L. If Eq. (4-199) is satisfied, b_G must be zero, which implies that the two-port is oscillating. Since $|\Gamma_G|$ is normally less than or equal to unity, this requires that $|S'_{11}|$ be greater than or equal to unity.

The oscillator designer must simply guarantee a stability factor less than unity and resonate the input port by satisfying Eq. (4-199), which implies that Eq. (4-200) has also been satisfied. Another way of expressing the resonance condition of Eq. (4-199) is the following:

$$R_{\text{in}} + R_G = 0 \qquad (4\text{-}206)$$

$$X_{\text{in}} + X_G = 0 \qquad (4\text{-}207)$$

386 LOOP COMPONENTS

This follows from substituting

$$S'_{11} = \frac{R_{in} + jX_{in} - Z_0}{R_{in} + jX_{in} + Z_0} \tag{4-208}$$

$$\Gamma_G = \frac{R_G + jX_G - Z_0}{R_G + jX_G + Z_0}$$
$$= \frac{-R_{in} - Z_0 - jX_{in}}{-R_{in} + Z_0 - jX_{in}} \tag{4-209}$$

into Eq. (4-199), giving

$$\Gamma_G S'_{11} = \frac{-R_{in} - Z_0 - jX_{in}}{-R_{in} + Z_0 - jX_{in}} \times \frac{R_{in} + jX_{in} - Z_0}{R_{in} + Z_0 + jX_{in}} = 1$$

which proves the equivalence of Eq. (4-199) to Eqs. (4-206) and (4-207).

Before proceeding with the oscillator design procedures, some typical oscillator specifications are given in Table 4-10 for the major types of oscillators. The high-Q or cavity-tuned oscillators usually have better spectral purity than do the low-Q VCOs (voltage-controlled oscillators), which have faster tuning speeds. The FM

Table 4-10 Typical oscillator specifications

Parameter	High-Q or Cavity-Tuned (e.g., YIG)	Low-Q or Varactor-Tuned VCO
Frequency	2–4 GHz	2–4 GHz
Power	+10 dBm	+10 dBm
Power variation versus f	±2 dB	±2 dB
Temperature stability versus f	±10 ppm/°C	±500 ppm/°C
Power versus temperature (−30 to 60 °C)	±2 dB	±2 dB
Modulation sensitivity	10–20 MHz/mA	50–200 MHz/V
FM noise	−110 dBc/Hz at 100 kHz	−100 dBc/Hz at 100 kHz
AM noise	−140 dBc/Hz at 100 kHz	−140 dBc/Hz at 100 kHz
FM noise floor	−150 dBc/Hz at 100 MHz	−150 dBc/Hz at 100 MHz
All harmonics	−20 dBc	−20 dBc
Short-term post Tuning drift	±2 MHz 1 μs	±2 MHz 1–100 μs
Long-term post Tuning drift	±2 MHz 5–30 s	±2 MHz 5–30 s
Pulling of f for all phases of 12-dB return loss	±1 MHz	±20 MHz
Pushing of f with change of bias voltage	5 MHz/V	5 MHz/V

noise is usually measured at about 100 kHz from the carrier in units of dBc, which means decibels below the carrier level, in a specified bandwidth of 1 Hz. If the measurement bandwidth is 1 kHz, the specification changes by 10^3.

In selecting a transistor to meet the specifications, the amplifier transistors with the same frequency and power performance are usually suitable. Lower close-in noise can be achieved from silicon bipolar transistors compared to GaAs MESFETs because of the $1/f$ noise difference.

4-8-2 The Compressed Smith Chart

The normal Smith chart is a plot of the reflection coefficient for $|\Gamma| \leq 1$. The compressed Smith chart includes $|\Gamma| > 1$, and the chart is given in Figure 4-164 for $|\Gamma| \leq 3.16$ (10 dB of return gain). This chart is useful for plotting the variation of S'_{11} and S'_{22} for oscillator design. The impedance and admittance properties of the Smith chart are retained for the compressed chart. For example, a Γ_{in} of $1.2\underline{/150°}$ gives the following values of Z and Y normalized to $Z_0 = 50\,\Omega$:

$$Z_{in}/Z_0 = -0.10 + j0.25$$
$$Z^*_{in}/Z_0 = -0.10 - j0.25$$
$$Y_{in}/Y_0 = -1.0 - j3.0$$
$$Y^*_{in}/Y_0 = -1.0 + j3.0$$

These values are plotted in Figure 4-164 for illustration.

A frequency resonance condition simply requires the circuit imaginary term be zero. If the impedance resonance is on the left-hand real axis, this is a series resonance; that is, at frequencies above resonance the impedance is inductive and below resonance the impedance is capacitive. If the impedance resonance is on the right-hand real axis, the resonance is a parallel resonance; that is, at frequencies above resonance the impedance is capacitive and below resonance the impedance is inductive.

An oscillator resonance condition implies that both the circuit imaginary term and the circuit real term are zero, as given by Eqs. (4-206) and (4-207). Impedances and admittances can be transformed on the compressed Smith chart by the methods discussed; however, when $|\Gamma|$ is greater than unity, the goal of impedance transformation is usually to achieve either a series or a parallel resonance condition. Another method for visualizing negative resistance is to plot $1/S_{11}$ and multiply the result by -1. This allows the designer readily to use available Smith charts, with $|\Gamma| \leq 1$, to analyze circuits with $|\Gamma| \geq 1$. The proof of this concept can be shown by expressing the reflection coefficient of a one-port by

$$S_{11} = \frac{Z_s - Z_0}{Z_s + Z_0} \qquad (4\text{-}210)$$

$$\frac{1}{S_{11}} = \frac{Z_s + Z_0}{Z_s - Z_0} = \frac{Z_1 - Z_0}{Z_1 + Z_0} \qquad (4\text{-}211)$$

388 LOOP COMPONENTS

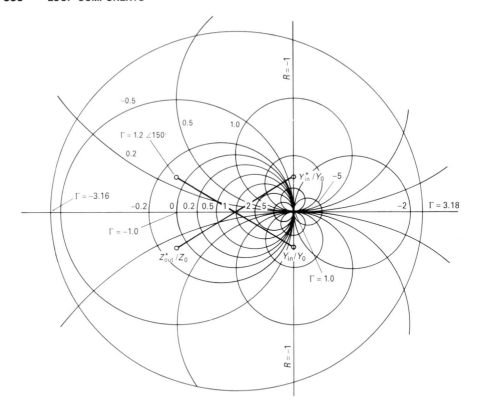

Figure 4-164 Compressed Smith chart.

where $Z_1 = -Z_s$, which gives a negative resistance on Smith chart coordinates. For example, using the case in Figure 4-164, we have

$$S_{11} = 1.2\underline{/150°}$$

$$\frac{1}{S_{11}} = 0.833\underline{/-150°}$$

$$\frac{Z_1}{Z_0} = 0.10 - j0.25$$

$$\frac{Z_s}{Z_0} = -0.10 + j0.25$$

The impedance of the one-port is plotted at Z_1 but understood to be Z_s.

4-8-3 Series or Parallel Resonance

Oscillators can be classified into two types—series-resonant or parallel-resonant—as shown in Figure 4-165. The equivalent circuit of the active device is chosen from

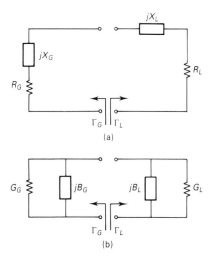

Figure 4-165 Oscillator equivalent circuits: (a) series-resonant and (b) parallel-resonant.

the frequency response of the output port, that is, the frequency response of Γ_G. For the series-resonant condition, the negative resistance of the active device must exceed the load resistance Γ_L at start-up of oscillation by about 20%. As the oscillation builds up to a steady-state value, the resonance condition will be reached as a result of limiting effects, which cause a reduction of Γ_G under large-signal drive.

For start-up of oscillation,

$$|R_G| > 1.2 R_L \tag{4-212}$$

For resonance,

$$R_G + R_L = 0 \tag{4-213}$$

$$X_G + X_L = 0 \tag{4-214}$$

For the parallel-resonant condition, the negative conductance of the active device must exceed the load conductance G_L at start-up of oscillation by about 20%. The parallel-resonant oscillator is simply the dual of the series-resonant case. For start-up of oscillation,

$$|G_G| > 1.2 G_L \tag{4-215}$$

For resonance,

$$G_G + G_L = 0 \tag{4-216}$$

$$B_G + B_L = 0 \tag{4-217}$$

To design the oscillator for series resonance, the reflection coefficient of the active transistor is moved to an angle of 180° (i.e., the left-hand real axis of the compressed Smith chart). Keeping in mind Eq. (4-198) for the input resonating port, we see that a nearly lossless reactance will resonate the transistor. For the example in Figure 4-165, we have

$$\Gamma = 1.2 \underline{/150°} = S'_{11}$$
$$\Gamma_G = 0.83 \underline{/-150°} \simeq 1.0 \underline{/-150°}$$

The large-signal drive of the transistor will reduce S'_{11} to about $1.0\underline{/150°}$. For parallel-resonant oscillator design, the reflection coefficient of the active transistor is moved to an angle of 0° (i.e., the right-hand real axis of the compressed Smith chart). Alternatively, the reflection coefficient associated with impedance can be inverted to an admittance point, and the admittance can be moved to an angle of 180° (i.e., the left-hand real axis of the compressed Smith chart).

4-8-4 Two-Port Oscillator Design

A common method for designing oscillators is to resonate the input port with a passive high-Q circuit at the desired frequency of resonance. It will be shown that if this is achieved with a load connected on the output port, the transistor is oscillating at both ports and is thus delivering power to the load port. The oscillator may be considered a two-port structure, where M_3 is the lossless resonating port and M_4 provides lossless matching such that all of the external RF power is delivered to the load. The resonating network has been described. Nominally, only parasitic resistance is present at the resonating port, since a high-Q resonance is desirable for minimizing oscillator noise. It is possible to have loads at both the input and the output ports if such an application occurs, since the oscillator is oscillating at both ports simultaneously.

The simultaneous oscillation condition is proved as follows. Assume that the oscillation condition is satisfied at port 1:

$$1/S'_{11} = \Gamma_G \tag{4-218}$$

Thus,

$$S'_{11} = S_{11} + \frac{S_{12}S_{21}\Gamma_L}{1 - S_{22}\Gamma_L} = \frac{S_{11} - D\Gamma_L}{1 - S_{22}\Gamma_L} \tag{4-219}$$

$$\frac{1}{S'_{11}} = \frac{1 - S_{22}\Gamma_L}{S_{11} - D\Gamma_L} = \Gamma_G \tag{4-220}$$

By expanding Eq. (4-220), we find

$$\Gamma_G S_{11} - D\Gamma_L \Gamma_G = 1 - S_{22}\Gamma_L$$
$$\Gamma_L(S_{22} - D\Gamma_G) = 1 - S_{11}\Gamma_G$$
$$\Gamma_L = \frac{1 - S_{11}\Gamma_G}{S_{22} - D\Gamma_G} \tag{4-221}$$

Thus,

$$S'_{22} = S_{22} + \frac{S_{12}S_{21}\Gamma_G}{1 - S_{11}\Gamma_G} = \frac{S_{22} - D\Gamma_G}{1 - S_{11}\Gamma_G} \quad (4\text{-}222)$$

$$\frac{1}{S'_{22}} = \frac{1 - S_{11}\Gamma_G}{S_{22} - D\Gamma_G} \quad (4\text{-}223)$$

Comparing Eqs. (4-221) and (4-223), we find

$$1/S'_{22} = \Gamma_L \quad (4\text{-}224)$$

which means that the oscillation condition is also satisfied at port 2; this completes the proof. Thus, if either port is oscillating, the other port must be oscillating as well. A load may appear at either or both ports, but normally the load is in Γ_L, the output termination. This result can be generalized to an n-port oscillator by showing that the oscillator is simultaneously oscillating at each port:

$$\Gamma_1 S'_{11} = \Gamma_2 S'_{22} = \Gamma_3 S'_{33} = \cdots = \Gamma_n S'_{nn} \quad (4\text{-}225)$$

Before concluding this section on two-port oscillator design, the buffered oscillator shown in Figure 4-166 must be considered. This design approach is used to provide the following:

1. A reduction in loading-pulling, which is the change in oscillator frequency when the load reflection coefficient changes.
2. A load impedance that is more suitable to wideband applications, Eq. (4-198).
3. A higher output power from a working design, although the higher output power can also be achieved by using a larger oscillator transistor.

Buffered oscillator designs are quite common in wideband YIG applications, where changes in the load impedance must not change the generator frequency.

Two-port oscillator design may be summarized as follows:

1. Select a transistor with sufficient gain and output power capability for the frequency of operation. This may be based on oscillator data sheets, amplifier performance, or S-parameter calculation.
2. Select a topology that gives $k < 1$ at the operating frequency. Add feedback if $k < 1$ has not been achieved.

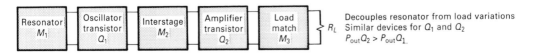

Figure 4-166 Buffered oscillator design.

Table 4-11 HP2001 bipolar chip common base (V_{CE} = 15 V, I_C = 25 mA)

$L_B = 0$	$L_B = 0.5$ nH
$S_{11} = 0.94\ \underline{/174°}$	1.04 $\underline{/173°}$
$S_{21} = 1.90\ \underline{/-28°}$	2.00 $\underline{/-30°}$
$S_{12} = 0.013\ \underline{/98°}$	0.043 $\underline{/153°}$
$S_{22} = 1.01\ \underline{/-17°}$	1.05 $\underline{/-18°}$
$k = -0.09$	-0.83

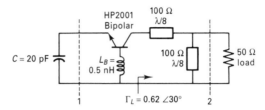

Figure 4-167 Oscillator example at 2 GHz.

3. Select an output load matching circuit that gives $|S'_{11}| > 1$ over the desired frequency range. In the simplest case this could be a 50-Ω load.
4. Resonate the input port with a lossless termination so that $\Gamma_G S'_{11} = 1$. The value of S'_{22} will be greater than unity with the input properly resonated.

In all cases, the transistor delivers power to a load and the input of the transistor. Practical considerations of realizability and dc biasing will determine the best design.

For both bipolar and FET oscillators, a common topology is common-base or common-gate, since a common-lead inductance can be used to raise S_{22} to a large value, usually greater than unity even with a 50-Ω generator resistor. However, it is not necessary for the transistor S_{22} to be greater than unity, since the 50-Ω generator is not present in the oscillator design. The requirement for oscillation is $k < 1$; then resonating the input with a lossless termination will provide that $|S'_{11}| > 1$.

A simple example will clarify the design procedure. A common-base bipolar transistor (HP2001) was selected to design a fixed-tuned oscillator at 2 GHz. The common-base S-parameters and stability factor are given in Table 4-11. Using the load circuit in Figure 4-167, we see that the reflection coefficients are

$$\Gamma_L = 0.62\underline{/30°}$$
$$S'_{11} = 1.18\underline{/173°}$$

Thus, a resonating capacitance of $G = 20$ pF resonates the input port. In a

MICROWAVE OSCILLATOR DESIGN 393

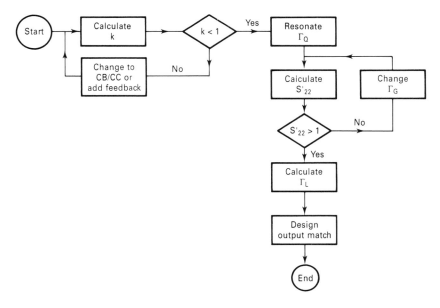

Figure 4-168 Oscillator design flowchart.

YIG-tuned oscillator, this reactive element could be provided by the high-Q YIG element. For a dielectric resonator oscillator (DRO), the puck would be placed to give $\Gamma_G \approx 1.0 \underline{/-173°}$.

Another two-port design procedure is to resonate the Γ_G port and calculate S'_{22} until $|S'_{22}| > 1$, then design the load port to satisfy. This design procedure is summarized in Figure 4-168.

An example using this procedure at 4 GHz is given in Figure 4-169 using an AT 41400 silicon bipolar chip in the common-base configuration with a convenient value of base and emitter inductance of 0.5 nH. The feedback parameter is the base inductance, which can be varied if needed.

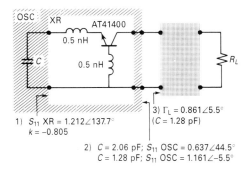

1) S_{11} XR = 1.212∠137.7°
   $k = -0.805$
2) $C = 2.06$ pF; S_{11} OSC = 0.637∠44.5°
   $C = 1.28$ pF; S_{11} OSC = 1.161∠−5.5°
3) $\Gamma_L = 0.861∠5.5°$
   ($C = 1.28$ pF)

Figure 4-169 A 4-GHz lumped resonator oscillator using AT41400.

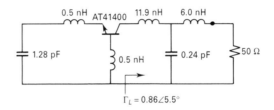

Figure 4-170 Completed lumped resonator oscillator (LRO).

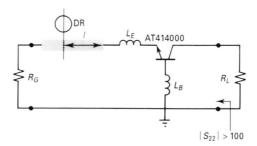

Figure 4-171 Transmission line oscillator with dielectric resonator.

The two-port common-base S-parameters were used to give

$$k = -0.805$$
$$S'_{11} = 1.212 \underline{/137.7°}$$

Since a lossless capacitor at 4 GHz of 2.06 pF gives $\Gamma_G = 1 - 0\underline{/-137.7°}$, this input termination is used to calculate S'_{22}, giving $S'_{22} = 0.637\underline{/44.5°}$. This circuit will not oscillate into any passive load. Varying the emitter capacitor about 20° on the Smith chart to 1.28 pF gives $S'_{22} = 1.16\underline{/-5.5°}$, which will oscillate into a load of $\Gamma_L = 0.861\underline{/5.5°}$. The completed lumped element design is given in Figure 4-170.

We now switch from the lumped design to a microstrip design that incorporates a dielectric resonator. This oscillator circuit is given in Figure 4-171, where the dielectric resonator (DR) will serve the function of the emitter capacitor. This element is usually coupled to the 50-Ω microstripline to present about 1000 Ω of loading ($\beta = 20$) at f_0, the lowest resonant frequency of the dielectric puck, at the correct position on the line. The load circuit will be simplified to 50 Ω ($\Gamma_L = 0$), so the oscillator must have an output reflection coefficient of greater than 100, thus presenting a negative resistance between -49 and -51 Ω. The computer file for analyzing this design is given in Table 4-12, where the variables are the puck resistance, the 50-Ω microstripline length, and the base feedback inductance. The final design is given in Figure 4-172, where the 10-μH coils are present for the dc bias connections that need to be added to the design. It is important to check the stability of this circuit with the DR removed. The input 50-Ω termination will usually guarantee unconditional stability at all frequencies. The phase noise of this oscillator is very low at -117 dBc/Hz at 10-kHz frequency offset.

Table 4-12 Super-compact file for DRO design in Figure 4-172

```
*
—
* AT41400 AT 7.5V, 30mA IN DRO
* OSCILLATOR By Vendelin et al. Microwave Journal June
1986 pp. 151-152
BLK
     TRL          1  2       Z=50 P=250MIL K=6.6
     RES          2  3       R=?955.06?
     TRL          3  4       Z=50 P=?224.16MIL?
                                 K=6.6
     IND          4  0       L=1E4NH
     IND          4  5       L=.5NH
     TWO          6  7  5    Q1
     IND          6  0       L=?.33843NH?
     IND          7  0       L=1E4NH
     OSC:2POR     1  7
END
*
FREQ
     4GHZ
END
OUT
     PRI OSC S
END
OPT
     OSC
     MS22 = 100 GT
END
DATA
  Q1:S
  4  .8057 −176.14 2.5990 74.77 .0316 56.54 .4306
     −22.94
END
```

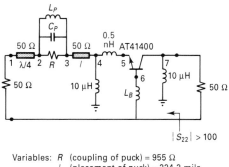

Variables: R (coupling of puck) = 955 Ω
  l (placement of puck) = 224.2 mils
  (ε_L = 10, h = 25 mils)
  L_B (base inductance) = 0.34 nH

Figure 4-172 Equivalent circuit for dielectric resonator oscillator (DRO).

396 LOOP COMPONENTS

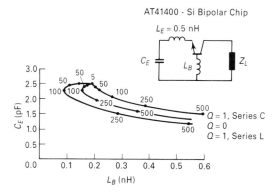

Figure 4-173 Tuning parameters for a 4-GHz oscillator versus load impedance as the load varies from 50 Ω.

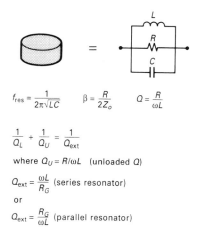

Figure 4-174 Simple equivalent circuit for the dielectric resonator.

For simple oscillators with no isolating stage, one can expect a certain amount of pulling. Figure 4-173 shows the tuning parameters as the load varies from 50 Ω. The load $C_L = R + jX$ influences the required input capacitance C_E and the base inductor L_B. The numbers in the graph are the resonant portion of the load impedance and the ratio X/R determines the Q line. It is obvious that such a circuit is quite interactive. As to the model for the dielectric resonator, the valid relationship is shown in Figure 4-174.

In Section 4-9 on microwave resonators, we will look at a more physical model.

Finally, Table 4-13 describes the same DRO in the familiar Spice format. This particular Compact Software Inc. Spice model uses transmission elements T1 and T2 and the resonant frequency of the oscillator is determined by both the dielectric resonator and its position relative to the transmission line. In the equivalent circuit

Table 4-13 Spice format

```
Compact Software - SUPER-SPICE 1.1 08/09/95 13:38:56
File: C:\SPICE\CIR\DRO.cir

Dielectric Resonator Oscillator with a BJT
Q1 1 2 3 Q2NXXXX
C1 2 4 100pf
L1 4 0 0.3384nh
L2 1 100 1uh
L3 3 6 0.5nh
lb1 6 0 1uh
T1 6 0 7 0 Z0=50 TD=5.4378e-11
cdro 7 8 .0397p
ldro 7 8 40nh
rdro 7 8 955
T2 8 0 9 0 Z0=50 TD=4.876e-11
R1 9 0 50
C4 1 10 100pf
P1 10 0 PNR=1 ZL=50
*Biasing
R3 100 2 3.6k
R4 2 0 1.2k
V1 100 0 7.5V
.model Q2NXXXX NPN(Is=1.65e-18 Vaf=20 Bf=50 Nf=1.03
+     Ise=5f lkf=.1 Xtb=1.818 Br=5 cjc=.75p
+     Fc=.5 Cje=.75p Mje=.6 Vje=1.01 xcjc=.5
+     Tf=14p Itf=.3 Vtf=6 Xtf=4 Ptf=35)
.IC V(2)=.001
.TRAN 2N 500N
.AC LIN 500 3GHZ 5GHZ
.opt itl5=0
.PROBE
.END
```

of the transistor, no values for a base-spreading resistor have been assumed. This modeling is done for demonstration purposes and does not relate to an actual transistor. A more practical circuit will follow.

4-9 MICROWAVE RESONATORS

For microwave applications, one is rapidly moving away from lumped to distributed elements. In the previous section, we looked at the case of a transmission line-based oscillator, which by itself has a low Q and was shown only for descriptive and design purposes. In similar fashion, we looked at the simplified description of a dielectric resonator-based oscillator.

From a practical design point of view, most relevant applications are SAW resonators, dielectric resonators, and YIG oscillators. These are the three types of resonators we will cover in this section.

398 LOOP COMPONENTS

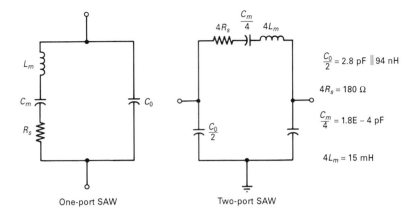

Figure 4-175 Appropriate capacitance to ground for the SAW resonator.

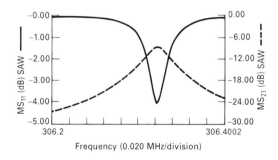

Figure 4-176 Frequency response of a SAW oscillator.

4-9-1 SAW Oscillators

The SAW oscillator has an equivalent circuit similar to a crystal but should be enhanced by adding the appropriate capacitance to ground. Figure 4-175 shows this. SAW oscillators are frequently used in synthesizers and provide a low phase noise, highly stable source, as can be seen in Figure 4-175. The SAW oscillator comes as either a one-port or two-port device.

The SAW resonator has fairly high insertion loss, as can be seen from Figure 4-176. The actual circuit of a high-performance SAW oscillator, as shown in Figure 4-177, consists of a bipolar transistor with a dc stabilizing circuit, SAW oscillator, and a feedback loop, which allows the phase to be adjusted. The SAW oscillator provides very good phase noise. The measured phase noise of such an oscillator is shown in Figure 4-178. The actual measured phase noise agrees quite well with this prediction [1].

MICROWAVE RESONATORS   399

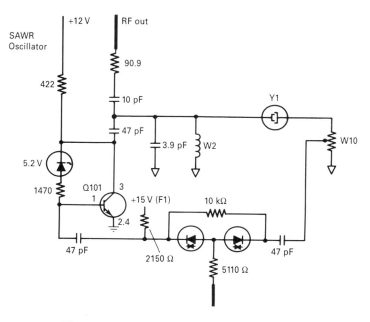

Figure 4-177 Schematic of a SAW oscillator.

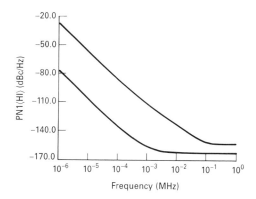

Figure 4-178 Phase noise as determined by the initial start-up values and after optimization.

4-9-2 Dielectric Resonators

In designing dielectric resonator-based oscillators, several methods of frequency stabilization are available that have been proposed by various authors. Figure 4-179 shows some recommended methods of frequency stabilization for dielectric resonator oscillators. The dielectric resonator consists of some high dielectric material coupled to a transmission line or microstrip structure.

400 LOOP COMPONENTS

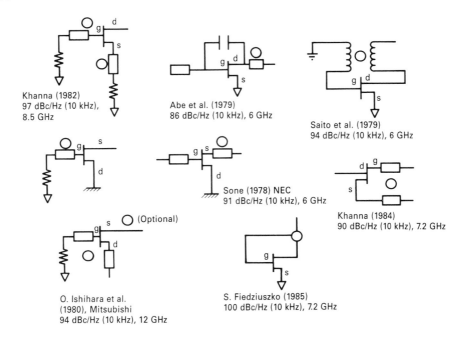

Figure 4-179 Recommended methods of frequency stabilization for DROs.

Table 4-14 Physical dimensions of DR

```
BLK

DRMS 1  2 D=6.12e-3 HD-2.45e-3 ER=38 HT=1.5e-3 S=.5e-3;
        + W=1.1e-3 L=4e-3 SRD=1e-4 BPF SUB;
        trf 2 0 3 N=1
        pug: 2POR 1 3
END
DATA
        SUB: MS er=2.4 h=0.380e-3 met1=cu 3.175e-6
        and=0.0001
END
```

Figure 4-180 shows the field distribution and interaction between the microstrip and the dielectric resonator. The two resulting applications, BandStop and BandPass filters, are displayed. Modeling this type of resonator is done by describing the resonator in the form of its physical dimensions.

Table 4-14 shows the physical dimensions of the dielectric resonator in Super-Compact/Microwave Harmonica format.

A practical example of a 6-GHz dielectric resonator-based oscillator is shown in Figure 4-181 and its predicted phase noise is shown in Figure 4-182.

For calibration purposes, it may be useful to plot the phase noise of different oscillators, including YIG oscillators, as shown in Figure 4-183, but normalized

MICROWAVE RESONATORS 401

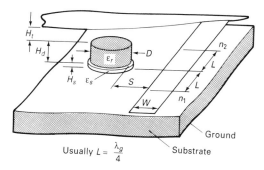

Figure 4-180a DRO on microstrip as BandStop filter.

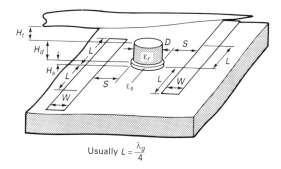

Figure 4-180b DRO on microstrip as BandPass filter.

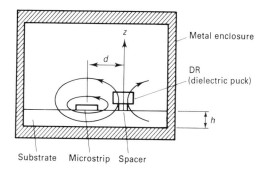

Figure 4-180c Field distribution and interaction between the microstrip and the DRO.

402 LOOP COMPONENTS

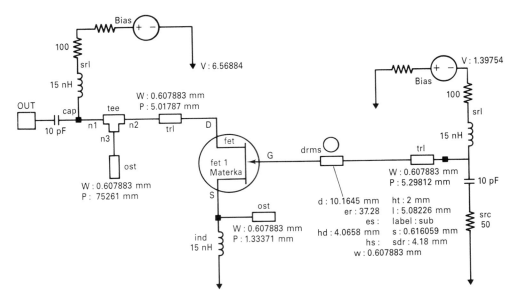

Figure 4-181 Schematic of 6-GHz DRO.

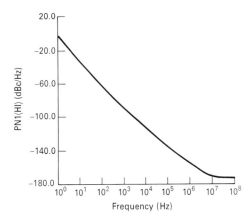

Figure 4-182 Predicted phase noise of the 6-GHz DRO pictured in Figure 4-181.

to a center frequency of 6 GHz. Another way of plotting this is to show the phase noise of silicon bipolar transistors versus FETs at 10 kHz offset from the carrier, as shown in Figure 4-184. This plot does not incorporate for heterojunction bipolar transistors because they are not yet readily or commercially available.

4-9-3 YIG Oscillators

For wideband electrically tunable oscillators, we use either a YIG or a varactor resonator. The YIG resonator is a high-Q, ferrite sphere of yittrium ion garnet,

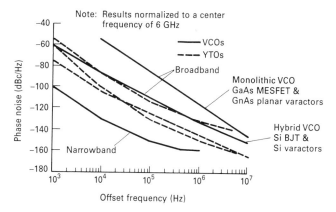

Figure 4-183 Phase noise comparison of different YIG and varactor tuned oscillators.

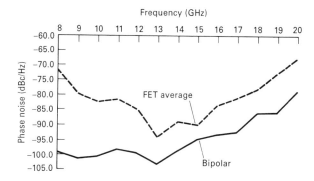

Figure 4-184 Phase noise at 10 kHz off the carrier of silicon bipolar transistors versus FETs.

$Y_2Fe_2(FeO_4)_3$, that can be tuned over a wide band by varying the biasing dc magnetic field. Its high performance and convenient size for applications in microwave integrated circuits make it an excellent choice in a large number of applications, such as filters, multipliers, discriminators, limiters, and oscillators. A YIG resonator makes use of the ferrimagnetic resonance, which, depending on the material composition, size, and applied field, can be achieved from 500 MHz to 50 GHz. An unloaded Q greater than 1000 is usually achieved with typical YIG material.

Figure 4-185 shows the mechanical drawing of a YIG oscillator assembly. The drawing is somewhat simplified and the actual construction is actually more difficult to do. Its actual circuit diagram is shown in Figure 4-186.

4-9-4 Varactor Resonators

The dual of the current-tuned YIG resonator is the voltage-tuned varactor, which is a variable reactance achieved from a low-loss, reverse-biased semiconductor PN

404 LOOP COMPONENTS

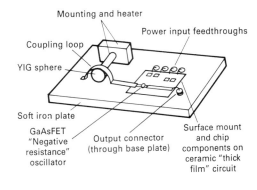

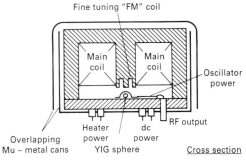

Figure 4-185 The yttrium–iron–garnet (YIG) sphere serves as the resonator in the sweep oscillators used in many spectrum analyzers.

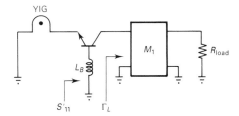

Figure 4-186 Actual circuit diagram for YIG-tuned oscillator depicted in Figure 4-185.

junction. These diodes are designed to have very low loss and therefore high Q. The silicon varactors have the fastest settling time in fast-tuning applications, but the gallium arsenide varactors have higher Q values. The cutoff frequency of the varactor is defined as the frequency where $Q_v = 1$. For a simple series RC equivalent circuit, we have

$$Q_v = \frac{1}{\omega R C_v} \qquad (4\text{-}226)$$

$$f_{c0} = \frac{1}{2\pi R C_v} \qquad (4\text{-}227)$$

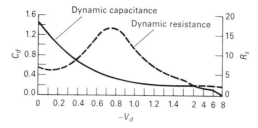

Figure 4-187 Dynamic capacitance and dynamic resistors as a function of tuning voltage for GaAs varactor.

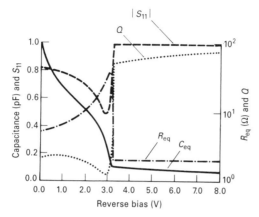

Figure 4-188 Varactor parameters: capactance, equivalent resistor, and Q, as well as the magnitude of S_{11}, as a function of reverse voltage.

The tuning range of the varactor will be determined by the capacitance ratio C_{max}/C_{min}, which can be 12 or higher for hyper-abrupt varactors. Since R is a function of bias, the maximum cutoff frequency occurs at a bias near breakdown, where both R and C_v have minimum values. Tuning diodes or GaAs varactors for microwave and millimeter-wave applications are frequently obtained by using a GaAs FET and connecting source and drain together. Figure 4-187 shows the dynamic capacitance and dynamic resistors as a function of tuning voltage. In using a transistor instead of a diode, the parameters become more complicated. Figure 4-188 shows the capacitance, equivalent resistor, and Q, as well as the magnitude of S_{11}, as a function of reverse voltage. This is due to the breakdown effects of the GaAs FET.

Previously, we had discussed in great detail the tuning diode applications. The major differences between these applications and microwave applications have to do with the resulting low Q and different technology. This is the reason why discussions of both applications were separated.

4-9-5 Ceramic Resonators

An important application for a new class of resonators called ceramic resonators (CRs) has emerged for wireless applications. The CRs are similar to shielded coaxial cable, where the center controller is connected at the end to the outside of the cable. These resonators are generally operating in quarter-wavelength mode and their characteristic impedance is approximately 10 Ω. Because their coaxial assemblies are made for a high-ϵ low-loss material with good silver plating throughout, the electromagnetic field is internally contained and therefore provides very little radiation. These resonators are therefore ideally suited for high-Q, high-density oscillators. The typical application for this resonator is VCOs ranging from not much more than 200 MHz up to about 3 or 4 GHz. At these high frequencies, the mechanical dimensions of the resonator become too tiny to offer any advantage. One of the principal requirements is that the physical length is considerably larger than the diameter. If the frequency increases, this can no longer be maintained.

Calculation of Equivalent Circuit. The equivalent parallel-resonant circuit has a resistance at resonant frequency of

$$R_p = \frac{2(Z_0)^2}{R^* l}$$

where Z_0 = characteristic impedance of the resonator
l = mechanical length of the resonator
R^* = equivalent resistor due to metalization and other losses

As an example, one can calculate

$$C^* = \frac{2\pi\epsilon_0\epsilon_r}{\log_e(D/d)} = 55.61 \times 10^{-12} \frac{\epsilon_r}{\log_e(D/d)} \qquad (4\text{-}228)$$

and

$$L^* = \frac{\mu_r\mu_0}{2\pi} = \log_e\left(\frac{D}{d}\right) = 2 \times 10^{-7} \log_e\left(\frac{D}{d}\right) \qquad (4\text{-}229)$$

$$Z_0 = 60\,\Omega \frac{1}{\sqrt{\epsilon_r}} \log_e\left(\frac{D}{d}\right) \qquad (4\text{-}230)$$

A practical example for $\epsilon_r = 88$ and 450 MHz is

$$C_p = \frac{C^* l}{2} = 49.7\,\text{pF} \qquad (4\text{-}231)$$

$$L_p = 8L^* l = 2.52\,\text{nH} \qquad (4\text{-}232)$$

$$R_p = 2.5\,\text{k}\Omega \qquad (4\text{-}233)$$

Manufacturers supply these resonators on a prefabricated basis. Figure 4-189 shows the standard round/square packaging available and the typical dimensions for a ceramic resonator.

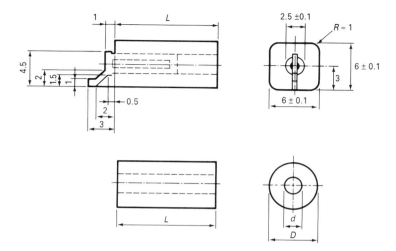

Figure 4-189 Standard round/square packaging.

The available material has a dielectric constant of 88 and is recommended for use in the 400- to 1500-MHz range. The next higher frequency range (800 MHz to 2.5 GHz) uses an ϵ of 38, while the top range (1 to 4.5 GHz) uses an ϵ of 21. Given the fact that ceramic resonators are prefabricated and have standard outside dimensions, the following quick calculation applies:

Relative dielectric constant of resonator material	$\epsilon_r = 21$	$\epsilon_r = 38$	$\epsilon_r = 88$
Resonator length in millimeters	$l = \dfrac{16.6}{f}$	$l = \dfrac{12.6}{f}$	$l = \dfrac{8.2}{f}$
Temperature coefficient (ppm/°C)	10	6.5	8.5
Available temperature coefficients	−3 to +12	−3 to +12	−3 to +12
Typical resonator Q	800	500	400

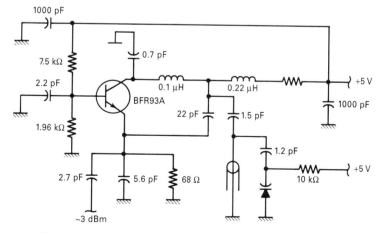

Figure 4-190 Schematic of ceramic resonator-based oscillator.

408 LOOP COMPONENTS

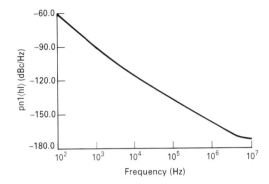

Figure 4-191 Simulated phase noise of an NPN bipolar 1-GHz ceramic resonator-based oscillator.

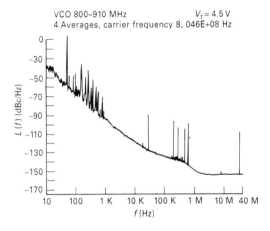

Figure 4-192 Measured phase noise of a ceramic resonator-based oscillator.

Figure 4-190 shows the schematic of such an oscillator. Figures 4-191 and 4-192 show the simulated and measured phase noise of the ceramic-resonator-based oscillator.

By using ceramic-resonator-based oscillators in conjunction with miniature synthesizer chips, it is possible to build extremely small phase-locked loop systems for cellular telephone operation. Figure 4-193 shows one of the smallest currently available PLL-based synthesizers manufactured by Synergy Microwave Corporation. Because of the high-Q resonator, these types of oscillators exhibit extremely low phase noise. Values of better than 150 dB/Hz, 1 MHz off the carrier, are achievable. The ceramic resonator reduces the sensitivity toward microphonic effects and proximity effects caused by other components.

Figure 4-193 Miniature PLL-based synthesizer manufactured by Synergy Microwave Corporation.

REFERENCES

Section 4-1

1. *Capacitance Diodes, Tuner Diodes, Diode Switches, PIN Diodes, Basics and Applications*, ITT Semiconductors, System-Druck GmbH & Co., Freiburg, Germany, 1976/10.
2. H. Keller, "Properties and Applications of the Silicon Capacitance Diode," *Ionen + Elektronen*, April 1961, pp. 15–17.
3. H. Keller, M. Lehmann, and L. Micic, "Diffused Silicon Capacitance Diodes," *Radio Mentor*, Vol. 28, No. 8, 1962, pp. 661–667.
4. H. Keller, "An FM Receiver with Electronic Tuning and Automatic Station Tracking," *Funk-Technik*, Vol. 18, No. 22, 1963, pp. 827–828.
5. A. Gilly and L. Micic, "DC Amplifier with Capacitance Diodes for Low Power Input Signals," *Elektronik*, Vol. 12, No. 9, 1963, p. 263.
6. DIN 41791, sheet 8; DIN 41785, sheet 20 (German standards).
7. H. Keller, "Electronic UHF Tuning in TV Receivers," *Radio-Fernseh-Phono-Praxis*, No. 3, 1967.
8. L. Micic, "The Tuner Diode," *Radio Mentor Electronic*, Vol. 32, No. 5, 1966, pp. 404–405.
9. B. Dietrich and M. Lehmann, "Epitaxial Planar Silicon Transistors—Technology and Properties," *Radio Mentor*, Vol. 29, No. 10, 1963, pp. 851–855.
10. W. Pruin and A. Swamy, "Diode Switches BA243 and BA244," *Funk-Technik*, Vol. 24, No. 1, 1969, pp. 11–14.

11. U. Dolega, "Semiconductor Diodes," *Funkschau*, 1974: No. 20, pp. 789–791; No. 21, pp. 819–820; No. 22, pp. 857–858.
12. K. Reinarz, "AF Signal Switching by Means of Diodes," *Funkschau*, Vol. 43, No. 23, 1971, pp. 769–772.
13. "Diodes," ITT Intermetall data manual.
14. H. Sarkowski, *Dimensioning Semiconductor Circuits*, Lexika Verlag, 7031 Grafenau Doffingen, L, 1973.
15. H. Keller, "Radio and TV Receiver Tuning by Diodes," *Elektronik Anzeiger*, Vol. 1, No. 1/2, 1969, pp. 45–48.
16. H. Keller, "The Capacitance Diode in Parallel Resonant Circuits," *Funkschau*, Vol. 39, No. 7, 1967, pp. 185–188.
17. O. Dietrich and H. Keller, "Non-linear Distortion in Capacitance Diodes," *Radio Mentor Electronic*, Vol. 33, No. 4, 1967, pp. 266–269.
18. U. Dolega, "Temperature-Compensated Zener Diodes," ITT Technical Information Semiconductors, Freiburg, Germany, 1966 (Order No. 6200-73-1E).
19. H. Keller, "Station Selector Circuits for Receivers with Capacitance Diode Tuning," *Radio-Fernseh-Phono-Praxis*, No. 5, 1966, pp. 151–154.
20. P. Flamm, "Ultrasonic Remote Control Circuits with New IC's," *Funkschau*, 1975: No. 8, pp. 81–84; No. 9, pp. 67–69.
21. "Funktechnische Arbeitsblatter Re 91," *Funkschau*, 1973, No. 1.
22. *Reference Data for Radio Engineers*, Howard W. Sams, Indianapolis, IN, 1972.
23. K. Schroter, "VHF Tuner for Low Tuning Voltage," *Radio-Fernseh-Phono-Praxis*, No. 10, 1974, p. 5.
24. O. Dietrich and F. Lowel, "Electronically Tuned and Switched TV Tuners with Diodes BA141, BA142, and BA143," *Funk-Technik*, Vol. 22, No. 7, 1967, pp. 209–211.
25. E. Kinne, "A Survey on Tuners for TV Receivers," *Funk-Technik*, Vol. 25, 1970: No. 23, pp. 927–928; No. 24, pp. 961–964; Vol. 26, 1971: No. 1, pp. 16–18; No. 2, pp. 51–52.
26. K. Schurig, "VHF Tuner Containing Field Effect Transistors," *Funk-Technik*, Vol. 29, No. 21, 1974, pp. 743–745.
27. G. Bernstein, "Capacitance Diodes Employed as Diode Switches; A Combined CCIR and OIRT TV Tuner," *Funkschau*, Vol. 43, No. 7, 1971, pp. 189–190.
28. H. Keller, "VHF Tuner with Diode Tuning," *Funk-Technik*, Vol. 21, No. 8, 1966, pp. 266–267.
29. J. Backwinkel, "From the Combi Tuner to the Strip Line Tuner," *Funk-Technik*, Vol. 26, No. 13, 1971, pp. 489–492.
30. H. Bender and K. Schurig, "An All-Channel Tuner with Only Two Transistors," *Funkschau*, Vol. 38, No. 10, 1966, pp. 313–316.
31. W. Klein, "Interference-Proof Universal Tuner with Tuned VHF Input," *Funk-Technik*, Vol. 24, No. 5, 1969, pp. 163–164.
32. J. Novotny, "Measurements on Capacitance Diodes," *Messen und Prufen*, No. 1, 1969, pp. 28–32.
33. H. Dahlmann, "Automatic High Speed 'Jumbo' Tester for the Computer Controlled Sorting of Tuner Diodes in 1200 Groups," *Funkschau*, No. 24, 1974, pp. 939–940.
34. L. Micic, "Diode Tuned Resonant Circuit," *Internationale Elektronische Rundschau*, Vol. 22, No. 6, 1968, pp. 138–140.
35. Ulrich L. Rohde, "Mathematical Analysis and Design of an Ultra Stable Low Noise 100 MHz Crystal Oscillator with Differential Limiter and Its Possibilities in Frequency

Standards." Presented at the 32nd Annual Frequency Symposium, Fort Monmouth, NJ, June 1978.

36. Hiroyuki Abe et al., "A Highly Stabilized Low-Noise GaAs FET Integrated Oscillator with a Dielectric Resonator in the C Band," *IEEE Transactions on Microwave Theory and Techniques*, Vol. MTT-26, No. 3, March 1978, pp. 156–162.
37. Robert Adler, "A Study of Locking Phenomena in Oscillators," *Proceedings of the IEEE*, Vol. 61, No. 10, October 1973, pp. 1380–1385.
38. Richard M. Beach, "Hyperabrupt Varactor Tuned Oscillators," *Tech-Notes*, Vol. 5, No. 4, July–August 1976, Watkins-Johnson Co., Palo Alto, CA.
39. R. Buswell, "Linear VCO's," *Tech-Notes*, Vol. 3, No. 2, March–April 1976, Watkins-Johnson Co., Palo Alto, CA.
40. R. Buswell, "Voltage Controlled Oscillators in Modern ECM Systems," *Tech-Notes*, Vol. 1, No. 6, December 1974, Watkins-Johnson Co., Palo Alto, CA.
41. R. J. Clark and D. B. Swartz, "Take a Fresh Look at YIG-Tuned Sources," *Microwaves*, February 1972.
42. A. Goodman, "Increasing the Band Range of a Voltage-Controlled Oscillator," *Electronic Design*, September 28, 1964, pp. 28–35.
43. C. Herbert and J. Chernega, "Broadband Varactor Tuning of Transistor Oscillators," *Microwaves*, March 1967, pp. 28–32.
44. T. E. Parker, "SAW Controlled Oscillators," *Microwave Journal*, October 1978, pp. 66–67.
45. P. Penfield and R. P. Rafuse, *Varactor Applications*, MIT Press, Cambridge, MA, 1962, Chap. 9.
46. *Solid-State Microwave Voltage Controlled Oscillators*, Frequency Sources, Inc., Chelmsford, MA, 1974.
47. S. Hamilton and R. Hall, "Shunt-Mode Harmonic Generation Using Step Recovery Diodes," *Microwave Journal*, April 1967, pp. 69–78.
48. "How to Select Varactors for Harmonic Generation," *Micronotes*, Vol. 10, No. 1, May 1973, Microwave Associates, Inc., Burlington, MA.
49. Kaneyuki Kurokawa, "Injection Locking of Microwave Solid-State Oscillators," *Proceedings of the IEEE*, Vol. 61, No. 10, October 1973, pp. 1386–1410.
50. F. S. Barnes and G. F. Eiber, "An Ideal Harmonic Generator," *Proceedings of the IEEE*, Vol. 53, 1973, pp. 693–695.
51. B. E. Keiser, "The Cycle Splitter—A Wide-band Precision Frequency Multiplier," *IRE National Conference Record*, Vol. 7, 1959, pp. 4, 275–281.
52. V. E. Van Duzer, "500 kc/s-500 Mc/s Frequency Doubler," *Hewlett-Packard Journal*, Vol. 17, October 1965.
53. D. Koehler, "The Charge-Control Concept in the Form of Equivalent Circuits, Representing a Link Between the Classic Large Signal Diode and Transistor Models," *Bell System Technical Journal*, Vol. 46, March 1967, pp. 523–576.
54. D. O. Scanlan and M. A. Laybourn, "Analysis of Varactor Harmonic Generators, with Arbitrary Drive Levels," *Proceedings of the IEE*, Vol. 114, 1967, pp. 1598–1604.
55. H. A. Watson, ed., *Microwave Semiconductor Devices and Their Circuit Applications*, McGraw-Hill, New York, 1969.
56. R. H. Johnston and A. R. Boothroyd, "Charge Storage Frequency Multipliers," *Proceedings of the IEEE*, Vol. 56, 1968, pp. 167–176.
57. J. J. Ebers and T. L. Moll, "Large Signal Behavior of Junction Transistors," *Proceedings of the IRE*, Vol. 42, 1954, pp. 1761–1772.

58. R. G. Harrison, "A Nonlinear Theory of Class C Transistor Amplifiers and Frequency Multipliers," *IEEE Journal of Solid-State Circuits*, Vol. SC-2, 1967, pp. 93–102.
59. M. Caulton et al., "Generation of Microwave Power by Parametric Frequency Multiplication in a Single Transistor," *RCA Review*, Vol. 26, June 1965, pp. 286–311.
60. D. Halford, A. E. Wainright, and J. A. Barnes, "Flicker Noise of Phase in RF Amplifiers and Frequency Multipliers: Characterization, Cause, and Cure," *Proceedings of the 22nd Annual Frequency Control Symposium*, Fort Monmouth, NJ, 1968.
61. T. Watanabe and F. Yoshiharu, "Characteristics of Semiconductor Noise Generated in Varactor Frequency Multipliers," *Review of the Electrical Communication Laboratories* (*Tokyo*), Vol. 15, November–December 1967, pp. 752–768.
62. Y. Saburi, Y. Yasuda, and K. Harada, "Phase Variations in the Frequency Multiplier," *Journal of the Radio Research Laboratories* (*Tokyo*), Vol. 10, 1968, pp. 137–175.
63. Hewlett Packard Associates, "Step Recovery Diode Frequency Multiplier Design," *Application Note 913*, Palo Alto, CA, May 15, 1967.
64. Hewlett Packard Associates, "Harmonic Generation Using Step Recovery Diodes and SRD Modules," *Application Note 920*, Palo Alto, CA, June 1968.
65. D. G. Tucker, "The Synchronization of Oscillators," *Electronic Engineering*, Part I, Vol. 15, March 1943, pp. 412–418; Part II, Vol. 15, April 1943, pp. 457–461; Part III, Vol. 16, June 1943, pp. 26–30.
66. R. Adler, "A Study of Locking Phenomena in Oscillators," *Proceedings of the IRE*, June 1946, pp. 351–357.
67. L. J. Paciorek, "Injection Locking of Oscillators," *Proceedings of the IEEE*, November 1965, pp. 1723–1727.

Section 4-2

1. *Precision Time and Frequency Handbook*, 7th ed., Ball, Efratom Division, Irvine, CA, 1989.
2. United States Army LABCOM Staff, "MIL-O-55310, Rev. B, Military Specification, Oscillators, Crystal, General Specification," Dayton, OH, Defense Logistics Agency, 1988.
3. Marvin E. Frerking, *Crystal Oscillator Design and Temperature Compensation*, Van Nostrand Reinhold, New York, 1978.
4. C. Erikson and H. Pak, "A Digitally Compensated Crystal Oscillator," *Proceedings of the 10th Quartz Devices Conference and Exhibition*, 1988.
5. R. L. Clark, "Reducing TCXO Error After Aging Adjustment," *Proceedings of the 39th AFCS*, 1985, pp. 166–170.
6. V. Rosati, S. Schodowski, and R. L. Filler, "Temperature Compensated Crystal Oscillator and Test Results," *Proceedings of the 37th AFCS*, 1983.
7. A. Benjaminson, "A Microprocessor Compensated Crystal Oscillator Using A Dual-Mode Resonator," Frequency Control Symposium, 1982.
8. V. E. Bottom, *Introduction to Quartz Crystal Unit Design*, Van Nostrand Reinhold, New York, 1982.
9. E. Hafner, "The Piezoelectric Crystal Unit—Definitions and Methods of Measurements," *Proceedings of the IEEE*, Vol. 57, February 1969, pp. 179–201.
10. Benjamin Parzen, *Design of Crystal and Other Harmonic Oscillators*, Wiley, New York, 1983.

11. A. Ballato, "Static and Dynamic Behavior of Quartz Resonators," *IEEE Transactions on Sonics and Ultrasonics*, Vol. SU-26, July 1979, pp. 299–306.
12. R. Filler, "The Aging of Resonators and Oscillators Under Various Test Conditions," *Proceedings of the 41st AFCS*, 1987, pp. 444–451.
13. J. Kusters, "The SC Cut Crystal—An Overview," *Proceedings of the IEEE Ultrasonics Symposium*, 1981, pp. 402–409.
14. R. L. Filler, "The Effect of Vibration on Frequency Standards and Clocks," *Proceedings of the 41st AFCS*, 1987, pp. 398–408.
15. J. Gleick, *Chaos*, Penguin Books, New York, 1987.
16. R. J. Gilmore and M. B. Steer, "Nonlinear Circuit Analysis Using the Method of Harmonic Balance—A Review of the Art. Part I: Introductory Concepts; Part II: Advanced Concepts," *International Journal of Microwave and Millimeterwave Computer-Aided Engineering*, January 1991, April 1991.
17. J. Matthys, *Crystal Oscillator Circuits*, Wiley, New York, 1983.
18. M. M. Driscoll, "Low Noise, Microwave Signal Generation Using Bulk and Surface Acoustic Wave Resonators," *Proceedings of the 42nd AFCS*, 1988, pp. 369–377.
19. P. Gray and R. Meyer, *Analysis and Design of Integrated Circuits*, Wiley, New York, 1984.
20. D. J. Heally III, "Flicker of Frequency and Phase and White Frequency and Phase Fluctuations in Frequency Sources," *Proceedings of the 26th AFCS*, June 1972, pp. 29–42.
21. L. A. Meacham, "Bridge-Stabilized Oscillator," *Proceedings of the IRE*, Vol. 26, No. 10, 1938, pp. 1278–1294.
22. A. Benjaminson, "Balanced Feedback Oscillators," *Proceedings of the 38th AFCS*, 1984, pp. 327–333.
23. A. Benjaminson, "Results of Continued Development of the Differential Crystal Oscillator," *Proceedings of the 39th AFCS*, 1985, pp. 171–175.
24. C. Adams and J. Kusters, "Improved Long-Term Aging in Deeply Etched SAW Resonators," *Proceedings of the 32nd AFCS*, 1978, pp. 74–76.

Section 4-3

1. R. S. Caruthers, "Copper Oxide Modulators in Carrier Telephone Repeaters," *Bell System Technical Journal*, Vol. 18, No. 2, April 1939, pp. 315–337.
2. J. C. Holgarrd, "Spurious Frequency Generation in Frequency Converters, Part 1," *Microwave Journal*, Vol. 10, No. 7, July 1967, pp. 61–64.
3. J. C. Holgarrd, "Spurious Frequency Generation in Frequency Converters, Part 2," *Microwave Journal*, Vol. 10, No. 8, August 1967, pp. 78–82.
4. R. B. Mouw and S. M. Fukuchi, "Broadband Double Balanced Mixer Modulators, Part 1," *Microwave Journal*, Vol. 12, No. 3, March 1969, pp. 131–134.
5. R. B. Mouw and S. M. Fukuchi, "Broadband Double Balanced Mixer Modulators, Part 2," *Microwave Journal*, Vol. 12, No. 5, May 1969, pp. 71–76.
6. U. L. Rohde, "Optimum Design for High-Frequency Communications Receivers," *Ham Radio*, October 1976.
7. U. L. Rohde, "Performance Capability of Active Mixers," presented at Wescon/81, September 16, 1981.
8. Doug DeMaw and George Collings, "Modern Receiver Mixers for High Dynamic Range," *QST*, January 1981, p. 19.

9. U. L. Rohde, "Zur optimalen Dimensionerung von UKW-Eingangsteilen," *Internationale Elektronische Rundschau*, Vol. 27, No. 5, 1973, pp. 103–108.
10. U. L. Rohde, "High Dynamic Range Receiver Input Stages," *Ham Radio*, October 1975.
11. "Reactive Loads—The Big Mixer Menace," *Anzac Electronics Technical Note*.

Section 4-4

1. W. Egan and E. Clark, "Test Your Charge-Pump Phase Detectors," *Electronic Design*, Vol. 26, No. 12, June 7, 1978, pp. 134–137.
2. U. L. Rohde, "Modern Design of Frequency Synthesizers," *Ham Radio*, July 1976.
3. G. Alonzo, "Considerations in the Design of Sampling-Based Phase-Lock-Loops," *WESCON/66 Technical Papers*, Session 23, Western Electronic Show and Convention, 1966, Part 23/2.
4. C. J. Byrne, "Properties and Design of the Phase Controlled Oscillator with a Sawtooth Comparator," *The Bell System Technical Journal*, March 1962, pp. 559–602.
5. Fairchild Data Sheet: "Phase/Frequency Detector, 11C44," Fairchild Semiconductor, Mountain View, CA.
6. Fairchild Preliminary Data Sheet: "SH8096 Programmable Divider—Fairchild Integrated Microsystems," April 1970.
7. J. D. Fogarty, "Digital Synthesizers . . .," *Computer Design*, July 1975, pp. 100–102.
8. R. Funk, "Low-Power Digital Frequency Synthesizers Utilizing COS/MOS IC's," *Application Note ICAN-6716*, RCA Solid State Division, Somerville, NJ, March 1973.
9. A. Jay Goldstein, "Analysis of the Phase-Controlled Loop with a Sawtooth Comparator," *The Bell System Technical Journal*, March 1962, pp. 603–633.
10. Wayne M. Grove, "A D.C. to 12 GHz Feedthrough Sampler for Oscilloscopes and Other R.F. Systems," *Hewlett-Packard Journal*, October 1966, pp. 12–15.
11. S. Krishnan, "Diode Phase Detectors," *The Electronic and Radio Engineer*, February 1959, pp. 45–50.
12. Venceslav Kroupa, *Frequency Synthesis Theory Design and Applications*, Wiley, New York, 1973.
13. Stephan R. Kurtz, "Mixers as Phase Detectors," *Tech-Notes*, Vol. 5, No. 1, January–February 1978, Watkins-Johnson Co., Palo Alto, CA.
14. Stephan R. Kurtz, "Specifying Mixers as Phase Detectors," *Microwaves*, January 1978, pp. 80–87.
15. Motorola Data Sheet, MC12012, 1973. Motorola Semiconductor Products, Inc., Phoenix, AZ 85036.
16. Motorola Data Sheet: "Phase-Frequency Detector, MC4344, MC4044."
17. D. Richman, "Color-Carrier Reference Phase Synchronization Accuracy in NTSC Color Television," *Proceedings of the IRE*, January 1954, p. 125.
18. J. M. Cohen, "Sample-and-Hold Circuits Using FET Analog Gates," *EEE*, January 1971, pp. 34–37.
19. Roland Best, *Theorie und Anwendungen des Phase-locked Loops*, Fachschriftenverlag Aargauer Tagblatt AG, Aarau, Switzerland, 1976.
20. U.S. Patent, Fritze, Rohde, and Schwarz, Munich.
21. Floyd M. Gardner, "Charge Pump Phase-Lock Loops," *IEEE Transactions on Communications*, Vol. COM-28, No. 11, November 1980.

Section 4-5

1. Rohde & Schwarz, Operating and Repair Manual for the SMS/SMS2 Synthesizer.
2. Rohde & Schwarz, Operating and Repair Manual for the ESH2 and ESVN Test Receiver.
3. David Norton, U.S. Patent 3,891,934.

Section 4-6

1. U. L. Rohde, "Modern Design of Frequency Synthesizers," *Ham Radio*, July 1976.
2. E. Horrman, "The Inductance–Capacitance Oscillator as a Frequency Divider," *Proceedings of the IRE*, Vol. 34, 1946, pp. 799–803.
3. B. Chance et al., eds., *Waveforms*, McGraw-Hill, New York, 1949.
4. A. L. Plevy and E. N. Monacchio, "Fail-Safe Frequency Divider," *Electronics*, Vol. 39, September 1966, p. 127.
5. S. Plotkin and O. Lumpkin, "Regenerative Fractional Frequency Generators," *Proceedings of the IRE*, Vol. 48, 1960, pp. 1988–1997.
6. Y. Kamp, "Amorcage des diviseurs de fréquence à capacité non linéaire," *L'Onde électrique*, Vol. 48, September 1968, pp. 787–793.
7. B. Preston, "A Microelectronic Frequency Divider with a Variable Division Ratio," *Electronic Engineering*, Vol. 37, April 1965, pp. 240–244.
8. H. G. Jungmeister, "Eine bistabile Kippschaltung fur den Gigahertz-Bereich," *Archiv der elektrischen Ubertragung*, Vol. 21, September 1967, pp. 447–458.
9. E. J. Kench, ed., *Electronic Counting: Circuits, Techniques, Devices*, Mullard, London, 1967.
10. W. E. Wickes, *Logic Design with Integrated Circuits*, Wiley, New York, 1968.
11. E. J. Kench, ed., *Integrated Logic Circuit Applications: Mullard FC Range*, Mullard, London, 1967.
12. L. F. Blachovicz, "Dial Any Channel to 500 MHz," *Electronics*, Vol. 39, May 2, 1966, pp. 60–69.
13. J. Stinehelfer and J. Nichols, "A Digital Frequency Synthesizer for an AM and FM Receiver," *Transactions of the IEEE*, Vol. BTR-15, No. 3, 1969, pp. 235–243.
14. S. Jannazzo and G. Rustichelli, "A Variable-Ratio Frequency Divider Using Micrologic Elements," *Electronic Engineering*, Vol. 39, July 1967, p. 419.
15. R. W. Frank, "The Digital Divider," *General Radio Experimenter*, Vol. 43, January–February 1969, pp. 3–7.
16. J. L. Hughes, *Computer Lab Workbook*, Digital Equipment Corp., Maynard, MA, 1969.
17. "A 1 GHz Prescaler Using GPD Series Thin-Film Amplifier Modules," *Microwave Component Applications, ATP-1036*, Avantek, Inc., Santa Clara, CA, 1977.
18. S. Bearse, "TED Triode Performs Frequency Division," *Microwaves*, November 1975, p. 9.
19. W. R. Blood, Jr., *MECL System Designer's Handbook*, 2nd ed., Motorola Semiconductor Products, Inc., Mesa, AZ, 1972.
20. W. J. Goldwasser, "Design Shortcuts for Microwave Frequency Dividers," *The Electronic Engineer*, May 1970, pp. 61–65.
21. W. D. Kasperkovitz, "Frequency-Dividers for Ultra-High Frequencies," *Philips Technical Review (Netherlands)*, Vol. 38, No. 2, 1978–1979, pp. 54–68.

22. S. Lee, *Digital Circuits and Logic Design*, Prentice Hall, Englewood Cliffs, NJ, 1976.
23. R. L. Miller, "Fractional-Frequency Generators Utilizing Regenerative Modulation," *Proceedings of the IRE*, July 1939, pp. 446–457.
24. J. Nicholds and C. Shinn, "Pulse Swallowing," *EDN*, October 1, 1970, pp. 39–42.
25. SP8750–8752 Data Sheets, Plessey Semiconductors, 1641 Kaiser Avenue, Irvine, CA 92714.
26. M. J. Underhill et al., "A General Purpose LSI Frequency Synthesizer System," *Proceedings of the 32nd Annual Symposium on Frequency Control*, 1978, pp. 366–367.
27. Data sheet for the HEF 4750/51, Philips, Mullard, London.
28. M. J. Underhill, "Wide Range Frequency Synthesizers with Improved Dynamic Performance," private communication.
29. M. J. Underhill and R. I. H. Scott, "FM Models of Frequency Synthesizers," private communication.

Section 4-7

1. Anatol I. Zverev, *Handbook of Filter Synthesis*, Wiley, New York, 1967.
2. Floyd M. Gardner, *Phaselock Techniques*, 2nd ed., Wiley, New York, 1980.
3. Roland Best, *Theorie und Anwendungen des phase-locked Loops*, Fachschriftenverlag Aargauer Tagblatt AG, Aarau, Switzerland, 1976.

Section 4-8

1. George Vendelin, Anthony M. Pavio, and Ulrich L. Rohde, *Microwave Circuit Design Using Linear and Nonlinear Techniques*, Wiley, New York, 1990.

Section 4-9

1. Ulrich L. Rohde, "All About Phase Noise in Oscillators," *QEX*, December 1993, January 1994, and February 1994.
2. J. Cheah, "Analysis of Phase Noise in Oscillators," *RF Design*, November 1991.
3. V. F. Kroupa, "Noise Properties of PLL Systems," *IEEE Transactions on Communications*, October 1992.
4. M. R. McClure, "Residual Phase Noise of Digital Frequency Dividers," *Microwave Journal*, March 1992, pp. 124–130.
5. D. Scherer, "Design Principles and Measurement of Low Phase Noise RF and Microwave Sources," *Hewlett-Packard RF & Microwave Measurement Symposium*, April 1979.
6. M. Bomford, "Selection of Frequency Dividers for Microwave PLL Applications," *Microwave Journal*, November 1990.
7. J. A. Mezak and G. D. Vendelin, "CAD Design of YIG Tuned Oscillators," *Microwave Journal*, December 1992.
8. R. Kiefer and L. Ford, "CAD Tool Improves SAW Stabilized Oscillator Design," *Microwaves & RF*, December 1992.
9. B. Parzen, "Clarification and a Generalized Restatement of Leeson's Oscillator Noise Model," *42nd Annual Frequency Symposium*, 1988.

10. J. K. A. Everard, "Minimum Sideband Noise in Oscillators," *40th Annual Frequency Control Symposium*, 1986, pp. 336–339.
11. R. Muat and A. Upham, "Low Noise Oscillator Design," *Hewlett-Packard RF & Microwave Measurement Symposium*, 1995.
12. R. Muat, "Designing Oscillators for Spectral Purity," *Microwaves & RF*, August 1984.
13. R. Muat, "Choosing Devices for Quiet Oscillators," *Microwaves & RF*, August 1984.
14. R. Muat, "Computer Analysis Aids Oscillator Designers," *Microwaves & RF*, September 1984.
15. F. Pergal, "Detail a Colpitts VCO as a Tuned One-Port," *Microwaves*, April 1979.
16. R. G. Rogers, "Theory and Design of Low Noise Microwave Oscillators," *42nd Annual Frequency Control Symposium*, 1988, pp. 301–303.
17. R. Spence, "A Theory of Maximally Loaded Oscillators," *IEEE Transactions on Circuit Theory*, June 1966.
18. K. L. Kotzebue and W. J. Parrish, "The Use of Large Signal S-Parameters in Microwave Oscillator Design," *Proceedings of the IEEE International Symposium on Circuits and Systems*, 1975.
19. R. G. Meyer and M. L. Stephens, "Distortion in Variable Capacitance Diodes," *IEEE Journal on Solid-State Circuits*, February 1975.
20. D. F. Peterson, "Varactor Properties for Wideband Linear Tuning Microwave VCOs," *IEEE Transactions on Microwave Theory and Techniques*, February 1980.

Suggested Readings

Bell, D. A., *Noise and the Solid State*, Wiley, New York, 1985.

Curtis, G. Stephen, "The Relationship Between Resonator and Oscillator Noise, and Resonator Noise Measurement Techniques," *Proceedings of the 41st AFCS*, 1987.

Driscoll, M. M., "Two-Stage Self-Limiting Series Mode Type Quartz Oscillator Exhibiting Improved Short-Term Stability," *Proceedings of the 26th AFCS*, 1972, pp. 43–49.

Driscoll, M. M., "Low Noise VHF Crystal-Controlled Oscillator Utilizing Dual, SC-Cut Resonators," *Proceedings of the 39th AFCS*, 1985, pp. 197–201.

Filler, R. L., "The Effect of Vibration on Frequency Standards and Clocks," *Proceedings of the 35th AFCS*, 1981.

Filler, R. L., Kosinski, J. A., Rosati, V. J., and Vig, J. R., "Aging Studies on Quartz Resonators and Oscillators," *Proceedings of the 38th AFCS*, 1984, pp. 225–231.

Halford, D., Wainwright, A., and Barnes, J., "Flicker Noise of Phase in RF Amplifiers and Frequency Multipliers: Characterization, Cause, and Cure," *Proceedings of the 22nd AFCS*, 1968, pp. 340–341.

Ho, J., "Hybrid Miniature Oven Quartz Crystal Oscillator," *Proceedings of the 38th AFCS*, 1984, pp. 193–196.

Parker, T. E., "1/f Frequency Fluctuations in Acoustic and Other Stable Oscillators," *Proceedings of the 39th AFCS*, 1985, pp. 97–106.

Rohde, Ulrich L., Whitaker, Jerry, and Bucher, T. T. N., *Communications Receivers*, 2nd ed., McGraw Hill, 1997, pp. 319–448.

Rosati, V. and Thompson, P., "Further Results of Temperature Compensated Crystal Oscillator Testing," *Proceedings of the 38th AFCS*, 1984, pp. 507–509.

Stein, S. R., Manney, C. M. Jr., Walls, F. L., Gray, J. E., and Besson, R. J., "A Systems Approach to High Performance Oscillators," *Proceedings of the 32nd AFCS*, 1978, pp. 527–541.

van der Ziel, Aldert, *Noise in Solid State Devices and Circuits*, Wiley, New York, 1986.

Vergers, Charles A., *Handbook of Electrical Noise Measurement and Technology*, 2nd ed., TAB Books, Blue Ridge Summit, PA, 1987.

5

DIGITAL PLL SYNTHESIZERS

5-1 MULTILOOP SYNTHESIZERS USING DIFFERENT TECHNIQUES

By now, we have accumulated a large amount of knowledge about single-loop synthesizers. In Chapter 1 a loop with a mixer was described, which probably represents the most simple dual-loop synthesizer.

Adding an auxiliary frequency is the first step toward building a two-loop synthesizer, with this auxiliary frequency generated by another loop rather than by multiplying the reference frequency to mix down the VCO frequency to a lower frequency range for convenience of being able to use lower dividers. The frequency resolution is then equal to the reference frequency unless special techniques are used.

We have heard already about the fractional division N synthesizer, and we have seen the sequential phase shifter that enabled us to get additional resolution. In addition, the pure digital frequency synthesizer was explained where the waveform is generated with the aid of a lookup table. Multiloop synthesizers use a combination of these techniques.

Modern frequency synthesizers no longer use designs where each decade uses phase-locked loops that operate at the same frequency with the output divided by 10. These *mix-and-divide systems*, or *triple mix systems* with cancellation of drift, are seldom used, as they require an enormous amount of filtering, shielding, and power consumption. However, to be able to decide which building blocks to use, some of them have to be discussed here, and we will start with direct frequency synthesis showing various degrees of resolution.

5-1-1 Direct Frequency Synthesis

Direct frequency synthesis refers to the generation of new frequencies from one or more reference frequencies using a combination of multipliers, dividers,

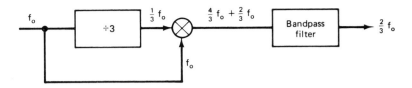

Figure 5-1 Direct frequency generation using the mix-and-divide principle. It requires excessive filtering.

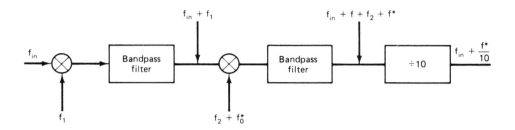

Figure 5-2 Direct frequency synthesizer using a mix-and-divide technique to obtain identical modules for high resolution.

bandpass filters, and mixers. A simple example of direct synthesis is shown in Figure 5-1. The new frequency $\frac{2}{3}f_0$ is realized from f_0 by using a divide-by-3 circuit, a mixer, and a bandpass filter. In this example $\frac{2}{3}f_0$ has been synthesized by operating directly on f_0.

Figure 5-2 illustrates the form of direct synthesis module most frequently used in commercial frequency synthesizers of the direct form. The method is referred to as the "double-mix-divide" approach. An input frequency f_{in} is combined with a frequency f_1, and the upper frequency $f_1 + f_{\text{in}}$ is selected by the bandpass filter. This frequency is then mixed with a switch-selectable frequency $f_2 + f^*$. (In the following, f^* refers to any one of 10 switch-selectable frequencies.) The output of the second mixer consists of the two frequencies $f_{\text{in}} + f_1 + f_2 + f^*$ and $f_{\text{in}} + f_1 - f_2 - f^*$; only the higher-frequency term appears at the output of the bandpass filter. If the frequencies f_{in}, f_1, and f_2 are selected so that

$$f_{\text{in}} + f_1 + f_2 = 10 f_{\text{in}} \qquad (5\text{-}1)$$

then the frequency at the output of the divide by 10 will be

$$f_{\text{out}} = f_{\text{in}} + \frac{f^*}{10} \qquad (5\text{-}2)$$

The double-mix-divide module has increased the input frequency by the switch-selectable frequency increment $f^*/10$. These double-mix-divide modules can be cascaded to form a frequency synthesizer with any degree of resolution. The

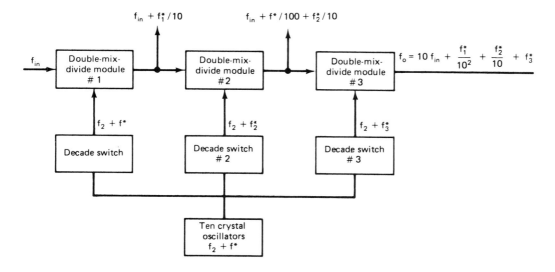

Figure 5-3 Phase incoherent frequency synthesizer with three-digit resolution.

double-mix-divide modular approach has the additional advantage that the frequencies f_1, f_2, and f_{in} can be the same in each module, so that all modules can contain identical components.

A direct frequency synthesizer with three digits of resolution is shown in Figure 5-3. Each decade switch selects one of 10 frequencies $f_2 + f^*$. In this example the output of the third module is taken before the decade divider. For example, it is possible to generate the frequencies between 10 and 19.99 MHz (in 10-kHz increments), using the three module synthesizer, by selecting

$$f_{in} = 1 \text{ MHz}$$
$$f_1 = 4 \text{ MHz}$$
$$f_2 = 5 \text{ MHz}$$

Since

$$f_{in} + f_1 + f_2 = 10 f_{in}$$

the output frequency will be

$$f_0 = 10 f_{in} = f_3^* + \frac{f_2^*}{10} + \frac{f_1^*}{100} \qquad (5\text{-}3)$$

Since f^* occurs in 1-MHz increments, $f_1^*/100$ will provide the desired 10-kHz frequency increments.

Theoretically, either f_1 or f_2 could be eliminated provided that

$$f_{in} + f_1(\text{or } f_2) = 10 f_{in} \qquad (5\text{-}4)$$

but the additional frequency is used in practice to provide additional frequency separation at the mixer output. This frequency separation eases the bandpass filter requirements. For example, if f_2 is eliminated, $f_1 + f_{in}$ must equal $10 f_{in}$ or 10 MHz. If an f_1^* of 1 MHz is selected, the output of the first mixer will consist of the two frequencies 9 and 11 MHz. The lower of these closely spaced frequencies must be removed by the filter. The filter required would be extremely complex. If, instead, a 5-MHz signal f_2 is also used so that $f_{in} + f_1 + f_2 = 10$ MHz, the two frequencies at the first mixer output will (for an f_1^* of 1 MHz) be 1 and 11 MHz. In this case the two frequencies will be much easier to separate with a bandpass filter. The auxiliary frequencies f_1 and f_2 can only be selected in each design after considering all possible frequency products at the mixer output.

Direct synthesis can produce fast frequency switching, almost arbitrarily fine frequency resolution, low phase noise, and the highest-frequency operation of any of the methods. Direct frequency synthesis requires considerably more hardware (oscillators, mixers, and bandpass filters) than the two other synthesis techniques to be described. The hardware requirements result in direct synthesizers being larger and more expensive. Another disadvantage of the direct synthesis technique is that unwanted frequencies (spurious) can appear at the output. The wider the frequency range, the more likely that spurious components will appear in the output. These disadvantages are offset by the versatility, speed, and flexibility of direct synthesis.

5-1-2 Multiple Loops

Multiple-loop synthesizers, as found in signal generators and in communication equipment, are probably best understood when examining their block diagrams. Let us take a look at Figure 5-4, which provides us with the information about the frequency generation of a shortwave receiver; several multiloop synthesizers are being used here. This block diagram shows the various methods that are currently being used.

The shortwave receiver operating from 10 kHz to 29.99999 MHz has a first IF of 81.4 MHz. The oscillator injection therefore requires operating from 81.465 to 111.45499 MHz, as seen in the block diagram.

The oscillator marked "G" in the block diagram uses an auxiliary frequency of 69.255 to 69.35499 MHz to down-convert the output loop to an IF from 11.2 to 41.4 MHz. Note that a bandpass filter is used to avoid any feedthrough of the higher frequencies in the mixer. A programmable divider divides this frequency band down to the reference frequency of 100 kHz, switching the output loop in 100-kHz increments.

The master standard is multiplied up to 80 MHz by using a PLL at 80 MHz to generate the auxiliary frequency, which, together with the fine-resolution synthesizer portion on the left, is used to generate the 69.255 to 69.35499 MHz window.

The fine resolution is achieved by operating a single-loop synthesizer from 64.501 to 74.5 MHz in 1-kHz steps and then dividing it by 100. The division by 100 gives a step size of 10 Hz, while good switching speed is offered by operating at 1 kHz reference. The output of about 700 kHz is mixed with the 10-MHz frequency standard to a 10.645- to 10.745-MHz IF. A crystal filter can be used

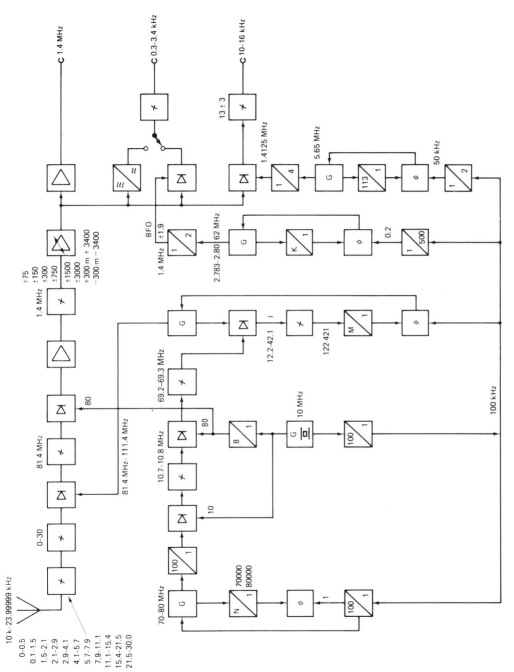

Figure 5-4 Block diagram of the frequency synthesizer of the Rohde & Schwarz EK070 shortwave receiver.

to take out all unwanted frequencies, and this 10.7-MHz signal, together with the 80 MHz generated from the 10-MHz standard, then results in the auxiliary frequency to be mixed into the output loop.

This system has several advantages. The output loop is extremely fast, and the division ratio inside the loop is fairly small. The mixer inside the loop reduces the division ratio from approximately 1104 to about 421 at the most, and therefore the noise generated because of the multiplication is kept small relative to a single-loop approach. However, the divider ratio is now 4:1; without the mixing, the ratio would have been 11:8. Therefore, the loop has to cope with higher gain variations, and the loop filter has to incorporate a mechanism that changed the loop gain corresponding somewhat with the gain variation.

We have learned that the VCO, when switching diodes are used to add capacitance, has a lower loop gain at the lower frequency, where more capacitance is added than at the higher frequency. This provides a simple method to adjust the gain variation inside the loop. With a 100-kHz reference, a loop bandwidth of 2 or 3 kHz will provide enough suppression of reference, and a settling time in the vicinity of several milliseconds is achievable.

The fine-resolution loop that provides the 10-Hz increments now limits the switching time. Most likely, the loop filter will be in the vicinity of 10 Hz or 1% of reference, which will provide 40 dB of reference suppression. The division by 100 at the output increases the reference suppression by another 100, so that the reference at the output is suppressed by at least 80 dB. The use of special LC filters can easily increase this to 100 dB.

The noise sideband of the 70–80 MHz oscillator depends mainly on the VCO. Even a very simple LC oscillator should provide 120 dB/Hz 20 kHz off the carrier, and the additional 40-dB improvement based on the division by 100 will theoretically increase the noise to 160 dB. This is not very likely, and the noise floor is now determined by the noise floor of the dividers, the mixers, and postamplifiers and will be in the vicinity of 150 dB.

The output loop operating with about 2 kHz of bandwidth will, outside the loop bandwidth, reproduce the noise performance of the oscillator. We have learned a great deal about low-noise oscillators in this book, and it would be considered standard practice to divide the output loop into at least three oscillators of 10-MHz range so that relative bandwidth $\Delta f/f$ is about 10% or less.

The RF has to pass an 81.4-MHz crystal filter ±6 kHz wide and then is mixed down to the second IF of 1.4 MHz. The second LO is derived from the same 80 MHz that is used inside the synthesizer loop. This avoids another PLL because of the clever combination of frequencies.

The IF of 1.4 MHz offers the choice of different band filters, as can be seen from the block diagram, and in CW and single-sideband modes, a BFO is required. In addition, this receiver offers, as a novel approach, a recorder output where the IF frequency is mixed to a frequency band from 10 to 16 kHz. As a result of this, additional synthesizers are required. For reasons of short-term stability and noise, the BFO synthesizer is operated at twice the frequency and the output is divided by 2 to obtain the final frequency. A similar approach is used to generate the 468 kHz required to obtain the 10 to 16 kHz of output.

Let us assume that the frequency resolution of this synthesizer has to be increased by the factor of 10. What would be the easiest approach? The easiest approach would be to take advantage of the recently developed HEF4750 and HEF4751 synthesizer ICs made by Philips.

The fractional offset portion of this single-loop synthesizer would allow a 100-Hz step size with the 1-kHz reference, and therefore the same switching speed would be maintained. The division by 100 is sufficient to suppress any possible reference problems if the loop filter is changed. This simple change would allow the required resolution.

A much higher resolution would be gained by substituting the 64.5–74.5 MHz single-loop synthesizer with a high-resolution fractional division synthesizer, which then could give almost any arbitrary resolution. In doing so, it would be possible to increase the reference frequency to the 100 kHz used in the output loop, and as a result, the entire switching speed of the synthesizer would be a few milliseconds while at 1 kHz reference, and the 10-Hz loop filter would currently dictate a switching time of about 100 ms. This approach, because of the high division ratio at the output, guaranteees the necessary cleanliness. The multiloop synthesizers require a certain amount of hand-holding as far as the construction is concerned. It is highly desirable to provide adequate shielding. Figure 5-5 shows the mechanical construction of an output loop similar to the one described. The metal can on the left contains the VCO. All voltages are fed to the VCO via feed through capacitors. On the top right side of the PC board, one can see the crystal filter marked 20.095 MHz. In this case, the 40–70 MHz oscillator is being mixed down to a 50–20 MHz IF, and the crystal filter shown on the PC board assembly is used to clean up the output from the fine-resolution synthesizer. Although this picture was taken from a lab model rather than a production unit, it indicates that it has to be built extremely carefully to give any meaningful results. Note the solid copper surface of the PC board, with all the wire connections underneath the PC board in printed form. Shielding is the next important thing for good reference suppression, and very frequently the design goal will not be met if the shielding is not optimized.

Figure 5-5 Photograph of the output loop of a multiloop frequency synthesizer. The shielded box contains the VCO; the double-balanced mixer and the monolithic crystal filter can also be seen.

426 DIGITAL PLL SYNTHESIZERS

Figure 5-6 Photograph of the frequency-divider chain of a multiloop synthesizer. Note the ground-plane construction and the filtering to the input connector.

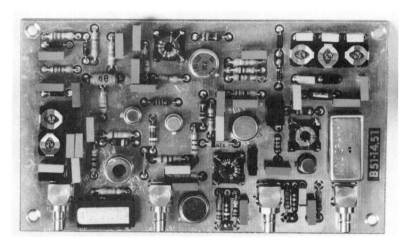

Figure 5-7 Photograph of the input stage of a receiver. RF connections are accomplished via coaxial connectors. A solid ground plane avoids ground loops even on the first experimental layout.

It is advisable to separate RF and logic circuits as much as possible. Figure 5-6 shows the digital frequency divider that is required for the dual-loop synthesizer shown in Figure 5-5. The PC board has its own regulator to minimize transient crosstalk on the power supply terminals, and this PC board uses both ECL and CMOS dividers. On the left, the input connector provides the switching information to the dividers. Because of this arrangement, a minimum of shielding between the dividers and this connector is required. RF circuitry requires similar careful layout. Figure 5-7 shows a PC board containing the RF input stage of an experimental receiver. The top side of the board is again a solid surface, and therefore ground loops are minimized. The various input and output connections are achieved through the four connectors shown, and the switching is done under dc control with switching diodes. The dc information is fed through the connector on the lower left-hand side of the PC board.

5-2 SYSTEM ANALYSIS

During our various discussions, it has become apparent that the single-loop synthesizer really is somewhat limited in its application.

Unless the fractional division N principle or other methods are used, it is really not possible to build a clean frequency synthesizer at a high frequency output, say, 100 to 150 MHz, in small increments such as 100 Hz or even 1 kHz with good switching time, reference suppression, and other important parameters.

This is probably most easily understood when we analyze various systems. Let us start with a single-loop synthesizer operating from 260.7 to 460.7 MHz, as may be used for a receiver (see Figure 5-8). These are the requirements:

Frequency range	260.7 to 460.7 MHz
Frequency increments	1 kHz
Frequency stability	1×10^{-8} per day
Spurious outputs	-70 dB
Switching time	20 ms
Phase noise	120 dB/Hz, 20 kHz off the carrier

These are the six most important requirements that have to be analyzed and kept track of in a system. The single-loop synthesizer is not really a system but a single phase-locked loop with a number of inherent limitations.

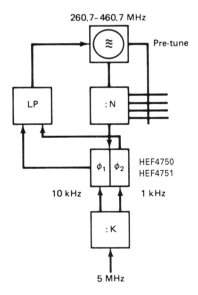

Figure 5-8 Single-loop synthesizer operating from 260.7 to 460.7 MHz in 1-kHz steps.

428 DIGITAL PLL SYNTHESIZERS

The frequency range is determined by the VCO. The phase noise of the loop outside the loop bandwidth is determined by the influence of the tuning diodes and the question of whether or not the oscillator is coarse-tuned and whether or not several oscillators are used to cover the range. The following table shows the noise typically found in a free-running oscillator in this area of operation.

Phase Noise, $\mathscr{L}(f)$ (dB/Hz)	Offset from the Carrier
-55	10 Hz
-75	100 Hz
-95	1000 Hz
-120	10 kHz
-140	100 kHz
-160	1 MHz

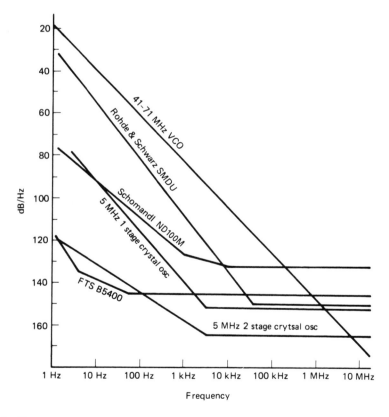

Figure 5-9 Measured noise sideband performance of a 41–71 MHz VCO; Rohde & Schwarz SMDU signal generator, Schomandl ND100M frequency synthesizer; frequency and time services (FTS) BS5400 modulator; and single- and double-stage 5-MHz crystal oscillators.

Figure 5-9 shows a graph that compares the noise sideband of several different oscillators. It becomes apparent from this that similar designs have quite different noise performance if the design is not carefully analyzed and the purpose of the synthesizer is not fully understood from the beginning. Both the 41–71 MHz VCO and the Rohde & Schwarz SMDU signal generator use free-running oscillators. The 41–71 MHz VCO is divided into three subranges, and the SMDU uses mechanical tuning, while the maximum electronic tuning is about 1 MHz. This explains the difference in the noise performance, and, in addition, the higher slope of the SMDU indicates also the higher Q of the circuit. The Schomandl ND100M is a frequency synthesizer constructed from many loops, and the improvement in noise there is due to division inside the loops, as we will see later in the chapter.

Two crystal oscillators are shown: the 5-MHz one-stage crystal oscillator shows fairly high noise below 10 Hz compared to the 5-MHz two-stage crystal oscillator. The phase noise discussions in Chapters 1 and 4 have explained the reason for the different performance.

Let us assume for a moment that our one-loop synthesizer, as shown in Figure 5-8, uses a 5-MHz two-stage crystal oscillator and a tuned-cavity oscillator ranging from 260.7 to 460.7 MHz.

Because of the high Q of the cavity, the VCO noise of this oscillator will be substantially better than that of the 41–71 MHz VCO shown in Figure 5-9.

In order to multiply the 1-kHz reference up to an average frequency of 300 MHz, a division ratio of 300,000 or a multiplication of 300,000 is required. Assuming that a -160-dB/Hz reference signal is present at the output of the reference divider chain at 1 kHz, we can calculate the noise at 300 MHz from this multiplication. The multiplication of 300,000 is equivalent to 109.54 dB, and if we subtract this from 160 dB, the noise floor, the resulting signal-to-noise ratio is 50.46 dB/Hz, 1 kHz off the carrier.

However, if a 1-kHz reference is used, the loop filter has to be narrower than 1 kHz in order to get, say, 70-dB reference suppression.

It can be assumed that a carefully built tri-state phase/frequency comparator will have 40-dB reference suppression by itself, while at least an additional 30 dB of reference suppression has to be provided by the loop filter.

This roughly leads to a natural loop frequency of the PLL in the vicinity of 50 Hz. The reference noise removed more than 50 Hz from the carrier is then reproduced in the output, and we have to reduce our calculation and take 50 Hz rather than 1 kHz off the 5-MHz reference. For 50 Hz our frequency standard shows a noise sideband of -140 dB/Hz, and we have to do the same calculation and deduct 109.54 dB from 140 dB, resulting in a signal-to-noise ratio of 30.46 dB. This signal-to-noise ratio is now less than the VCO would have had by itself, which means that we are actually making the VCO noisier than it would be by itself.

In order to have less influence from the loop, it would theoretically be better to use a wider-loop bandwidth to take advantage of the lower noise of the VCO, since the multiplication inside the loop is so tremendous. However, the reference suppression will then suffer.

Taking a 50-Hz loop bandwidth into consideration, lockup time will be in the vicinity of 60 to 100 ms, and we are not going to meet our target as far as switching time is concerned. We learned in Chapter 1 that the type 2 third-order loop

430   DIGITAL PLL SYNTHESIZERS

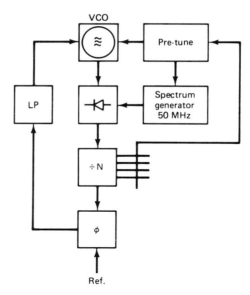

Figure 5-10   A 260.7–460.7 MHz dual-loop approach with a comb generated to obtain a low IF for the dividers.

provides faster lock and higher reference suppression than the type 2 second-order loop. Much-higher-order loops, such as fifth order or higher, have an advantage only if the reference frequency is much higher than the loop bandwidth, as the additional phase shift that is being introduced if both frequencies get too close will make the loop unstable; then the simple type 3 second-order loop is better.

Another way to overcome this problem is to use a sample/hold comparator, where the phase shift seems to be smaller.

The fractional division N principle with the zero averaging detector allows an extension of resolution. This method was explained previously and will not be treated again here.

The noise sideband performance outside the loop bandwidth is determined by the VCO, and the switching time by the loop filter. In order to increase the switching speed and improve the noise performance of the oscillator, let us use a design as shown in Figure 5-10. Here we split the range 260.7 to 460.7 MHz into a number of 50-MHz subbands by selecting the appropriate harmonic of a 50-MHz comb spectrum generated by the 5-MHz frequency standard with the help of a times-10 multiplier and a comb generator (using snap-off or MESFET varactors).

Let us take the same 300-MHz center frequency and use a 350-MHz comb line to beat the 300 MHz down to an IF of 50 MHz. By doing this, we have decreased the division ratio inside the loop by 10 or 20 dB, and by using 10 discrete oscillators covering the range 260.7 to 460.7 MHz, we have decreased the noise sideband performance of the VCO by 20 dB. In doing so we have achieved both goals, increasing the close-in noise performance as well as the noise outside the loop bandwidth at the expense of additional circuits.

The additional circuits incorporate the following:

1. A large number of VCOs (can be simulated by the coarse switching range in increments of 50 MHz).
2. Designing a 5–50 MHz multiplier and a comb generator to mix frequency ranges down to a lower IF.
3. Developing circuitry selecting the appropriate harmonic of the comb and steering the VCO to prevent lockup against the wrong comb harmonic.

This is a somewhat drastic but effective method.

Figure 5-11 shows another way of achieving this. This dual-loop synthesizer now takes advantage of a high-gain loop using a reference frequency of 1 MHz; omitting the influence of the mixing for a moment, the multiplication in the coarse loop is now only 300 or the reduction in noise relative to the 5-MHz reference is about 50 dB.

As the noise floor-out is about 160 dB/Hz for the particular crystal oscillator, the noise floor is increased to -110 dB/Hz and we now can choose our loop filter to cross over with the VCO noise at this point. From the table we used to determine the noise performance of our UHF VCO, the 100-dB noise of the VCO can be measured at 5 kHz off the carrier.

It will therefore be reasonable to use a filter of 5-kHz bandwidth, as the VCO above this cutoff has less noise than the noise generated by the multiplication inside the loop.

Our reference frequency of 1 MHz would be suppressed by more than 60 dB from a filter having 1-kHz loop bandwidth, not taking the reference suppression of a tri-state and/or sample/hold comparator into consideration.

It is barely possible to build a 1-MHz sample/hold discriminator with low leakage, and the best possible choice will be a combination of some discrete flip-flops optimized in layout forming a flip-flop phase/frequency comparator. A reference suppression of 30 to 40 dB can be expected here, which adds to a total of 100 dB. The lockup time in this case will be in the vicinity of 1 ms, depending on the type of loop filter.

In Section 1-10 we learned that it is possible to use a dual-time-constant filter, where the frequency acquisition is speeded up by a factor of 20, and therefore the settling time is determined by phase lock rather than by frequency lock. The auxiliary synthesizer mixed into the loop is now responsible for the final resolution.

In our example, we have used an auxiliary synthesizer that has a 10-kHz reference rather than 1 kHz, and its output is also divided by 100. As a result of this, the switching speed of the auxiliary loop is now 10 times higher than the switching speed of our initial one-loop design, taking the same reference suppression into consideration, and the output noise from the VCO, even using the initial crude design where one VCO had to cover the entire frequency range, now permits 20-dB-better phase noise. Figure 5-12 shows the resulting phase noise (A + B) for the two frequency synthesizers as shown in Figures 5-8 and 5-11.

If finer frequency resolution is required and the digibase system, for which Hewlett-Packard and Racal seem to have patents, has to be avoided or if a microwave frequency synthesizer has to be designed, the number of loops has to be increased.

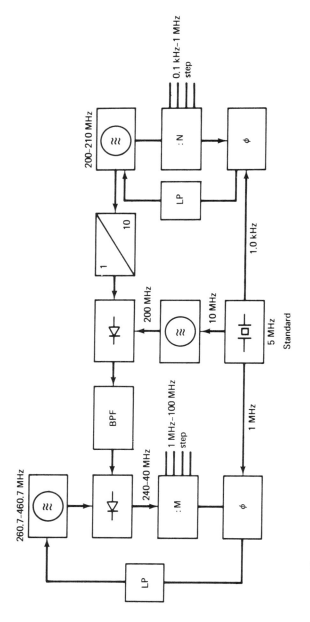

Figure 5-11 Dual-loop frequency synthesizer operating from 260.7 to 460.7 MHz with 100-Hz resolution.

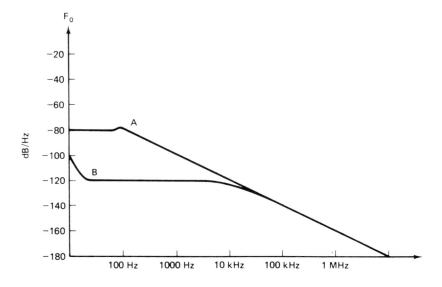

Figure 5-12 Noise sideband performance of synthesizer in Figures 5-8 and 5-11.

The introduction of the mixer, however, causes two problems:

1. The required filter is designed with a variable divider. The mixer does not have constant delay and the change in delay can introduce loop stability problems. This filter must be optimized for flat delay inside the passband characteristic.
2. The mixer has a large number of spurious products, as we learned in Section 4-3.

Besides the question of proper drive and termination, proper bandpass filters at the output of the mixer are important, and a proper choice of frequencies is similarly important. The phase/frequency detector by itself is a highly nonlinear device capable of mixing actions, which may cause problems when such output is fed into the programmable divider. The programmable divider has only a limited suppression of its input frequency, and therefore the phase comparator will receive not only the output frequency but also, with some limited suppression, unwanted mixer products. It is therefore vital to incorporate a low-pass filter at the output of the divider chain, unless a slow divider chain such as a CMOS is used. I have found that combinations of swallow counters in ECL with CMOS dividers do not suffer from this difficulty, whereas ECL/TTL divider chains definitely require the additional low-pass filter. Similar difficulties have occurred in the past where the input signal from the fine-resolution loop, after being divided down by 10, was mixed into the main loop. It is absolutely necessary to incorporate a filter between the divide-by-10 stage and the mixer and to drive the mixer with a sine wave rather than a square wave.

The next important question is which of the two inputs of the mixer, the LO

434 DIGITAL PLL SYNTHESIZERS

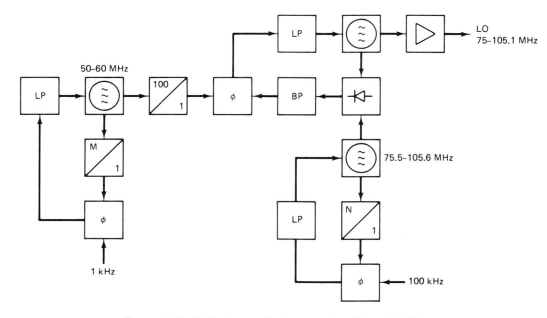

Figure 5-13 Triple-loop synthesizer covering 75 to 105 MHz.

and RF portion, is being driven by the VCO output. This will determine the spurious response. In our particular case, where we suspect some spurious output to be generated because of the mixer action, it is advisable to use the fine-resolution loop as the LO and have the UHF VCO be at the RF input level.

Because of the losses inside the mixer, a postamplifier will be required that can be included in the bandpass filter driving the programmable counter for the output loop.

Another way to reduce output noise, avoid spurious response at the output, and use a triple-loop synthesizer to achieve high resolution is shown in Figure 5-13. The output loop uses three VCOs covering the range from 75 to 105 MHz in 10-MHz increments. Each range has about 10% variations, where $\Delta f/f$ equals 10 MHz/85 MHz as the first range.

A second set of VCOs of identical design is locked in a single-loop synthesizer in increments of 100 kHz, and therefore, the programmable divider requires a division ratio between 750 and 1050. The fine resolution is achieved.

Let us take a look at the noise. The highest frequency, 105.1 MHz, dictates a multiplication of 1051 or reduces the noise relative to the reference at 100 kHz by 60.43 dB.

Let us assume that the VCO noise at 1 kHz is about −100 dB/Hz and about −130 dB/Hz at 10 kHz. If the 100-kHz reference noise is −160 dB/Hz (determined by the reference divider noise rather than the standard), the reference noise multiplied up would reduce the signal-to-noise ratio to about −100 dB/Hz, equivalent to the 1-kHz noise of the VCO. It is therefore advisable to set the loop bandwidth of the synthesizer at 1 kHz. Inside the loop bandwidth, the noise will now stay approximately −100 dB/Hz at 1 kHz, deteriorating to −60 dB/Hz at about

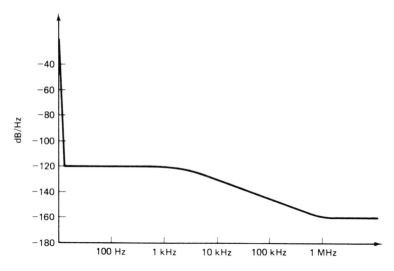

Figure 5-14 Noise sideband performance of the step loop of synthesizer in Figure 5-13.

1 Hz off the carrier. Outside the loop bandwidth, the VCO determines the noise, and Figure 5-14 shows the resulting noise of this section of the synthesizer. We will call this the *coarse-tuning loop* or *step loop*, as we step through the entire frequency range in increments of 100 kHz. If those oscillator sections would be totally identical and mixed against each other, the resulting difference frequency would be zero.

We can, however, use a third loop, a single-loop synthesizer as the fine-resolution loop, and therefore compare the output of the mixing of the two loops with the fine-resolution loop.

The fine-resolution loop uses a 50–60 MHz VCO inside a 1-kHz reference loop.

If this loop is divided by 100 at the output, the resulting output frequency is 500 to 600 kHz in increments of 10-Hz steps.

The triple-loop synthesizer has several unique features:

1. There is no divider at the output loop, and therefore the noise present at the phase comparator is not multiplied at the output.

2. The output noise is equal to the geometric average of the noises between the fine-resolution loop and the step loop.

3. The noise of the output loop is determined outside the loop bandwidth by the performance of the VCO and inside the loop bandwidth by the 500- to 600-kHz reference, which is improved by 40 dB because of the division and the step loop, which has a low-noise performance because of the low division ratio, where N remains less than 1100.

The 100-kHz loop can be designed in such a way that the loop filter, together with the phase/frequency discriminator, achieves more than 90-dB suppression with enough safety margin for stability, and the switching time is in the vicinity of 1 ms.

4. The settling time of the fine-resolution loop can be made much faster due to the fact that the output frequency is divided by 100, and therefore, the reference suppression is increased by an additional 40 dB.

Let us assume that the required reference suppression of the 1-kHz reference is 100 dB. We know that the division by 100 at the output reduce1s the rference by 40 dB, so we have to achieve an additional 60 dB between the loop filter and the phase/frequency comparator.

A tri-state phase/frequency comparator enables us to obtain at least 40 dB of reference suppression, so that the output filter only has to supply an additional 20 dB. As a result of this, the loop filter can be set to a loop bandwidth of approximately 100 Hz. In practice, however, one would drop the requirement of the reference suppression of 100 dB, setting it at 90 dB, and then a loop bandwidth of 300 Hz is sufficient. In doing this, a settling time in the vicinity of 6 ms is achievable, providing a total system's settling time in this vicinity, as the output loop and the step loop are much faster.

5. There is, however, a potential hazard. As both VCOs operate very close at such a high frequency, care has to be taken that one VCO always remains higher than the other to avoid an image problem. Such an image problem would definitely allow false lock, and therefore make the loop unstable, and would give the wrong output frequency. To avoid this, the output loop is receiving coarse-steering information from the step loop, and the step loop by itself is coarse set by a 100-kHz, 1-MHz, and 10-MHz activated D/A converter.

An additional auxiliary circuit is provided, which assures that the one frequency always remains higher than the other, and a set of operational amplifiers, together with a frequency detector, takes care of this problem.

At first, this type of circuit may appear difficult, but this principle allows the design of an extremely low-noise synthesizer together with a substantial reduction in spurious signal inside the loop, as two large frequencies are mixed against each other down to a low IF, which in our case is 500 to 600 kHz. Other combinations may have some advantages from certain design points but definitely have more spurious outputs and have worse noise performance. Table 5-1 shows the performance of this multi-loop synthesizer.

Table 5-1 Performance of 75–105 MHz multiloop synthesizer, 10-Hz step size

Stability	Depends on standard
Phase noise	−90 dB/Hz
	1 kHz off the carrier
	−135 dB/Hz
	20 kHz off the carrier
	−140 dB/Hz
	100 kHz off the carrier
	−85 dB/Hz
	60 Hz off the carrier
Switching speed	6 ms
Spurious output	−90 dB

5-3 LOW-NOISE MICROWAVE SYNTHESIZERS

Low-noise microwave synthesizers, although they operate above 1 GHz, consist of a number of different building blocks. These blocks are either analog or digital in nature. The digital interfaces such as microprocessors will not be addressed here but are necessary to perform some of the number crunching involved in controlling the internal synthesizer auxiliary stages. Synthesizer building blocks are comprised of oscillators, dividers, and various loops, such as translation loops. We will first look at a number of block diagrams and then proceed from the more traditional approach toward the very latest technology. Despite progress made in the various disciplines, fundamental performance as far as phase noise is concerned is still determined by the loop and its components (such as transistors and tuned circuits). The achievable figure of merit, or Q, depends solely on mechanical size and materials used. I do not believe that we will see the development of any additional high-Q resonators, that is, crystals, dielectric resonators, SAW resonators, ceramic resonators, YIG oscillators, and LC oscillators.

For microwave applications, the YIG oscillator (while temperamental in nature) combines the best tuneability, linearity, and widest frequency range. The electronic equipment used to provide coarse and fine steering, however, is complex and costly. Modern CAD tools will permit us to look at the loop response for phase noise, gain (stability), and lock-in time.

We will examine critical stages of various oscillator types, as available, and look at CAD applications versus actual measurements and different technological application for clean signals.

Moving toward the millimeter-wave range, 40 GHz and higher, I will show a MMIC oscillator used for "smart" ammunition purposes. It turns out that even these applications require low-noise high-performance synthesizers. Finally, I will present a quick look at the transient response of oscillators for the purpose of examining the actual time it takes the oscillator to settle, which is a limiting factor generally not considered by designers.

5-3-1 Building Blocks

Microwave synthesizers are essentially an extension of the RF synthesizers, which are found in test and communication equipment. The traditional approach for building synthesizers with fairly simple structures pretty much ends at 1 GHz. There are many reasons for this. The number one reason, of course, is that the resulting division ratio becomes very high. As a result of such high division ratios, the output phase noise can change quite drastically. By tightening the loop bandwidth, one can use the output VCO to be the dominant noise source outside the loop bandwidth; however, this has its limits. One of the early high-performance microwave synthesizers designed by California Microwave is shown in Figure 5-15 [85]. It shows all the typical building blocks one must consider when looking into microwave synthesizer design. A voltage-controlled crystal oscillator, which is locked to a stable reference, is first multiplied by 2 and then by 10. This provides an output frequency between 1280 and 1380 MHz and the phase detector is typically a harmonic sampler consisting of two diodes as a microstrip discriminator. The output VCO can be locked with a fairly wide bandwidth (up to several hundred

438 DIGITAL PLL SYNTHESIZERS

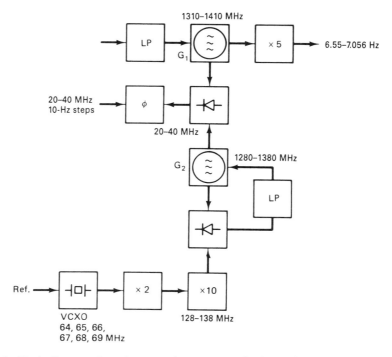

Figure 5-15 Block diagram of a microwave frequency synthesizer using an internal IF of 20 to 40 MHz. G_1 and G_2 are cavity oscillators. If a wider frequency range is required, YIG oscillators may be used to replace those oscillators and the ×5 multiplier in the output may not be necessary.

kHz) and will have sufficient suppression for the subharmonic frequencies generated in the loop (80 to 90 dB).

Actually, one can replace this subsystem with a comb generator, but then it becomes quite tricky to filter out the appropriate spectral line and sufficiently suppress the adjacent unwanted line. If the output frequency has only a fairly narrow bandwidth requirement, the actual output loop can be mixed down to a low IF, in this case 20 to 40 MHz, and then the fine resolution can be obtained here. A ×5 multiplier at the output then brings the signal up to the desired value. The actual tuning range is about 8% and if both G_1 and G_2 are cavity-tuned oscillators with very high Q, the resulting phase noise is quite good. Figure 5-16 shows the noise sideband performance of the model CV3595 microwave downconverter measured at 7 GHz and Figure 5-17 shows the actual block diagram of the total system.

This early type of microwave synthesizer, while achieving quite good phase noise, has a large number of building blocks and is both bulky and expensive. On the other hand, if the output is divided down (for comparison purposes) into a VHF frequency like 150 MHz, the phase noise can be reduced by 33.4 dB. In two-way communications a spacing of 30 kHz off the carrier is always a critical number and the resulting phase noise would be roughly 148 dB. Compared to modern signal generators, this is not an extremely high performance, but even

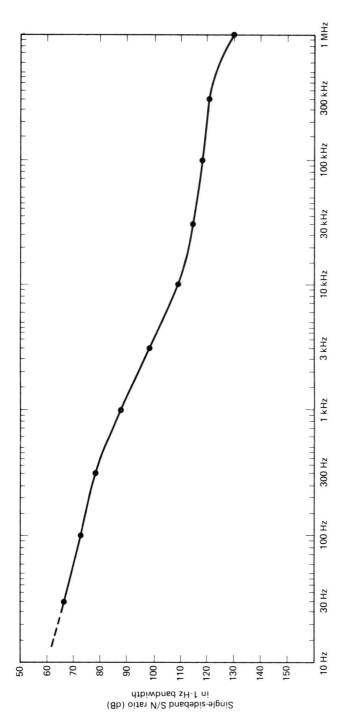

Figure 5-16 Noise sideband performance of model CV3594 microwave down-converter measured at 7000 MHz.

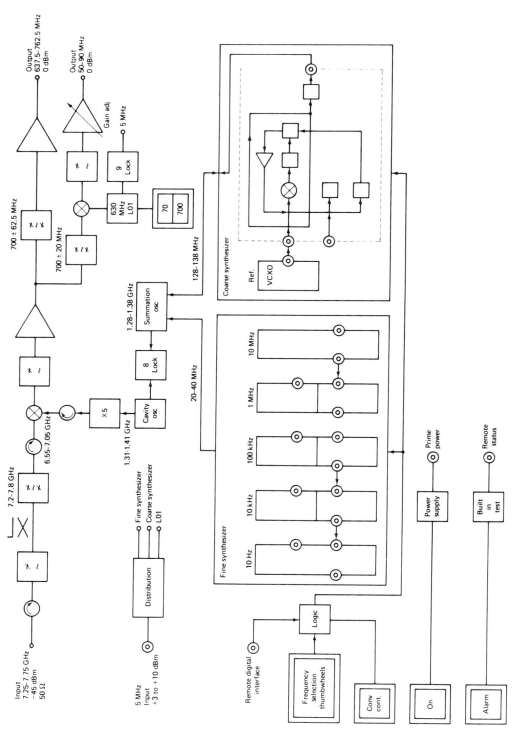

Figure 5-17 Block diagram of model CV3594 microwave down-converter.

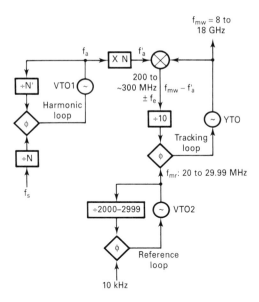

Figure 5-18 Wideband microwave frequency synthesizer, 8 to 10 GHz.

today's state of the art in this frequency range is approximately 150 dB and the 2-dB variance is not significant. The major drawback of this design, of course, lies in the fact that it only shows a very narrow frequency, which for general applications is not very useful.

This method can be extended as shown in Figure 5-18 [86]. The phase noise analysis, however, shows that the resulting phase noise is in a similar category, which is obvious since the division ratios are fairly high. The resulting phase noise depends strongly on the quality of the oscillators labeled VTO1 and VTO2 and because of the fixed divide-by-10 ratio, the phase noise is worsened by 20 dB (multiplied inside the loop). The fact that the oscillator VTO2 is divided by a fairly large number also means that one should keep the loop bandwidth fairly narrow, otherwise the phase noise is multiplied up into the VCO. A narrow loop bandwidth's drawbacks are that the switching speed is slow and the VCO is subject to microphonic effects.

When building a 20–30 MHz oscillator, there are no particular high-Q resonators available and such an LC oscillator will have a general operating $Q_L \lesssim 200$. This is small compared to the Q of ceramic resonator oscillators (CROs) or dielectric resonator oscillators (DROs), which would be used in a different frequency scheme. The same applies to VTO1 as well, since its output is multiplied inside the loop. The sample in Ref. 86 uses approximately 1 GHz for the VTO1. This is an ideal frequency for using a CRO.

A different approach, which results in overall better performance, is shown in Figure 5-19. Figure 5-19 shows the block diagram of the YIG oscillator-based first local oscillator (LO) of a spectrum analyzer with very low phase noise. While the basic approach of the synthesizer is similar to the previous two examples, there are some exceptions to the rule. First, in order to have a very low phase noise oscillator, a 200-MHz CRO is used as input for the multiplier.

442 DIGITAL PLL SYNTHESIZERS

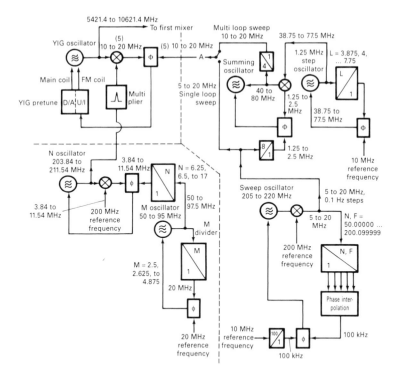

Figure 5-19 Interaction of the frequency-determining modules of the first local oscillator of a microwave spectrum analyzer.

At a frequency of 200 MHz, an operating Q_L of 600 is obtainable for such an oscillator and because of the pulling range of <5%, there is very little noise contribution from the tuning diode. By mixing the oscillator down to an IF of 3.84 to 11.54 MHz, using a 200-MHz VXCO-based reference frequency, the actual division ratio is 1. Therefore, the added phase noise caused by a divider chain is avoided. The phase noise of the 200-MHz oscillator itself is determined by the operating Q_L and by the output power of the transistor oscillator. At these frequencies, one has the option to use either FETs or low phase noise bipolar transistors (BIPTs). Device and oscillator topology will be addressed later. Fine resolution for this oscillator is achieved by a fractional division N dual-loop system, whereby both the divide by N and divide by M use both integer and fractional values. This particular principle has two distinct advantages. One is that despite the fairly low division ratio one obtains fine resolution, and the second is that the actual spurs that occur are at least 500 kHz away. Since their location is predetermined, notch filters can remove those discrete spurs while maintaining the widest possible loop bandwidth. This method does not require all the complicated housekeeping mathematics required by the full fractional division N. The 50–95 MHz oscillator can be made quite clean and the divide-by-N stage, which feeds the phase detector, has an extremely low phase noise floor, in the vicinity of -170 dBc/Hz operating with several hundred kHz loop bandwidth. The

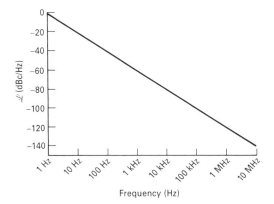

Figure 5-20 Predicted worst case single-sideband phase noise of a YIG oscillator operating at 7 GHz.

divide-by-M loop, which controls the 50–95 MHz oscillator, is also very clean and its phase noise depends on the actual oscillator.

In many cases, it is desirable not only to have fixed frequencies but also to have the ability to have very fast sweeps. The 5–20 MHz signal required for the phase detector can either by generated from a very fast fractional divide-by-N synthesizer or from a double-loop fine-resolution synthesizer, as shown in the block diagram. The decision to use either the fractional divide-by-N synthesizer or a direct digital synthesizer (DDS) will be discussed shortly.

5-3-2 Output Loop Response

For the purpose of analyzing the output loop, we are going to look into the VCO (YIG) oscillator and other contributors to get a better understanding of what the overall response is. Figure 5-20 shows the phase noise of a free-running high-Q YIG oscillator at 7 GHz. The particular YIG is based on the latest Siemens 15-GHz f_t silicon transistor and uses a combination of coarse and fine steering. The topic of YIG stabilization will be addressed later. This phase noise performance is based on stabilization with very narrow bandwidth, and free-running measurement even with a delay line discriminator would be difficult. The approach selected here allows a switch in the loop bandwidth, which results in a speed-versus-phase-noise comparison.

Figures 5-21 and 5-22 show the single-sideband phase noise as a function of the loop bandwidth. The loop bandwidth certainly influences the locking of the phase lock loop. The higher-order loop, as applicable here, uses complex filters and the stability border diagram as shown in Figure 5-23 indicates such a design. The open-loop gain curve indicates a sharper roll-off above 500 kHz and the response components will be suppressed by 12 more than 90 dB. It is very important to select the proper time constants in the loop filter. Figure 5-24 shows an optimized response and Figure 5-25 shows the ringing that occurs if the postfilters do not provide the desirable 45° of phase margin. Figure 5-26 shows the phase noise achieved by a fine-resolution system.

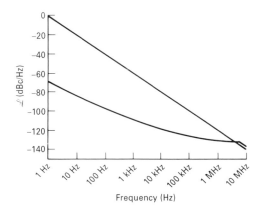

Figure 5-21 Predicted worst case open- and closed-loop phase noise of a YIG oscillator with loop filter frequency of 4 MHz.

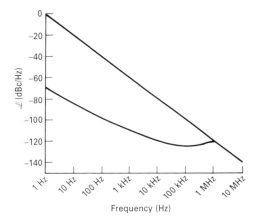

Figure 5-22 Predicted worst case open- and closed-loop phase noise of a YIG oscillator with 1-MHz loop bandwidth.

5-3-3 Low Phase Noise References: Frequency Standards

The synthesizer's architecture requires a number of auxiliary frequencies that are all traceable to a master standard. However, requirements for long-term stability and low phase noise are diametrically opposed. Long-term stability means low aging, and, therefore, the operating mode for the crystal oscillator is different from that used in a high signal-to-noise ratio operation. A similar case is found in a rubidium atomic frequency standard (second standard) or in a primary cesium standard. These features are combined, but their cost may exceed the budget for the appropriate time/frequency standard. Therefore, an acceptable approach is to dedicate crystal oscillators at 10 MHz or 100 MHz, which will serve sufficiently as low phase noise auxiliary sources.

LOW-NOISE MICROWAVE SYNTHESIZERS 445

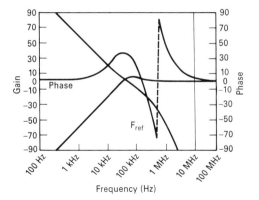

Figure 5-23 Bode plot of the fifth-order PLL system for a microwave synthesizer. The theoretical reference suppression is better than 90 dB.

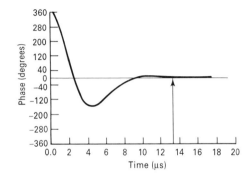

Figure 5-24 Lock-in function of the fifth-order PLL. Note that the phase lock time is approximately 13.3 μs.

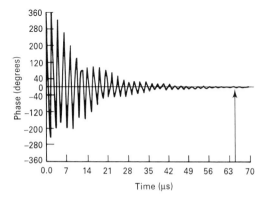

Figure 5-25 Lock-in function of the fifth-order PLL. Note that the phase margin has been reduced to 33° from the ideal 45°. This results in a much longer settling time of 62 μs.

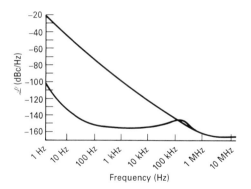

Figure 5-26 Predicted worst case phase noise as achieved by the fine-resolution system (open- and closed-loop).

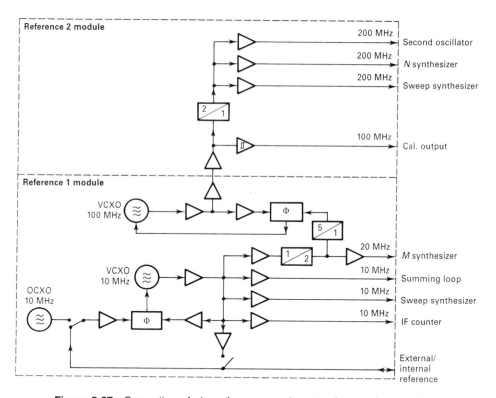

Figure 5-27 Generation of phase/frequency coherent reference frequencies.

Figure 5-27 shows the reference source generation for fixed frequencies as used in a complex synthesizer. Another item one needs to consider is the pulling effect or interaction between the different outputs. For high-performance synthesizers, one needs not only frequency but also phase stability, and therefore, the change of load must not change any output phase relationship.

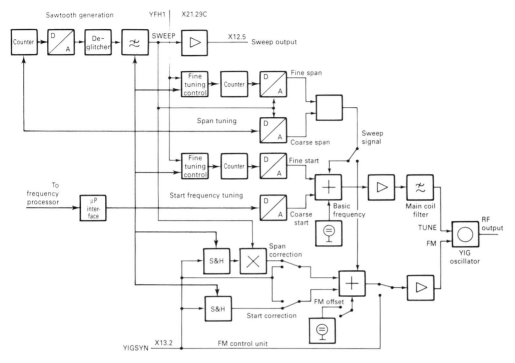

Figure 5-28 Coarse/fine steering of the YIG oscillator including sweeping capability.

A typical case is where the frequency divider is driven by a reference frequency as a time-dependent loading, and therefore, the phase will jump as a function of the toggle occurring at the input of the integrated circuit. To prevent these effects, one has to design very special isolation stages.

In a similar fashion, one has difficulties with the YIG oscillator relative to the comb generator. The tuning sensitivity of the main coil is extremely high and complex circuitry is required to properly position the YIG oscillator prior to when the frequency/phase lock will occur.

Figure 5-28 shows the block diagram of this pretuning circuit for the YIG oscillator. By analyzing the circuit, it becomes apparent that the level of effort to do this properly is quite high. There are also provisions for modulating the system, and therefore, there are also inputs for FM. Figure 5-29 shows the YIG's synchronization in greater detail.

5-3-4 Critical Stages

Oscillators. While YIG oscillators are specialty items and must be purchased from select manufacturers, such as Hewlett-Packard, Avantek, and Watkins-Johnson, the general tendency is to custom-build VCOs for each application. At the lower frequencies, below 200 MHz, there are two types of voltage-controlled oscillators. One is actually a voltage-controlled crystal oscillator and the other is a high-performance LC oscillator.

448 DIGITAL PLL SYNTHESIZERS

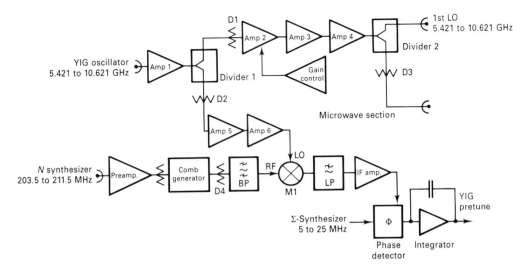

Figure 5-29 Block diagram of the YIG synchronization.

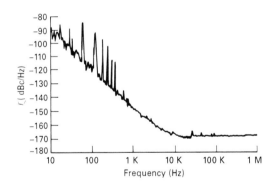

Figure 5-30 VCXO short-term frequency stability for 200 μW crystal dissipation.

Figure 5-30 shows the low phase noise of a 10-MHz voltage-controlled crystal oscillator, where its phase noise is about 8 to 10 dB better than the HP10811A frequency standard; however, its long-term stability obviously is not as good.

Figure 5-31 shows the actual circuit diagram. The oscillator's configuration is similar to what is referred to as a Butler oscillator in common literature. This means the frequency-selective device is located between two emitters and the tuned circuit is in the collector of one of the transistors.

The PNP transistors type 2N5160 are 2-watt plus PNP transistors with an f_t above 1 GHz. By having a two-stage crystal oscillator, the feedback loop can be made high enough so dissipation in the crystal remains reasonable.

Figure 5-32 shows the circuit of the 100-MHz voltage-controlled crystal oscillator. Modern CAD tools like harmonic balance simulators allow accurate

LOW-NOISE MICROWAVE SYNTHESIZERS

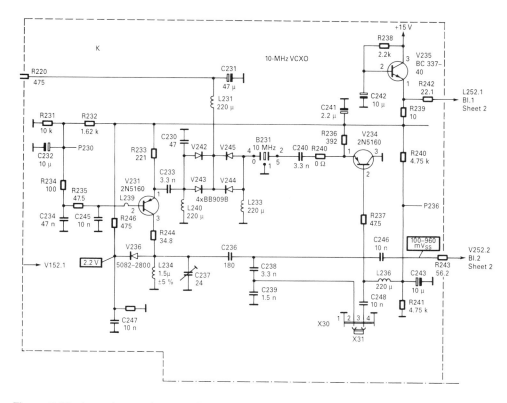

Figure 5-31 Low phase noise crystal oscillator for 10-MHz generation will be locked against the master standard.

phase noise predictions for any type of oscillator and can also handle optimization.

Figure 5-33 depicts a screen capture of the simulator used for predicting the phase noise. The simulator shows the circuit file including the nonlinear parameters for the bipolar transistor and, after performing the harmonic balance simulation, shows the harmonic contents of the oscillator signal in the frequency domain and in the time domain (distorted waveform rather than sinusoidal curve). Finally, it also shows the SSB phase noise. My experience with these types of circuits has been that accuracy is within 1 or 2 dB compared to the measured data. By allowing the component's values to vary and by varying the dc bias point, one can optimize the circuit for the best phase noise.

Figure 5-34 shows that the close-in phase noise of the free-running oscillator was improved by 32 dB. By introducing negative resistance feedback, we change the feedback loop gain and therefore the loading of the tuned circuit. The "cleanup" of the close-in phase noise is done at the expense of the far-out noise. This oscillator example shows that for frequencies greater than 20 MHz away, the noise source is now marked to 160 dBc/Hz compared to 168 dBc/Hz. However, this is still far better than needed for practical application.

450 DIGITAL PLL SYNTHESIZERS

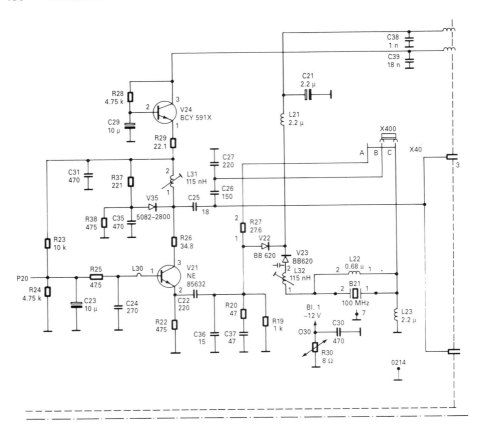

Figure 5-32 A 100-MHz VCXO with extremely low phase noise.

Other Key Components. Other key components for low phase noise are the use of SAW devices and DROs. These oscillators cannot be pulled too much but can be phase locked fairly narrow against a frequency standard. This is useful in obtaining auxiliary frequencies that are very clean. Figure 5-35 shows the measured phase noise spectrum for one laboratory prototype 675-MHz SAW delay line VCO.

Figure 5-36 shows the measured phase noise spectrum for one L-band (982 MHz) dielectric resonator oscillator. The measured data for Figures 5-35 and 5-36 were provided by Don Parker, NIST, formerly of Raytheon Corp.

Now we turn to voltage-controlled oscillators. Figure 5-37 shows a cavity-based low phase noise oscillator with a decoupling stage. Its performance is similar to the HP8640 signal generator or the Rohde & Schwarz SMDU. The phase noise at 400 MHz, 25 kHz off the carrier, is generally 145 to 148 dBc/Hz. A hybrid for the very-high-Q resonator oscillator and LC oscillator is the ceramic resonator oscillator (CRO). This type of oscillator uses a ceramic resonator that is electric quarter-wavelength and uses dielectric material for an ε_r of 38 to 88.

The advantage of this type of oscillator is that it combines small mechanical

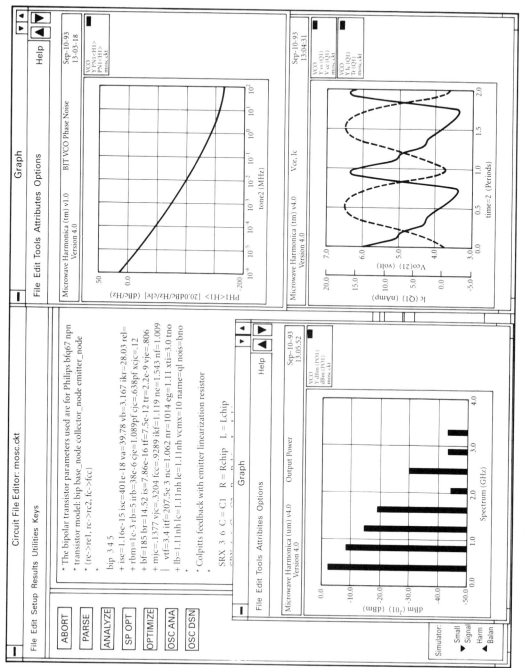

Figure 5-33 Screen capture of the simulator used for predicting the phase noise.

452 DIGITAL PLL SYNTHESIZERS

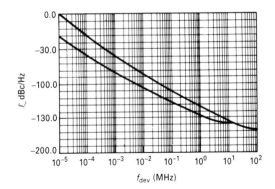

Figure 5-34 Phase noise before and after automatic optimization, which improved close-in phase noise by 32 dB.

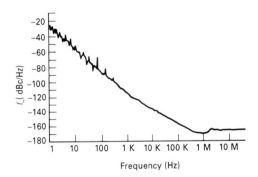

Figure 5-35 Measured phase noise spectrum for 675-MHz SAW delay line VCO.

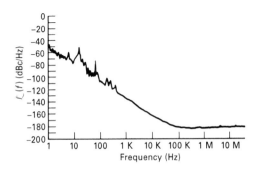

Figure 5-36 Measured phase noise spectrum for one L-band (982-MHz) dielectric resonator oscillator.

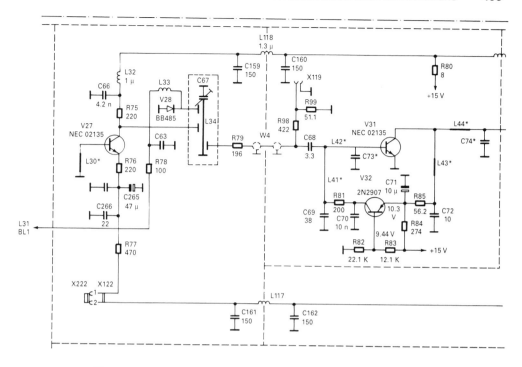

Figure 5-37 Schematic of low phase noise cavity stabilized VCO.

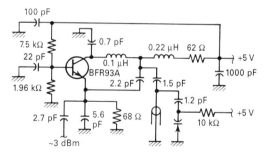

Figure 5-38 Typical test circuit for use in a ceramic resonator. These resonators are available in the 500-MHz to 2-GHz range. For higher frequencies, dielectric resonators are recommended.

size with high Q and low cost. Figure 5-38 shows the schematic of such a CRO. In modeling this circuit, the CRO should be modeled with a cable for high dielectric constant.

Figure 5-39 shows the measured phase noise of the oscillator shown in Figure 5-38.

Figure 5-40 shows the predicted phase noise of the 1-GHz ceramic resonator VCO without the tuning diode, and Figure 5-41 shows the predicted phase noise of the 1-GHz ceramic resonator VCO with the tuning diode attached. Note the good agreement between the measured and predicted phase noise.

454 DIGITAL PLL SYNTHESIZERS

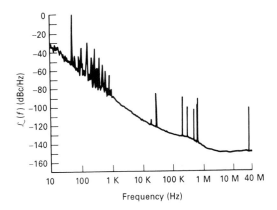

Figure 5-39 Measured phase noise of the oscillator shown in Figure 5-38.

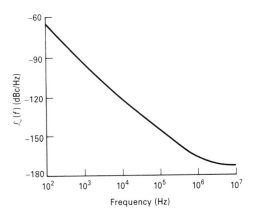

Figure 5-40 Predicted phase noise of the 1-GHz ceramic resonator VCO without the tuning diode.

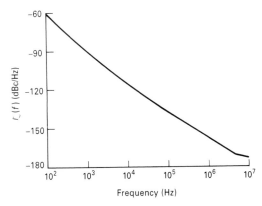

Figure 5-41 Predicted phase noise of the 1-GHz ceramic resonator VCO with the tuning diode attached. Note the good agreement between the measured and predicted phase noise.

LOW-NOISE MICROWAVE SYNTHESIZERS 455

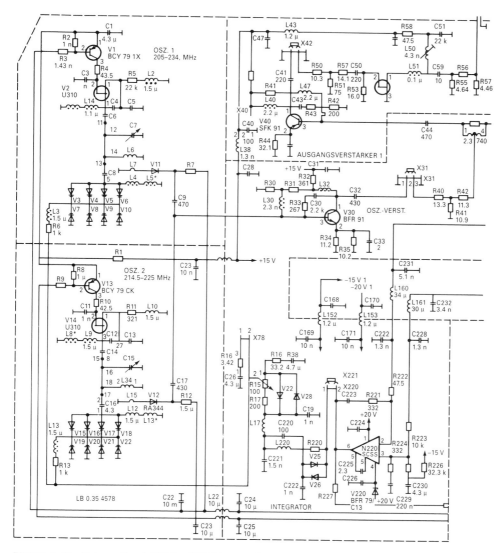

Figure 5-42 Schematic of 205 to 225 MHz very low phase noise oscillator system. It uses FETs for low flicker noise contribution and multiple-diode arrangement to reduce the diode noise.

For lower frequency application and wider tuning range, a different approach is necessary. The previous example has shown that the tuning diode adds a lot to the phase noise and, therefore, the pulling range should be kept minimal. Figure 5-40 shows a narrowband VCO that has been optimized for low phase noise operation at 200 MHz.

In order to maintain very low phase noise, the decoupling is done by taking energy off the tuned circuit with magnetic coupling and making the output part of the resonator circuit. This improves the phase noise and the harmonic content.

Figure 5-42 depicts the schematic of a 205 to 225 MHz very low phase noise

456 DIGITAL PLL SYNTHESIZERS

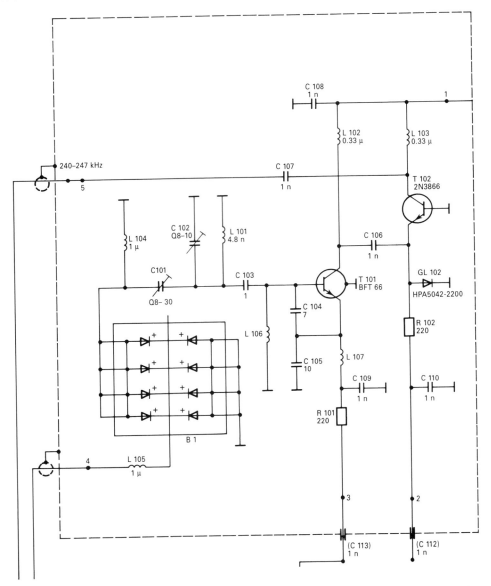

Figure 5-43 Bipolar implementation of the low phase noise 205 to 225 MHz oscillator. Note that the tuning range is much smaller and set from 240 to 247 MHz.

oscillator system. It uses FETs for low flicker noise contribution and multiple-diode arrangement to reduce the diode noise.

Figure 5-43 shows the bipolar equivalent of this circuit operating in the vicinity of 200 MHz. In instances where a much wider tuning range like 1:2 is required, a single circuit with multiple parallel diodes of high voltage gain must be used, as shown in Figure 5-44. The diode clamping circuit marked V9 is responsible for

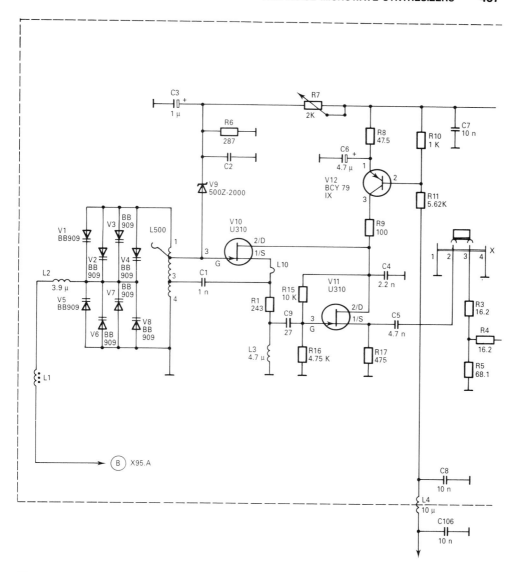

Figure 5-44 Very wideband low phase noise oscillator for frequency range 40 to 80 MHz. The design takes advantage of the multiple-diode arrangement and clamping diode V9 for good signal purity.

preventing the RF voltage from exceeding certain values and from allowing the gate source area of the U310 to become conductive.

Isolation Stage. As previously mentioned, the success of low phase noise operation depends highly on preventing any load changes resulting in phase jumps. Figure 5-45 shows a combination of a power splitter and low feedback buffer stage. The neutralization of this circuit makes it possible to minimize loading effects on different channels.

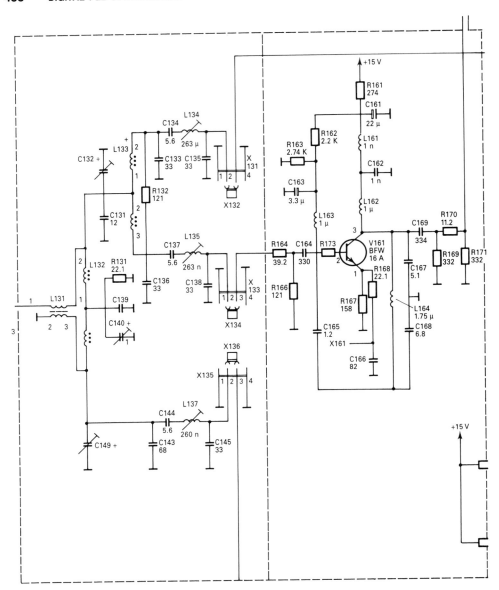

Figure 5-45 Distribution amplifier system that combines the input power splitter and a feedback amplifier with neutralization.

Harmonic Generators. Figure 5-46 shows a harmonic generation circuit. The BFW16 transistor drives the snap-off diode and the output filter in tune selects the appropriate output frequency. Such a filter can be made tunable and can track the appropriate desired harmonic. The multiplier circuit is extremely critical because any noise from additional unwanted nonlinear effects will deteriorate the performance.

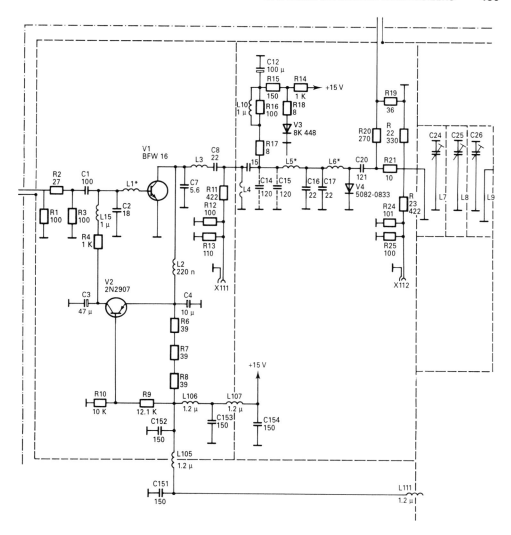

Figure 5-46 Comb generator and postselection filter for reference oscillator. Note that the biasing of the step recovery diode or snap-off diode is very important for good phase noise performance.

The harmonic multiplier diodes have flicker noise and there must be a good compromise between the flicker noise contribution and efficiency.

Millimeter-Wave Oscillators. At frequencies above 20 GHz the bipolar technology runs out of steam. One is faced with two options: (1) using a frequency doubler, which requires additional volume and power, or (2) using GaAs FETs.

Depending on the application, either of the two options may be chosen; however, at this point, I would like to highlight the performance of millimeter-wave oscillators built with GaAs FETs. Figure 5-47 shows a Texas Instrument VCO.

460 DIGITAL PLL SYNTHESIZERS

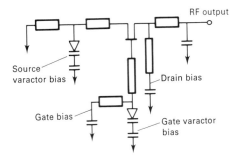

Figure 5-47 Texas Instrument 8132 VCO topology.

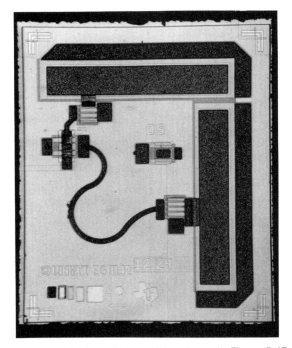

Figure 5-48 Layout of oscillator shown in Figure 5-47.

This arrangement uses two tuning diodes to increase the tuning range. This basic approach can be extended to frequencies as high as 100 GHz. The initial implementation of the oscillator was pushed up in frequency and its layout is shown in Figure 5-48. It is obvious that in this case the tuning diodes also exert a major influence on the phase noise and a number of studies have been done to improve the performance of those diodes.

In most cases these diodes are built by using an FET and connecting the gate and drain together. A nonlinear junction is then used as a tuning diode.

Figure 5-49 shows a 39-GHz oscillator design that uses a symmetrical ring type of arrangement as a test vehicle. This oscillator was developed under MIMIC

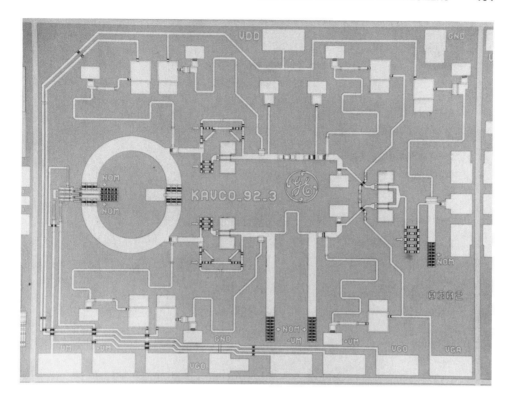

Figure 5-49 A 39-GHz oscillator design that uses a symmetrical ring type of arrangement.

activities by General Electric (now Martin Marietta/Sanders). Please note that two of the transmission lines have extensions for possible laser trimmings. Again, the highest reported approach for this is about 100 GHz on the fundamental sources. Consistent with previous statements, one can use modern CAD tools to predict the output power and the phase noise of such a circuit (Figures 5-50 and 5-51). Because of the unavailability of measured data for the Texas Instruments design, the performance of the 39-GHz oscillator is still under evaluation. However, indications are that simulation and predictions are quite close.

5-3-5 Time Domain Analysis

One of the requirements for the synthesizers is a very fast switching time and I would like to highlight that there is a trade-off between the high-Q oscillator and switching time or start-up condition for the oscillator. Figure 5-52 shows a dielectric resonator oscillator (DRO), which is being utilized for the purpose of examining the switching time but otherwise shows no particular performance advantages. The operating Q of the DRO is kept as high as possible in order to obtain low phase noise and the active elements determine the phase noise of the transistor and resonator.

462 DIGITAL PLL SYNTHESIZERS

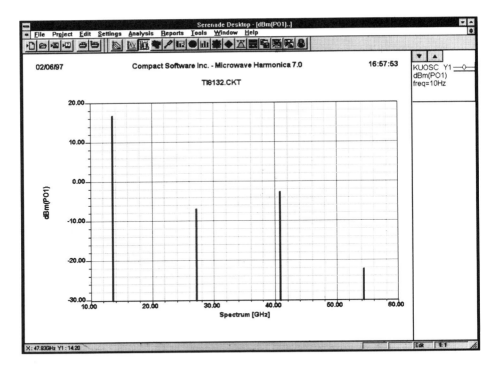

Figure 5-50 Harmonic output power.

For time domain analysis, one needs to use a Spice program as harmonic balance programs used for phase noise analysis only consider steady-state conditions.

Consistent with the linear approach, one should first determine the presence of negative resistance or the equivalent of $S_{11} > 1$ to guarantee start-up of the oscillator, which is shown in Figures 5-53a and 5-53b. Unless this is established, one cannot determine the output power shown in Figure 5-54.

Finally, following an examination of the start-up condition of the oscillator shown in Figure 5-55, it becomes apparent that it requires approximately 180 ns for the oscillator to start and it is fair to assume that the oscillator has started after 500 ns.

In order to establish the feedback pass, the DRO uses inductor feedback. If one simulates the current in the inductor, a pattern consistent with the output voltage develops. Fifty nanoseconds after the switch-on time, an initial current surge occurs that gets the oscillator started. Figure 5-56 depicts the initial current surge occurring 52 ns after switch-on time.

5-3-6 Summary

I have shown both the multiloop approaches and the contribution from dependent building blocks as they affect the overall performance of millimeter-wave synthesizers. Table 5-2 gives a list of the key elements.

LOW-NOISE MICROWAVE SYNTHESIZERS 463

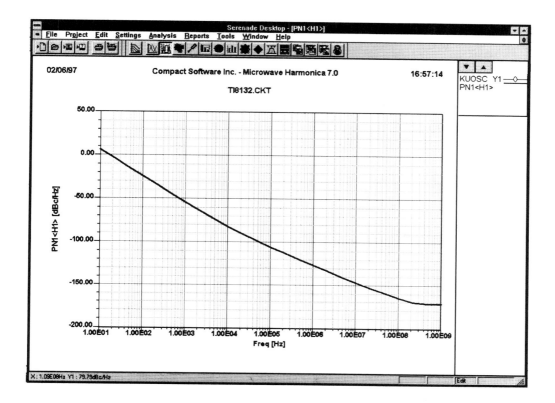

Figure 5-51 Phase noise simulation of Figure 5-49.

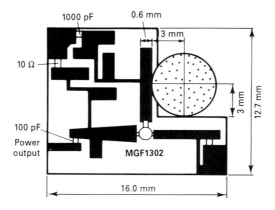

Figure 5-52 A 10-GHz DRO.

464 DIGITAL PLL SYNTHESIZERS

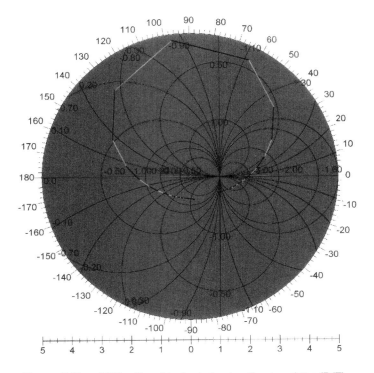

Figure 5-53a DRO with a bipolar heterojunction transistor (BJT).

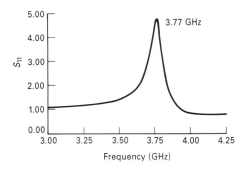

Figure 5-53b Calculation of S_{11} as a function of frequency. Note that the resonant frequency occurs at roughly 3.77 GHz.

Figure 5-57 shows the single-sideband (SSB) of a 10-GHz oscillator made by Rohde & Schwarz.

Finally, Figure 5-58 shows the measured phase noise of a 47.104-GHz frequency source as advertised by Fujitsu Limited.

These types of microwave circuits are being used in both signal generators, such as the Rohde & Schwarz SMP22, as shown in Figure 5-59, and in the high end

Table 5-2 Key elements for millimeter-wave synthesizers

Techniques

- Basic PLL principles for digital and analog loops including use of delay line stabilizers and variable reference frequency
- Fractional division N with high-resolution counters and accumulators (using gate arrays)
- Direct digital synthesizers having arbitrary resolution and picosecond access time
- Selection of low noise summing loops with high bandwidth and fast response
- Availability of computer program for evaluating SSB noise for different VCOs (SONATA available through Compact Software, Inc., PLL DESIGNKIT also available through Compact Software, Inc.)
- Selection of low-noise transistors: bipolar transistors (N-junction FETs) and bipolar heterojunction transistors (GaAs FETs)

Sources

- Crystal oscillator—designed for low aging
- Use of buffer oscillator at 10 MHz and 100 MHz for auxiliary frequencies (VCXOs and DROs)
- Choice of lowest possible phase noise design for all VCOs (modern low-gain YIG oscillators)
- Buffer amplifiers selected for highest isolation and low AM-to-PM conversion
- Adaptive loop bandwidth for fast locking and low noise operation
- Selection of low noise dividers with low spike operation
- All opamps in loops must be of low noise design
- Choice of harmonic sampling over division due to $1/f$ noise
- Use of analog phase/frequency detectors

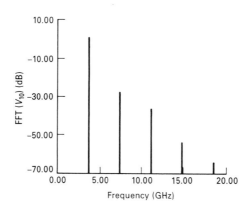

Figure 5-54 Output power of DRO with a BJT.

466 DIGITAL PLL SYNTHESIZERS

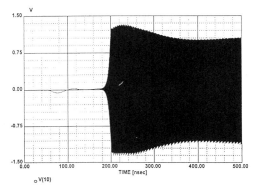

Figure 5-55 Start-up condition of the DRO with a BJT.

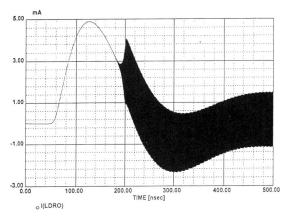

Figure 5-56 Initial current surge occurring 52 ns afater switch-on-time.

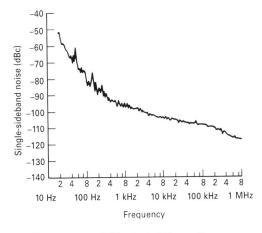

Figure 5-57 SSB of 10-GHz oscillator.

LOW-NOISE MICROWAVE SYNTHESIZERS 467

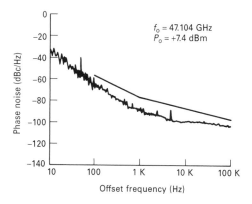

Figure 5-58 Measured phase noise of 47.104-GHz frequency source.

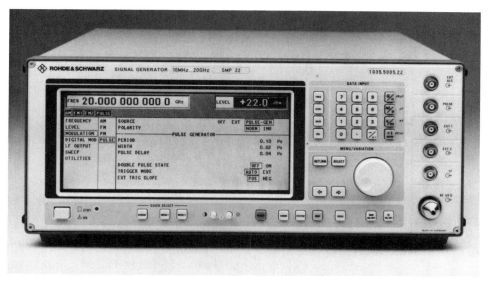

Figure 5-59 Rohde & Schwarz signal generator SMP 22.

Rohde & Schwarz Spectrum Analyzer Series FSEA-30, as shown in Figure 5-60.

5-3-7 Two Commercial Synthesizer Examples

The previously detailed microwave synthesizer example was based on the Rohde & Schwarz FSB spectrum analysis synthesizer, which operates from 100 Hz to 5 GHz at the input.

For test signal generators, the requirements are slightly different. All signal generators require modulation capabilities. An interesting approach implemented in the HP8642B signal generator is that its synthesizer is a combination of a number

468 DIGITAL PLL SYNTHESIZERS

Figure 5-60 Rohde & Schwarz spectrum analyzer Series FSEA-30.

of reference signals generated by SAW oscillators and based on the mixing scheme shown in Figure 5-61, using various oscillator frequency images ranging between 607.5 and 967.5 MHz. These frequencies are a combination of mixing the SAW frequencies with a very clean 135-MHz signal, which contains the FM components. The SAW oscillators are stabilized against a 45-MHz reference. Please note that 135 MHz is the third harmonic of 45 MHz.

A fractional N division synthesizer with a window of 45 to 90 MHz is then used as a fine-resolution synthesizer to generate the output frequency from dividers and higher frequencies from a frequency doubler. While the block diagram looks fairly simple, a great deal of care must be taken to generate clean signals and the spurious-free requirements for its reference oscillator is very high. The arrangement shown allows the use of a very wide loop bandwidth, as can be shown in Figure 5-62. The noise pedestal between 80 kHz and 6 MHz indicates that the loop bandwidth is somewhere below 100 kHz and the phase noise of 20 kHz of better than 140 dB is quite good. This is possible because there is no multiplication within the loop and the close-in phase noise between 10 Hz and 10 kHz is typically that of a high-Q oscillator rather than that of a synthesizer. However, for a low-cost instrument this is quite acceptable.

Much higher performance at much higher costs is achieved from the multiloop

LOW-NOISE MICROWAVE SYNTHESIZERS

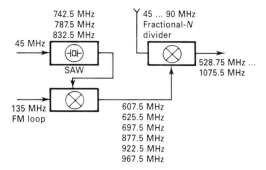

Figure 5-61 Mixing scheme using various oscillator frequency images ranging between 607.5 and 967.5 MHz for the HP8642 generator.

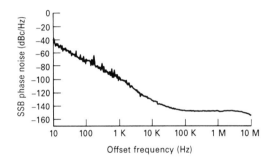

Figure 5-62 Noise pedestal between 80 kHz and 6 MHz indicates the loop bandwidth is somewhere below 100 kHz and the phase noise of 20 kHz of better than 140 dB is quite good.

approach found in the Rohde & Schwarz SMHU85 signal generator, which covers 100 kHz to 4.320 GHz. The following provides an overview of its multiloop architecture. Figure 5-63 shows the RF oscillator assembly, which is housed in module A11. It consists of three oscillators covering the range from 1000 to 2160 MHz in three ranges. To achieve the output frequency of 4.320 GHz, an additional frequency doubler is used. The output phase-locked loop takes advantage of the separate oscillators, which receive pretuning and are locked against the appropriate harmonic of the 40–41.575 MHz reference loop. The module underneath labeled A1 shows a block diagram, which explains the generation of the various reference frequencies. Three crystal oscillators, operating at 10 MHz, 40 MHz, and 130 MHz, are used to produce extremely clean, low phase noise signals, which are used in the auxiliary loops. The 10-MHz crystal oscillator is the internal reference, which can also be replaced by an external frequency standard. However, both the 103-MHz and the 40-MHz crystal oscillator are phase locked against the master standard. The modulation required for modern signal generators is fed into the input of module A8 (called step synthesis FM), which handles both frequency and phase modulation. The AM modulation is applied

470 DIGITAL PLL SYNTHESIZERS

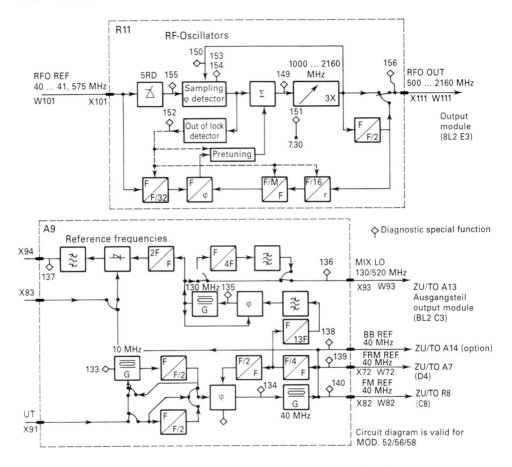

Figure 5-63 RF oscillator assembly, housed in module A11.

to the output module. The step synthesizer, which generates output between 23.125 and 29.375 MHz, uses the 40 MHz from the reference generator and also provides an FM output that goes back into the reference portion. The reference frequency output labeled X94 generates a 300-MHz modulated output, which is fed into the summing loop synthesizer, as seen in Figure 5-64. The fine-resolution signal of down to 1-Hz step size is obtained in module A7 (FRN synthesizer) of Figure 5-65, which internally operates from 38 to 58 MHz and also gets a 40-MHz reference from the reference portion. It is divided down into the frequency range from 3 to 3.625 MHz. This frequency output is then fed into the summing loop portion A10 (Figure 5-64) and each successive stage has approximately a 10 × higher input frequency but achieves this by a mixing scheme rather than a multiplying scheme. The microprocessor system is extremely busy, finding all the right combinations; the phase noise of the FRO reference between 40 and 41.575 MHz within the loop bandwidth determines the output phase noise of the system. Figure 5-66 shows the measured phase noise of the SMHU synthesizer.

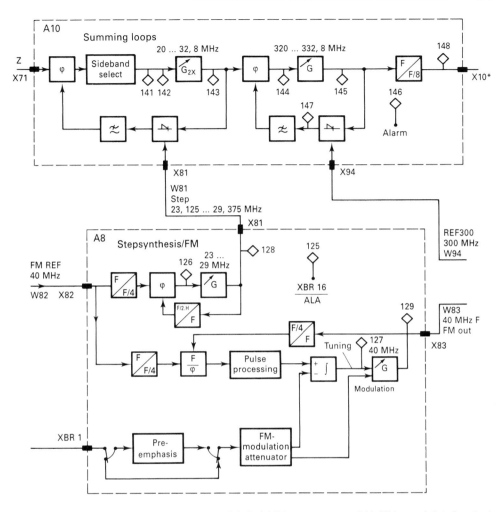

Figure 5-64 Reference frequency output labeled X94 generates a 300-MHz modulated output, which is fed into the summing loop synthesizer.

As we are always interested in higher frequencies, Figure 5-67 shows the single-sideband (SSB) phase noise of the 10-GHz synthesizer SMP made by Rohde & Schwarz.

Finally, Figure 5-68 shows the measured phase noise of a 47.104-GHz frequency source as advertised by Fujitsu Limited.

5-4 MICROPROCESSOR APPLICATIONS IN SYNTHESIZERS

Today's technology is changing at a fast pace, and it may be dangerous to go into great detail about microprocessor applications using specific devices, as constant improvements require the manufacturers to come out with new types of

472 DIGITAL PLL SYNTHESIZERS

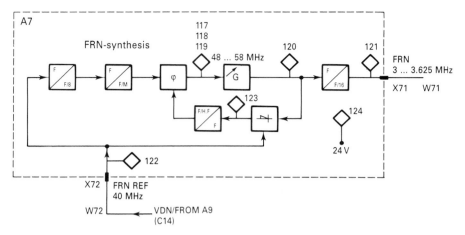

Figure 5-65 FRN synthesizer labeled module A7 operates from 38 to 58 MHz and also gets a 40-MHz reference from the reference point.

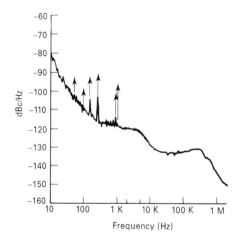

Figure 5-66 Measured phase noise of the SMHU synthesizer at 800 MHz.

microprocessors. However, there are certain fundamentals that are independent of the particular manufacturer or device.

1. Modern frequency synthesizers have a certain intelligence. This is accomplished by incorporating a number of routines in the system. The most frequently used is a scanning routine where a start frequency, a stop frequency, and a frequency increment or step size can be defined. In addition, modern signal generators can be programmed in output power (dBm), output voltage (μV, mV, V) or dB above 1 μV. Different users of signal generators will use different specifications in their system, and to avoid conversion tables and possible errors in translating one figure into the other, the built-in intelligence of the signal

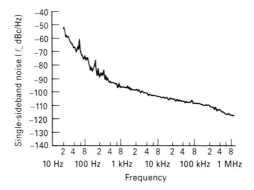

Figure 5-67 SSB phase noise of the SMP synthesizer at 10 GHz.

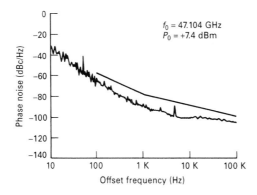

Figure 5-68 Measured phase noise of 47.104-GHz frequency source.

generator via the microprocessor is capable of converting one value into another or receiving commands in different format.

2. Frequency synthesizers found in signal generators are typically multiloop synthesizers. In Section 1-10 we have seen that, depending on a change of loop gain and change of frequency range, certain compensations have to be done within the loop, causing the loop to go out of lock for a certain time. If the out-of-lock sensor used in all superior circuits gives an error command to the microprocessor responsible for the housekeeping, the microprocessor will then either wait until lock is achieved, or, if this is not done within a reasonable time determined by the program, it will alert the user that the frequency synthesizer is out of lock. This so-called built-in self-check, sometimes referred to as BITE (for Built-In Test Equipment), refers to the housekeeping capability of a microprocessor whereby, under software control, certain routines are made available to verify the system operation. This can occur immediately after switching on the instrument or by pressing a check button that activates the relevant circuitry.

3. A number of loops may be used with what is called *offset*, which means that

the actual command value given to the loop does not correspond to the value shown on the display. Therefore, the microprocessor has to perform certain arithmetic, offsetting certain frequencies. Again, this can be called housekeeping and is an essential part of the system. In addition to this, some loops are being mixed, and by determining which sideband is to be chosen from this mixing process, different output frequencies can be made available using the same oscillators. The microprocessor can keep track of the system's requirements, such as which oscillator range has to be operated, which actual programming has to be done with the various loops, what output filters have to be activated, what modulation capabilities have to be considered, and so on.

4. Advanced technology allows construction of synthesizers with large-scale integrated circuits. Because of the high complexity, several commands are required by the frequency divider. Supplied in parallel, the number of lines would be excessive. It is therefore a simplification to address the frequency dividers in serial format rather than parallel, and the microprocessor again has to keep track of the proper format. The HEF4750/4751 is a typical example where both approaches are used, which means that there is a 4-bit parallel six-, seven-, or eight-digit serial input required. The number of digits depends on the number of external frequency dividers used and therefore may vary. If this had to be done in discrete logic, a large number of additional integrated circuits would be necessary.

What does this lead to? The various details I have just listed are most likely to be found in a modern frequency synthesizer and are all necessary at the same time. A microprocessor is essentially a serial device. This means that it performs one task after the other following a certain set of instructions. The programmed microprocessor sends certain commands to the ROMs and IO ports, which have latch circuits in them. Therefore, certain information can be initiated and held in latches. Updating is done by changing the contents of the latches and counters. It is apparent that once housekeeping, arithmetic, verifications, and switching exceed a certain amount, the microprocessor will be extremely busy.

Microprocessors and microcontrollers are available through well-established companies, such as Intel and Motorola to name the most dominant ones. Customarily, one uses microprocessors, such as the ones being used in IBM-compatible computers, for instrumentation control. The main reason for this is the incorporation of high-powered devices in the hardware, like the 80486 and Pentium processor where one obtains a PC as part of the instrumentation. For some applications, fairly simple calculations have to be performed; however, the very moment graphics become involved, the execution speed requirements change drastically. While this topic is not a subject of this book, it should be pointed out that a computer-like display requiring fast graphic and high throughput has become the industry standard for complicated test equipment. Over the years, 16-bit processors have been replaced by 32-bit, and in many cases, the actual calculation is done internally in 64 bits. Also as higher integration occurs, the math coprocessors have become part of the "chip" and in fact are frequently referred to as floating-point calculation devices.

Consistent with experience in computers, where one uses a keyboard processor, graphics processor, memory manager, and CPU (central processing unit), modern communications test equipment with many synthesizers will also break down activities. Typical interface between the user and the instrument such as a signal generator are either keyboard entry from the front panel or parallel addressing via a parallel bus such as the IEEE 488 or 526 bus with a serial RS 232 bus. Chip manufacturers continue to develop dedicated chips to ease programming for these parallel and serial interfaces.

Having mentioned the most powerful Intel devices, I should like to mention two Motorola products. One is the MC68060 Superscalar 32-bit Microprocessor, which is fully compatible with all previous members of the M68000 family. The MC68060 features dual on-chip caches, full independent demand-paged memory management units (MMUs) for both instructions and data, dual integer execution pipelines, on-chip floating-point unit (FPU), and a branch target cache. A high degree of instruction/execution parallelism is achieved through the use of a full internal Harvard architecture, multiple internal busses, independent execution units, and dual instruction issue within the instruction controller. Power management is also a key part of the MC68060 architecture. The MC68060 offers a low-power mode of operation that is accessed through the LPSTOP instruction, allowing for full power-down capability. The MC68060 design is fully static so that when circuits are not in use, they do not draw power. Each unit can be disabled so that power is used only when the unit is enabled and executing an instruction.

Complete code compatibility with the MC68000 family allows the designer to draw on existing code and past experience to bring products to market quickly. There is also a broad base of established development tools, including real-time kernels, operating systems, languages, and applications, to assist in product design. The functionality provided by the MC68060 makes it the ideal choice for a range of high-performance computing applications as well as many portable applications that require low power and high performance.

Relative to the controllers, Motorola now offers the MC68EC040, 32-bit high-performance embedded controller. It is the performance leader for top-of-the-line embedded applications. The EC040 is capable of delivering 29 MIPS of sustained performance at 1.2 cycles per instruction with a system cost that is unattainable by competing architectures.

This impressive performance is a result of a six-level pipelined integer unit, independent four-way set-associated instruction and data caches, and a very high level of on-chip parallelism. The EC040 also supports multimaster and multiprocessor systems with bus snooping.

By integrating all these features into the EC040, the microprocessor is able to perform the vast majority of work on-chip, limiting external memory accesses to allow for higher system performance with less expensive DRAMs. The result is virtual immunity to the effects of memory wait states.

Sometimes, if a microprocessor is incorporated as part of a system such as a signal generator or communication equipment, the microprocessor section will be handled by a different design engineer, who takes care of the software. It is, however, important in the early stage that those two people coordinate their tasks, because even for fairly simple systems, where certain arithmetic processing, digital-to-analog converting, supplying some mathematical offsets, or other simple tasks are required, it may be wise initially to make the microprocessor portion flexible enough to address the synthesizer in a highly flexible way for testing. As

476 DIGITAL PLL SYNTHESIZERS

frequency synthesizers are being built, and a large number of frequencies are available, it becomes fairly difficult to test the performance of the synthesizer at all frequencies. If the microprocessor accepts a serial input for testing purposes, an automatic test system can be developed whereby the extremely large number of possible channels are being actually scanned and measured, and therefore the synthesizer can be evaluated. In multiloop synthesizers, there are some critical ranges where if the wrong tolerances in the various loops come together, loop stability may not be guaranteed, and therefore it may be a good idea to develop a test program and verification very early. This is the reason why these ideas have to be incorporated into the microprocessor and synthesizer design.

Let us now, however, after these more-or-less philosophical approaches, go back to some hard-core requirements, and we will use the Rohde & Schwarz SMS synthesized signal generator as a typical example of how to build an extremely flexible frequency synthesizer with the help of a microprocessor, without which it would not be possible. Because of the large amount of arithmetic and housekeeping involved, there is a distinct trade-off between software and microprocessor application versus discrete gates, decoding circuits, and switching of such a synthesizer system. Somewhere in the middle between the microprocessor and discrete decoding are the so-called programmable logic arrays, which are large-scale integrated circuits and, similar to ROMs, are customized for one particular application.

Typical examples of a microprocessor-controlled low cost synthesizer/signal generator are the Rohde & Schwarz SMS and SMS2 signal generators. As shown in Figure 5-69, it covers the frequency range from 400 kHz to 520 MHz, and with the help of a frequency doubler, can be extended to 1040 MHz. In the base frequency range, the frequency resolution is 100 Hz. Frequency stability and drift depend on the reference standard used. Figure 5-70 gives short, condensed specifications.

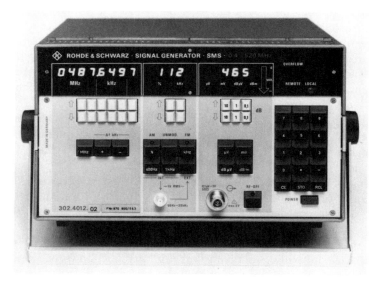

Figure 5-69 Photograph of the Rohde & Schwarz model SMS synthesized generator.

Specifications
Frequency

Frequency range, Model 22	0.4 to 520/1040 MHz	Option SMS-B2 see Specfications "Options"
Model 24	0.1 to 520/1040 MHz	

Frequency readout .... 8-digit LED display ; in MHz
  Resolution .... 100 Hz

Frequency error with reference oscillator

	Standard	Option
Aging	$< \pm 1 \times 10^{-6}$/month	$< \pm 5 \times 10^{-8}$/month
Temperature effect	$< \pm 1 \times 10^{-6}$/°C	$< \pm 1 \times 10^{-7}$ (5 to 45° C)
Warm-up period		15 min

Output/input for internal/external reference frequency, 10 MHz (single connector)
Output .... TTL level
Input .... > 0.5 V (sinewave) or TTL level

Spectral purity

Harmonics .... down ≥ 30 dBc[1])
Non-harmonic spurious
responses .... down ≥ 60 dBc[1]) (≥ 5 kHz from carrier)
Spurious deviation, rms
  0.3 to 3 kHz ... ≤ 4 Hz
                  (weighted in accordance with CCITT)
  0.3 to 20 kHz ≤ 16 Hz
Spurious AM, rms
  0.3 to 20 kHz .. down ≥ 70 dBc[1])
Single-sideband phase noise
(see also diagram below) ...... typ. down 120 dBc[1])
                  (test bandwidth 1 Hz; 20 kHz from carrier)
Single-sideband broadband
noise .... typ. down 145 dBc[1])
                  (test bandwidth 1 Hz; 1 MHz from carrier)

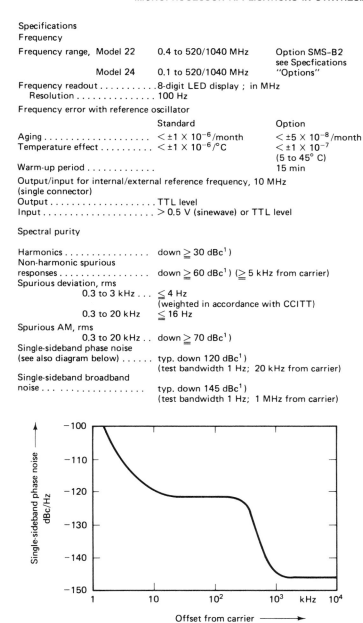

Typical single-sideband phase noise of signal generator SMS ($f_{carrier}$ = 360 MHz).

Figure 5-70 Specifications and plot of noise performance of the Rohde & Schwarz model SMS synthesized signal generator.

Figure 5-71 shows the simplified block diagram. FM modulation is accomplished by modulating the 80/40 MHz PLL, which is locked to a 10-kHz reference. Because of the narrow loop, the average frequency will be correct, but the actual frequency then depends on the modulation.

AM modulation is not shown but is accomplished with the help of a pin diode-modulator at the output. A 10-MHz frequency standard is used and can be stabilized against a more accurate external 10-MHz source.

Several auxiliary frequencies are required. Because of the modulation capability, the 80-MHz loop, which under certain conditions is divided down to 40 MHz, is mixed with a fixed frequency of 380 MHz.

Depending on the ranges, as can be seen in Table 5-3, certain mixing combinations of the 380 MHz and the 80 and 40 MHz are used. For instance, 380 MHz plus 80 MHz results in 460 MHz, 380 MHz plus 40 MHz equals 420 MHz, 380 MHz minus 80 MHz equals 300 MHz, and finally, 380 MHz minus 40 MHz equals 340 MHz. Range selection and filter switching are controlled by the microprocessor.

Several programmable dividers are used; the M/1 divider is found twice, and a somewhat complex formula can be developed to describe the output frequency as a function of divider ratio settings. Let us first take a look at the output. The basic frequency range is 100 kHz (special version) to 525 MHz, and a switchable frequency doubler that has compensation for the change in amplitude expands the frequency range up to 1040 MHz. The main range of 260 to 520 MHz is obtained by taking the output of either of the two switchable oscillators of the output loop and feeding this to the output of the signal generator. The range 130 to 260 MHz is achieved by using the divide-by-2 stage, as shown in the block diagram. The noise performance of this frequency range will therefore be 6 dB better than that of the main range. The third range, 400 kHz (100 kHz) to 130 MHz, is obtained by mixing the signal with the auxiliary 250–380 MHz output.

The output loop receives a variable input frequency of about 2 to 2.2 MHz at the input, and by selecting the proper frequency between 300 and 460 MHz, the output of the two oscillators is converted down to an internal IF of 20 to 26 MHz. The variable divider therefore has a division ratio between 10 and 30. Without this mixing, the division ratio would have been substantially larger, and therefore the noise multiplication inside the loop would have been much higher.

The 20–22 MHz loop acts as a translation loop and receives the two inputs A and B. Input B covers the frequency range 100 to 135 kHz, which is generated from the fine-resolution or interpolation oscillator, and input A is generated from the 50-kHz interpolation oscillator with an output frequency of 19.88 to 21.88 MHz. The programmable divider of the loop generating the A signal covers the range 796 to 1592, and as this signal is mixed into the translation loop, the noise contribution of this loop appears with only 10% at the output since the output of the translation loop is divided by 10 to drive the output loop.

The fine-resolution or interpolation loop, with the divider P ranging from 1000 to 3498, is divided by 250 and through the translation loop by 10, and therefore a total division ratio of 2500 is used. Thus, the noise contribution of this loop is extremely small. Only the 380-MHz and 80/40-MHz signals appear unchanged at the output and therefore determine the noise sideband performance. If the modulation requirement for the signal generator were to be omitted, the

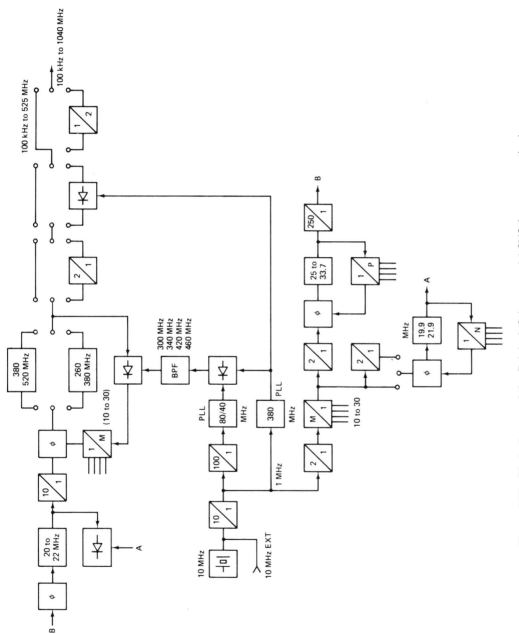

Figure 5-71 Block diagram of the Rohde & Schwarz model SMS frequency synthesizer.

Table 5-3 Microprocessor-selected frequency ranges in the Rohde & Schwarz SMS, used to obtain the proper output frequencies

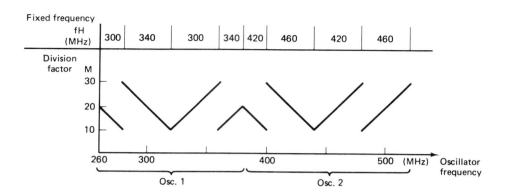

80/40-MHz stage could be replaced by a crystal oscillator, and therefore the noise could be improved there; the same is correct for the 380-MHz loop, where the noise contribution of the oscillator appears without any improvement of the loop.

As the output loop is being operated at a fairly narrow bandwidth, the ultimate signal-to-noise ratio is then determined by those two VCOs outside the loop bandwidth, as can be seen from Figure 5-10, showing the noise sideband performance of the oscillator.

Table 5-3 shows the relationship among the four fixed frequencies, the division factor M, and the frequency of the main oscillator. The output frequency of the synthesizer can be expressed as

$$F_0 = \left| \pm F_A + R\left(\frac{P}{10^4} + \frac{N}{20}\right) \right| \qquad (5\text{-}5)$$

$$F_A = \left.\begin{matrix}300,\ 340\\420,\ 460\end{matrix}\right\} \quad \begin{matrix}P = 1000 \text{ to } 3498\\N = 796 \text{ to } 1592\end{matrix}$$

$$R = 10\,\text{MHz}$$

Because so many frequencies change during the course of stepping this synthesizer through its range, it would be next to impossible to do this with discrete logic; and to obtain the same frequency range and the same performance with a mix-and-divide approach, such a large number of filters and auxiliary circuits would be required that it would not be feasible to build a device of the same performance. Because the SMS, in addition, offers a memory capacity where the signal generator remembers 10 channels and the settings for modulation and output level, the microprocessor also updates the memory chips and makes sure that, as certain frequencies are recalled from memory, the other settings are adjusted. Thanks to the microprocessor, these things are possible today.

Similar approaches can be used for microwave synthesizers, where as in the case of our four frequencies (300, 340, 420, and 460 MHz) either selected discrete frequencies or a special comb will be generated derived from the master standard. With the help of these down-conversion systems, it is possible to build frequency synthesizers up to 10 GHz or more, which provide excellent output phase noise. Figure 5-00 showed the phase noise of a microwave up/down-converter made by California Microwave. Here oscillators are locked against such combs using fairly wide bandwidth, and high-Q cavity oscillators ensure very low noise performance outside the loop bandwidth.

Recent developments in new material other than quartz produce resonators to be operated at frequencies up to several thousand megahertz with extremely high Q, and therefore low noise. Again, combinations of these techniques require the use of microprocessors.

One final word on 16/32-bit microprocessors: the 68000 series microprocessor is probably the most popular, being used by such companies as Hewlett-Packard, Tektronix, and Fluke as main processors. The other microprocessors seem to have applications in areas other than controllers, such as for test equipment or synthesizers.

5-5 TRANSCEIVER APPLICATIONS

Modern short-wave transceivers use the most advanced digital implementation both in the RF/IF section as well as in the synthesizer. These generally allow simplification of the frequency selection. In the case of this example, we will look at a short-wave transceiver (Model XK2100 by Rohde & Schwarz) that operates from 10 kHz to 1.5 MHz in receive and 1.5 to 30 MHz in transceive. The first IF

482 DIGITAL PLL SYNTHESIZERS

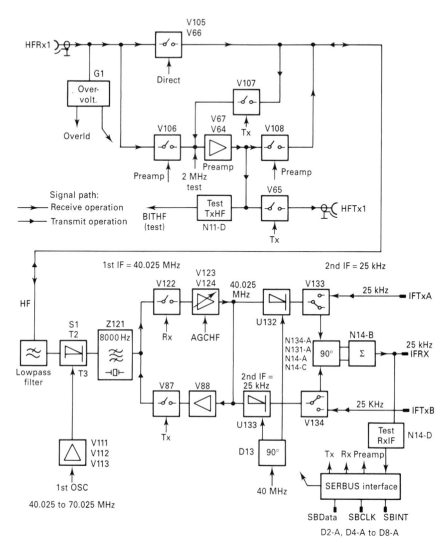

Figure 5-72 HF Unit, block diagram.

is 40.025 MHz and the second IF is 25 kHz using a digital signal processing (DSP) implementation. Therefore, the transceiver is a dual conversion system that requires a first oscillator, commonly referred to as a local oscillator (LO), which generates frequencies between 40.025 and 70.025 MHz in small steps such as 1 Hz. Also, an additional auxiliary frequency of 40 MHz is required to translate the first IF down to the 25-kHz second IF. Figure 5-72 shows the block diagram of the RF input down to the second IF and backward for transmit.

The synthesizer as shown in Figure 5-73 consists of various loops. The internal reference of 10 MHz or an external reference of 1, 5, or 10 MHz is used to generate the first auxiliary frequency of 40 MHz. A VCO operating at 40 MHz is locked against the 1-MHz reference. Outputs from its programmable divider are used to

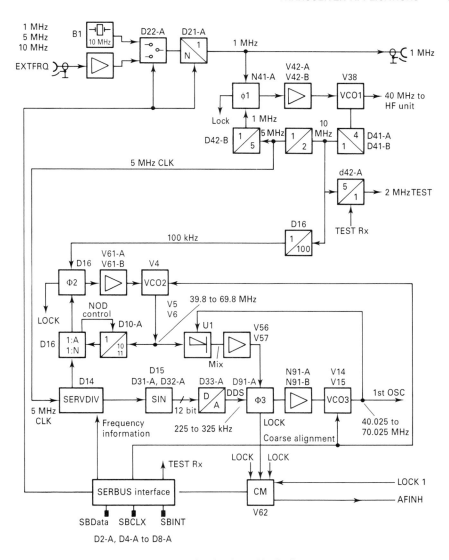

Figure 5-73 Synthesizer, block diagram.

obtain auxiliary frequencies such as 2 MHz, 5 MHz, and 100 kHz. The main RF synthesizer consists of two loops and a fine-resolution synthesizer. The output signal of VCO III, operating from 40.025 to 70.025 MHz is mixed with the output signal of VCO II, resulting in an internal synthesizer IF of 39.8 to 69.8 MHz. The interface controller marked "SERBUS-interface" provides frequency information to the SERVDEV unit, which is clocked by the 5-MHz reference and obtains the frequency information sent from the front panel. As part of the DDS, the sinus lookup table generates a fine-resolution signal for 10-kHz steps down to 1-Hz steps and provides this to Phase Detector III. At the same time, a D/A converter scheme is used for coarse presteering of both VCO II and VCO III. This two-loop synthesizer with an embedded DDS minimizes the division ratio. The loop around VCO II is stepped in 100-kHz increments.

An example of a frequency calculation is presented next. The receive or transmit frequency set to 12.34567 MHz requires a local oscillator frequency of 52.37067 MHz. The division factor for PLL II (VCO II) must be 123 + 398 = 521. VCO II therefore operates at 51.2 MHz. The DDS generates a frequency of 45.67 kHz + 225 kHz or 270.67 kHz. This occurs at the output frequency of VCO II of 52.1 MHz + 0.277067 MHz or the required 52.37067. In Chapter 6, we will look at a more advanced concept of a hybrid synthesizer.

Some of the tricky questions that must be addressed are the issue of switching speed, the combination of spurious products and shielding, and power consumption. In the hybrid synthesizer's case, seen in Chapter 6, we will see that instead of using the possible 0.007-Hz resolution, we will verify it to 1-Hz steps. The resolution is much finer than the customary 1 Hz and the microprocessor must now reduce the resolution and ensure the proper steps are selected. In this case, the DDS operating from 225 to 325 kHz is simply added to the auxiliary frequencies and therefore no offset calculation or compensation for multiplications has to be made. The penalty, however, is an overall more complex scheme. As new integrated circuits appear, some of these trade-offs will have to be reconsidered. We need to be reminded that in a mixing scheme, such as this, many noise sources are adding up and each mixer produces spurious signals. The choice of whether or not to use a DDS or fractional division N synthesizer is mostly determined by cost, power consumption, and shielding. In applications involving high-performance FM, there is some merit to modulating the division ratio, and this may result in a fractional division synthesis preference. Also, in many instances, established in-house technologies, short development times, and the ability to bring the product quickly to market determine its selection.

ACKNOWLEDGMENTS

I would like to again acknowledge the contributions of Texas Instruments, General Electric—Martin Marietta/Sanders, Rohde & Schwarz, and other industry sources, which assisted me in compiling the overview of Section 5-3.

REFERENCES

1. Eric G. Breeze, "High Frequency Digital PLL Synthesizer," *Fairchild Journal of Semiconductor Progress*, November–December 1977, Fairchild Semiconductor, Mountain View, CA, pp. 11–14.
2. W. Byers et al., "A 500 MHz Low-Noise General Purpose Frequency Synthesizer," *Proceedings of the Twentieth Annual Frequency Control Symposium*, U.S. Army Electronic Command, Fort Monmouth, NJ, 1973.
3. H. W. Cooper, "Why Complicate Frequency Synthesis?" *Electronic Design*, July 19, 1974, pp. 80–84.
4. W. F. Egan, "LOs Share Circuitry to Synthesize 4 Frequencies," *Microwaves*, May 1979, pp. 52–65.

5. R. Papaiech and R. Coe, "New Technique Yields Superior Frequency Synthesis at Lower Cost," *EDN*, October 20, 1975, pp. 73–79.
6. Racal Technical Manual RA6790, HF Receiver RCI 84244, Racal Communications, Inc., 5 Research Place, Rockville, MD 20850, June 1979, pp. 4-11 to 4-22.
7. P. G. Tipon, "New Microwave-Frequency Synthesizers that Exhibit Broad Bandwidths and Increased Spectral Purity," *IEEE Transactions on Microwave Theory and Techniques*, Vol. MTT-22, December 1974, p. 1251.
8. V. E. Van Duzer, "A 0–50 Mc Frequency Synthesizer with Excellent Stability . . .," *Hewlett-Packard Journal*, Vol. 15, No. 9, May 1964, pp. 1–6.
9. H. J. Finden, "The Frequency Synthesizer," *Journal of the IEEE*, Part III, Vol. 90, 1943, pp. 165–180.
10. V. Kroupa, "Theory of Frequency Synthesis," *IEEE Transactions on Instrumentation and Measurement*, Vol. IM-17, 1968, pp. 56–68.
11. O. Perron, *Irrationalzahlen*, 3rd ed., Walter de Gruyter, Berlin, 1947.
12. L. Essen, E. G. Hope, and J. V. L. Parry, "Circuits Employed in the N. P. L. Caesium Standard," *Proceedings of the IEE*, Vol. 106, Part B, 1959, pp. 240–244.
13. J. Holloway et al., "Comparison and Evaluation of Cesium Atomic Beam Frequency Standards," *Proceedings of the IRE*, Vol. 47, 1959, pp. 1730–1736.
14. W. E. Montgomery, "Application of Integrated Electronics to Military Communications and Radar Systems," *Proceedings of the IEEE*, Vol. 52, 1964, pp. 1721–1731.
15. J. Gerhold, "Dekadischer HF-Messender SMDH," *Neues von Rohde und Schwarz*, Vol. 8, January 1968, pp. 5–12.
16. B. M. Wojciechowski, "Theory of a Frequency-Synthesizing Network," *Bell System Technical Journal*, Vol. 39, May 1960, pp. 649–673.
17. M. Colas, "Le Stabilidyne," *L'Onde Electrique*, Vol. 36, February 1956, pp. 83–93.
18. H. Flicker, "Stand der Frequenzmesstechnik nach dem Überlagerungsverfahren," *Handbuch fur Hochfrequenz- und Electro-Techniker*, Vol. 6, Verlag für Radio-Foto-Kinotechnik GMBH, Berlin, 1960, pp. 349–392.
19. The Marconi Company, "Hydrus HF Receiver," *Telecommunications*, Vol. 2, July 1968, pp. 40–41.
20. M. Boella, "Generatore di frequenze campione per misure di alta precisione," *Alta Frequenza*, Vol. 14, September–December 1945, pp. 183–194.
21. R. Leonhardt and H. Flicker, "Eine Neuentwicklung: Dekadische Frequenzmessanlage 10 Hz bis 30 MHz mit Absolutkontrolle," *Rohde und Schwarz-Mitteilungen*, August 1952, p. 69.
22. R. J. Breiding and C. Vammen, "RADA Frequency Synthesizer," *Proceedings of the 21st Annual Frequency Control Symposium*, Fort Monmouth, NJ, 1967, pp. 308–330.
23. J. K. Clapp and F. D. Lewis, "A Unique Standard-Frequency Multiplier," *IRE National Convention Record*, 1957, Part 5, pp. 131–136.
24. M. L. Stitch, N. O. Robinson, and W. Silvey, "Parametric Diodes in a Maser Phase-locked Frequency Divider," *IRE Transactions on Microwave Theory and Technique*, Vol. MTT-8, March 1960, pp. 218–221.
25. R. Vessot et al., "An Intercomparison of Hydrogen and Caesium Frequency Standards," *IEEE Transactions on Instrumentation and Measurement*, Vol. IM-15, December 1966, pp. 165–176.
26. H. Valdorf and R. Klinger, "Die Entwicklung einer hochkonstanten dekadischen Kurzwellensteurstufe fur den Bereich 1,5 . . . 30 MHz," *Frequenz*, Vol. 14, October

1960, pp. 335–343.
27. L. Mooser, "Precision Offset Exciter Equipment XZO for Suppression of TV Common Channel Interference," *News from Rohde-Schwarz*, Vol. 7, May 1967, pp. 40–43.
28. R. J. Hughes and R. J. Sacha, "The LOHAP Frequency Synthesizer," *Frequency*, Vol. 6, August 1968, pp. 12–21.
29. J. Noordanus, "Frequency Synthesizers—A Survey of Techniques," *IEEE Transactions on Communication Technology*, Vol. COM-17, April 1969, pp. 257–271.
30. J. R. Woodbury, "Phase-Locked Loop Pull-in Range," *IEEE Transactions on Communication Technology*, Vol. COM-16, February 1968, pp. 184–186.
31. L. Sokoloff, "IC Voltage Variable Capacitors (VVC)," *IEEE Transactions on Broadcast and Television Receivers*, Vol. BTR-15, February 1969, pp. 33–40.
32. A. Noyes, Jr., W. F. Byers, and G. H. Lohrer, "Coherent Decade Frequency Synthesizers," a set of articles in *General Radio Experimenter*, September 1964, May 1965, November–December 1965, September 1966 (summarized in the G. R. Company reprint E119), May–June 1969, January–February 1970.
33. Adret-Electronique, "2 MHz Signal Generator-Synthesizer," Codasyn 201, Instruction Manual, Trappes, July 1969.
34. G. A. G. Rowlandson, "Frequency Synthesis Techniques," *Industrial Electronics*, Vol. 6, August 1968, pp. 320–323; and September 1968, pp. 355–359.
35. G. C. Gillette, "The Digiphase Synthesizer," *Frequency Technology*, Vol. 7, August 1969, pp. 25–29.
36. J. C. Shanahan, "Uniting Signal Generation and Signal Synthesis," *Hewlett-Packard Journal*, Vol. 23, December 1971, pp. 2–13.
37. R. L. Allen, "Frequency Divider Extends Automatic Digital Frequency Measurements to 12.4 GHz," *Hewlett-Packard Journal*, Vol. 18, April 1967, pp. 2–7.
38. A. Noyes, Jr., "The Use of Frequency Synthesizer for Precision Measurements of Frequency Stability and Phase Noise," *General Radio Experimenter*, Vol. 41, January 1967, pp. 15–21.
39. D. E. Maxwell, "A 5 to 50 MHz Direct-Reading Phase Meter with Hundredth-Degree Precision," *IEEE Transactions on Instrumentation and Measurement*, Vol. IM-15, December 1966, pp. 304–310.
40. R. L. Moynihan, "A Sweeper for GR Synthesizers," *General Radio Experimenter*, Vol. 41, January 1967, pp. 15–21.
41. Adret-Electronique, "Nouveu générateurs de signaux électriques programmables," *Note d'information 03*.
42. V. Kroupa, "An All-Band 'Single-frequency' Synthesizer," International Broadcasting Convention, London, *IEE Conference Publication No. 69*, September 1970, pp. 117–119.
43. G. J. McDonald and C. S. Burnham, "Review of Progress in Mercantile-Marine Radiocommunication," *Proceedings of the IEE*, Vol. 116, 1969, pp. 1807–1820.
44. A. Ruhrmann, "The Remotely Controlled Transmitter Center at Elmshorn," *Telefunken-Zeitung*, Vol. 35, December 1962, pp. 284–298.
45. D. W. Watt-Carter et al., "The New Leafield Radio Station," set of articles in *The Post Office Electrical Engineers' Journal*, Vol. 59, 1966, pp. 130–134, 178–181, 196–198, 267–270, 283–287.
46. "Unconventional Communications Receiver," *Wireless World*, Vol. 63, August 1957, pp. 388–389.
47. J. J. Muller and J. Lisimaque, "Portable Single-Sideband High-Frequency Transceiver

with Military Applications," *Electrical Communication*, Vol. 43, December 1968, pp. 360–368.
48. J. Gerhold and G. Pilz, "The EK 47—A Communications Receiver with Digital Tuning Facilities for the Range 10 kHz to 30 MHz," *News from Rohde-Schwarz*, Vol. 9, October–November 1969, pp. 8–12.
49. D. H. Throne, "A Report of the Performance Characteristics of a New Rubidium Vapor Frequency Standard," *Frequency Technology*, Vol. 8, January 1970, pp. 16–19.
50. Y. Yasuda, K. Yoshimura, and Y. Saito, "One of the Methods of Frequency Offsetting," *Journal of the Radio Research Laboratories* (Tokyo), Vol. 13, July–September 1966, pp. 211–225.
51. V. Kroupa, "Single-Frequency Synthesis and Frequency-Coherent Communications Systems," Conference on Frequency Generation and Control for Radio Systems, London, *IEEE Conference Publication No. 31*, May 1967, pp. 96–99.
52. R. Morrison, *Grounding and Shielding Techniques in Instrumentation*, Wiley, New York, 1967.
53. H. Ott, *Noise Reduction Techniques in Electronic Systems*, Wiley, New York, 1976.
54. H. P. Westman, ed., *Reference Data for Radio Engineers*, 5th ed., Howard W. Sams, Indianapolis, IN, 1968.
55. D. White, *Electromagnetic Interference and Compatibility*, Vol. 3, Don White Consultants Inc., 14800 Springfield Rd., Germantown, MA 20267, 1973, pp. 4.1–8.30, 10.1–12.14.
56. Rohde & Schwarz Operating and Repair Manual for the SMS Synthesizer.
57. Rohde & Schwarz Operating and Repair Manual for the EK070 Shortwave Receiver.
58. Operating Manual for the Model CV3594, California Microwave.
59. U.S. Patent 3,588,732, Robert D. Tollefson, Richardson, TX., June 28, 1971.
60. K. Fukui et al., "A Portable All-Band Radio Receiver Using Microcomputer Controlled PLL Synthesizer," *IEEE Transactions on Consumer Electronics*, Vol. CE-26, October 1980.
61. M. E. Peterson, "The Design and Performance of an Ultra Low-Noise Digital Frequency Synthesizer for Use in VLF Receivers," *Proceedings of the 26th Frequency Control Symposium*, 1972, pp. 55–70.
62. J. Tierney, C. M. Rader, and B. Gold, "A Digital Frequency Synthesizer," *IEEE Transactions on Audio and Electroacoustics*, Vol. AU-19, 1971, pp. 48–57.
63. B. E. Bjerede and G. D. Fisher, "A New Phase Accumulator Approach for Frequency Synthesis," *Proceedings of the IEEE NAECON '76*, May 1976, pp. 928–932.
64. J. Stinehelfer and J. Nichols, "A Digital Frequency Synthesizer for an AM and FM Receiver," *IEEE Transactions on Broadcast and Television Receivers*, Vol. BTR-15, 1969, pp. 235–243.
65. L. F. Blachowicz, "Dial Any Channel to 500 MHz," *Electronics*, May 2, 1966, pp. 60–69.
66. U. L. Rohde, "Modern Design of Frequency Synthesizer," *Ham Radio*, July 1976, pp. 10–22.
67. B. Bjerede and G. Fisher, "An Efficient Hardware Implementation for High Resolution Frequency Synthesis," *Proceedings of the 31st Frequency Control Symposium*, 1977, pp. 318–321.
68. J. Gibbs and R. Temple, "Frequency Domain Yields Its Data to Phase-Locked Synthesizer," *Electronics*, April 27, 1978, pp. 107–113.

69. J. Gorski-Popiel, *Frequency Synthesis: Techniques and Applications*, IEEE, New York, 1975.
70. V. F. Kroupa, *Frequency Synthesis: Theory, Design and Applications*, Wiley, New York, 1973.
71. V. Manassewitsch, *Frequency Synthesizers: Theory and Deisgn*, Wiley, New York, 1976.
72. W. F. Egan, *Frequency Synthesis by Phase Lock*, Wiley, New York, 1981.
73. L. Sample, "A Linear CB Synthesizer," *IEEE Transactions on Consumer Electronics*, Vol. CE-23, No. 3, August 1977, pp. 200–206.
74. I. Dayoff and B. Kirschner, "A Bulk CMOS 40-Channel CB Frequency Synthesizer," *IEEE Transactions on Consumer Electronics*, Vol. CE-23, No. 4, November 1977, pp. 440–446.
75. G. W. M. Yuen, "An Analog-Tuned Digital Frequency Synthesizer Tuning System for AM/FM Tuner," *IEEE Transactions on Consumer Electronics*, Vol. CE-23, No. 4, November 1977, pp. 440–446.
76. B. E. Beyers, "Frequency Synthesis Tuning Systems with Automatic Offset Tuning," *IEEE Transactions on Consumer Electronics*, Vol. CE-24, No. 3, August 1978, pp. 419–428.
77. T. B. Mills, "An AM/FM Digital Tuning System," *IEEE Transactions on Consumer Electronics*, Vol. CE-24, No. 4, November 1978, pp. 507–513.
78. K. Ichinose, "One Chip AM/FM Digital Tuning System," *IEEE Transactions on Consumer Electronics*, Vol. CE-26, August 1980, pp. 282–288.
79. T. Yamada, "A High Speed NMOS PLL-Synthesizer LSI with On-Chip Prescaler for AM/FM Receivers, *IEEE Transactions on Consumer Electronics*, Vol. CE-26, August 1980, pp. 289–298.
80. T. Rzezewski and T. Kawasaki, "A Microcomputer Controlled Frequency Synthesizer for TV," *IEEE Transactions on Consumer Electronics*, Vol. CE-24, No. 2, May 1978, pp. 145–153.
81. K. J. Mueller and C. P. Wu, "A Monolithic ECL/I^{2}L Phase-Locked Loop Frequency Synthesizer for AM/FM TV," *IEEE Transactions on Consumer Electronics*, Vol. CE-25, No. 3, August 1979, pp. 670–676.
82. G. d'Andrea, V. Libal, and G. Weil, "Frequency Synthesis for Color TV-Receivers with a New Dedicated μ-Computer," *IEEE Transactions on Consumer Electronics*, Vol. 27, 1981, pp. 272–283.
83. B. Apetz, B. Scheckel, and G. Weil, "A 120 MHz AM/FM PLL-IC with Dual On-Chip Programmable Charge Pump/Filter Op-Amp," *IEEE Transactions on Consumer Electronics*, Vol. 27, 1981, pp. 234–242.
84. K. Tanaka, S. Ikeguchi, Y. Nakayama, and Osamu Ikeda, "New Digital Synthesizer LSI for FM/AM Receivers," *IEEE Transactions on Consumer Electronics*, Vol. 27, 1981, pp. 210–219.
85. James A. Crawford, *Frequency Synthesizer Design Handbook*, Artech House, Norwood, MA, 1994.
86. Ulrich L. Rohde, "Low Noise Microwave Synthesizers, WFFDS: Advances in Microwave and Millimeter-Wave Synthesizer Technology," *IEEE MTT—Symposium*, Orlando, FL, May 19, 1995.

6

A HIGH-PERFORMANCE HYBRID SYNTHESIZER

The previous chapters have dealt with the design principles of frequency synthesizers and the effect that parameters have on the loop performance. It is impossible to show all details relevant to the design of frequency synthesizers, especially regarding the selection of components, PC board layouts, and which principle to use over another, as sometimes they are equally good and the choice is very difficult. Engineers typically want to reinvent everything themselves. This is not a very economical way to do research, and inasmuch as one relies on literature, it is also good to take a look at proven designs. A nonworking novel approach is more difficult to digest than looking at a reliable and working approach and trying to improve this and also to understand why it has been done the way it has been done.

In this chapter we will look at a high-performance hybrid synthesizer as a combination of most of the technologies we have analyzed thus far.

Frequency agile synthesizers can now be built easily, essentially by using off-the-shelf available integrated circuits. If very fine resolution is required, one either ends up with a multiloop synthesizer or a combination of a phase-locked loop (PLL) synthesizer and a fine-resolution loop. The two commonly used choices for fine resolution are fractional division N synthesizers and direct digital synthesizers (DDSs). Fractional division N synthesizers, as previously described, are usually found in test equipment and offer some distinct advantages due to their predictability of spurious sidebands. Also, they can be frequency modulated by a fast change of the division ratio. Conventional analog cancellation circuits, which are responsible for suppressing the unwanted sidebands, are both temperature and component sensitive. Therefore, a digital version of this is needed. For most communication circuits, where high-resolution synthesizers are required, the direct digital frequency synthesizers are attractive because all necessary components are located on the chip and a custom-tailored synthesizer using both a PLL and DDSs can easily be put together.

490 A HIGH-PERFORMANCE HYBRID SYNTHESIZER

To demonstrate the ease of application, I would like to show a high-performance hybrid synthesizer the uses a unique combination of generally available techniques. Such an implementation, for which a patent had been issued to Qualcom in the United States, had been done initially in 1979 in the Rohde & Schwarz XPC synthesizer. The interpolation synthesizer portion of the overall approach, shown in detail, has two chips. However, if the application is less cost sensitive, this can be replaced by a higher integrated version.

6-1 BASIC SYNTHESIZER APPROACH

Figure 6-1 shows an application example using the Motorola Series MC1451XX. Because of their low power consumption and silent operation mode capability,

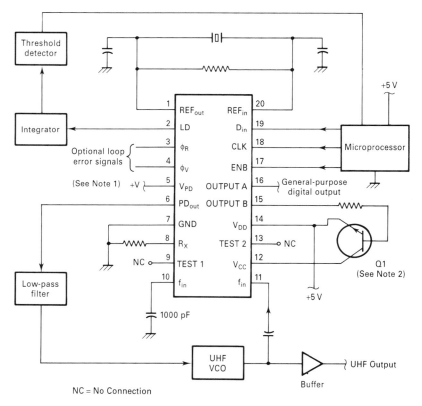

Figure 6-1 Example application. Notes: (1) When used, the ϕ_R and ϕ_V outputs are fed to an external combiner/loop filter. (2) Transistor Q_1 is required only if the standby feature is needed. Q1 permits the bipolar section of the device to be shut down via use of the general-purpose digital pin, OUTPUT B. If the standby feature is not needed, tie pin 12 directly to the power supply. (3) For optimum performance, bypass the V_{CC}, V_{DD}, and V_{PD} pins to ND with low-inductance capacitors. (4) The R counter is programmed for a divide value = REF_{in}/f_R. Typically, f_R is the tuning resolution required for the VCO. Also, the VCO frequency divided by $f_R - NT = N \times 64 + A$; this determines the values (N, A) that must be programmed into the N and A counters, respectively.

BASIC SYNTHESIZER APPROACH

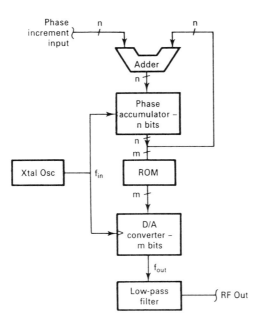

Figure 6-2 Block diagram of a DDS system.

these synthesizers allow the design of very powerful single-loop synthesizers. The required interface is fairly simple to handle and, of course, makes intensive use of microprocessors.

On the other hand, the required quasi-arbitrary resolution is obtained from the direct digital synthesizer. These DDSs are available through several sources and late arrivals, such as the Analog Devices model AD7008, and have a built-in digital/analog converter to provide the necessary output.

Figure 6-2 shows the functional block diagram of a DDS system. In analyzing both the resolution and signal-to-noise ratio (or rather signal to spurious performance) of the DDS, one has to know the resolution and input frequencies. As an example, if the input frequency is approximately 35 MHz and the implementation is for a 32-bit device, the frequency resolution compared to the input frequency is $35E6 \div 2^{32} = 35E6 \div 4.294967296E9$ or $0.00815 \, \text{Hz} \approx 0.01 \, \text{Hz}$. Given the fact that modern shortwave radios with a first IF of about 75 MHz will have an oscillator between 75 and 105 MHz, the resolution at the output range is more than adequate. In practice, one would use the microprocessor to round it to the next increment of 1 Hz relative to the output frequency.

As to the spurious response, the worst-case spurious response is approximately $20 \log = 2^R$, where R is the resolution of the digital/analog converter. For an 8-bit A/D converter, this would mean approximately 48 dB down (worst case), as the output loop would have an analog filter to suppress close-in spurious noise. In our application, we will use an 8-bit external D/A converter. However, devices such as the Analog Devices AD7008 DDS modulator have a 10-bit resolution, as shown in Figure 6-3. Ten bits of resolution can translate into $20 \log 2^{10}$ or 60 dB of suppression. The actual spurious response would be much better. The current

492 A HIGH-PERFORMANCE HYBRID SYNTHESIZER

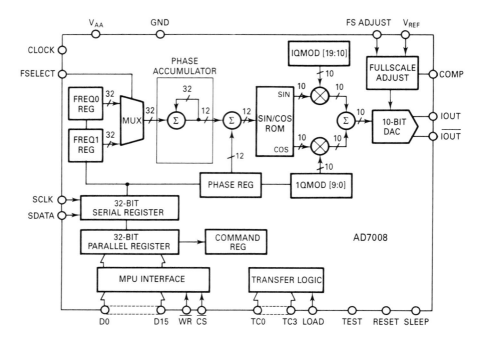

Figure 6-3 Functional block diagram of the Analog Devices AD7008 DDS modulator.

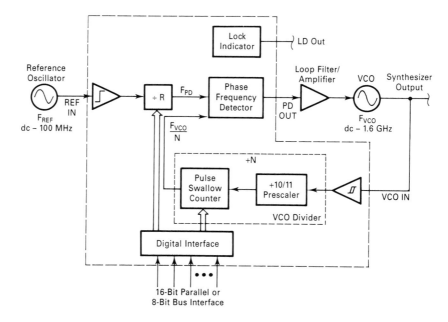

Figure 6-4 Block diagram of a single-loop PLL synthesizer showing all the necessary components for microwave and RF application.

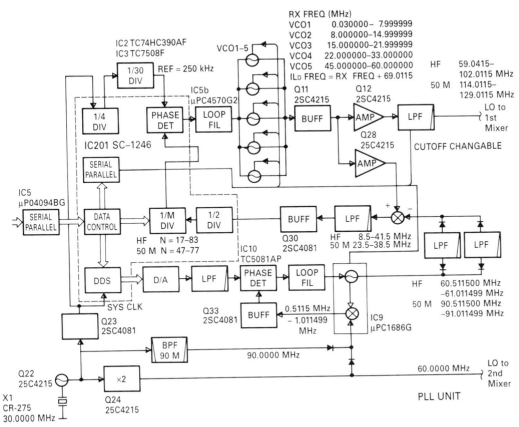

Figure 6-5 Synthesizer used in the ICOM IC 736, 6-meter transceiver. The IC 736 combines both the DDS and PLL approaches.

production designs for communication applications, such as shortwave transceivers, despite the fact that they are resorting to a combination of PLLs and DDSs, still end up somewhat complicated.

Figure 6-4 shows the necessary components of a single PLL system, which are hidden in the chip approach outlined in Figure 6-1. Figure 6-5 shows the combination of a standard PLL and a DDS, as implemented in the ICOM IC 736 HF/6m transceiver. This approach uses the DDS in a frequency range between 500 kHz and 1 MHz. This frequency gets converted up to either 60 MHz at the shortwave band or 90 MHz at the 6-meter ham band. The resulting frequency is used as an auxiliary frequency to convert the frequency of the first local oscillator (LO) (69.0415 to 102.0115 MHz) down to the synthesizer IF between 8.5 and 41.5 MHz. There is an additional divide-by-2 stage in the loop, which therefore requires a reference frequency of 250 kHz instead of 500 kHz. This is done to extend the operating range of the synthesizer chip, including its prescaler's capability of operating at much higher frequencies, although it does not have such a hybrid DDS incorporated (Figure 6-6). While this approach obtains a fairly small

494 A HIGH-PERFORMANCE HYBRID SYNTHESIZER

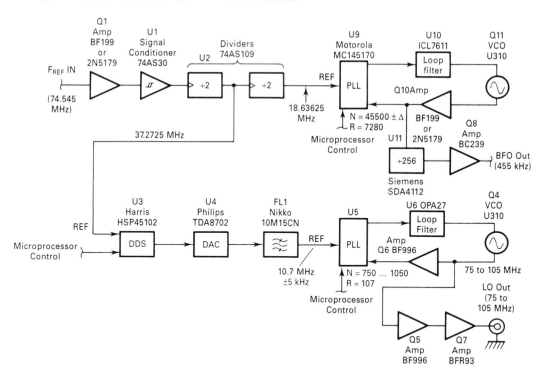

Figure 6-6 Hybrid synthesizer with output frequency of about 455 kHz, which provides the 75 to 105 MHz at approximately 0.01-Hz resolution. This synthesizer uses a combination of a standard PLL and DDS.

division ratio, it is still a four-loop synthesizer. One loop is the DDS itself. The second is the translator loop that mixes the DDS up to 60 MHz. The third is the main loop responsible for the desired output frequency. The fourth loop, so to speak, is the generation of the auxiliary frequencies of 60 and 90 MHz, which are derived from the 30-MHz frequency standard. For reasons of good phase noise, it employs a total of five VCOs. The fact that the division ratio varies between 80 and 17 also indicates that the loop gain will change considerably (Figure 6-7).

The 10.7-MHz signal from the crystal filter now goes to the single chip MC145170 shown in Figure 6-8, which contains all the necessary dividers and the phase/frequency discriminator. The operational amplifier is driven from a 28-V source and the negative supply of the OPA27 is connected to a voltage doubler, which receives its ac voltage from the synthesizer IC. This trick allows extension of the operating AGC voltage. The resistive filter following the op-amp is a spike suppression filter. The actual VCO consists of an arrangement of 2×6 tuning diodes BB805. The inductor is 92 nH and consists of four turns, with the taps on turns 2 and 3. The oscillator also has a clamping circuit as opposed to a diode, similar to a grid leakage current detector. This circuit provides the cleanest output from a phase noise point of view.

BASIC SYNTHESIZER APPROACH 495

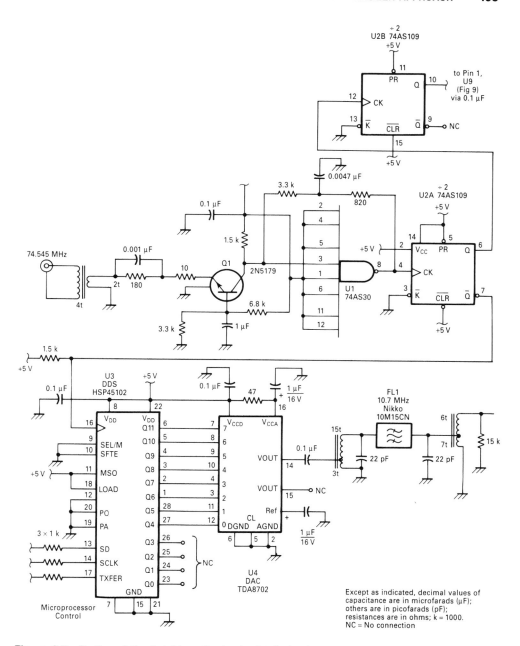

Figure 6-7 Portion of the hybrid synthesizer's detailed schematic. It takes 74.55 MHz from the second LO and drives the BFO, PLL, and DDS systems.

496 A HIGH-PERFORMANCE HYBRID SYNTHESIZER

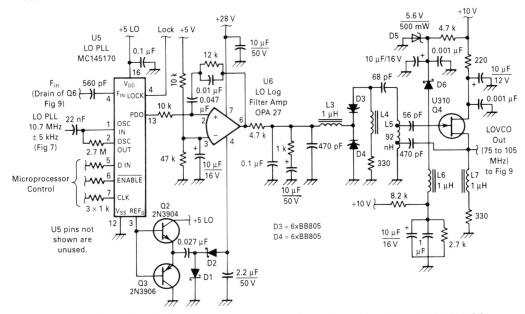

Figure 6-8 Single-loop PLL synthesizer of the main loop LO and the 75 to 105 MHz VCO.

The output from this VCO is then applied to a distribution amplifier system as shown in Figure 6-9. One dual-gate MOSFET provides the output for the PLL IC and the other dual-gate MOSFET drives a feedback stage, which, in turn, supplies 17 dBm output power for the first mixer. Both the BFO and the local oscillator (LO) synthesizer have their own regulator.

Finally, the BFO synthesizer, as shown in Figure 6-10, follows the same principal pattern since the voltage swing for the tuning diode can be much smaller: it operates off 10 V. Also, the BFO oscillator is much simpler in that its output gets divided down to 455 kHz, which is done by using a fixed divide-by-256 divider. Both the synthesizer chips and the DDS are driven by an appropriate microprocessor. The microprocessor system is then responsible for all the housekeeping activities.

6-2 LOOP FILTER DESIGN

The synthesizer uses a type 2 third-order loop, which is sufficient in both reference suppression and switching speed. In order to accomplish this, we have to look at both the free-running phase noise and the phase noise under closed-loop conditions. For the calculation of the open loop, we assume an equivalent noise resistor of R_n for the tuning diode of about 3 kΩ and a large signal-to-noise figure of the transistor of 10 dB. The definition of R_n was explained earlier. We will calculate the noise at 100-MHz frequency and assume a loaded resonator Q of 120. The flicker frequency noise, because of using an N-junction FET, is assumed to be 50 Hz. The resulting phase noise is -134 dBc/Hz. Since the VCO gain is 1 MHz per volt, it can be shown that the noise contribution is mostly from the

LOOP FILTER DESIGN 497

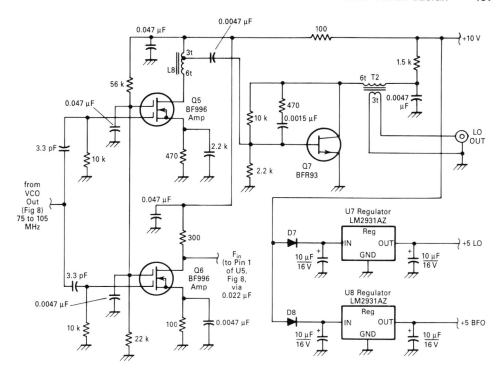

Figure 6-9 Isolation and driver for the first LO. Each synthesizer stage is driven by a separate regulated power supply.

tuning diode. A change in the Q value will have no contribution because the flicker noise of the tuning diode gets modulated on the oscillator. The phase noise for the free-running oscillator is around 134 dBc/Hz at 20 kHz off the carrier. Table 6-1 shows all the values used to calculate the SSB phase noise of the oscillator. This was done with Compact Software's PLL Design Kit.

The next step to consider is the difference between switching time and phase noise. Also, because of the up-multiplication of the phase noise into the loop, the loop frequency has to be carefully selected. As a first example, we set the natural loop frequency at 1 kHz; this results in a phase noise deterioration from 107 dBc/Hz down to about 90 dBc/Hz. Figure 6-11 shows the comparison between open- and closed-loop phase noise prediction. Note the overshoot around 1 kHz off the carrier. Also, because of the many dividers of the loop, the phase noise below 10 Hz increases dramatically. In order to improve the single-loop synthesizer, one has to allow for a loop bandwidth at around 300 Hz. Why is this important in practical use? It is important because the CW operation of a commercial receiver would be poorer in signal-to-noise ratio than with a different filter bandwidth. A re-run of the same analysis with a 300-Hz bandwidth shows a significant reduction of the phase noise compared to the previous example. Figure 6-12 shows the comparison between the open- and closed-loop phase noise

498 A HIGH-PERFORMANCE HYBRID SYNTHESIZER

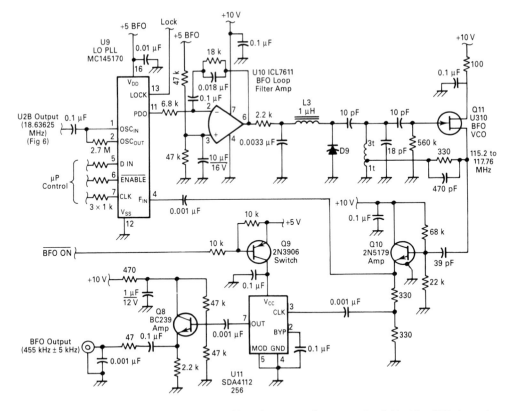

Figure 6-10 Single-loop BFO synthesizer. Note the output frequency is divided by 256 down to 455 kHz.

Table 6-1 Values used in calculation of SSB phase noise

Equivalent tuning-diode noise resistance	3000 Ω
Transistor noise figure	10 dB
RMS noise per SQR (1 Hz) bandwidth	7.04 nV
VCO gain in Hz/volt	1.E6
SSB noise at frequency offset	25.e3 Hz
Enter VCO center frequency	100 MHz
Loaded resonator Q	120
Flicker frequency (1 Hz to 100 MHz)	50
LO output power	1.0 mW
The phase noise in 25-kHz offset is −134 dBc/Hz	

prediction. Note the overshoot around 300 Hz off the carrier. This results in much less deterioration of the free-running oscillator.

The values for the active element have been computed using Compact Software's PLL Design Kit and Figure 6-13 shows the Bode diagram for the one-loop synthesizer.

Figures 6-14 and 6-15 show the spectral analysis plots of the first LO loop, at

LOOP FILTER DESIGN   499

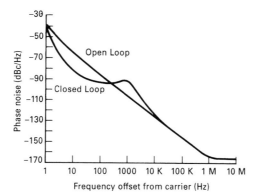

Figure 6-11 Comparison between open- and closed-loop noise prediction. Note the overshoot of around 1 kHz off the carrier.

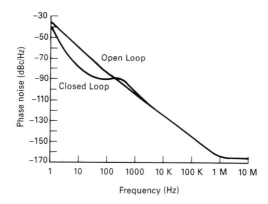

Figure 6-12 Comparison between open- and closed-loop noise prediction. Note the overshoot around 300 Hz off the carrier.

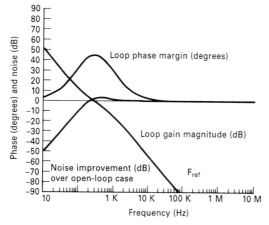

Figure 6-13 Bode diagram of a type 2 third-order loop. It is used in the main loop of our hybrid synthesizer. The predicted reference suppression is 90 dB.

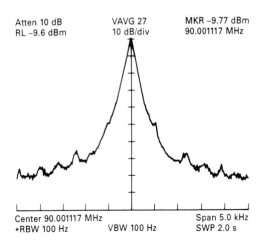

Figure 6-14 Spur analysis of the hybrid synthesizer in which there are no close-in discrete spurs within ±2 kHz.

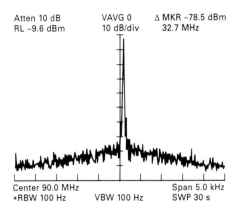

Figure 6-15 Spur analysis of the main loop synthesizer in the 100-kHz regime. Note that there are two discrete spurs approximately ±32 kHz and 78.5 dB down. These are due to the DDS contribution and other pick-ups.

a speed of ±2.5 kHz. We have not found any discrete spurs close in. There are two spurs located at approximately ±32 kHz, which seem to come from radiation and not directly from the DDS. We can then calculate the transient response, which is approximately 5 ms. (This is the time it takes the synthesizer to lock.)

The measured phase noise for the system in Figure 6-16 shows quite good agreement with the prediction outside the loop bandwidth. It shows a hump between 200 and 300 Hz at about 75 dBc/Hz and a phase noise of approximately 105 dBc/Hz at 1 kHz. The reason the phase noise values on the left side between 1 and 300 Hz differ from the measurement has to do with the fact that the designer does not have enough insight into all the noise contributions, including the one provided by the DDS system. The simulation is optimistic by approximately 10 dB,

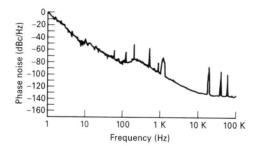

Figure 6-16 SSB phase noise of the hybrid frequency synthesizers, which is the topic of this discussion.

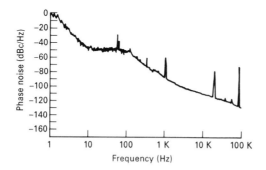

Figure 6-17 Phase noise of the English-made Lowe model HF150 single-loop synthesizer as it drives the first mixer. Its phase noise is significantly higher than the approach demonstrated here. Note that the discrete spurs below 100 Hz are small compared to previous measurements.

but for practical operating points, the phase noise at distances of 500 Hz off the carrier are still quite acceptable and can be tweaked by changing the phase of the filter. At 1 kHz off the carrier, the simulation is off by about 5 dB, meaning that the simulation is slightly too optimistic. At 10 kHz, the simulation is too pessimistic. At 20 kHz, the simulation agrees with the measurement, and further out, the measurement was limited by the test equipment. The measured area between 1 and 10 Hz is questionable because of the 50-dB jump in once decade. This area is referred to as random walk. The area between 10 and 100 Hz still has 40-dB decay, which is also on the high side, while the area between 100 Hz and 10 kHz seems reasonable. It is useful to compare this with another synthesizer approach. We look at the Lowe HF150 receiver's phase noise in Figure 6-17. It is significantly worse in all areas and also uses only a single VCO design. Figure 6-18 shows a Rohde & Schwarz multiloop synthesizer model SMK, which sold for more than $10,000. Its measured phase noise is not that far from the synthesizer approach used in the hybrid synthesizer detailed here.

After "polishing" the loop filter, Figure 6-19 shows the synthesizer phase noise of our hybrid frequency synthesizer. For those interested in the transient response of the synthesizer, Figure 6-20 shows the switching time of the synthesizer. After 5.11 ms we can estimate a locking error of less than 1°. There is also no ringing on the response curve. Finally, since the "power supply" is an important issue, Figure 6-21 shows a dc voltage supply for the synthesizer including the generation

502 A HIGH-PERFORMANCE HYBRID SYNTHESIZER

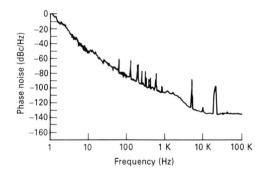

Figure 6-18 Measured SSB phase noise of the Rohde & Schwarz high-performance multiloop synthesizer model SMK. *Note:* In all cases, the reference oscillator was the Hewlett-Packard HP8662. Therefore, measurements above 20 kHz off the carrier were limited by the test setup. The actual phase noise further out may be better.

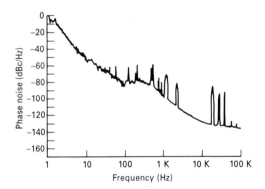

Figure 6-19 Measured phase noise of our hybrid synthesizer. The combination of the filters was optimized to reduce the overshoot, as shown in Figure 6-15. This "correction" has changed the far-out phase noise somewhat, and due to the higher VCO gain, the diodes are slightly more noisy.

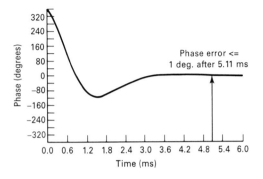

Figure 6-20 Switching time of the synthesizer. After 5.11 ms, phase/frequency error can be estimated to be less than 1°.

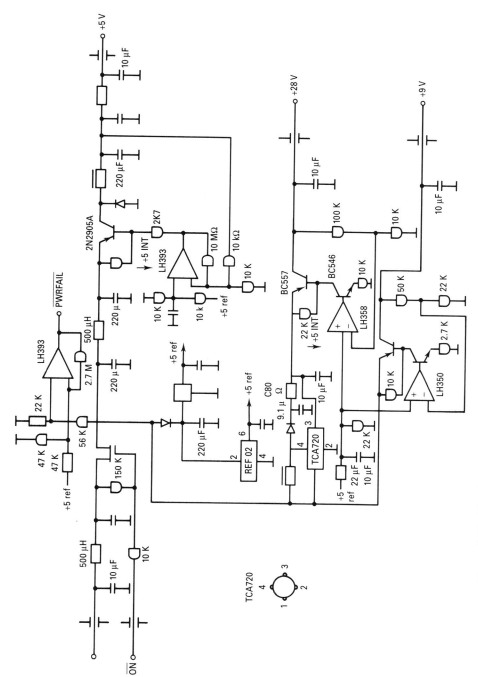

Figure 6-21 The dc power supply voltage for the synthesizer, including generation of the 28 V. It is based on an ITT TV IC, which generates a stabilized, regulated, and temperature-compensated voltage.

of the 28 V. It is based on an ITT TV IC, which generates a stabilized, regulated, and temperature-compensated voltage.

6-3 SUMMARY

In designing this hybrid synthesizer, a combination of two technologies has made it possible to design and build a high-performance synthesizer that meets today's requirement for purity and acceptable switching speed. The switching speed can be increased by a factor of 10 to 30 by going to a type 2 fifth-order loop at the expense of more components.

REFERENCES

1. Philips Semiconductors, *Integrated Circuits for Frequency Synthesizers*, May 1994.
2. J. Stilwell, Philips Semiconductors, "A Flexible Fractional-*N* Frequency Synthesizer for Digital RF Communication," *RF Design*, February 1993, pp. 39–43.
3. U. L. Rohde, "Key Components of Modern Receiver Design, Parts I, II and III," *QST*, May 1994, pp. 29–32; June 1994, pp. 27–30; July 1994, pp. 43, 45, respectively.
4. U. L. Rohde, "All About Phase Noise in Oscillators," *QEX*, December 1993, January 1994, February 1994.
5. Ulrich L. Rohde, "A High Performance Hybrid Synthesizer," *QST*, March 1995, pp. 30–38.

APPENDICES

The following appendices provide in-depth detail in several important areas. For those readers who would like to become acquainted with more of the mathematical details, Appendix A contains insight into the computations that form the foundation for the programs used to calculate loop response. In addition, I have prepared additional detailed information relative to the design of low-noise oscillators. Aside from providing information relative to the calculation of the tuning range of transmission line-based oscillators, I have included information relative to the amplitude stabilization mechanism. The stabilization of the oscillator is actually handled by modern computer-aided design (CAD) programs such as Microwave Harmonica by Compact Software, Inc.

Appendix B provides detailed insight into this area using CAD along the lines of the oscillator. We will also look into an electronic feedback system that dramatically improves the phase noise and a novel "push–push" oscillator, which has been proposed for various wireless applications. Appendix B includes a complete introduction to nonlinear phase noise analysis of oscillators with validation. This is not intended for duplication by engineers writing their own programs; its purpose is to explain this very modern and unique approach for solving problems. In essence, this is the exact solution for the linearized model developed in 1966 by Leeson.

Appendix C contains a reproduction of two state-of-the-art synthesizer data sheets for wireless applications. The current trend is toward smaller, faster, and less expensive devices. However, the basic application will not change. Again, the three technologies in this area are (1) single-loop synthesizers based on digital PLL detectors, (2) hybrid synthesizers involving DDS or computations including fractional division N, and (3) MMIC-based synthesizers. The fractional division N synthesizer approach is of particular interest because of its clean output. It is the most difficult to describe in light of current patent situations and the need for large custom-made gate array circuits.

Appendix D provides additional insight into MMIC-based synthesizers. The information is a compilation of personal communications with and publications by Dr. Takashi Ohira of NTT Radio Communication Systems Labs in Japan. The basic VCO shown in Figure D-1 uses a transmission line resonator.

APPENDIX A

MATHEMATICAL REVIEW

A-1 FUNCTIONS OF A COMPLEX VARIABLE

Definition:

$$\sqrt{-1} = i \text{ or } j \qquad i^2 = -1 \quad j^2 = -1$$

i or j is also used to indicate reactive components in electrical circuits; i is used in nonelectrical work. A complex number such as

$$p = x + jy \qquad (A\text{-}1)$$

or

$$p = 4 + j5$$

can be shown in the complex plane as in Figure A-1. This is called the *rectangular form*. The magnitude M, as well as the direction or angle θ, can be computed from

$$M = \sqrt{x^2 + y^2} \qquad (A\text{-}2)$$

or

$$M = \sqrt{16 + 25} = 6.40$$

and

$$\theta = \arctan \frac{y}{x} \qquad (A\text{-}3)$$

MATHEMATICAL REVIEW

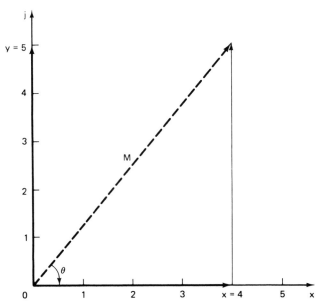

Figure A1 Complex plane showing a complex number as the sum of two components.

or

$$\theta = \arctan \tfrac{5}{4} = 51.34°$$

This is called the *polar form*. The conversion from one form to the other is achieved by

$$x = M \cos \theta \qquad \text{(A-4)}$$

$$y = M \sin \theta \qquad \text{(A-5)}$$

and

$$p = M(\cos \theta + j \sin \theta) \qquad \text{(A-6)}$$

This is the same as

$$p = M \exp(j\theta) \qquad \text{(A-7)}$$

as

$$p = M(\cos \theta + j \sin \theta)$$

We can expand both $\sin \theta$ and $\cos \theta$ into a series.

$$\sin \theta = \theta - \frac{\theta^3}{3!} + \frac{\theta^5}{5!} - \frac{\theta^7}{7!} + \cdots \qquad |x| < \infty, \quad x = \text{radians} \qquad \text{(A-8)}$$

$$\cos \theta = 1 - \frac{\theta^2}{2!} + \frac{\theta^4}{4!} - \frac{\theta^6}{6!} + \cdots \qquad |x| < \infty, \quad x = \text{radians} \qquad \text{(A-9)}$$

Table A-1 Operations for complex mathematics

Rectangular to Polar

100 R = SQR(A*A + B*B)
110 S = 0
120 IF R = 0 Then 170
130 S = ACS(A/R)
140 IF B < 0 Then 170
150 B = S
160 Go to 180
170 B = −S
180 A = R
190 Return

Note: X,Y are positional parameters
X − A
Y − B
A/B → Input
R/S → Output → A/B

Polar to Rectangular

100 A = R*COS(S)
110 B = R*SIN(S)
120 R = A
130 S = B
140 Return

Note: R,S are positional parameters

R − R
S − S

Multiplication

100 A3 = A1*A2 − B1*B2
110 B3 = A1*B2 + A2*B1
120 Return

Arrays A(1,2)B(1,2)C(1,2)

Note: A,B,C are positional parameters
A(1,1) − A1 R
A(1,2) − B1 I rectangular
B(1,1) − A2 R
B(1,2) − B2 I rectangular
C(1,1) − A3 Real ⎫ Product
C(1,2) − B3 Imag ⎭ rectangular

Division

100 B3 = A2*A2 + B2*B2
110 IF B3 < > Then 140
120 Print "Error − Denominator = 0"
130 End
140 A3 = (A1*A2 + B1*B2)/B3
150 B3 = (A2*B1 − A1*B2)/B3
160 Return

Arrays A(1,2)B(1,2)C(1,2)

Note: A,B,C are positional parameters

A(1,1) − A1 ⎫
A(1,2) − B1 ⎬ both inputs
B(1,1) − A2 ⎨ rectangular
B(1,2) − B2 ⎭
C(1,1) − A3 ⎫ output
C(1,2) − B3 ⎭ rectangular

Table A-1 *(Continued)*

Complex Number Raised to a Complex Power

```
100 A = A1
110 B = B1
120 GOSUB 240
130 IF R = 0 Then 210
140 R = LOG(R)
150 Z1 = A2*R − B2*S
160 Z2 = A2*S + B2*R
170 Z1 = Exp(Z1)
180 A3 = Z1*COS(Z2)
190 B3 = Z1*SIN(Z2)
200 Return
210 A3 = 0
220 B3 = 0
230 Return
240 R = SQR(A*A + B&B)
250 S = 0
260 IF R = 0 Then 300
270 S = ACS(A/R)
280 IF > B = >0 Then 300
290 S = − S
300 Return
```

Arrays A(1,2) B(1,2) C(1,2)

Note: A,B,C are positional parameters

A(1,1) − A1 ⎫
A(1,2) − B1 ⎬ both inputs
B(1,1) − A2 ⎨ rectangular
B(1,2) − B2 ⎭

C(1,1) − A3 ⎫ output
C(1,2) − B3 ⎬ rectangular

Logarithm of a Complex Number to a Complex Base

```
100 GOSUB 190
110 IF R1 < >0 or R2 < >0 Then 140
120 Print "Error − Complex No. = 0"
130 End
140 R1 = LOG(R1)
150 R2 = LOG(R2)
160 R3 = R1/R2
170 S3 = S1 − S2
180 Return
190 R1 = SQR(A1*A1 + B1*B1)
200 R2 = SQR(A2*A2 + B2*B2)
210 S1 = 0
220 S2 = 0
230 IF R1 = 0 or R2 = 0 Then 290
240 S1 = ACS(A1/R1)
250 S2 = ACS(A2/R2)
260 IF B1 = >0 or B2 = >0 Then 290
270 S1 = −S1
280 S2 = −S2
290 Return
```

Arrays A(1,2) B(1,2)

Note: A,B,X,Y are positional parameters

A(1,1) − A1   X − R3
A(1,2) − B1   Y − S3
B(1,1) − A2
B(1,2) − B2

Both inputs: rectangular
Output: polar

Table A-1 *(Continued)*

Sinh

100 B2 = Exp(−A1)
110 A2 = −0.5*COS(B1)*(B2 − 1/B2)
120 B2 = 0.5*SIN(B1)*(B2 + 1/B2)
130 Return

Arrays A(1,2) B(1,2)

Note: A,B are positional parameters
A(1,1) − A1
A(1,2) − B1
B(1,1) − A2
B(1,2) − B2

Cosh

100 B2 = Esp (A1)
110 A2 = 0.5*COS(B1)*(1/B2 + B2)
120 B2 = −0.5*SIN(B1)*(1/B2 − B2)
130 Return

Arrays A(1,2) B(1,2)

Note: A,B are positional parameters
A(1,1) − A1
A(1,2) − B1
B(1,1) − A2
B(1,2) − B2

Tanh

100 Z1 = Exp(2*A1)
110 Z2 = 1/Z1
120 B2 = (Z1 + Z2)*0.5 + COS(2*B1)
130 IF ABS(B2) > 1.0E − 12 Then 160
140 Print "Error − TANH is infinite"
150 End
160 A2 = (Z1 − Z2)*0.5/B2
170 B2 = SIN(2*B1)/B2
180 Return

Arrays A(1,2) B(1,2)

Note: A,B are positional parameters

A(1,1) − A1 Input
A(1,2) − B1 rectangular
B(1,1) − A2 Output
B(1,2) − B2 rectangular

Arc sinh

100 A2 = (1 − B1)*(1 − B1) + A1*A1
110 B2 = SQR(A2 + 4*B1)
120 A2 = SQR(A2)
130 Z1 = 0.5*(A2 + B2)
140 Z2 = 0.5*(A2 − B2)
150 B2 = −ASN(Z2)
160 A2 = LOG(Z1 + SQR(ABS(Z1*Z1 − 1)))
170 Return

Arrays A(1,2) B(1,2)

Note: A,B are positional parameters

A(1,1) − A1
A(1,2) − B1
B(1,1) − A2
B(1,2) − B2

Table A-1 *(Continued)*

Arc cosh

100 GOSUB 150	Arrays A(1,2) B(1,2)
110 Z2 = A2	
120 A2 = −B2 Note:	A,B are positional parameters
130 B2 = Z2	
140 Return	A(1,1) − A1
150 A2 = (A1 + 1)*(A1 + 1) + B1*B1	A(1,2) − B1
160 B2 = SQR(A2 − 4*A1)	B(1,1) − A2
170 A2 = SQR(A2)	B(1,2) − B2
180 Z1 = 0.5*(A2 + B2)	
190 Z2 = 0.5*(A2 − B2)	
200 A2 = ACS(Z2)	
210 B2 = −LOG(Z1 + SQR(ABS(Z1*Z4 − 1)))	
220 Return	

Arc tanh

100 A2 = A1*A1 + B1*B1	Arrays A(1,2) B(1,2)
110 IF A2 < > 1 Then 170	
120 IF ABS(A1) < > Then 150	
Note:	A,B are positional parameters
130 Print "Error − ARCTANH not defined for Complex No. =1 or −1"	A(1,1) − A1
	A(1,2) − B1
140 End	B(1,1) − A2
150 B2 = PI/4	B(1,2) − B2
160 Go to 180	
170 B2 = 0.5*ATN(2*B1/(1 − A2))	
180 A2 = −0.25*LOG((A2 − 2*A1 + 1)/(A2 + 2*A1 + 1))	
190 Return	

Using Euler's theorem, it can be shown that adding the power series together results in the expansion series for e^θ.

Some other useful conversion equations are

$$\sin\theta = \frac{y}{\sqrt{x^2 + y^2}} = \frac{\exp(j\theta) - \exp(-j\theta)}{2j} = \text{Im}\{\exp[j(\theta)]\} \quad \text{(A-10)}$$

$$\cos\theta = \frac{x}{\sqrt{x^2 + y^2}} = \frac{\exp(j\theta) + \exp(-j\theta)}{2} = \text{Re}\{\exp[j(\theta)]\} \quad \text{(A-11)}$$

The FORTRAN computer language allows the use of complex mathematics, whereas the BASIC language does not permit such easy conversions.

Table A-1 lists operations that are useful in dealing with complex mathematics.

A-2 COMPLEX PLANES

The complex number

$$p = x + jy$$

was pictured as a point located x units to the right and y units up from the zero point. Any complex number whose real and imaginary parts are given can be located as a point in the xy-plane. It has become conventional in mathematics to call the variables x and y and call z the resulting complex number. As this book deals mainly with engineering problems rather than mathematics, we have substituted p for z.

Any complex number z consists of a real part x and an imaginary part y and occupies a definite point in the complex plane. The particular plane used to plot z values will therefore be called the *z-plane* or *complex impedance plane*. If we had chosen to stay with the previously used p, we could have called this the p-plane.

As we have more values of x and y as they are connected together with a mathematical function, we will see that any line in the z plane is actually a chain of connective points. If the function $F(x,y)$ is known, the graph of the function can be presented.

In generating maps, several methods of projection are used, some of which are to show the real distance between any given points and some of which are chosen to give the correct surface of an area. Depending on the projection used, the resulting image looks different while still providing the same basic information. If one is not familiar with a particular area because of the difference in projection methods, it will seem that there is no similarity, and the same area can look so different that it is hard to think of the two projections as being the same.

Apparently, it is desirable to have more than one particular projection or plane. We can now use this assumption in mathematics, and we will create an additional plane, the *s-plane*. The *s*-plane will be drawn with vertical and horizontal lines in the same way as is done for the *z*-plane, but the coordinates of the point locations will be given as σ units to the right or left and ω units up or down from the zero reference. In mathematics, the letters u and v are used at times, but since we want to solve engineering problems rather than mathematics, and since this is only a question of whether or not one agrees to a certain abbreviation, we will use the term "*s*-plane" and stay with the nomenclature, since this technique will be used for the Laplace transformation, where it is used this way. We therefore define

$$s = \sigma + j\omega \qquad (A\text{-}12)$$

By using the identity

$$s = f(z) \qquad (A\text{-}13)$$

we can transform one plane to the other once we know the particular function.

Let us try a simple example. If

$$s = \frac{1}{z} \tag{A-14}$$

or

$$s = \frac{1}{x+jy} = \frac{x-jy}{(x+jy)(x-jy)} \tag{A-15}$$

and

$$s = \sigma + j\omega = \frac{x-jy}{x^2+y^2} \tag{A-16}$$

then

$$\sigma = \frac{x}{x^2+y^2} \qquad \omega = \frac{-y}{x^2+y^2} \tag{A-17}$$

This transform was fairly simple and straightforward and for any given pair of values for x and y we can find the corresponding σ and ω values.

An example using an LC oscillator is given below.

Functions in the Complex Frequency Plane. The frequency response of a network or the steady-state response to a sinusoidal input is directly related to the transfer function of the network. It is important to make sure that the waveform of the signal applied to the electrical circuit is really sinusoidal. The steady-state response assumes sinusoidal waveforms, and an analysis of the response to a nonsinusoidal waveform is better analyzed with the mathematical aid of the Fourier series and integral, which then leads to the Laplace transformation and the inverse Laplace transformation. The following discussion covers the transfer functions of several useful networks with $s = j\omega$ and $\sigma = 0$. Probably the most interesting transfer characteristics for our phase-locked loop applications are the ones for the simple RC network and the ones for the compensated RC network. The simple RC network shown in Figure A-2 is described by

$$F(s) = \frac{1}{sCR + 1} \tag{A-18}$$

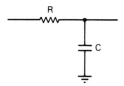

Figure A-2 Simple RC network.

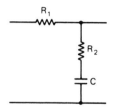

Figure A-3 RC lag filter.

The magnitude of the frequency response is

$$|F(j\omega)| = \frac{1}{\sqrt{(1+\omega^2 R^2 C^2)}} \quad \text{(A-19)}$$

and the phase is

$$\theta(\omega) = -\arctan(\omega C R) \quad \text{(A-20)}$$

For the type 2 second-order loop with an active filter, the RC lag network shown in Figure A-3 is commonly used. Its frequency response is

$$F(j\omega) = \frac{1 + j\omega R_2 C}{1 + j\omega C(R_1 + R_2)} \quad \text{(A-21)}$$

and the magnitude of the frequency response is

$$|F(j\omega)|^2 = \frac{1 + \omega^2 R_2^2 C^2}{1 + \omega^2 C^2 (R_1 + R_2)^2} \quad \text{(A-22)}$$

The phase is

$$\theta(\omega) = \arctan(\omega R_2 C) - \arctan[\omega C(R_1 + R_2)] \quad \text{(A-23)}$$

The phase and frequency response are sketched in Figure A-4.

The general transfer characteristics of the networks we are dealing with are defined as

$$F(s) = \frac{A(s)}{B(s)} \quad \text{(A-24)}$$

and $F(s)$ is the ratio of two polynomials in s. In an expanded form this reads

$$F(s) = \frac{a_m s^m + a_{m-1} s^{m-1} + \cdots + a_0}{b_n s^n + b_{n-1} s^{n-1} + \cdots + b_0} \quad \text{(A-25)}$$

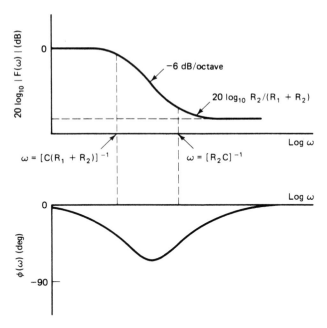

Figure A-4 Phase and frequency response of the lag filter in Figure A-3.

$m < n$ is a practical network. A polynomial may be factored and expressed as a product of binomials.

$$F(s) = \frac{a_m(s - z_m)(s - z_{m-1}) \cdots (s - z_1)}{b_n(s - p_n)(s - p_{n-1}) \cdots (s - p_1)} \quad \text{(A-26)}$$

Roots of the numerator are called *zeros*, whereas roots of the denominator are called *poles*. A zero occurs at a frequency where no power is transmitted through the complex network; a pole occurs at a frequency where no power is absorbed by the network. There are m zeros and n poles. The network is said to be of nth order; the order is equal to the number of poles, which is the same as the degree of the denominator.

Now let us try an example. Figure A-5 shows a tuned parallel circuit consisting of the capacitor C, the inductance L, the loss resistor R, and the negative resistor R_n, which is generated by an amplifier as dealt with in Chapter 4. The equation for this can be written

$$I = V \frac{1 + sCR + s^2LC}{R - R_n + s(L - CRR_n) - s^2 R_n CL} \quad \text{(A-27)}$$

The denominator provides us with the characteristic equation, which will be set to zero.

$$R - R_n + s(L - CRR_n) - s^2 R_n CL = 0 \quad \text{(A-28)}$$

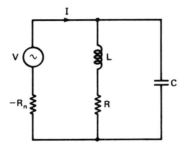

Figure A-5 Tuned circuit with negative resistor.

This is a quadratic equation, and its roots are

$$p_{1,2} = \frac{L - CRR_n}{2R_n CL} \pm \sqrt{\left(\frac{L - CRR_n}{2R_n CL}\right)^2 + \frac{R - R_n}{R_n CL}} \qquad \text{(A-29)}$$

and as $s = \sigma \pm j\omega$, we finally obtain

$$\sigma = -\frac{R}{2L} + \frac{1}{2R_n C} \qquad \text{(A-30)}$$

and

$$\omega_0 = \sqrt{\frac{1}{LC} - \frac{R}{R_n LC} + \sigma^2} \qquad \text{(A-31)}$$

We can define this result in three cases:

1. $\sigma > 1$; any initial oscillation will cease rapidly.
2. $\sigma = 0$; this is the case of a lossless circuit, in which the losses generated by R are exactly compensated by R_n.
3. $\sigma < 1$; in this case we have oscillation that will grow in amplitude until some saturation or limiting effect occurs in the device that produces the negative resistance. Circuits of this kind are called *negative resistance oscillators*.

Let us return now to the general transfer characteristic of the system and determine the stability from the Bode diagram as a graphical method rather than from the characteristic equation of the system under closed-loop conditions, because in complicated systems this will become very difficult. In Chapter 1 we analyzed higher-order loops. However, we will now think what the transfer characteristic of a network in the form of a polynomial expression can be. For phase-locked-loop circuits this function would describe an nth-order PLL.

To review: roots of the numerator are called zeros and roots of the denominator are called poles. There are m zeros and n poles. The network is said to be of nth order. The order is equal to the number of poles, which is the same as the degree

518 MATHEMATICAL REVIEW

of the denominator. What does this mean for phase-locked-loop circuits? Phase-locked-loop circuits are generally categorized into systems of a certain type and of a certain order. The type of phase-locked loop indicates the number of integrators, and as we have seen before, a type 1 first-order loop is a loop in which the filter is omitted and the loop therefore has only one integrator, the VCO. For good tracking, a large dc gain is needed and as the type 1 first-order loop has no filter, the bandwidth also must be large. It is apparent that narrow bandwidth and good tracking are incompatible for first-order loops—the principal reason why they are not used very often.

This network has only one pole. If we use a phase-locked loop with an active integrator, we now cascade two integrators, the VCO and the active integrator, and the loop automatically becomes a type 2 loop. Depending on the filter, we will have a type 2 second-order, third-order, or up to nth-order system. The consequences of the higher order are explained in the next section, where we deal with stability and use the Bode diagram to analyze the stability. On very rare occasions, loops with three integrators have been built, but since they find no application in frequency synthesizers, they are not dealt with here.

A-3 BODE DIAGRAM

In Chapter 1, in dealing with the question of stability, the Bode diagram was used. It is an aid to determining loop stability by plotting the amplitude and phase characteristic of the transfer function of a system and applying several criteria.

There are several ways in which stability can be analyzed. The Nyquist stability analysis requires a fairly large amount of calculation, and as most of the information available about phase-locked loops is based on an approximation, it is difficult to obtain all the necessary information.

Considerable information about the behavior of a phase lock can be obtained by determining the location of poles in the closed-loop response. These poles change their locations as the loop gain changes. The path that the pole traces in its migrations in the s-plane is known as the root-locus plot. This method again requires substantial mathematical effort because the roots of the denominator have to be determined with a digital computer. We will see that applying the Bode diagram is a fairly easy and convenient way of forecasting the stability of a loop by analyzing the open-loop gain.

First let us take a look at Figure A-6, which contains all the necessary loop components.

The loop, according to this block diagram, consists of the VCO, a dc amplifier with gain K_2, two lag filters called F_3 and F_4, which are determining τ_1 and τ_2, and the two cutoff frequencies F_6 and F_7. F_6 refers to the cutoff frequency of the operational amplifier used for the active filter, and F_7 is the 3-dB bandwidth that is generated by possible series resistors and bypass capacitors in the system. These various frequencies allow the simulation of influences as they actually occur. In addition, we saw in Chapter 1 that it is possible to use elliptic LC filters for high attenuation, and we also dealt with them in Section 4-7. The Cauer or elliptic low-pass filter can be described by providing the poles and zeros, the order, and the cutoff frequency of the filter. M refers to the order, F_5 refers to the cutoff

CHECK OF INPUT STATEMENTS : FIRST ORDER LOOP

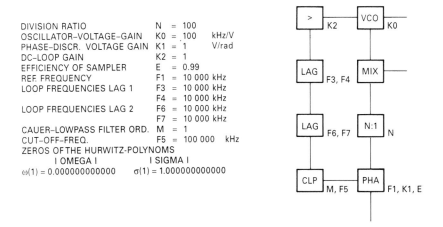

DIVISION RATIO	N	=	100	
OSCILLATOR-VOLTAGE-GAIN	K0	=	100	kHz/V
PHASE-DISCR. VOLTAGE GAIN	K1	=	1	V/rad
DC-LOOP GAIN	K2	=	1	
EFFICIENCY OF SAMPLER	E	=	0.99	
REF. FREQUENCY	F1	=	10 000	kHz
LOOP FREQUENCIES LAG 1	F3	=	10 000	kHz
	F4	=	10 000	kHz
LOOP FREQUENCIES LAG 2	F6	=	10 000	kHz
	F7	=	10 000	kHz
CAUER-LOWPASS FILTER ORD.	M	=	1	
CUT-OFF-FREQ.	F5	=	100 000	kHz

ZEROS OF THE HURWITZ-POLYNOMS
| OMEGA | | SIGMA |
$\omega(1) = 0.000000000000$ $\sigma(1) = 1.000000000000$

Figure A-6 Block diagram of a universal phase-locked loop system used in the computer program for high-order phase-locked loops.

frequency, and ω and σ refer to the transfer function of the filter. In order to be able to describe a complex system, we will allow for a mixer, and we also have the divider and the phase detector included.

In our first example, we are trying to simulate a first-order loop. The first-order loop has no active integrator and the loop bandwidth is determined by $K_0 K_\theta / N$; all other values, F_1 and F_7, are set so high that they will have no influence. Therefore, our next drawing, Figure A-7, shows the ideal first-order open-loop frequency response. The phase is $-90°$ and constant as a function of frequency, the gain marked V on the plot has a slope of -6 dB/octave, and our open-loop bandwidth is 1 kHz. Note that the frequency display on the x-axis is logarithmic, and the gain is expressed in decibels. This is an ideal situation, and there is no question of stability, as there is only 90° phase shift. The very moment we add a simple filter to the first-order loop, it becomes a type 1 second-order loop, which refers to one integrator and a simple RC network. As long as the following requirements are fulfilled, there is no problem with stability.

1. The open-loop gain $A(s)$ as plotted must fall below 0 dB before the phase shift reaches 180°. A typical gain margin of $+10$ dB is desirable for $-180°$.
2. A phase shift of less than 180° must be provided at the gain crossover frequency for $A(s)$. This is called *phase margin*. A typical phase margin of 45° is desirable.

It is possible that a loop is conditionally stable and violates the Bode criteria. However, once it meets the Bode criteria, the loop is unconditionally stable. As the type 1 first-order loop phase stays at $-90°$, it will always remain stable. The type 1 second-order loop has only one element for phase shifting, as seen in Figure A-8, and the phase margin at $A(s) = 0$ is sufficient. The gain margin at $-180°$ phase

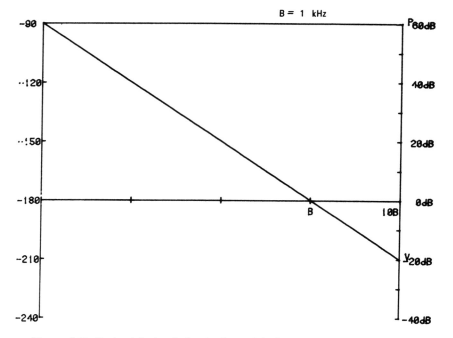

Figure A-7 Bode plot of a first-order loop; P is for phase and V is for gain.

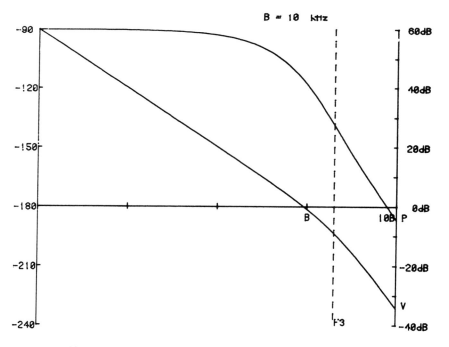

Figure A-8 Type 1 second-order loop with simple RC filter.

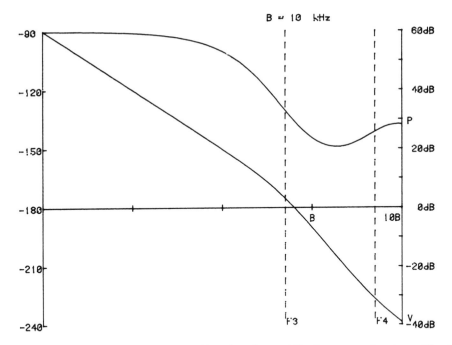

Figure A-9 Type 1 second-order loop taking the phase shift of an operational amplifier into consideration.

is about 35 dB. Therefore, the type 1 second-order loop, as plotted, is unconditionally stable.

Next, we look at a type 1 second-order loop that has phase compensation. In the block diagram of the loop, we made allowance to indicate the phase shift introduced by various components. Rather than use the simple RC network, we now use a lag filter corresponding to the two time constants τ_1 and τ_2. The cutoff frequency determined by F_3 is set below the open-loop bandwidth of 10 kHz, and it is evident that the phase is being compensated by the introduction of the time constant calculated from F_4. This lag filter therefore increases the phase margin. At the point of 0-dB gain, we have sufficient phase margin, and even at -40-dB gain, the phase is still at about $-130°$. This is equal to a phase margin of about $50°$ (see Figure A-9).

Next, we take into consideration the finite cutoff frequency of the operational amplifier. Figure A-10 shows the Bode diagram in which the operational amplifier, used as a dc amplifier (gain $K_2 = 1$), introduces considerable phase shift. The system is still stable, and for $-180°$ phase, the gain is about -20 dB, resulting in a 20-dB gain margin. At 0 dB, about $45°$ phase margin is available. The operational amplifier is responsible for a $180°$ phase shift.

We will now look at the influence of more parameters and will plot a type 2 second-order loop. Figure A-11 shows a type 2 high-order loop with an open-loop bandwidth of 200 kHz using the lag filter with the two cutoff frequencies F_3 and

522 MATHEMATICAL REVIEW

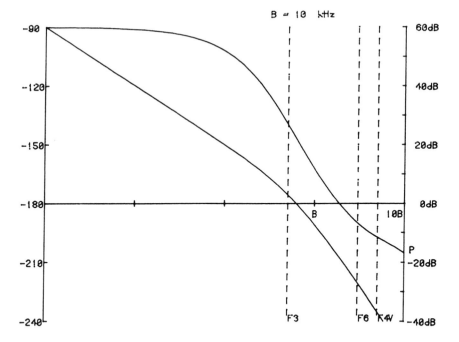

Figure A-10 Type 1 second-order loop with the time constants expressed in frequencies F_3 and F_4, as well as the additional phase shift caused by an operational amplifier.

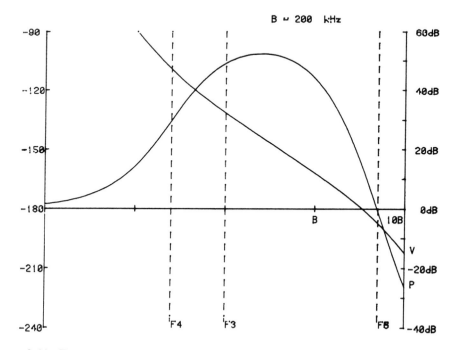

Figure A-11 Type 2 second-order loop showing the influence of the operational amplifier (F_6).

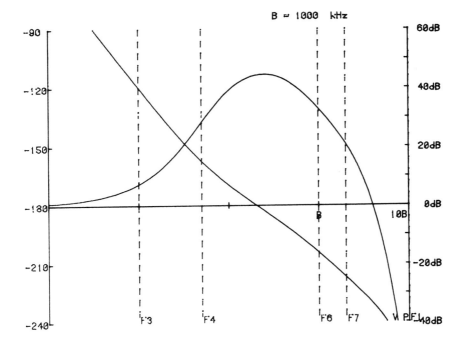

Figure A-12 Type 2 second-order loop in which the two cutoff frequencies F_3 and F_4, the phase shift and cutoff frequencies of the operational amplifier F_6, and an additional RC network (F_7) are incorporated.

F_4 (note that F_4 is smaller than F_3). F_6 describes the cutoff frequency of the operational amplifier used for the active filter.

The gain curve marked V starts off with 12 dB/octave due to two integrators and then, because of the effect of the lag filter, decays with 6 dB/octave. The phase margin at 0 dB gain is about 30°, and the gain margin at $-180°$ of phase is about 7 or 8 dB. This is a stable loop.

Next, we make allowance for the low-pass filter action of the RC network generated by bypass capacitors, cutoff frequency F_7. This loop, by choosing the right F_3 and F_4 values, is stable as the phase margin at 0 dB gain is 60° and the gain margin at $-180°$ of phase is about 40 dB (see Figure A-12).

Finally, let us take a look at Figure A-13, which shows the open-loop performance of a type 2 nth-order loop that contains allowance for the phase shift of the operational amplifier RC filtering and shows the effect of a first-order elliptic filter. This loop is no longer stable, as the gain does not fall to 0 dB while the phase is still less than $-180°$. It is very convenient to use a digital computer to generate these plots because once all the parameters are known, the Bode diagram instantaneously reveals whether a loop is stable and what parameters have to be changed to obtain the necessary phase and gain margins.

524 MATHEMATICAL REVIEW

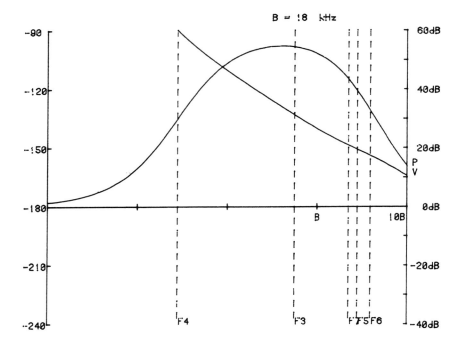

Figure A-13 Type 2 nth-order loop in which several elements are incorporated. This is an unstable loop.

A-4 LAPLACE TRANSFORMATION

Introduction. The Laplace transformation is a convenient mathematical way to analyze and synthesize electronic circuitry with much less effort and far more accuracy than the conventional method by solving differential equations. The Laplace transformation is based on a method described by Pierre Simon de Laplace, a great French mathematician who developed the foundation of potential theory and made important contributions to celestial mechanics and probability theory. The word "transformation" in this case means that functions in time are converted to functions in frequency, and vice versa. Let us look at Figure A-14.

Figure A-14 shows a square wave generated by a suitable generator. We all know that square waves contain harmonics up to very high orders and that the Fourier analysis can be used to synthesize the square waveform. The Laplace transformation allows the direct transformation of the square wave into the Fourier spectrum. This method is used in engineering to analyze the performance of an electrical circuit where an electrical short pulse, a single event, or a periodic event that is not merely a sine or cosine function excites this circuit. Therefore, the Laplace transformation is used as a final method of solving differential equations and will provide an algebraic method of obtaining a *particular* solution of a differential equation from stated initial conditions. Since this is often what is desired in

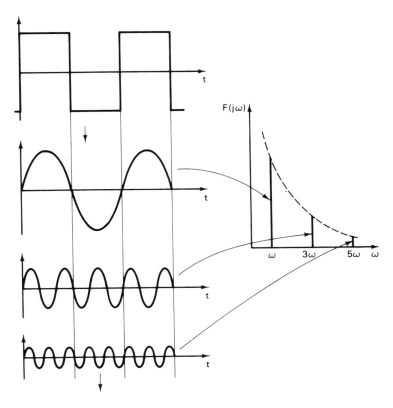

Figure A-14 Square wave showing its sine-wave contents.

practice, the Laplace transformation is preferred for the solution of differential equations for electronic engineering.

Let us assume that $f(t)$ is a given function like the one shown in Figure A-15 and is defined for all $t \geq 0$. This function $f(t)$ is multiplied by e^{-st} and integrated with respect to t from 0 to infinity. Provided that the resulting integral exists, we can write

$$F(s) = \int_0^\infty e^{-st} f(t)\, dt \tag{A-32}$$

The function $F(s)$ is called the *Laplace transform* of the original function $f(s)$ and will be written

$$F(s) = \mathscr{L}(f) = \int_0^\infty e^{-st} f(t)\, dt \tag{A-33}$$

Let us assume that we start with the Laplace transform and want to get the

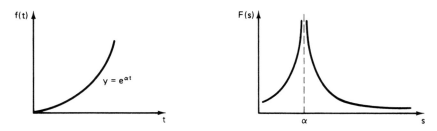

Figure A-15 Function f(t) to be transformed into F(s).

resulting time function. Mathematically this would be done with the inverse Laplace transformation and will be denoted by $\mathscr{L}^{-1}\{F(s)\}$. We shall write

$$f(t) = \mathscr{L}^{-1}[F(s)] \tag{A-34}$$

Rather than get scared, it may be nice to use it.

The Step Function. Let us assume that we have a step function

$$f(t) = 0 \quad \text{for } t < 1$$

and

$$f(t) = 1 \quad \text{for } t \geq 0$$

We want to determine $F(s)$

We obtain by integration

$$\mathscr{L}(f) = \mathscr{L}(1) = \int_0^\infty e^{-st} dt = -\frac{1}{s} e^{-st} \Big|_0^\infty \tag{A-35}$$

Hence, when $s > 0$,

$$\mathscr{L}(1) = \frac{1}{s} \tag{A-36}$$

The Ramp. Accordingly, for a ramp

$$\mathscr{L}(f') = \int_0^\infty t e^{-st} dt = \lim_{c \to \infty} \int_0^c t e^{-st} dt$$

$$= \lim_{c \to \infty} \frac{e^{-st}(-st-1)}{s^2} \Big|_0^c = \frac{1}{s^2} \tag{A-37}$$

We will use the ramp function as well as the step function in analyzing the loop performance of initial disturbance. In using actual Laplace transformation, the linearity theorem is important.

Linearity Theorem. Because the Laplace transformation is a linear operation, we can state that for any given functions $f(t)$ and $g(t)$ whose Laplace transforms exist, and any constants a and b, we have

$$\mathscr{L}[af(t) + bg(t)] = a\mathscr{L}[f(t)] + b\mathscr{L}[g(t)] \tag{A-38}$$

In addition, we have to know about the derivatives and integrals.

Differentiation and Integration. The differentiation is made very simple by the fact that differentiation of a function $f(t)$ corresponds simply to multiplication of the transform $F(s)$ by s. This permits replacing operations of calculus by simple algebraic operations on transforms. Furthermore, since integration is the inverse operation of differentiation, we expect it to correspond to division of transforms by s. This means that

$$\mathscr{L}(f') = s\mathscr{L}(f) - f(0) \tag{A-39}$$

and

$$\mathscr{L}\left[\int_0^t f(\tau)\,d\tau\right] = \frac{1}{s}\mathscr{L}[f(t)] \tag{A-40}$$

Table A-2 shows some functions $f(t)$ and their Laplace transforms.

Initial Value Theorem. If we apply a nonsinusoidal signal to an electrical circuit, we are interested in obtaining the value of $f(t)$ at the time $t = 0$, and this can be determined from the Laplace transform by

$$\lim_{t \to 0} f(t) = \lim_{s \to \infty} sF(s) \tag{A-41}$$

After the initial start condition, we are interested in determining the final value.

Final Value Theorem. The final value can be determined accordingly,

$$\lim_{t \to \infty} f(t) = \lim_{s \to 0} sF(s) \quad \text{(provided that such a limit exists)}$$

Let us now use our knowledge and the integration table for one particular case, the active integrator.

The Active Integrator. Figure A-16 shows the circuit of an active RC integrator being driven with a step; because of the integration, the output voltage has to be a ramp. Let us prove this. The differential equation can be written

$$v_2(t) = -\frac{1}{C}\int_0^t i\,dt = \frac{-1}{RC}\int_0^t v_1\,dt \tag{A-42}$$

Table A-2 Functions $f(t)$ and their Laplace transforms $F(s)$

	$F(s)$	$f(t)$
1.	$sF(s) - f(0)$	$\dfrac{df(t)}{dt}$
2.	$\dfrac{F(s)}{s} + \dfrac{f^{(-1)}(0)}{s}$	$\int_0^t f(t)\,dt$
3.	$F(s)e^{-s\tau}$	$f(t-\tau)$
4.	$kF(s)$	$kf(t)$
5.	$F_1(s)F_2(s)$	$f_1(t)f_2(t)$
6.	$F_1(s)F_2(s)$	$f_1(t)f_2(t)$
7.	0	0
8.	$\dfrac{1}{s}$	$u(t)$
9.	1	$\delta(t)$
10.	$\dfrac{1}{s^2}$	t
11.	$\dfrac{1}{s^3}$	$\dfrac{t^2}{2}$
12.	$\dfrac{1}{s^n}\ n>0$	$\dfrac{t^{n-1}}{(n-1)!}$
13.	$\dfrac{1}{s-\alpha}$	$e^{\alpha t}$
14.	$\dfrac{1}{s(s-\alpha)}$	$\dfrac{1}{\alpha}(e^{\alpha t}-1)$
15.	$\dfrac{1}{s(s+\alpha)}$	$\dfrac{1}{\alpha}(1-e^{-\alpha t})$
16.	$\dfrac{1}{(s-\alpha)^n}\ n>0$	$\dfrac{t^{n-1}}{(n-1)!}e^{\alpha t}$
17.	$\dfrac{1}{s^2+\alpha^2}$	$\dfrac{1}{\alpha}\sin\alpha t$
18.	$\dfrac{s}{s^2+\alpha^2}$	$\cos\alpha t$
19.	$\dfrac{1}{s(s^2+\alpha^2)}$	$\dfrac{1}{\alpha^2}(1-\cos\alpha t)$
20.	$\dfrac{1}{s^2-\alpha^2}$	$\dfrac{1}{\alpha}\sinh\alpha t$
21.	$\dfrac{s}{s^2-\alpha^2}$	$\cosh\alpha t$
22.	$\dfrac{1}{s(s^2-\alpha^2)}$	$\dfrac{1}{\alpha^2}(\cosh\alpha t-1)$
23.	$\dfrac{1}{(s-\alpha)(s-\beta)}$	$\dfrac{e^{\beta t}-e^{\alpha t}}{\beta-\alpha}$

Table A-2 (Continued)

24.	$\dfrac{s}{(s-\alpha)(s-\beta)}$	$\dfrac{\beta e^{\beta t} - \alpha e^{\alpha t}}{\beta - \alpha}$
25.	$\dfrac{1}{s(s-\alpha)(s-\beta)}$	$\dfrac{\beta e^{\alpha t} - \alpha e^{\beta t}}{\alpha\beta(\alpha - \beta)} + \dfrac{1}{\alpha\beta}$
26.	$\dfrac{1}{s^2 + 2s\zeta\omega_n + \omega_n^2}$	$\dfrac{e^{-\zeta\omega_n t} \sin\sqrt{1-\zeta^2}\,\omega_n t}{\sqrt{1-\zeta^2}\,\omega_n}$
27.	$\dfrac{s}{s^2 + 2s\zeta\omega_n + \omega_n^2}$	$\left[\cos\sqrt{1-\zeta^2}\,\omega_n t - \dfrac{\zeta}{\sqrt{1-\zeta^2}} \sin\sqrt{1-\zeta^2}\,\omega_n t\right]e^{-\zeta\omega_n t}$
28.	$\dfrac{1}{s(s^2 + 2s\zeta\omega_n + \omega_n^2)}$	$\dfrac{1}{\omega_n^2}\left[1 - \left(\cos\sqrt{1-\zeta^2}\,\omega_n t + \dfrac{\zeta}{\sqrt{1-\zeta^2}} \sin\sqrt{1-\zeta^2}\,\omega_n t\right)e^{-\zeta\omega_n t}\right]$
29.	$\dfrac{1}{(s-\alpha)(s-\beta)^2}$	$\dfrac{e^{\alpha t} - [1 + (\alpha - \beta)t]e^{\beta t}}{(\alpha - \beta)^2}$
30.	$\dfrac{s}{(s-\alpha)(s-\beta)^2}$	$\dfrac{\alpha e^{\alpha t} - [\alpha + \beta(\alpha - \beta)t]e^{\beta t}}{(\alpha - \beta)^2}$
31.	$\dfrac{s^2}{(s-\alpha)(s-\beta)^2}$	$\dfrac{\alpha^2 e^{\alpha t} - [2\alpha - \beta + \beta(\alpha - \beta)t]\beta e^{\beta t}}{(\alpha - \beta)^2}$
32.	$\dfrac{1}{(s-\alpha)(s-\beta)(s-\gamma)}$	$\dfrac{(\beta-\gamma)e^{\alpha t} + (\gamma-\alpha)e^{\beta t} + (\alpha-\beta)e^{\gamma t}}{(\alpha-\beta)(\beta-\gamma)(\gamma-\alpha)}$
33.	$\dfrac{1}{(s^2+\alpha^2)(s^2+\beta^2)}$	$\dfrac{\alpha \sin\beta t - \beta\sin\alpha t}{\alpha\beta(\alpha^2 - \beta^2)}$
34.	$\dfrac{s}{(s^2+\alpha^2)(s^2+\beta^2)}$	$\dfrac{\cos\beta t - \cos\alpha t}{\alpha^2 - \beta^2}$
35.	$\dfrac{1}{\sqrt{s}}$	$\dfrac{1}{\sqrt{\pi t}}$
36.	$\dfrac{1}{s\sqrt{s}}$	$2\sqrt{\dfrac{t}{\pi}}$
37.	$\dfrac{1}{s^n\sqrt{s}}$	$\dfrac{n!}{(2n)!}\dfrac{4^n}{\sqrt{\pi}} t^{n-1/2}$
38.	$\dfrac{1}{\sqrt{s-\alpha}}$	$\dfrac{1}{\sqrt{\pi t}} e^{\alpha t}$
39.	$\dfrac{1}{s\sqrt{s+\alpha}}$	$\dfrac{2}{\sqrt{\alpha\pi}}\int_0^{\sqrt{\alpha t}} e^{-\xi^2}\,d\xi$
40.	$\dfrac{1}{(s+\alpha)\sqrt{s+\beta}}$	$\dfrac{2e^{-\alpha t}}{\sqrt{\pi(\beta-\alpha)}}\int_0^{\sqrt{(\beta-\alpha)t}} e^{-\xi^2}\,d\xi$
41.	$\dfrac{\sqrt{s+\alpha}}{s}$	$\dfrac{e^{-\alpha t}}{\sqrt{\pi t}} + 2\sqrt{\dfrac{\alpha}{\pi}}\int_0^{\sqrt{\alpha t}} e^{-\xi^2}\,d\xi$
42.	$\dfrac{1}{\sqrt{s^2+\alpha^2}}$	$I_0(\alpha t)$
43.	$\dfrac{1}{\sqrt{s^2-\alpha^2}}$	$J_0(\alpha t)$

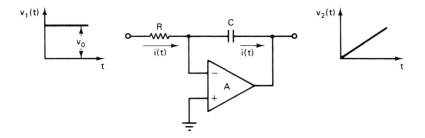

Figure A-16 Active RC integrator being driven with a step function.

Using $\tau = RC$, we obtain in Laplace notation

$$V_2(s) = \frac{-1}{s\tau} V_1(s) \tag{A-43}$$

We assume that the capacitor at $t = 0$ has no charge. The step function $v_1(t)$ rises to the value v_0 and

$$V_1(s) = \frac{v_0}{s} \tag{A-44}$$

Therefore,

$$V_2(s) = \frac{-v_0}{\tau} \frac{1}{s^2} \tag{A-45}$$

According to Table A-2,

$$v_2(t) = \mathcal{L}^{-1}[V_2(s)] = -v_0 \frac{t}{\tau} \tag{A-46}$$

This is the equation of a linear ramp.

Let us now become more challenging and determine the locking behavior of a phase-locked loop, using a lag filter as shown in Figure A-17.

Locking Behavior of the PLL. The transfer function of the lag filter is

$$F(s) = \frac{1 + s\tau_2}{s\tau_1} \tag{A-47}$$

and the phase detector voltage is

$$v_\phi(t) = K_\theta \theta \tag{A-48}$$

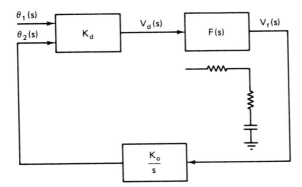

Figure A-17 PLL with lag filter.

and in Laplace notation

$$V_\phi(s) = K_\theta \theta(s) \tag{A-49}$$

The output frequency of the VCO is

$$\omega_o = K_o v(t) \tag{A-50}$$

$$\theta_o = K_o \int_0^t v(t)\,dt \tag{A-51}$$

and in Laplace notation

$$\theta(s) = K_o \frac{V(s)}{s} \tag{A-52}$$

There are three building blocks for which we define the following functions:

1. *Phase comparator*:

$$\frac{V_\phi(s)}{\theta_e(s)} = K_o \tag{A-53}$$

2. *Low-pass filter*:

$$\frac{V(s)}{\theta(s)} = F(s) = \frac{1 + s\tau_2}{s\tau_1} \tag{A-54}$$

3. *VCO*:

$$\theta_2(s) = \theta_1(s) - \theta_e(s) = K_o \frac{V(s)}{s} \tag{A-55}$$

This can be rearranged to give

$$\theta_2(s) = \theta_1(s) \frac{s^2}{s^2 + sK_o K_\theta(\tau_2/\tau_1) + (K_o K_\theta/\tau_1)} \qquad \text{(A-56)}$$

Using similar abbreviations,

$$\omega_n = \left(\frac{K_o K_\theta}{\tau_1}\right)^{1/2} \qquad \text{(A-57)}$$

$$\zeta = \frac{\tau_2}{2}\left(\frac{K_o K_\theta}{\tau_1}\right)^{1/2} \qquad \text{(A-58)}$$

we can rearrange the equation above in the form

$$\theta_e(s) = \theta_1(s) \frac{s^2}{s^2 + 2s\zeta\omega_n + \omega_n^2} \qquad \text{(A-59)}$$

Applying a step to the input,

$$\theta_1(s) = \frac{\Delta\phi}{s} \qquad \text{(A-60)}$$

we obtain

$$\theta_e(s) = \frac{s\Delta\phi}{s^2 + 2s\zeta\omega_n + \omega_n^2} \qquad \text{(A-61)}$$

We apply the initial value theorem,

$$\lim_{t \to 0} \theta_e(t) = \lim_{s \to \infty} s\theta_e(s) = \frac{s^2 \Delta\phi}{s^2 + 2s\zeta\omega_n + \omega_n^2} = \Delta\phi \qquad \text{(A-62)}$$

This means that the initial phase error is equal to the step in phase $\Delta\phi$.
Using the final value theorem, we find

$$\lim_{t \to \infty} \theta_e(t) = \lim_{s \to 0} s\theta_e(s) = \frac{s^2 \Delta\phi}{s^2 + 2s\zeta\omega_n + \omega_n^2} = 0 \qquad \text{(A-63)}$$

This means that, if we wait long enough, the phase error will be zero. The final remaining task is to look up the equation above in our table of Laplace transform functions, and we find the required transform in No. 27:

$$f(t) = \theta_e(t)$$
$$= \Delta\phi \left[\cos\sqrt{1-\zeta^2}\,\omega_n t - \frac{\zeta}{\sqrt{1-\zeta^2}} \sin\sqrt{1-\zeta^2}\,\omega_n t\right] e^{-\zeta\omega_n t} \qquad \text{(A-64)}$$

A-5 LOW-NOISE OSCILLATOR DESIGN

The design of low-noise oscillators is based on various principles.

1. We have learned that one way of reducing the noise is to keep as much energy storage in the capacitor as possible. We can assign for any tuned circuit an equivalent transmission impedance $C_o = \sqrt{L/C}$. This would indicate that the larger the C, the lower the transmission impedance. In addition, such a circuit is less sensitive to circuit board capacitance and should provide better performance.
2. We have learned that the noise outside the loop bandwidth of an oscillator is determined by the Q of the LC network—the highest possible Q that can be obtained in an LC circuit when the losses are minimized. High-Q tuned circuits can be built with transmission lines, and quarter-wavelength transmission lines are specifically used for this purpose. The easiest way of accomplishing this is to take a mechanical cavity that is adjusted to odd numbers of quarter-wavelengths, whereby any material inside the cavity has to be taken into consideration. The wavelengths of a quarter-wave transmission line can be determined from $\lambda_0 = 300/f_0$. If the frequency is inserted in megahertz, the resulting wavelength is in meters. In the event that a dielectric material is used, as in the case of coaxial cable as a cavity oscillator, the wavelengths electrically and mechanically differ:

$$\lambda = \frac{\lambda_o}{\sqrt{\varepsilon_r}} \quad \text{(A-65)}$$

For Teflon, $\varepsilon_r = 2$.

This second principle is used in the Hewlett-Packard HP8940 signal generator, where a cavity is mechanically tuned. This cavity has a high Q of about 600 to 800, and therefore the noise sideband is very low.

Let us design such an oscillator.

Example A quarter-wavelength oscillator using a rigid coaxial line will be built covering the frequency range from 250 to 450 MHz. We have to use the equation

$$\frac{dz}{\lambda} = \frac{1}{2\pi} \arctan \omega CZ \quad \text{(A-66)}$$

where dz is the amount by which the cavity is reduced in size relative to quarter-wavelength. The highest frequency of our oscillator is 450 MHz, and therefore $\lambda_o = 66.6$ cm. Quarter-wavelength is $\lambda_o/4 = 16.66$ cm. For reasons of available mechanical space, we have decided to make the transmission line quarter-wavelength cable 5 cm long. Therefore,

$$L = \frac{\lambda_o}{4} - dz = 16.66 - 5 = 11.66$$

We now rearrange the equation above and solve it for C.

$$C = \frac{1}{\omega Z} \tan \frac{2\pi dz}{\lambda_0} \quad (A\text{-}67)$$

or

$$C_1 = \frac{1}{2\pi \times 450 \times 10^6 \times 50} \tan \frac{2\pi \times 11.66}{66.6}$$

$$C_1 = 7.736 \times 10^{-12} \times \tan 1.1 \text{ (rad)}$$

$$C_1 = 7.736 \times 10^{-12} \times 1.9649$$

$$C_1 = 13.899 \times 10^{-12} = 13.899 \text{ pF}$$

Electrically, the transmission line, which is now operating as a quarter-wavelength resonator, is an inductance that requires an external capacitor of about 14 pF to be in resonance for 450 MHz. For 250 MHz we will get a new value for the capacitance. First, we determine λ. $\lambda_o = 1.2$ m and $\lambda_o/4 = 30$ cm. Because the mechanical length of our quarter-wavelength is 5 cm,

$$L = \frac{\lambda_o}{4} - dz = 30^{-5} = 25 \text{ cm}$$

We now compute

$$C_2 = \frac{1}{2\pi \times 250 \times 10^6 \times 50} \tan \frac{2\pi \times 25}{120}$$

$$C_1 = 12.732 \times 10^{-12} \tan 1.309$$

$$= 12.732 \times 10^{-12} \times 3.7321 = 47.156 \text{ pF}$$

These are the two values required for the oscillator to cover the frequency range. If one compares these two capacitance values with values obtained with conventional high-Q inductors, it is apparent that those values are substantially larger. This is due to the fact that we have chosen a 50-Ω transmission line. The use of a low-impedance transmission line has several advantages.

1. It can be shown mathematically that the optimum Q of a coaxial transmission line occurs at about 70 Ω. All higher impedances exhibit more losses and lower Q.
2. If a rigid line or its equivalent mechanical arrangement is used, the low-impedance version will have fewer microphonic effects due to mechanical vibration than a high-impedance transmission line and is therefore electrically much more stable.

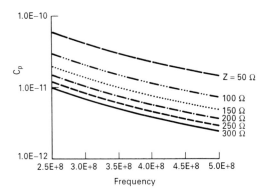

Figure A-18 Capacitance required to tune a quarter-wavelength resonator oscillator from 250 to 500 MHz as a function of the impedance of the quarter-wavelength.

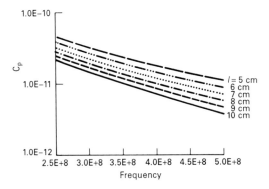

Figure A-19 Rigid cable used as a quarter-wave resonator at various lengths showing the external capacitance value required to tune it from 250 to 500 MHz as a function of length.

Figure A-18 shows an analysis of this done on a digital computer. We find that, as the impedance increases, the external shunt capacitance goes down in value. In this figure, the computer has plotted the curves from 50 to 300 Ω, and the necessary capacitance can be read from this drawing as a function of frequency and characteristic impedance. For a 300-Ω transmission line and 500 MHz, an external capacitance of about 2.5 pF is required. Most likely, circuit board and other stray capacitances will be around that magnitude. For a 100-Ω oscillator, about 7 pF is required. It is evident that the oscillator we have just calculated, which requires about 14 pF, is a better choice. Another interesting relationship is the required capacitance as a function of increase of resonator length. Figure A-19 shows a diagram in which the capacitance is plotted as a function of frequency and resonator length with a 50-Ω transmission line. If we use a 10-cm resonator, we need about 3.5 pF at 500 MHz and about 25 pF at 250 MHz. Again, this gives some interesting insight into the mechanism. Conventional LC circuits theoretically could be built using such low inductances. However, the stray field of this

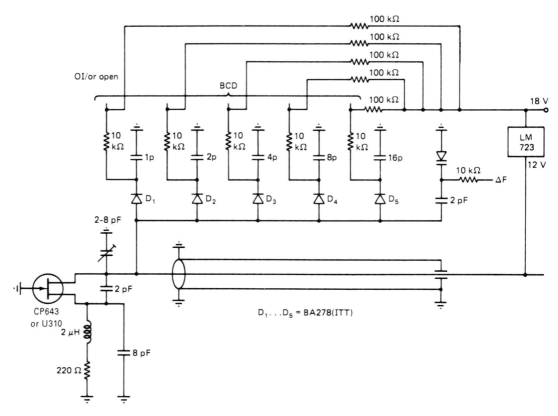

Figure A-20 Schematic of a quarter-wavelength oscillator, including switching diodes.

unconfined resonator would result in losses, consequently lowering the magnetic Q of the circuit. A similar principle is used in helical resonators, and the Rohde & Schwarz SMDU signal generator uses this principle. There is really no difference between the two approaches. In the case of Hewlett-Packard, the quarter-wave transmission line is mechanically adjusted in its length. As a result of this, a mechanically more elaborate system is required, whereas in the Rohde & Schwarz SMDU signal generator the helical resonator is loaded with a very large, low microphonic air-variable capacitor of large diameter. Both arrangements are electrically excellent. The air-variable capacitor has the advantage that there is no mechanical abrasion, and therefore the lifetime will be longer. The cavity, on the other hand, provides a somewhat more linear frequency versus tuning curve.

Figure A-20 shows the schematic of such an oscillator. It becomes apparent that a switching technique is used to coarse steer the oscillator within certain ranges. Since we have learned that the tuning diodes will introduce more noise than fixed capacitors switched in by diodes that are not sensitive to noise pickup and other radiation effects, this technique is used. Let us take a look at the possible resolution. The minimum additional capacitor that can be added is 1 pF. At 450 MHz, 1 pF will result in the following detuning:

$$\frac{f_1}{f_2} = \sqrt{\frac{13.951}{14.951}} = \frac{450}{434.69}$$

or a change of 15.3 MHz. At the low end of 225 MHz, this will result in

$$\frac{f_1}{f_2} = \sqrt{\frac{47.517}{48.517}} = \frac{250}{247.47}$$

or we obtain a frequency shift of 2.6 MHz.

Our highest resolution at the top is therefore about 15 MHz, with 2.6 MHz at the low-frequency end. Thus, we have to use a decoding circuit that selects the proper capacitor for the same step at the lower frequency range. In order to get 15.3 MHz, we calculate

$$\sqrt{\frac{265}{250}} = 1.0296$$

or

$$\frac{C_1^*}{C_1} = 1.06$$

Our starting value at the low end is 47.517 pF, which has to be reduced to 44.827 for a 15-MHz shift. The difference is about 2.7 pF. Therefore, following the first 1-pF capacitor, we must be able to switch in 2 pF, resulting in a total of 3 pF, which is a close approximation to the required 2.7 pF for the required 15-MHz step. We now follow this binary system, and therefore our next capacitances are 4, 8, 16 pF.

Our binary switch requires a 5-bit data command. If we add all the capacitors together, we obtain a total capacitance of 31 pF. Since the initial starting capacitance at 500 MHz was set to be about 14 pF, which is found by the feedback network as well as the stray capacitance and a coarse-tuning capacitor, the additional 30 pF, if all five capacitors are switched in, will result in 43 pF. We have to take into consideration the fact that these capacitors have some tolerances and therefore, by selecting the proper values with slightly larger amounts, we can easily make the total 33 pF to obtain the 47.5 pF required. This oscillator exhibits superior performance relative to the normal LC oscillator.

Some authors have found it useful to build a $\lambda/2$ oscillator, which then has twice the mechanical length we have currently used, and this may be helpful at higher frequencies. In addition, because of the transmission properties of a half-wavelength cable, a capacitor used at the output of the cable is transformed into an inductor. The drawback of this method, however, is that the resonant impedance for constant Q at the transistor varies as a function of frequency, whereby for higher frequencies where the gain is lower, the impedance gets lower.

Figure A-21 Photograph of the Rohde & Schwarz SMDU oscillator.

This is opposite to the quarter-wavelength system and, in my opinion, less desirable.

The tuning diode in the quarter-wave oscillator is responsible for the fine tuning and will cover about 20 MHz of range. At 250 MHz, this is less than 10%, and as we have seen previously, the noise influence under these circumstances is extremely small.

As this oscillator is highly useful, in the next section we will analyze the feedback circuit to determine the amplitude stabilization and harmonic contents with the aid of some nonlinear analysis. Figure A-21 shows a picture of the Rohde & Schwarz SMDU oscillator.

A-6 OSCILLATOR AMPLITUDE STABILIZATION

In Chapter 4 we mentioned briefly that the oscillator amplitude stabilizes due to some nonlinear performance of the transistor. There are various mechanisms involved and, depending on the circuit, several of them are simultaneously responsible for the performance of an oscillator. Under most circumstances, the transistor is operated in an area where the dc bias voltages are substantially larger than the ac voltages. Therefore, the theory describing the transistor performance under these conditions is called *small-signal theory*. In a transistor oscillator, however, we are dealing with a feedback circuit that applies positive feedback. The energy that is being generated by the initial switch-on of the circuit is being fed back to the input of the circuit, amplified, and returned to the input again until oscillation starts. The oscillation would theoretically increase in value indefinitely unless some limiting or stabilization occurs. In transistor circuits, we have two basic phenomena responsible for limiting the amplitude of oscillation.

OSCILLATOR AMPLITUDE STABILIZATION

1. Limiting because of gain saturation and reduction of open-loop gain.
2. Automatic bias generated by the rectifying mechanism of either the diode in the bipolar transistor or in the junction field-effect transistor. In MOSFETs an external diode is sometimes used for this biasing.

A third phenomenon would be external AGC, but it will not be considered here.

The oscillators we discuss here are self-limiting oscillators.

The self-limiting process, which by generating a dc offset bias moves the operating point into a region of less gain, is generally noisy. For very low noise oscillators, this operation is not recommended. After dealing with the quarter-wavelength oscillator in the preceding section, we will deal here only with the negative resistance oscillator, in which, through a mechanism explained in Chapter 4, a negative resistance is generated due to feedback and is used to start oscillation with the passive device. Here we look at what is happening inside the transistor that is responsible for amplitude stabilization, and we will thus be in a position to make a prediction regarding the available energy and the harmonic contents.

Figure A-22 shows the quarter-wavelength oscillator redrawn in such a way that the source electrode is now at ground potential while the gate and drain electrode are electrically hot. The reason for doing this is because we will look at the gate-to-source transfer characteristic and use its nonlinearities as a tool to describe what is happening. The same analysis can be applied to a transistor circuit, provided that the resistors used for dc bias are small enough not to cause any dc offset. The field-effect transistor characteristic follows a square law and therefore can be expressed as

$$i_2 = I_{DSS}\left(1 - \frac{v_1}{V_p}\right)^2 \tag{A-68}$$

For any other device, we have to take the necessary transfer characteristic into

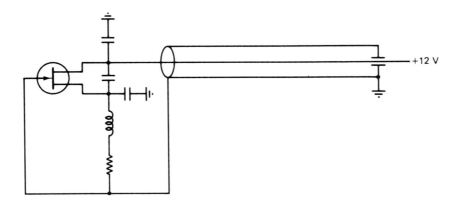

Figure A-22 Quarter-wavelength oscillator with grounded source electrode.

consideration, and this could theoretically be done by changing the square law into nth order. The voltage v_1 will be in the form

$$v_1 = V_b + V_1 \cos \omega t \tag{A-69}$$

This is the voltage that is being generated due to the selectivity of the tuned circuit at which there is a resonant frequency. Inserting this into the above equation and using

$$V_x = V_p - V_b \tag{A-70}$$

we obtain

$$i_2 = \frac{I_{DSS}}{V_p^2}(V_x^2 - 2V_x V_1 \cos \omega t + V_1^2 \cos^2 \omega t) \tag{A-71}$$

Once we know the peak value of i_2, we can expand this into a Fourier series. In this case a Fourier series expansion for i_2 has only three terms; that is,

$$i_2(t) = I_o + I_1 \cos \omega t + I_2 \cos 2\omega t \tag{A-72}$$

$$I_o = \frac{I_{DSS}}{V_p^2} V_x^2 + \frac{V_1^2}{2} \tag{A-73}$$

$$I_1 = -2 \frac{I_{DSS}}{V_p^2} V_x V_1 \tag{A-74}$$

$$I_2 = \frac{I_{DSS}}{V_p^2} \frac{V_1^2}{2} \tag{A-75}$$

Because of the square-law characteristic, I_1 is a linear function of V_1 and we can define a large-signal average transconductance G_m:

$$G_m = \frac{I_1}{V_1} = -2 \frac{I_{DSS}}{V_p^2} V_x \tag{A-76}$$

In the case of the square-law characteristic, we find the interesting property that the small-signal transconductance g_m at any particular point is equal to the large-signal average transconductance G_m at the same point. The second harmonic distortion in the output current is given by

$$\frac{I_2}{I_1} = \frac{V_1}{4V_x} = \frac{V_1}{4V_p} \frac{g_{mo}}{g_m} \tag{A-77}$$

The transconductance G_m can be defined in such a way that it indicates the gain for a particular frequency relative to the fundamental, which means that there is a certain G_m for the fundamental frequency and one for the second harmonic, and in the general case, a G_{mn} for the nth-order harmonic. In the more general form, we rewrite our equation

$$i_d = C_n(-V_b + V_1 \cos x)^n \tag{A-78}$$

As this current will exist only during the period from $-\alpha$ to $+\alpha$, the equation

$$-\alpha < x < +\alpha$$

exists only for

$$i_2 = 0$$
$$x = \pm\alpha$$
$$\cos\alpha = \frac{V_b}{V_1}$$

We can rewrite our equation for the drain current or collector current of a transistor:

$$i_d = C_n V_1^n (\cos x - \cos \omega)^n \qquad \text{(A-79)}$$

The dc value of the current is

$$I_d = \frac{1}{\pi} \int_0^\alpha i_d \, dx \qquad \text{(A-80)}$$

or

$$I_d = \frac{C_n V_1}{\pi} \int_0^\alpha (\cos x - \cos \alpha)^n \, dx \qquad \text{(A-81)}$$

The amplitude of the fundamental frequency is

$$I_1 = \frac{2}{\pi} \int_0^\alpha i_d \cos x \, dx \qquad \text{(A-82)}$$

or

$$I_1 = \frac{2 C_n V_1^n}{\pi} \int_0^\alpha (\cos x - \cos \alpha)^n \cos x \, dx \qquad \text{(A-83)}$$

For $n = 1$, the collector current is

$$I_d = C_1 V_1 A_1 \qquad \text{(A-84)}$$

and the amplitude of the fundamental frequency is

$$I_1 = C_1 V_1 B_1 \qquad \text{(A-85)}$$

For $n = 2$, the collector current is therefore

$$I_d = C_2 V_1^2 A_2 \qquad \text{(A-86)}$$

Table A-3 Normalized Fourier coefficients

$\dfrac{V_b}{V_1}$	A_1	B_1	$\dfrac{B_1}{A_1}$	A_2	B_2	$\dfrac{B_2}{A_2}$
0	0.318	0.500	1.57	0.250	0.425	1.7
0.1	0.269	0.436	1.62	0.191	0.331	1.73
0.2	0.225	0.373	1.66	0.141	0.251	1.78
0.3	0.185	0.312	1.69	0.101	0.181	1.79
0.4	0.144	0.251	1.74	0.0674	0.126	1.87
0.5	0.109	0.195	1.79	0.0422	0.0802	1.90
0.6	0.077	0.141	1.83	0.0244	0.0458	~1.95
0.7	0.050	0.093	1.86	0.0118	0.0236	~2
0.8	0.027	0.052	1.92	0.0043	0.0082	~2
0.9	0.010	0.020	2	0.00074	0.00148	2
1.0	0	0	2	0	0	2

and the amplitude of the fundamental frequency is

$$I_1 = C_2 V_1^2 B_2 \qquad \text{(A-87)}$$

With the definition of the conduction angle, we find

$$\alpha = \operatorname{arc} \frac{V_b}{V_1} \qquad \text{(A-88)}$$

These values are listed in Table A-3.

These are the normalized Fourier coefficients as a function of n and the conduction angle. Theoretically, this has to be expanded to the order n of 3 or 4, depending on the particular device, and can be found from tables or by a digital computer.

For simplifications, let us go back to the case of our square-law device, where our transconductance is

$$G_m = \frac{I_1}{V_1} = -2 \frac{I_{DSS}}{V_p^2} V_x \qquad \text{(A-89)}$$

This can be rewritten in the form

$$G_m = -\frac{2 I_{DSS}}{V_p^2} (V_p - V_b + V_1 \cos \omega t) \qquad \text{(A-90)}$$

V_p is the pinch-off voltage of the field-effect transistor, V_b is the bias voltage that is measured between source and ground, and V_1 is the peak value of the voltage of the fundamental frequency. Figure A-23 shows the effect where the sine wave is driving the transfer characteristic, and the resulting output currents are narrow pulses. Based on the duration, the mutual conductance g_m becomes a fraction of

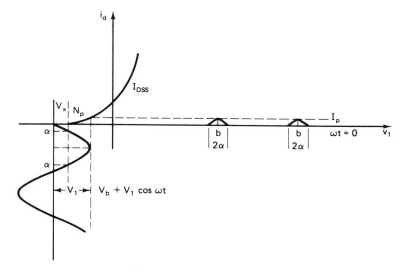

Figure A-23 Current tips as a function of narrow conduction angles in a square-wave transfer characteristic.

the dc transconductance G_m, and therefore the gain is reduced. For small conduction angles I_n/I_d, the mutual conductance can take very small values, and therefore the gain gets very small; this is the cause for stabilizing the amplitude in the oscillator. We note that the gain is being reduced as the amplitude causing the small conduction angle is increased.

Fourier analysis indicates that, for a small harmonic distortion, the RF voltage at the source or gate (depending on where it is grounded) has to be less than 80 mV. Now we can design the oscillator performance.

Let us assume that the saturation voltage of the active device is 2 V, battery voltage applied to the transistor is 12 V, and the transistor starts at a dc current of 10 mA with a source resistor of 200 Ω. This results in a voltage drop of 1 V at the source and 2 V in the device; therefore, 9 V is available. It can be assumed that the maximum voltage at the drain will be $9 \times \sqrt{2}$. The capacitor voltage divider from drain to voltage now depends on the gain. If we assume an I_n/I_d of 0.15 for about 50° conduction angle, 2α, and the dc conductance of the transistor at the starting dc operating point is 20 mA/V, the resulting transconductance is 3 mA/V.

Next, we need the output impedance provided by the quarter-wave resonator:

$$R_L = Q \frac{1}{\omega C} (250 \text{ MHz})$$

or

$$R_L = 600 \frac{1}{2\pi \times 47 \times 10^{-12} \times 250 \times 10^6} = 8127 \, \Omega$$

As we want 9 V rms at the output, we have to use the equation

$$\frac{V_{\text{out}}}{V_{\text{in}}} = A(\text{voltage gain}) = g_m R_L = 3 \times 10^{-3} \times 8.127 \times 10^3$$

$$A = 24.38$$

or

$$V_{\text{in}} = \frac{8\text{V}}{A} = 328\,\text{mV}$$

This would mean that the capacitance ratio of the feedback capacitors C_1 and C_2 would be 1:24.38. In practice, we will find that this is incorrect, and we need a 1:4 or 1:5 ratio. The reason for this is that the equations we have used so far are not accurate enough to represent the actual dc shifts and harmonic occurrences. As mentioned in Sections 4-1 and A-5 a certain amount of experimentation is required to obtain the proper value. To determine the actual ratio, it is recommended that one obtain from the transistor manufacturer the device with the lowest gain and build an oscillator testing it over the necessary temperature range. As the gain of the transistor changes as a function of temperature (gain increases as temperature decreases for field-effect transistors and acts in reverse for bipolar transistors), a voltage divider has to be chosen that is, on the one hand, high enough to prevent the device from going into saturation, which will cause noise, and, on the other hand, small enough to allow oscillation under worst-case conditions. Suitable values were determined for the U310 transistor and shown in the circuit for the field-effect quarter-wavelength transistor.

A-7 VERY LOW PHASE NOISE VCO FOR 800 MHz

The previous quarter-wavelength resonator-based oscillator, while covering a large frequency range, is subject to switching noise while moving from one frequency range to another. The circuit shown in Figure A-24 is a high-performance VCO in which the transmission line is part of the layout. While the circuit is not too different from previously described circuits, a constant-current generator is used in the emitter of the oscillator's transistor. This feedback, along with flicker noise feedback obtained by the 27-Ω resistor labeled R444, results in overall smaller phase noise than one would obtain in a conventional oscillator, eliminating the constant-current generator V436 by using a fixed resistor. The other transistors, V438 and V440, only serve as dc switches and are part of the power supply circuit. In the selection of this particular oscillator transistor, a high-dissipation device was used and is also operating at fairly high currents. However, the maximum current of the device is significantly higher and therefore, compared to this maximum value, the operating value is still smaller.

To minimize the noise in the tuning diodes, the series loss resistors are kept to 3.3 Ω, which reduces the Q of the inductors to the point where one does not observe spurious resistant frequencies and has a minimal noise contribution of its own.

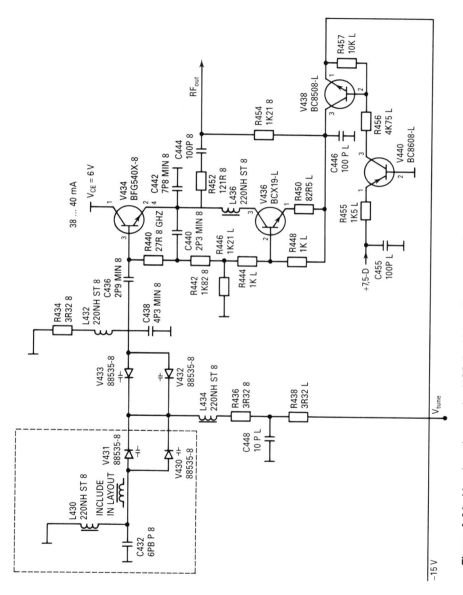

Figure A-24 Very low phase noise VCO for high-performance synthesizer application operating at 800 MHz.

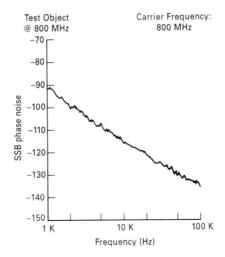

Figure A-25 Measured phase noise of the oscillator shown in Figure A-24.

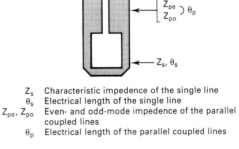

Z_s	Characteristic impedance of the single line
θ_s	Electrical length of the single line
Z_{pe}, Z_{po}	Even- and odd-mode impedence of the parallel coupled lines
θ_p	Electrical length of the parallel coupled lines

Figure A-26 Layout of the stepped impedance hair pin resonator with parallel coupled lines.

Figure A-25 shows the measured phase noise for this oscillator.

A voltage-controlled "push–push" oscillator using a hair pin resonator was recently described by Yabucki [1]. Figure A-26 shows the layout of the stepped impedance hair pin resonator with parallel coupled lines. To calculate such a structure, one needs to use a circuit simulator with electromagnetic models such as Super-Compact or an electromagnetic simulator such as Microwave Explorer, both made by Compact Software, Inc., Paterson, New Jersey.

Figure A-27 shows the frequency response of the resonator as a function of mode capacitor and Figure A-28 shows the voltage distribution of the hair pin resonator. One can build a push–push oscillator using this symmetrical resonator. This oscillator has certain similarities to a tuning fork oscillator using ceramic material described by my father, Lothar Rohde, around 1940 when he designed the world's first portable time/frequency standard. In the case of the time standard,

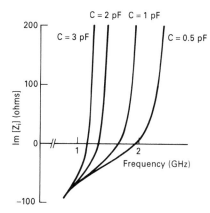

Figure A-27 Frequency response of the resonator.

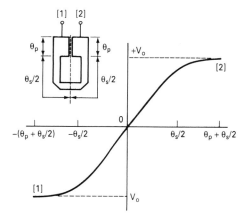

Figure A-28 Voltage distribution of the hair pin resonator.

the material chosen had a ±0 temperature coefficient and therefore did not require a proportional oven for maintaining constant temperature. In the case of the printed symmetrical hair pin resonator, the Q of the resonator depends solely on the chip material on which the circuit is assembled.

A push–push oscillator consists of two identical oscillators with one common resonator, where the two sides are 180° out of phase. This is applicable to the hair pin resonator and the authors [1] have shown that such an arrangement has 40-dB suppression of the second harmonic. While the third harmonic is only suppressed by about 10 dB, this is specifically due to the fact that the oscillator shows higher modes. When comparing the phase noise of a signal versus a push–push configuration (as shown in Figure A-29), an improvement of about 10 dB can be realized. A push–push arrangement helps to cancel noise from external devices; since the oscillators are also interface locked to each other, the overall performance

548 MATHEMATICAL REVIEW

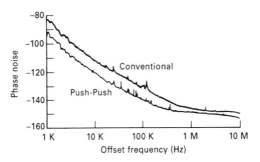

Figure A-29 Comparison between SSB phase noise of the conventional and push–push oscillators.

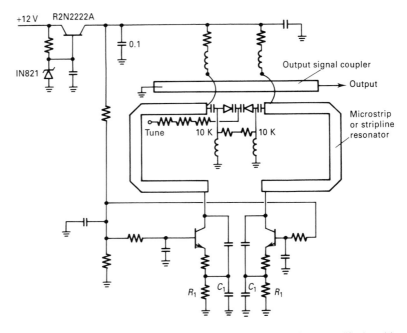

Figure A-30 State-of-the-art push–push VCO, based on the original [1] but modified and improved by James Crawford [2].

is improved. These types of oscillators are fairly recent developments and I believe more exciting results will be available in the future.

The actual circuit diagram is shown in Figure A-30. It is based on the original publication [1] as improved by James A. Crawford [2]. As shown in the original, the push–push arrangement has reduced the phase noise by at least 10 dB.

There are other circuits, such as feedback circuits, available, which reduce the phase noise by 10 dB or more. This is an important area where we expect to see more interesting contributions.

REFERENCES

1. H. Yabucki et al., "VCOs for Mobile Communications," *Applied Microwave*, Winter 91/92.
2. James A. Crawford, *Frequency Synthesizer Design Handbook*, Artech House, Boston, MA, 1994.

APPENDIX B

A GENERAL-PURPOSE NONLINEAR APPROACH TO THE COMPUTATION OF SIDEBAND PHASE NOISE IN FREE-RUNNING MICROWAVE AND RF OSCILLATORS

B-1 INTRODUCTION

We want to look at the contribution of the active devices like FETs and BIPs as well as a novel algorithm for the computation of near-carrier noise in free-running microwave oscillators by the nonlinear harmonic-balance (HB) technique [1].

The application of the HB methodology to nonlinear noise analysis is very effective, because frequency-domain analysis is well suited for describing the mechanism of noise generation in nonlinear circuits. In the last few years, this topic has received the interest of several research teams; however, until now, a rigorous treatment of noise analysis in autonomous circuits has not appeared in the technical literature.

The usual approach relying on a simple noise model of the active device and the frequency-conversion analysis is not sufficient to describe the complex physical behavior of a noisy oscillator. Instead, we apply the following approach:

- A complete bias-dependent noise model for bipolar transistors and FETs is developed.
- The frequency-conversion approach is reviewed and its limitations are pointed out.
- It is shown how the analysis procedure can be extended to include the case of autonomous circuits.
- The capabilities of the proposed algorithm are demonstrated by means of some application examples.

B-2 NOISE GENERATION IN OSCILLATORS

The qualitative picture of noise generation in oscillators is very well known. As previously outlined, Lesson had developed a linear model,

$$\mathcal{L}(f_m) = \frac{1}{2}\left[1 + \frac{1}{\omega_m^2}\left(\frac{\omega_o}{2Q_{\text{load}}}\right)^2\right]\frac{FkT}{P_{\text{sav}}}\left(1 + \frac{f_c}{f_m}\right)$$

that requires the following input parameters: (1) RF output power, (2) large signal noise figure, (3) loaded Q, and (4) flicker component. The harmonic-balance method is used to calculate the RF output power of the oscillator and at the same time calculate the loading of the tuned circuit as a function of the large-signal condition. The flicker frequency (flicker frequency corner) is a device-dependent parameter that has to be entered. A more complete expression is as follows:

$$\begin{aligned}S_\phi(f_m) = &[\alpha_R F_0^4 + \alpha_E(F_0/(2Q_L))^2]/f_m^3 \\ &+ [(2GFKT/P_0)(F_0/(2Q_L))^2]/f_m^2 \\ &+ (2\alpha_R Q_L F_0^3)/f_m^2 \\ &+ \alpha_E/f_m + 2GFKT/P_0\end{aligned}$$

where

G = compressed power gain of the loop amplifier
F = noise factor of the loop amplifier
K = Boltzmann's constant
T = temperature in °K
P_0 = carrier power level (in Watts) at the output of the loop amplifier
F_0 = carrier frequency in Hz
f_m = carrier offset frequency in Hz
$Q_L(= \pi F_0 \tau_g)$ = loaded Q of the resonator in the feedback loop
α_R and α_E = flicker noise constants for the resonator and loop amplifier, respectively

B-3 BIAS-DEPENDENT NOISE MODEL

Modeling the Device. Figure B-1 shows the traditional Gummel–Poon [1] model for bipolar transistors. For microwave applications, the model has to be transformed into a T-equivalent circuit, as shown in Figure B-2.

To be compatible, the equivalent circuit had to be updated by adding the appropriate resistor R_{ce} to it. For the bipolar transistor, a convenient starting solution to determine the intrinsic values of R_{bb}, R_e, and C_e is a set of equations that calculate the four noise parameters $F_{\min}$, Γ_{opt}, and R_n.

$$F_{\min} = a\frac{R_b + R_{\text{opt}}}{r_e} + \left(1 + \frac{f^2}{f_b^2}\right)\frac{1}{\alpha_0}$$

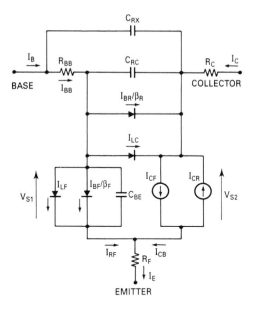

Figure B-1 Gummel–Poon bipolar transistor model.

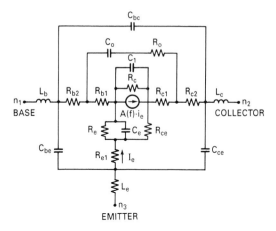

Figure B-2 Small-signal T-model derived from linear hybrid π model. Note the additional resistor R_{ce}, which is necessary for modeling reasons.

The optimum source resistance is

$$R_{\text{opt}} = \left\{ R_b^2 - X_{\text{opt}}^2 + \left(1 + \frac{f^2}{f_b^2}\right) \frac{r_e(2R_b + R_e)}{\alpha_0 a} \right\}^{1/2} \tag{B-2}$$

and optimum source reactance is

$$X_{\text{opt}} = \left(1 + \frac{f^2}{f_b^2}\right) \frac{2\pi f C_{Te} R_e^2}{\alpha_0 a}$$

where

$$a = \left[\left(1 + \frac{f^2}{f_b^2}\right)\left(1 + \frac{f^2}{f_e^2}\right) - \alpha_0\right]\frac{1}{\alpha_0} \quad \text{(B-3)}$$

$$R_n = R_b\left(A - \frac{1}{\beta_0}\right) + \frac{R_e}{2}\left(A + \left(\frac{R_b}{R_e}\right)^2\left\{1 - \alpha_0 + \left(\frac{f}{f_b}\right)^2 + \left(\frac{f}{f_e}\right)^2\right.\right.$$
$$\left.\left. + \left[\frac{1}{\beta_0} - \left(\frac{f}{f_b}\right)\left(\frac{f}{f_e}\right)\right]^2\right\}\right) \quad \text{(B-4)}$$

where

$$A = \frac{1 + \left(\frac{f}{f_b}\right)^2}{a_0^2}$$

and f_b denotes the *cutoff* frequency of the common base current gain $\alpha(f)$. The above provides a convenient set of equations for representing the low-frequency noise performance of a bipolar transistor. Unlike Fukui's formula, the new expression does not involve the unity current gain frequency f_T.

These equations are based on Refs. 2 and 3 and have been modified by us to reflect the modern geometry. These results have been published in the *IEEE-MTT Transactions* [4]. Further information can be found in Ref. 5. Based on actual noise measurements, they predict the starting values for the base spreading resistor R_{bb} and the input capacitor C_e in schematic, while the emitter diffusion resistor can be calculated directly from the dc bias point. These values have better accuracy than the traditional Gummel–Poon parameter extraction for the large-signal bipolar model. A typical set of parameters for a microwave resistor is shown in Table B-1. These are the results of a parameter extraction method using different dc bias points and ensuring that the noise measurements agree with the predictions. Subsequently, by invoking the transformation from the hybrid π to the T model, one obtains the dc or small-signal equivalent circuit for the transistor under the bias point chosen. The calculation of the four noise parameters, which is now based on the various dc bias points, generates a table that can be used for interpolation and can be translated into the equivalent noise correlation matrix.

Figure B-3 shows the measured and modeled noise based on the parameter extraction.

A similar approach is possible with GaAs FETs. Figure B-4 shows the linear equivalent circuit for GaAs FETs and Figure B-5 shows the large-signal equivalent circuit for which we have used the Materka model. An important feature is that the large-signal model using small-signal conditions generates the same set of S-parameters from the equivalent small-signal circuit.

Noise performance of microwave circuits is one of the major concerns of circuit design engineers. It is an important determining factor of receiver system sensitivity and dynamic range. Since the noise correlation matrix has been introduced, noise analysis of microwave linear circuits has been available and implemented in general-purpose CAD tools. However, in the practical world, most microwave circuits need to be analyzed using nonlinear analysis techniques.

Table B-1 Small- and large-signal parameters of an intrinsic transistor[a]

```
MICROWAVE HARMONICA PC V5.0   10-JUL-92   14:10:31
File: bfr965s.ckt

* Linear/NONlinear BIP description:
*
  BIP 53 56 58
+   ; LINEAR parameters:
+      LB = 0              LC = 0              LE = 0
+      RB2 = 0             RC2 = 0.931         CBE = 0
+      CCE = 0             CBC = 0             LBT = 0
+      ZBT = 50            LCT = 0             ZCT = 50
+      LET = 0             ZET = 50            CBEP = 0
+      CBCP = 0            CCEP = 0            RE1 = 0.53
+      RC1 = 0             RO = 0              T = 0
+      F = 8.868E+009      CO = 9.452E-013     RB1 = 7.869
+      A = 0.9686          RC = 1E+030         CI = 1.667E-013
+      RE = 0.4097         CE = 2.331E-010     RCE = 2337
+      TJ = 293
+   {
+   ; NONLINEAR parameters:
+      BF = 169            BR = 16.43          NF = 0.975
+      NE = 1.527          NR = 1.007          NC = 1.097
+      IS = 1.6E=016       ISE = 1.645E-014    ISC = 2.252E-015
+      VA = 117            VB = 1.782          IKR = 0.05949
+      IKF = 0.157         RE1 = 0.53          RC2 = 0.931
+      RBM = 0.025         RB = 10.24          IRB = 0.01051
+      TR = 0              TF = 1.8E-011       ITF = 1
+      XTF = 84            VTF = 0.7           FCC = 0.5
+      VJE = 1.079         MJE = 0.471         CJE = 5.46E-012
+      XCJC = 0.15         CJC = 2.631E-012    VJC = 0.106
+      MJC = 0.21          TJ = 293            XTB = 0
+      XTI = 3             TRE1 = 0            TRE2 = 0
+      TRB1 = 0            TRB2 = 0            TRM1 = 0
+      TRM2 = 0            TRC1 = 0            TRC2 = 0
+      TNOM = 293          VCMX = 11           IBMX = 9.973E+011
+      IBMN = 0            NPLT = 6            NAME = BIP_NPN
+      ANA = OFF           MODEL = NPN         RB2 = 0
+      LB = 0              LC = 0              LE = 0
+      CBE = 0             CCE = 0             CBC = 0
+      LBT = 0             ZBT = 50            LCT = 0
+      ZCT = 50            LET = 0             ZET = 50
+      CBEP = 0            CBCP = 0            CCEP = 0
+   }
```

[a] The small-signal parameters were generated from the large-signal model at the dc operating point of 90 mA.

This section illustrates how the noise performance of general mixer circuits can be simulated by using the harmonic-balance technique implemented in the enhanced version of Microwave Harmonica v4.0, a workstation product made by Compact Software, Inc.

Noise in a microwave FET is produced by sources intrinsic to the device. If

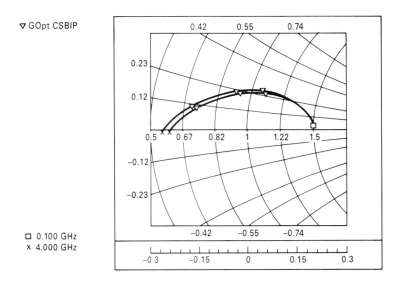

▽ GOpt CSBIP

□ 0.100 GHz
× 4.000 GHz

Figure B-3 Measured and modeled noise based on the parameter extraction.

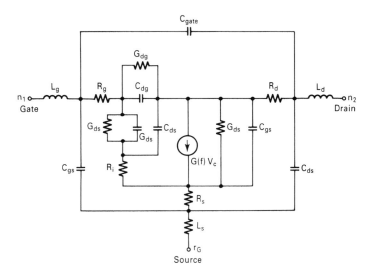

Figure B-4 Linear equivalent circuit of FET, which can be derived from Figure B-5, the large-signal equivalent circuit.

the equivalent noisy circuit of an intrinsic FET device is represented as in Figure B-6, the correlations of the gate and drain noise current sources are

$$\langle |I_d|^2 \rangle = 4K_B T \Delta f g_m P \tag{B-5}$$

$$\langle |I_g|^2 \rangle = 4K_B T \Delta f \frac{\omega^2 c_{gs}^2}{g_m} R \tag{B-6}$$

$$\langle I_g I_d^* \rangle = 4K_B T \Delta f j \omega c_{gs} \sqrt{PRC} \tag{B-7}$$

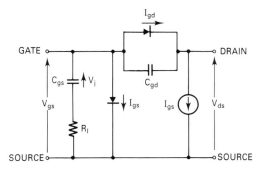

Figure B-5 Intrinsic Curtice–Ettenberg model for the MESFET. Other useful models are the Materka–Kacprzak model for the MESET (best for millimeter-wave applications), the Statz model, and the TOM model.

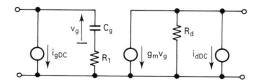

Figure B-6 Equivalent noise circuit of an intrinsic FET device.

and the correlation matrix of the noise current sources is

$$C_{dc}(\omega) = \frac{2}{\pi} K_B T d\omega \begin{bmatrix} \frac{\omega^2 C_{gs}}{g_m} R & -j\omega C_{gs}\sqrt{PR}\, C \\ j\omega C_{gs}\sqrt{PR}\, C & g_m P \end{bmatrix} \quad \text{(B-8)}$$

The gate and drain noise parameters R and P and the correlation coefficient C are related to the physical noise sources acting in the channel and are thus functions of the device structure and bias point. These noise parameters at a certain bias point can be calculated explicitly from measured device noise parameters using a noise de-embedding procedure. That is, by defining measured noise parameters, $F_{\min}$, R_n, and Γ_{opt}, and using the Super-Compact noise de-embedding procedure, the noise correlation matrix of an FET device can be determined.

The next step is to develop a bias-dependent model. This is necessary because one has to develop an analysis program that can handle the noise effect in conjunction with a general-purpose harmonic-balance simulator. The method to be used is concerned with an extension of the usual linearization adopted to apply the piecewise linear balance technique. For our purposes, the nonlinear subnetwork is a collection of intrinsic FET chips with all (linear) parasitic elements included in the linear subnetwork. These linear components are the time-averaged values as a function of the local oscillator pumping and the dc bias. It is therefore necessary to develop a bias-dependent model that can be used to obtain the necessary foundry coefficients. Using the noise correlation technique, they have

Table B-2 Corresponding correlation between the Gummel–Poon model predictions and the measurements of the chip: $V_C = 10$ V at 4 mA

Frequency (GHz)	F_{min} (dB) CSBIP	MG_{opt} (mag) CSBIP	PG_{opt} (deg) CSBIP	Run (ohm) CSBIP
Gummel–Poon Model Predictions				
0.500	1.16	0.138	41.7	8.363
1.000	1.30	0.165	79.8	8.381
1.500	1.49	0.214	106.8	8.411
2.000	1.72	0.272	123.8	8.454
2.500	1.97	0.329	134.8	8.510
3.000	2.23	0.379	142.2	8.578
3.500	2.50	0.423	147.6	8.660
4.000	2.77	0.461	151.6	8.756

Frequency (GHz)	F_{min} (dB) HPBIP	MG_{opt} (mag) HPBIP	PG_{opt} (deg) HPBIP	Run (ohm) HPBIP
Gummel–Poon Model Predictions				
0.500	1.20	0.148	37.2	8.886
1.000	1.34	0.164	74.5	8.907
1.500	1.54	0.205	103.4	8.943
2.000	1.78	0.259	122.0	8.993
2.500	2.05	0.314	133.7	9.057
3.000	2.32	0.364	141.6	9.137
3.500	2.60	0.408	147.2	9.232
4.000	2.88	0.446	151.4	9.343

been based on a de-embedding technique using measurements of the four noise parameters at a test frequency like 10 GHz. This de-embedding technique is the subject of another paper [6] while the general treatment of this had already been published [7].

This method is accurate enough that measured data and predicted data agree quite well. Table B-2 shows the corresponding correlation between the two. In the case of the bipolar transistor, the novel approach in generating the starting values for the Gummel–Poon model is using small-signal noise data as seed values first and then refining the values for different bias points. This is somewhat easier because the bipolar transistor in the Gummel–Poon approach is a physics-based model, while the Materka model is the result of a curve fit and has no physical equivalent.

Table B-3 shows the *R*, *P*, and *C* values of a MESFET as a function of bias, which are a result of calculation and measurement.

In generating large-signal models, a set of starting values is required. In the case of bipolar transistors, we use the values that are supported by the noise calculation. The standard parameter extraction programs, like in the HP TCAP

Table B-3

			P Values				
I_{ds}/I_{dss}							
0.85	2.742	2.850	2.860	2.851	2.846	2.857	
0.70	2.181	2.482	2.556	2.569	2.571	2.581	
0.50	1.576	1.939	2.051	2.078	2.083	2.089	
0.30	1.307	1.460	1.527	1.541	1.540	1.541	
0.15	1.358	1.263	1.254	1.246	1.240	1.242	
0.05	1.358	1.390	1.339	1.321	1.317	1.327	
	0.25 V	0.70 V	1.10 V	1.50 V	2.10 V	2.70 V	V_{ds}

			R Values				
I_{ds}/I_{dss}							
0.85	0.122	0.173	0.152	0.175	0.259	0.238	
0.70	0.131	0.193	0.180	0.211	0.306	0.294	
0.50	0.146	0.194	0.178	0.210	0.302	0.299	
0.30	0.192	0.215	0.200	0.231	0.315	0.332	
0.15	0.236	0.265	0.274	0.315	0.400	0.450	
0.05	0.236	0.283	0.328	0.380	0.464	0.538	
	0.25 V	0.70 V	1.10 V	1.50 V	2.10 V	2.70 V	V_{ds}

			C Values				
I_{ds}/I_{dss}							
0.85	0.211	0.430	0.537	0.592	0.621	0.650	
0.70	0.217	0.441	0.552	0.608	0.639	0.668	
0.50	0.230	0.456	0.570	0.628	0.661	0.691	
0.30	0.253	0.485	0.603	0.665	0.700	0.732	
0.15	0.279	0.535	0.665	0.735	0.775	0.810	
0.05	0.279	0.612	0.768	0.852	0.901	0.943	
	0.25 V	0.70 V	1.10 V	1.50 V	2.10 V	2.70 V	V_{ds}

program or the IC-CAP, are insufficient to obtain some of the microwave properties. In particular, the intrinisic delay times and, in the case of the bipolar transistors, the base-spreading resistor parameter extraction do not show enough sensitivity. We have seen variations of 5:1. In some cases, the standard large-signal parameter extraction program has provided a base-spreading resistor of 50 Ω, while the noise modeling approach has calculated 10 Ω with the correct value being 8 Ω. As an example, $R_n = 0.2$ transforms into $0.2 \times 50 = 10\,\Omega$. The equivalent circuit generation of the FET parameters, particularly the parasitics, is summarized in Ref. 7. The intrinisc value of the devices, however, can be obtained by the method shown.

Bias-Dependent Model. We now describe our bias-dependent small-signal FET model, which is used for the piecewise linear harmonic-balance technique. We stress the word "small" since the model is not a nonlinear one, but a linear one.

A bias-dependent small-signal model serves two purposes: (1) it permits "tweaking" of the MMIC by external (bias voltage) means and (2) it introduces another degree of freedom in "noise matching." The first application is important since MMICs, by their very nature, do not permit on-chip adjustments: indeed, this would be counter to the whole purpose of the MMIC approach. Therefore, only external means of adjustment are allowed. The second application is important because it allows one to achieve a better compromise between a noise match and a power–gain match than one could possibly reach by a matching circuit technique alone.

We believe very strongly that one cannot derive a bias-dependent linear FET model from a nonlinear FET model that would be sufficiently accurate to satisfy the critical MMIC designer. The reason for our belief is that whereas a nonlinear nonphysical based model is obtained by some "curve-fitting" technique, which will be adequate for large-signal excursions, this fitting procedure makes no attempt to match the *derivative* of the nonlinear function that one is fitting. But small-signal parameters are derivatives of a nonlinear function; therefore, one cannot ensure accuracy by this method by deriving the small-signal parameters. Rather, we believe that the required accuracy can be obtained by directly measuring the small-signal performance as a function of bias voltages and then *fitting* this dependence by some simple function. This simple function is no more complicated than a second-degree polynomial, that is, a quadratic function.

We restrict the model to "above the knee" operations, since it is a rare occasion that one would operate the FET as a linear device below the knee. We recognize, also, that the greatest dependence of the model parameters is on the gate–source voltage. The drain–source voltage plays a secondary role. Indeed, the dependence on the latter voltage is in most cases simply a linear one.

Derivation of the Model. For a selected set of drain–source bias voltages, one may represent any small-signal FET parameter in one of the following forms:

$$P(V_{gs}) = a + b(V_{gs} - V_p) + c(V_{gs} - V_p)^2 \tag{B-9}$$

or

$$P(I_{ds}) = a' + b I_{ds}^{0.5} + c' I_{ds} \tag{B-10}$$

where P represents any of the small-signal equivalent circuit parameters and the desired expansion coefficients. Here V_{gs}, V_p, and I_{ds} denote the gate–source bias voltage, the pinch-off voltage, and the drain–source bias current, respectively. Since the parameter P is temperature dependent, the expansion coefficients also are a function of temperature, albeit mild ones and most probably negligible. This has been verified, however, by analysis of the data taken on a group of devices.

Note that Eq. (B-10) is in reality a quadratic polynomial in the square root of I_{ds}. This form follows directly from Eq. (B-9) because of the nearly quadratic dependence of drain current on gate–source bias that we have observed with many devices. We shall show samples of this dependence for a small pinch-off device ($V_p = 1.8$ V).

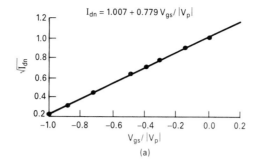

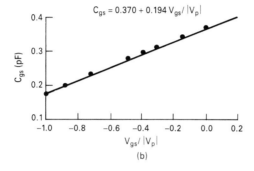

Figure B-7 Linear dependence of (a) normalized drain current and (b) gate–source capacitance on gate–source bias voltage.

Figure B-7 illustrates the quadratic dependence of I_{ds} on V_{gs} and the linear dependence of C_{gs} on V_{gs}. The fits are "perfect," as evidenced by the fact that the coefficients of the quadratic terms are zero. Figure B-8 shows, however, that in some cases quadratic terms are necessary. Although we have found that g_m usually can be fitted with a linear function for small pinch-off devices (this example being an exception), a quadratic term is usually necessary for large pinch-off devices (V on the order of 3 to 6 V). A quadratic term is always required for g_{ds}, however, regardless of pinch-off voltages. Some of the other model parameters such as the delay times usually require a quadratic term, although a linear approximation will probably suffice because of the insensitivity of the device performance to this quantity.

In a mixer or oscillator, the active device is not only dc biased but also pumped by the local oscillator. At the dc bias point of the device, the nonlinear noise sidebands are uncorrelated and are dependent on the bias point. When the device is pumped by the LO, the nonlinear noise sidebands are modulated accordingly and are partially correlated because each sideband is a combination of the original uncorrelated dc sidebands. During mixing, each sideband generates a correlated component in the vicinity of the IF. To determine the correct nonlinear noise power contribution, the correlation of the sidebands must be considered in the analysis. Similarly, contributions to the noise power at the IF load are made by

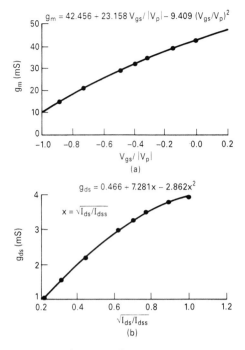

Figure B-8 Quadratic dependence of transconductance on gate–source voltage and drain–source conductance on the square root of the normalized drain current.

the thermal noise generated by the linear network through frequency conversion in the mixer. The thermal noise is not dependent on the LO excitation and, because it is uncorrelated, its noise power contribution is additive.

B-4 GENERAL CONCEPT OF NOISY CIRCUITS

In the evaluation of a two-port, it is important to know the amount of noise added to a signal passing through a network.

$$\frac{S_{in}}{N_{in}} \rightarrow \text{Network} \rightarrow \frac{S_{out}}{N_{out}} \tag{B-11}$$

An important parameter for expressing this characteristic is the noise factor (or noise figure).

$$\text{Noise factor} = F = \frac{S_{in}/N_{in}}{S_{out}/N_{out}}$$

$$\text{Noise figure} = NF = 10 \log (F) \tag{B-12}$$

The noise figure of cascaded networks can be calculated by the following

GENERAL CONCEPT OF NOISY CIRCUITS 563

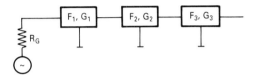

Figure B-9 Chain of amplifiers.

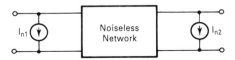

Figure B-10 Noiseless circuit noise sources at the input and output.

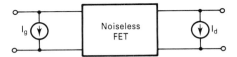

Figure B-11 Noiseless FET with noise sources at the input and output.

approximations shown in Figure B-9. The approximation assumes a 50-Ω resistive termination. The correct and frequently overlooked method for this, of course, is the noise correlation matrix.

$$F = F_1 + \frac{F_2 - 1}{G_1} + \frac{F_3 - 1}{G_1 G_2} + \frac{F_4 - 1}{G_1 G_2 G_3} + \cdots \quad \text{(B-13)}$$

The sources of the internal noise in a general circuit are described next.

Noise from Linear Elements. Thermal noise is related to the admittance of the elements:

$$C_n(\omega) = \frac{1}{\pi} K_B T \delta\omega [Y(\omega) + Y^*(\omega)] \quad \text{(B-14)}$$

A noise network can be treated as a noiseless network with equivalent noise current source at each external port.

The correlation of the noise current sources of a linear network is related to the Y matrix of this network. This is shown in Figure B-10.

The intrinsic noise sources of an active device (e.g., MESFET, BJT) are at the input and the output as shown in Figure B-11. The intrinsic noise model can be expressed by four measured parameters:

$F_{\min}$ Minimum noise figure
R_n Equivalent normalized noise resistance
MG_o Magnitude of the optimal noise reflection coefficient
PG_o Phase of the optimal noise reflection coefficient

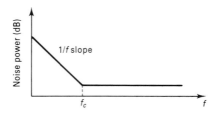

Figure B-12 The major parameter used to describe the flicker noise is f_c—corner frequency.

From these four parameters, the Van der Ziel noise model can be derived as

$$C_n(\omega) = \frac{2}{\pi} K_B T \delta\omega \begin{bmatrix} \dfrac{\omega^2 C_{gs}^2}{g_m} R & -j\omega C_{gs}\sqrt{PR}\,C \\ j\omega C_{gs}\sqrt{PR}\,C & g_m P \end{bmatrix} \quad \text{(B-15)}$$

This conversion, shown in Figure B-11 for all active devices, has been implemented in both Super-Compact and Microwave Harmonica.

In addition, we have to add the flicker noise of an active device (1/f noise), as displayed in Figure B-12.

We now look at the noise model of the active device when pumped by an LO. The noise sources and equivalent circuit model parameters are modulated by the LO. This is indicated in Figure B-13.

The noise correlation matrix of the device is modulated by the LO. This means

$$R, P, C, g_m, C_{gs}, \cdots = f(V_{gs}, V_{ds}) \quad \text{(B-16)}$$

In addition, the flicker noise is modulated by the drain current using the following equation:

$$\langle |I_f|^2 \rangle = 2 K_B T \delta\omega Q \frac{|I_D|^\beta}{f^\alpha} \quad \text{(B-17)}$$

If we consider an oscillator as a mixer driven by a noisy source, we have to consider

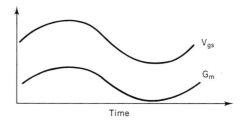

Figure B-13 The voltages and currents of devices are determined by harmonic-balance calculations.

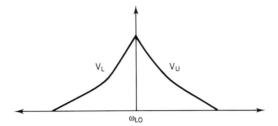

Figure B-14 LO signal with noise sidebands.

the noise contribution of the external sources (LO). The source noise is given by the single sideband (SSB) RF spectrum, and the amplitude. Noise or frequency fluctuation is a set of frequency deviations from the carrier. This is frequently referred to as the spectrodensity of a signal, as shown in Figure B-14.

B-5 NOISE FIGURE OF MIXER CIRCUITS

In order to calculate the noise figure of a mixer circuit, we need to calculate the total internal noise of the circuit at the IF frequency.

Noise Analysis Step 1. Do the harmonic-balance analysis to determine the steady state of the mixer. Figure B-15 shows the arrangement. The harmonic-balance calculation determines the Fourier coefficients of voltages and currents of the circuit. Any receiver configuration (e.g., LNAs, IF, AMP) may be considered.

Noise Analysis Step 2. Calculate the transfer functions of the sideband signals to the IF band signal. The noise at each sideband frequency contributes to the noise at the IF through frequency conversion, as shown in Figure B-16. The block diagram, Figure B-17, is a summary of the IF noise contributions in a general nonlinear mixer circuit. Please note the large number of contributing elements that make up the total noise at the output.

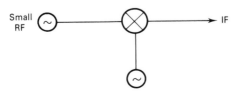

Figure B-15 Mixer arrangement.

566   A NONLINEAR APPROACH TO THE COMPUTATION OF SIDEBAND PHASE NOISE

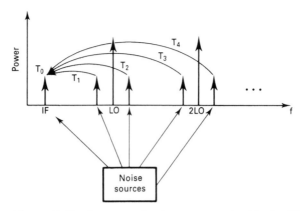

Figure B-16 Summary of noise sources mixed to the IF.

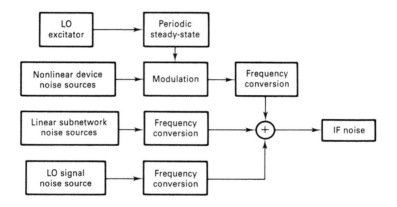

Figure B-17 Summary of IF noise contributions.

The calculation of *dN* is performed by Eq. (B-20), where the intermediate steps are given in Ref. 8.

$$\langle |\delta\Phi|^2(f_d)\rangle = \frac{\langle |V_l(f_d)|^2\rangle + \langle |V_u(f_d)|^2\rangle - 2\text{Re}\{\langle V_l^*(f_d)V_u^*(f_d)\rangle \exp(2j\Phi_0)\}}{|V_0|^2} \quad \text{(B-18)}$$

$$\langle |\delta A|^2(f_d)\rangle = 2\frac{\langle |V_l(f_d)|^2\rangle + \langle |V_u(f_d)|^2\rangle + 2\text{Re}\{\langle V_l^*(f_d)V_u^*(f_d)\rangle \exp(2j\Phi_0)\}}{|V_0|^2} \quad \text{(B-19)}$$

$$dN(\omega_{\text{IF}}) = R_{\text{IF}} \sum_p T_{0p} C_L(\omega_{\text{IF}} + p\omega_0) T_{0p}^*$$

$$+ R_{\text{IF}} \sum_{p,q} T_{0p} \left[\sum_s H_{p-s} C_{dc}(\omega_{\text{IF}} + s\omega_0) H_{s-q}\right] T_{0q}^*$$

$$+ R_{\text{IF}} \sum_{p,q} Y_p^s \begin{bmatrix} \langle |V_u(\omega_{\text{IF}})|^2\rangle & \langle V_u(\omega_{\text{IF}}) V_l^*(\omega_{\text{IF}})\rangle \\ \langle V_u^*(\omega_{\text{IF}}) V_l(\omega_{\text{IF}})\rangle & \langle |V_l(\omega_{\text{IF}})|^2\rangle \end{bmatrix} Y_q^{s*} \quad \text{(B-20)}$$

In Eq. (B-20), the T_{0x} terms are the sideband-to-IF conversion matrices; H_x terms are the spectral modulation components of the device; p, q, r, and s are sideband spectral indices; R_{IF} is the IF load; Y is a conversion admittance matrix between the LO noisy source and the IF load at the IF frequency; and ω_{IF} is a small frequency deviation in the neighborhood of the baseband frequency. The first term represents the noise contribution of the linear network, the second term is the noise contribution from the modulated nonlinear devices, and the third term is the noise contribution of the noisy LO.

B-6 OSCILLATOR NOISE ANALYSIS

The effect of the local oscillator (LO) frequency (f_o) and its noise can be determined through perturbation analysis.

The noise sources modulate the carrier frequency. Flicker noise is the predominant noise source. The arrangement in Figure B-18 looks similar to the mixer of Figure B-15. The near-carrier noise is proportional to the flicker, thermal, and device shot noise power and is inversely proportional to Q^2 of the oscillator. As shown in the block diagram, the noise contains the flicker components and the noise calculated via the correlation matrix.

The noise sources at sideband and baseband frequencies contribute to near-carrier noise through frequency conversion. The noise contribution is significant at deviations far from the carrier, as shown in Figure B-19.

The physical effects of random fluctuations taking place in the circuit are different depending on their spectral allocation with respect to the carrier:

Noise Components at Low-Frequency Deviations

- Frequency modulation of the carrier
- Mean square frequency fluctuation proportional to the available noise power

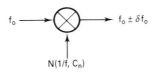

Figure B-18 Oscillator signal modulated by noise sources including flicker noise.

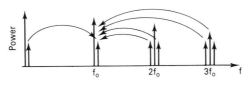

Figure B-19 Noise sources at different frequencies contribute to the near-carrier noise.

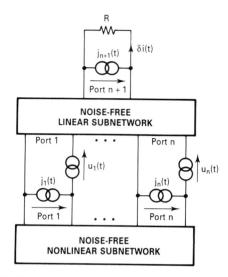

Figure B-20 Equivalent representation of a noisy nonlinear circuit.

Noise Components at High-Frequency Deviations

- Phase modulation of the carrier
- Mean square phase fluctuation proportional to the available noise power

The main purpose of this chapter is to show that the same results can be quantitatively derived from the nonlinear HB equations of the autonomous circuit and how the general nonlinear approach uses a nonlinear BIP or FET model for the noise calculation.

A general noise nonlinear network can be described by the circuit in Figure B-20. The circuit shown is then divided into linear and nonlinear subnetworks, represented as noise-free multiports. The noise generation is accounted for by connecting a set of noise voltage and noise current sources at the ports of the linear subnetwork.

For the frequency-conversion noise analysis, we now make the following assumptions: The circuit supports a large-signal time-periodic steady state of fundamental frequency ω_0 (carrier). Noise signals are small perturbations superimposed on the steady state, represented by families of pseudo-sinusoids located at the sidebands of the carrier harmonics. Therefore, the noise performance of the circuit is determined by the exchange of power among the sidebands of the unperturbed steady state through frequency conversion in the nonlinear subnetwork. Due to the perturbative assumption, the nonlinear subnetwork can be replaced by a multifrequency linear multiport described by a conversion matrix. The flow of noise signals can be computed by means of conventional linear circuit techniques.

B-7 LIMITATIONS OF THE FREQUENCY-CONVERSION APPROACH

The frequency-conversion approach is not sufficient to predict the noise performance of an autonomous circuit. The spectral density of the output noise power and, consequently, the PM noise computed by conversion analysis are proportional to the available power of the noise sources.

In the presence of both thermal and flicker noise sources, PM noise increases as ω^{-1} for $\omega \to 0$ tends to a finite limit for $\omega \to \infty$. While the frequency-conversion analysis correctly predicts the far-from-carrier noise behavior of an oscillator and, in particular, the oscillator noise floor, it does not provide results consistent with physical observations at low deviations from the carrier. The inconsistency is removed by the perturbation analysis of autonomous circuits.

Assumptions. The circuit supports a large-signal time-periodic autonomous regime. The circuit is perturbed by a set of small sources located at the carrier harmonics and at the sidebands at a deviation ω from carrier harmonics.

Now we can find the results of the perturbation of the HB equations. The perturbation of the circuit state $(\delta X_B, \delta X_H)$ is given by the uncoupled sets of equations:

$$\frac{\partial E_B}{\partial X_B} \delta X_B = J_B(\omega) \tag{B-21}$$

$$\frac{\partial E_H}{\partial X_H} \delta X_H = J_H(\omega) \tag{B-22}$$

where E_B, E_H = vectors of HB errors
X_B, X_H = vectors of state-variable (SV) harmonics (since the circuit is autonomous, one of the entries of X_H is replaced by the fundamental frequency ω_0)
J_B, J_H = vectors of forcing terms where the subscripts B and H denote sidebands and carrier harmonics, respectively

Conversion and Modulation Noise. For a spot noise analysis at a frequency deviation ω, the noise sources can be interpreted in either of two ways. For pseudo-sinusoids with random amplitude and phase located at the sidebands, noise generation is described by Eq. (B-21), which is essentially a frequency-conversion equation relating the sideband harmonics of the state variables and of the noise sources. This description is exactly equivalent to the one provided by the frequency-conversion approach. The mechanism is referred to as conversion noise. For sinusoids located at the carrier harmonics, randomly phase- and amplitude-modulated by pseudo-sinusoidal noise at frequency ω, noise generation is described by Eq. (B-22), which gives the noise-induced jitter of the circuit state, represented by the vector δX_H. The modulated perturbing signals are represented by replacing entries J_H with the complex modulation laws. This mechanism is referred to as modulation noise.

Properties of Modulation Noise. One of the entries in δX_H is $\delta\omega_0$. $\delta\omega_0(\omega)$ equals the phasor of the pseudo-sinusoidal components of the fundamental frequency fluctuations in a 1-Hz band at frequency ω. Equation (B-22) provides a frequency jitter with a mean square value proportional to the available noise power. In the presence of both thermal and flicker noise sources, PM noise increases as ω^{-3} for $\omega \to 0$ and tends to 0 for $\omega \to \infty$. Modulation-noise analysis correctly describes the noise behavior of an oscillator at low deviations from the carrier frequency but does not provide results consistent with physical observations at high deviations from the carrier frequency.

Noise Analysis of Autonomous Circuits. Conversion noise and modulation noise represent complementary descriptions of noise generation in autonomous circuits. The previous discussion has shown that very-near-carrier noise is essentially a modulation phenomenon, while very far-from-carrier noise is essentially a conversion phenomenon; also, Eqs. (B-21) and (B-22) necessarily yield the same evaluation of PM noise at some crossover frequency ωX. The computation of PM noise should be performed by modulation analysis below ωX and by conversion analysis above ω_X.

This criterion is not artificial since Eqs. (B-21) and (B-22) provide virtually identical results in a wide neighborhood of ω_X (usually more than two decades). The same criterion can be applied to AM noise. (In many practical cases, modulation and conversion analyses yield almost identical AM noise at all frequency deviations.)

Conversion Noise Analysis Results. After performing all the necessary calculations, we obtain the following:

- kth harmonic PM noise

$$\langle \Phi_k(\omega)|^2 \rangle = \frac{N_k(\omega) + N_{-k}(\omega) - 2\mathrm{Re}[C(\omega)]}{R|I_k^{SS}|^2} \tag{B-23}$$

- kth harmonic AM noise

$$\langle |\delta A_k(\omega)|^2 \rangle = 2\frac{N_k(\omega) + N_{-k}(\omega) + 2\mathrm{Re}[C_k(\omega)]}{R|I_k^{SS}|^2} \tag{B-24}$$

- kth harmonic PM–AM correlation coefficient

$$\begin{aligned}C_k^{\mathrm{PM\text{-}AM}}(\omega) &= \langle \delta\Phi_k(\omega)\,\delta A_k(\omega)^* \rangle \\ &= -\sqrt{2}\frac{2\,\mathrm{Im}[C_k(\omega)] + j[N_k(\omega) - N_{-k}(\omega)]}{R|I_k^{SS}|^2}\end{aligned} \tag{B-25}$$

where $N_k(\omega)$, $N_{-k}(\omega)$ = noise power spectral densities at the upper and lower sidebands of the kth carrier harmonic

$C_k(\omega)$ = normalized correlation coefficient of the upper and lower sidebands of the kth carrier harmonic
R = load resistance
I_k^{SS} = kth harmonic of the steady-state current through the load

Modulation Noise Analysis Results. Again, after performing all the necessary calculations, we obtain the following:

- kth harmonic PM noise

$$\langle|\delta\Phi_k(\omega)|^2\rangle = \frac{k^2}{\omega^2} T_F \langle J_H(\omega) J_H^\dagger(\omega)\rangle T_F^\dagger \qquad \text{(B-26)}$$

- kth harmonic AM noise

$$\langle|\delta A_k(\omega)|^2\rangle = \frac{2}{|I_k^{SS}|^2} T_{Ak} \langle J_H(\omega) J_H^\dagger(\omega)\rangle T_{Ak}^\dagger \qquad \text{(B-27)}$$

- kth harmonic PM–AM correlation coefficient

$$C_k^{\text{PM-AM}}(\omega) = \langle \delta\Phi_k(\omega)\,\delta A_k(\omega)^*\rangle$$
$$= \frac{k\sqrt{2}}{j\omega|I_k^{SS}|^2} T_F \langle J_H(\omega) J_H^\dagger(\omega)\rangle T_{Ak}^\dagger \qquad \text{(B-28)}$$

where $J_H(\omega)$ = vector of Norton equivalent of the noise sources
T_F = frequency transfer matrix
T_{Ak} = amplitude transfer matrix
R = load resistance
I_k^{SS} = kth harmonic of steady-state current through the load

B-8 SUMMARY OF THE PHASE NOISE SPECTRUM OF THE OSCILLATOR

The numerical approach is important in understanding that the oscillator phase noise is composed of two parts:

1. The near-carrier noise consists of contributions from the perturbation of the oscillating frequency caused by the noise sources at each sideband frequency. This part is the major noise source at the near-carrier frequencies.
2. The far-carrier noise consists of contributions from each sideband noise source through sideband-to-sideband transfer functions. As can be seen in Figure B-21, this part is similar to the mixer noise calculation and is the major noise source at frequencies far away from the carrier.

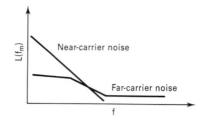

Figure B-21 Oscillator phase noise consisting of near- and far-carrier noise.

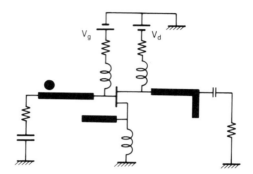

Figure B-22 Schematic topology of Microstrip DRO.

B-9 VERIFICATION EXAMPLES FOR THE CALCULATION OF PHASE NOISE IN OSCILLATORS USING NONLINEAR TECHNIQUES

Example 1: High-Q Case Microstrip DRO Figure B-22 shows the schematic topology of a Microstrip dielectric resonator oscillator (DRO). The specifications for this DRO are as follows:

Oscillation frequency	4.6 GHz
Output power	12.5 dBm
Q of the dielectric resonator	1700

Noise sources considered in the analysis include channel noise and flicker noise. Flicker noise is modeled by a voltage source connected in series to the gate terminal. The dc spectral density of this source is $3.35 \times 10^{-9}/\omega$ V^2/Hz.

This is probably the first time that the various noise contributions such as AM noise, PM noise, conversion contribution, and modulation contribution have been calculated and plotted, as shown in Figure B-23.

Example 2 The equivalent of a DRO at low frequency is a crystal oscillator. In the HP3048 phase noise measurement system, Hewlett-Packard uses the HP10811A 1-MHz frequency standard. For many years, this frequency standard has been the state of the art for low phase noise performance and similar crystal

CALCULATION OF PHASE NOISE IN OSCILLATORS USING NONLINEAR TECHNIQUES

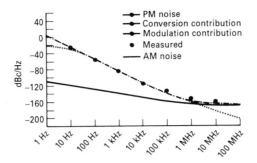

Figure B-23 Conversion contribution and modulation contribution have been calculated and plotted.

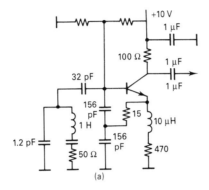

Figure B-24a Abbreviated circuit of a 10-MHz crystal oscillator, where $Q = 200{,}000$.

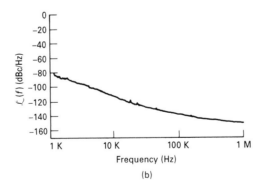

Figure B-24b Simulated phase noise of the oscillator shown in Figure B-24a.

oscillators are now built by a number of companies. Figure B-24a shows an abbreviated circuit of such a crystal oscillator, which uses an extremely high-Q crystal. Figure B-24b displays the simulated phase noise of this circuit.

Figure B-25 shows the phase noise measured and published by Hewlett-Packard and our findings using the Microwave Harmonica v.4.0 algorithm as previously outlined. The extremely good correlation between the two cases can be seen.

574 A NONLINEAR APPROACH TO THE COMPUTATION OF SIDEBAND PHASE NOISE

Example 3: The 1-GHz Ceramic Resonator VCO A number of companies have introduced resonators built from ceramic materials with an ε ranging from 20 to 80. The advantage of using this type of resonator is that they are a high-Q element that can be tuned by adding a varactor diode.

Figure B-26 shows a typical test circuit for use in a ceramic resonator. These resonators are available in the range of 500 MHz to 2 GHz. For higher frequencies, dielectric resonators are recommended. Figure B-27 shows the measured phase noise of the oscillator shown in Figure B-25. The noise pedestal above 100 kHz away from the carrier is due to the reference oscillator model HP8662.

Figure B-28 shows the predicted phase noise of the 1-GHz ceramic resonator VCO without a tuning diode and Figure B-29 shows the predicted phase noise of the 1-GHz ceramic VCO with a tuning diode attached. Please note the good agreement between the measured and predicted phase noise.

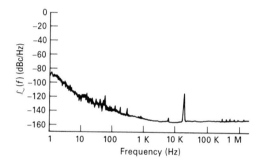

Figure B-25 Measured phase noise for this frequency standard by Hewlett-Packard.

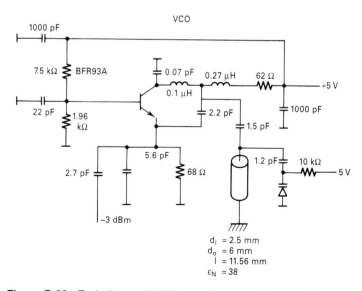

Figure B-26 Typical test circuit for use in a ceramic resonator.

CALCULATION OF PHASE NOISE IN OSCILLATORS USING NONLINEAR TECHNIQUES

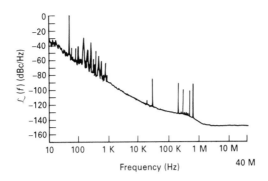

Figure B-27 Measured phase noise of the oscillator shown in Figure B-26.

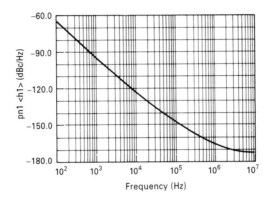

Figure B-28 Predicted phase noise of the 1-GHz ceramic resonator VCO without the tuning diode.

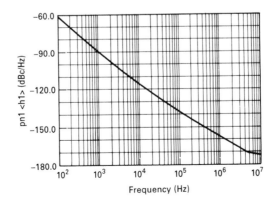

Figure B-29 Predicted phase noise of the 1-GHz ceramic resonator VCO with the tuning diode attached. Please note the good agreement between the measured and predicted phase noise.

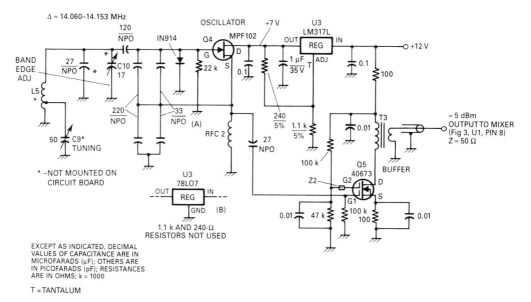

Figure B-30 A 20-meter VFO circuit from the 1993 *ARRL Handbook*.

The ceramic resonator has been modeled using the model "cable"; then the Spice type parameters of the BFR93A transistor data are shown at the beginning of the nonlinear program. The statement at the end of the last line "nois" = B noise indicates that the bias-dependent noise model of the bipolar transistor has been activated. In the data section, the bias-dependent flicker noise for the transistor (bnoise) and the bias-dependent noise of the tuning diode (dnoise) are defined. The center of the netlist shows the element values that are consistent with the previous shown circuit diagram. This noise analysis approach is unique because it takes all factors into consideration and therefore is referred to as the "exact" solution.

Example 4: Low Phase Noise FET Oscillator A number of authors recommend the use of a clipping diode to prevent the gate–source junction of an FET from becoming conductive and thereby lowering the phase noise. In reality, it turns out that this has been a misconception. A popular VCO circuit as described in the American Radio Relay League's (ARRL) manual and shown in Figure B-30 has been analyzed with and without the diode. Claims also have been made that the diode was necessary to obtain long-term stability.

Figure B-31 shows the measured phase noise of an oscillator of this type and Figures B-32 and B-33 show the simulated phase noise of the type of oscillator shown in Figure B-30, with and without a clipping diode. Please note the degradation of the phase noise if the diode is used.

An experimenter found that by removing the diode it did not change or degrade the stability. Additionally, the clipping diode did degrade the phase noise close-in. We have developed a VCO, however, that clips the negative peaks, in the sense that it prevents the oscillator from shutting off. This VCO, as shown in Figure

CALCULATION OF PHASE NOISE IN OSCILLATORS USING NONLINEAR TECHNIQUES 577

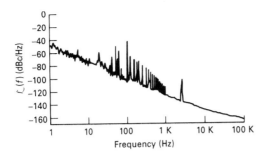

Figure B-31 Measured phase noise of a 10-MHz oscillator of the type shown in Figure B-30.

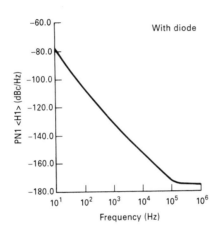

Figure B-32 Simulated phase noise of this type of oscillator with a clipping diode attached.

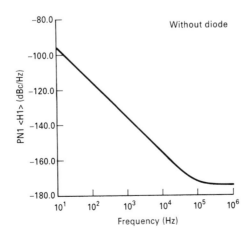

Figure B-33 Simulated phase noise of this type of oscillator without a clipping diode attached.

578 A NONLINEAR APPROACH TO THE COMPUTATION OF SIDEBAND PHASE NOISE

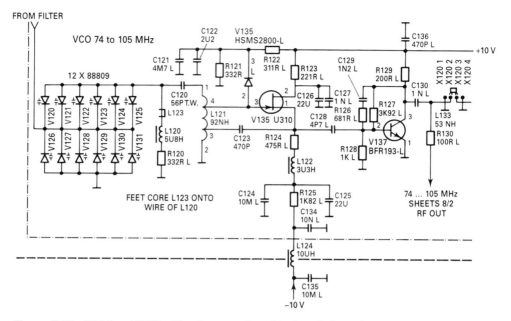

Figure B-34 Wideband VCO with a large number of tuning diodes to improve phase noise. Note that the diode is biased in reverse and does not follow the positive clipping as published by other authors.

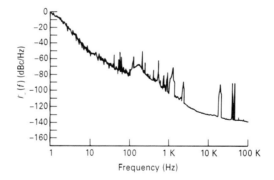

Figure B-35 Phase noise of the multidiode VCO in a PLL system.

B-34, was incorporated in a scheme with a digital direct synthesizer. This synthesizer will be the subject of a later publication.

The phase noise of the combined system was significantly improved. The phase noise of the oscillator shown in Figure B-35, which has only one VCO for the total range from 75 to 105 MHz, when compared to the phase noise of a very recent design like the synthesizer in the TS950, has a 10-dB better S/N ratio at 10 kHz (and further away). This is shown in Figure B-36.

CALCULATION OF PHASE NOISE IN OSCILLATORS USING NONLINEAR TECHNIQUES

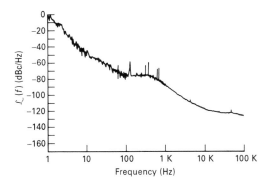

Figure B-36 Phase noise of the TS950, which is 10 dB worse than the multidiode system.

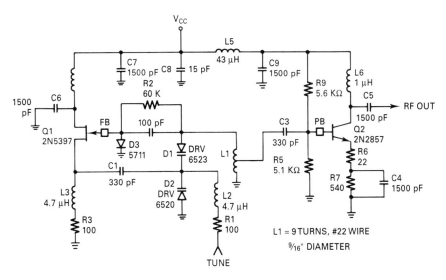

Figure B-37 41 MHz VCO that violates several rules of good design.

Previous authors have tried to build similar wideband oscillators with varying degrees of success. The VCO shown in Figure B-37 violates several rules of designing a good VCO. First, resistor R2 of 68 kΩ, together with C2, provides a time constant that gets close to the audio frequency range. This may result in a super-regenerative receiver, which, of course, is counterproductive. Second, the diode from gate to ground working as a clipping diode also generates more noise. This was outlined earlier. Finally, the feedback selected between the two tuning diodes reduces the operating Q of the resonator to unreasonably small values. If this particular circuit is favored, then the tuning diode D2 should be made out of several (three to five) diodes in parallel. It is therefore not surprising that the measured phase noise shown in Figure B-38 is significantly below state of the art.

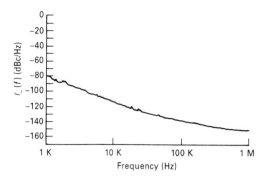

Figure B-38 Phase noise of a 41-MHz oscillator that is significantly below state of the art.

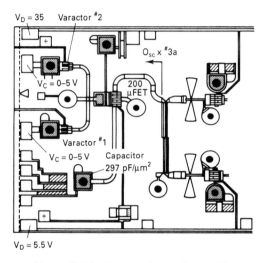

Figure B-39 Layout of a 39-GHz VCO.

Example 5: Millimeter-Wave Applications In millimeter-wave applications such as SMART weapons, which use small radar units for the tracking of enemy targets, MMICs with VCOs are used. One of the most severe tests of software is the combination of millimeter-wave accuracy and nonlinear phase noise calculation. As a last test, I am showing the layout of such a VCO that operates at 39 GHz. While a detailed circuit description of this proprietary design is beyond the scope of this presentation, it should be noted that it is now possible to analyze such complex structures.

Figure B-39 shows the actual layout of the 39-GHz VCO. Figure B-40 shows its schematic presentation. Both transmission lines are used as a resonator and the varactors have fairly low Q values. The resulting phase noise therefore is significantly below that seen in other examples. Even if we take low-frequency oscillators and multiply them up to 39 GHz, we would get better performance. VCOs like this are being used as part of PLL systems that "clean up" some of the noise.

CALCULATION OF PHASE NOISE IN OSCILLATORS USING NONLINEAR TECHNIQUES

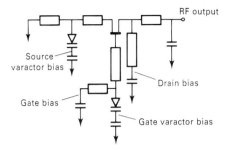

Figure B-40 Schematic presentation and topology of the millimeter-wave VCO.

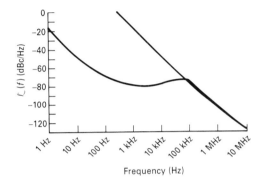

Figure B-41 "Clean-up" from a PLL, also showing the phase noise of the same oscillator in a free-running mode.

Figure B-41 shows the "clean-up" from a PLL and at the same time shows the phase noise of the same oscillator in free-running mode. The operating conditions were 10-MHz reference and 100-kHz loop bandwidth. The "clean-up" is dramatic if one considers the multiplication factor up to 39 GHz. The crystal oscillator's reference is the best low noise crystal type 10811A made by Hewlett-Packard.

Example 6: Discriminator Stabilized DRO Figure B-42 shows an advanced system rather than an oscillator where a DRO is stabilized by a discriminator. The oscillation frequency of the DRO is 6.161 GHz. The output signal is fed to a discriminator using a DR as the frequency selective element. The reflection and transmission coefficients of the DR are equal at the resonant frequency and change in opposite directions when frequency varies in the neighborhood of the resonance frequency. The reflected and transmitted waves are fed to a couple of detector diodes by two directional couplers and a 3-dB hybrid. The variation of the output voltage of the discriminator is roughly proportional to the frequency deviation of the oscillator from the resonance frequency of the DR.

The computed output voltage of the frequency discriminator is shown in Figure B-43. The phase fluctuations of the oscillator are compensated by feeding back

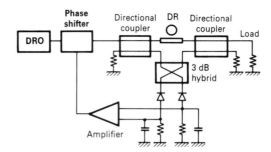

Figure B-42 Advanced system rather than an oscillator where a DRO is stabilized by a discriminator.

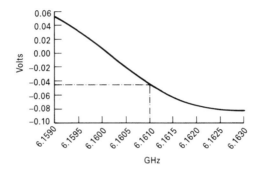

Figure B-43 Computed output voltage of the frequency discriminator shown in Figure B-42.

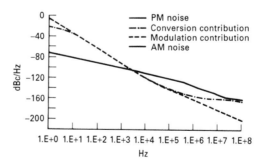

Figure B-44 Noise analysis results.

the output signal of the discriminator to a phase shifter placed between the oscillator and the DR. After properly dimensioning the feedback loop, the DRO PM noise is virtually cancelled, and the output noise is only determined by the noise contributions of the feedback amplifier and the detector diodes.

The noise analysis results are reported in Figure B-44.

Figure B-45 provides a comparison between the open-loop and closed-loop PM noise. An improvement of about 15 dB is obtained up to frequency deviations of

CALCULATION OF PHASE NOISE IN OSCILLATORS USING NONLINEAR TECHNIQUES

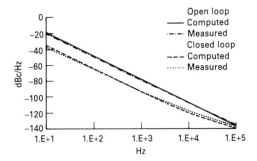

Figure B-45 Open-loop and closed-loop PM noise.

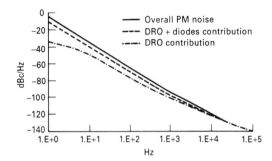

Figure B-46 Contributions to PM noise.

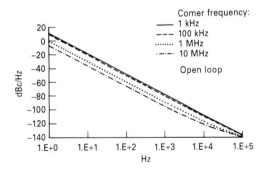

Figure B-47 PM noise versus diode corner frequency.

about 2 kHz. In the figure, measured data are also reported. Computed and experimental results are in very good agreement.

Figure B-46 shows the contribution of different noise sources to the output PM noise.

Finally, Figure B-47 shows the dependence of PM noise on the corner frequency of the detector diodes. (All previous results correspond to corner frequencies of 100 kHz for both the detector diodes and the feedback amplifier.)

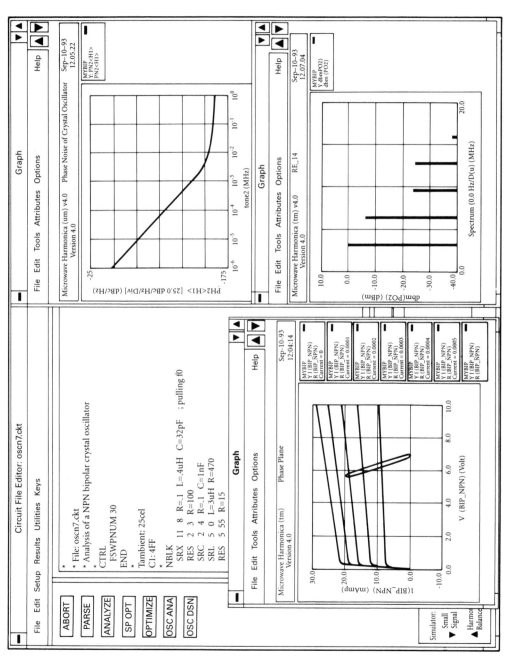

Figure B-48 Screen dump of Microwave Harmonica/Scope phase noise analysis.

B-10 SUMMARY

This combined mathematical and experimental discussion has shown that the new approach implemented in Compact Software's Microwave Harmonica and Scope workstation products provides a fast and accurate method for phase noise analysis. The results can be viewed on the workstation as shown in Figure B-48. This workstation approach is the result of a two-year cooperative effort between Professor Vittorio and his group at the University of Bologna and the engineering staff of Compact Software, Inc. and myself. An additional benefit is that this approach also handles mixer noise. A future enhancement will be the ability to optimize a circuit for output power and phase noise. References 8 to 12 refer to recent work done in this area.

REFERENCES

1. T. Antognefti and G. Massobdo, *Semi-conductor Device Modeling with SPICE*, McGraw-Hill, New York, 1988, p. 91.
2. R. J. Hawkins, "Limitations of Nielsen's and Related Noise Equations Applied to Microwave Bipolar Transistors, and a New Expression for the Frequency and Current Dependent Noise Figure," *Solid State Electronics*, Vol. 20, 1977, pp. 191–196.
3. Tzu-Hwa Hus, and Craig P. Snapp, "Low Noise Microwave Bipolar Transistor Sub-Half-Micrometer Emitter Width," *IEEE Transactions on Electron Devices*, Vol. ED-25, June 1978, pp. 723–730.
4. R. A. Pucel and U. L. Rohde, "An Accurate Expression for the Noise Resistance R_n of a Bipolar Transistor for Use with the Hawkins Noise Model," *IEEE Transactions on Microwave Theory and Techniques (submitted)*.
5. G. Vendelin, A. M. Pavio, and U. L. Rohde, *Microwave Circuit Design*, Wiley, New York, 1990.
6. R. A. Pucel, W. Struble, Robert Hallgren, and U. L. Rohde, "A General Noise De-embedding Procedure for Packaged Two-Port Linear Active Devices," *IEEE Transactions on Microwave Theory and Techniques*, Vol. 40, November 1992, pp. 2013–2025.
7. U. L. Rohde, "Improved Noise Modeling of GaAs FETS, Parts I and II: Using an Enhanced Equivalent Circuit Technique," *Microwave Journal*, November 1991, pp. 87–101; and December 1991, pp. 87–95, respectively.
8. V. Rizzoli, F. Mastri, and C. Cecchefti, "Computer-Aided Noise Analysis of MESFET and HEMT Mixers," *IEEE Transactions on Microwave Theory and Technique*, Vol. MTT-37, September 1989, pp. 1401–1410.
9. V. Rizzoli and A. Lippadni, "Computer-Aided Noise Analysis of Linear Multiport Networks of Arbitrary Topology," *IEEE Transactions on Microwave Theory and Techniques*, Vol. MTT-33, December 1985, pp. 1507–1512.
10. V. Rizzoli, F. Mastri, and D. Masofti, "General-Purpose Noise Analysis of Forced Nonlinear Microwave Circuits," *Military Microwave*, October, 1992, pp. 293–298.
11. C. R. Chang, "Mixer Noise Analysis Using the Enhanced Microwave Harmonics," Compact Software *Transmission Line News*, Vol. 6, No. 2, June 1992, pp. 4–9.

12. V. Rizzoli, F. Mastri, and D. Masotti, "A General Purpose Harmonic Balance Approach to the Computation of Near-Carrier Noise in Free-Running Microwave Oscillators," MTT-S, 1993, pp. 309–312.
13. R. Tayrani, J. E. Gerber, T. Daniel, R. Pengelly, and U. L. Rohde, "A New and Reliable Approach to Direct Parameter Extraction for MESFETs and HEMTs," European Microwave Conference, Madrid, Spain, September 1993.

APPENDIX C

EXAMPLE OF WIRELESS SYNTHESIZERS USING COMMERCIAL ICs

In this appendix I give two examples of synthesizers for wireless applications based on Motorola and Philips material.

MOTOROLA
SEMICONDUCTOR TECHNICAL DATA

Order this document
by MC145190EVK/D

MC145190EVK
MC145191EVK

Advance Information

1.1 GHz PLL Frequency Synthesizer Evaluation Kits

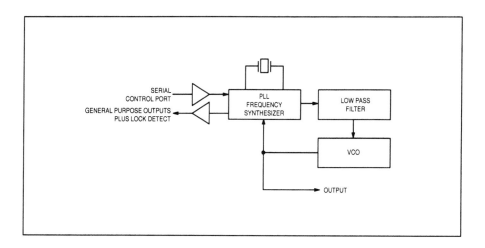

This document contains information on a new product. Specifications and information herein are subject to change without notice.

© Motorola, Inc. 1994

REV 1
10/94

Copyright of Motorola, Semiconductor Products Sector. Used with permission.

SECTION 1
GENERAL DESCRIPTION

1.1 ORGANIZATION OF DOCUMENT

This document is divided into four major sections and includes appendices. Section 1 introduces the MC145190/91EVK, lists its major features, and lists the contents of this evaluation kit. Section 2 contains a functional block diagram for the EVKs, and Section 3 outlines how to setup and use the board. Section 4 describes further operation of the board.

1.2 INTRODUCTION

The MC145190EVK and MC145191EVK are two versions of the same board that allow evaluation and demonstration of the Motorola MC145190 and MC145191 Frequency Synthesizers. The MC145190/91EVK is factory populated and tested with the MC145190F on the MC145190EVK and with the MC145191F on the MC145191EVK. A sample of the other device is included in this kit for those who wish to evaluate the alternate chip. In addition, the board may be modified to evaluate the MC145192.

Figure 1 is a simplified block diagram of the MC145190/91EVK.

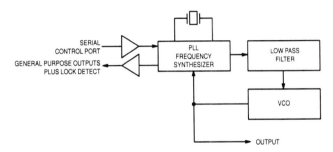

Figure 1. Simplified Block Diagram

1.3 ORDERING INFORMATION

These kits may be ordered through your local Motorola Semiconductor sales office or authorized distributor. Ask your Motorola representative to order the kits from the "chip warehouse", not the literature center. Request the part numbers shown below.

Part Number	Description
MC145190EVK	Kit with the MC145190 installed, also includes a MC145191 device and appropriate current–setting resistor.
MC145191EVK	Kit with the MC145191 installed, also includes a MC145190 device and appropriate current–setting resistor.

1.4 FEATURES

GENERAL FEATURES

- MC145190/91EVK Is a Complete Working Synthesizer, Including Voltage Controlled Oscillator (VCO), PLL Device, 5–Element Low–Pass Filter, and VCO.
- Boards Are Controlled by Any IBM PC Compatible Computer Through the Printer Port.
- Up to Three Boards Can Operate Independently From One Printer Port.
- Two or Three Board Cascades are Constructed with Only Three Output Lines (Clock, Data, and Load) from the Printer Card.
- A Prototype Area and Mounting Holes are Provided for VCOs, Mixers, and Amplifiers.
- External Reference Input Can Be Used.
- Lock Detect, Out A and Out B on Any Single Board Are Accessible Through the Printer Port.

SOFTWARE FEATURES

- Menu–Driven Control Program Written in Turbo Pascal.
- Source and Object Code Provided on a 3.5″ PC Compatible Diskette.
- Frequency Range of Operation, Step Size, and Reference Frequency Can Be Changed in the Control Program.

IBM PC is a trademark of International Business Machines Corp.

TYPICAL PERFORMANCE

Common to both kits, unless noted.

Supply voltage (J7)	11.5 – 12.5 V
Supply current (J7)	75 mA
Available current *	10 mA
Frequency Range ('190)	741.6 – 751.6 MHz
Frequency Range ('191)	733 – 743 MHz
Step size	100 kHz
Power output	6 – 8 dBm
Harmonic Level	> –10 dB
Frequency Accuracy	± 50 kHz
Reference Sidebands (100 kHz)	– 70 dB
Phase noise (10 kHz)	– 95 dBc/Hz
Switching time **	3.5 ms

1.5 CONTENTS OF THE EVALUATION KIT

- Fully Assembled and Tested MC145190/91 Evaluation Board
- Additional Device to Modify Board for Alternate Evaluation
- Flat Cable with 4 DB–25 Male Connectors
- 3.5″ PC Compatible Disk Containing Compiled Program and Turbo Pascal Source Code
- Complete Documentation and Support Materials, Including:
  — Schematic Diagram of MC145190/91EVK (See Appendix A)
  — Parts List (See Appendix B)
  — MC145190/91 Data Sheet
  — Printer Port Diagram
  — Typical MC145190EVK Signal Plot
  — Typical MC145191EVK Signal Plot

---

* Total current at 5 V and 8.5 V available to user in the prototype area. The 12 V supply is not regulated. Current at 12 V is limited by the power supply.
** 10 MHz step, within ± 2 kHz of final frequency.

592 EXAMPLE OF WIRELESS SYNTHESIZERS USING COMMERCIAL ICs

SECTION 2
FUNCTIONAL BLOCK DIAGRAM

Figure 2 is a basic functional block diagram for the MC145190/91EVK.

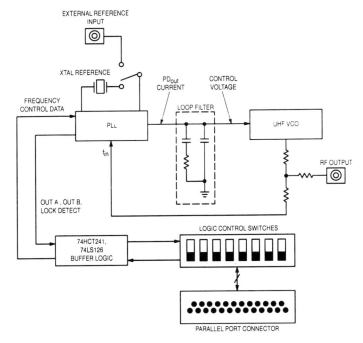

Figure 2. Functional Block Diagram

SECTION 3
GETTING STARTED

3.1 PRODUCTION TEST

After assembly is complete, the following alignment and test is performed on both the MC145190EVK and MC145191EVK before shipment:

(1) The control program is started in single board EVK mode.
(2) L menu item is selected.
(3) Power is applied to the board. DIP switch section 1 is closed circuit with all others being open circuit.
(4) After attaching computer cable, menu item B is selected.
(5) Trim capacitor VC1 is adjusted to obtain an output frequency at J8 of 741.58 – 741.62 MHz.
(6) Voltage at the control voltage test point is measured. It must be 3.0 – 3.8 V.
(7) When testing more than 1 board, steps 3 – 6 are repeated.

In step 5, if it isn't possible to obtain a signal on frequency, menu item O should be selected and the correct printer port address entered. Menu item B would then be selected to reload the data.

3.2 HARDWARE SETUP

(1) Remove the board from its conductive environment, carefully ensuring that you are not carrying static electricity on your body. You may discharge built-up static by touching a well–grounded, conductive body. Examine the board and its components to make certain nothing was damaged during shipment.
(2) Become familiar with the layout of the components and the various connectors on the board.
(3) Plug in 12 V at J6, observing the polarity marked on the board.
(4) Short circuit switch 1 of the DIP switch (S1) and open circuit all other switches (see Figure 3).
(5) Connect the supplied flat cable between the computer printer port and the DB–25 connector on the board (J5).
(6) Turn the PC on.

3.2.1 Reference Options

As shipped, the MC145190/91EVK is configured for a 10 MHz crystal (supplied). With the exception of J9 (a wire jumper), all components are present for either an external or crystal reference. To use an external reference:

(1) Remove X1, R3, C9, and VC1.
(2) Connect J9.
(3) Connect a reference signal at J10 which complies with data sheet requirements.
(4) Configure program for reference mode by selecting R in main menu.
(5) Modify reference frequency in program by selecting P in main menu, if it isn't 10 MHz.

3.3 SOFTWARE INSTALLATION

The MC145190/91EVK is controlled with a common program called PLL9091.EXE. Source code is in the file called PLL9091.PAS. It is in Turbo Pascal by Borland and is version independent.

(1) Copy the contents of the supplied disk onto a DOS compatible machine.
(2) To start the program, type PLL9091 at the DOS prompt.
(3) At startup the program asks how many boards (1 – 3 allowed) are connected and whether each is an MC145190EVK or an MC145191EVK. Answer all questions, and the main menu is displayed. Any main menu item can be selected by entering the first capitalized letter of the selection. Individual menu items when selected prompt the user for input in the correct format.

The main menu appears as shown in Table 1:

Table 1. Main Menu

Main Menu	
L	Set Frequency to Default MHz – Low Frequency
M	Set Frequency to Default MHz – Medium Frequency
H	Set Frequency to Default MHz – High Frequency
S	Set PLL Output Frequency
P	Change Clock (Reference) Frequency and Channel Spacing
A	Set Function of Output A *
C	Modify C Register *
I	Change the Default High Frequency
N	Change the Default Medium Frequency
W	Change the Default Low Frequency
O	Set Output Port Address
B	Initialize Board, Write all Registers
R	Modify Crystal Mode or Reference Mode *
T	Change Active Board
X	Exit to DOS

* = Contains Sub menu

If the printer port address is not $278, then the correct address must be entered under menu item O. Default frequencies at startup are those that are given in the specifications section.

Menu item B is used to reload data into a PLL board which has been turned on after the program started.

Menu item T is used to change the board being addressed in a cascade. The program remembers data for each board independently.

A, C, and R are the only main menu items with sub menus.

EXAMPLE OF WIRELESS SYNTHESIZERS USING COMMERCIAL ICs

The A sub menu is:

Table 2. A Sub Menu

A Sub Menu	
0	Port
1	Data Out
2	f_V
3	f_R

These selections correspond with features of the MC145190/91 device as described in the data sheet.

The C submenu is:

Table 3. C Sub Menu

C Sub Menu	
A	Output A
B	Output B
C	PD$_{out}$ Current
L	Lock Detect Enable
D	Phase Detector Select
P	Polarity

Items A and B above set Output A and Output B manually to logic 1 or 0.

The R submenu is:

Table 4. R Sub Menu

R Sub Menu	
0	Crystal Mode, Shutdown
1	Crystal Mode, Active
2	Reference Mode, REF$_{in}$ Enabled, REF$_{out}$ Low
3	Reference Mode, REF$_{out}$ = REF$_{in}$
4	Reference Mode, REF$_{out}$ = REF$_{in}$/2
5	Reference Mode, REF$_{out}$ = REF$_{in}$/4
6	Reference Mode, REF$_{out}$ = REF$_{in}$/8
7	Reference Mode, REF$_{out}$ = REF$_{in}$/16

Detailed information on the above modes is given in the MC145190/91 data sheet.

EXAMPLE OF WIRELESS SYNTHESIZERS USING COMMERCIAL ICs

SECTION 4
HARDWARE DESCRIPTION

4.1 JUMPER AND DIP SWITCH OPERATION

The following table shows all on–board jumpers and their corresponding functionality. Also listed are the major components. Please refer to the schematic for additional details.

Jumper/DIP Switch/Component	Functionality
D1	Data is sent to the PLL device through this DIP switch and 74HCT241 buffer.
D2	If power is properly connected, LED D2 should be lit.
J3	J3 is a cut–trace and jumper. On the MC145190EVK board, 8.5 V is routed through J3 to the charge pump supply and VCO. The MC145191EVK board uses 8.5 V only for VCO power. If it is desired to power the 8.5 V circuits directly, the trace under J3 can be cut, J3's shorting plug is removed and 8.5 V is applied through J7.
J5	Used to connect the computer to the board.
J6	A 12 V power supply should be used to power the board at J6 (Augat 2SV–02 connector). The 2SV–02 accepts 18–24 AWG bare copper power leads.
J7	2SV03 power connector.
M1	741 – 751 MHz VCO.
R10, R14	Power is fed to the PLL chip through a voltage divider R14 and R10. Two resistors terminate the PLL chip in 50 ohms and provide isolation.
U1	PLL Device.
U2	74LS126 Line Driver.
U3	78L08 regulator — configured as an 8.5 V regulator.
U4	Power for the 5 V logic is provided by U4 (78L05). U4 steps down the 8.5 V from U3.
U5	74HCT241 Buffer — provides isolation, logic translation, and a turn on delay for PLL input lines. Logic translation is needed from the TTL levels on the printer port to the CMOS levels on the MC145190/91EVK inputs. Turn on delay is used to ensure the power on reset functions properly. The clock line to the PLL must be held low during power up.

4.1.1 Modifying the MC145190EVK to Evaluate the MC145191F (and Vice–Versa)

Three modifications must be made to change the board between the MC145190EVK and the MC145191EVK:

	MC145190EVK	MC145191EVK
U1	MC145190	MC145191
R2	47 kΩ	22 kΩ
J10	Connect to 8.5 V	Connect to 5 V

EXAMPLE OF WIRELESS SYNTHESIZERS USING COMMERCIAL ICs

4.2 BOARD OPERATION

Please refer to Appendix A, MC145190/91EVK Schematic.

A computer is connected to the DB–25 connector J5.

A 12 V power supply should be used to power the board at J6 (Augat 2SV–02 connector). The 2SV–02 will accept 18–24 AWG bare copper power leads. No tools are needed for connection. If power is properly connected, LED D2 should be lit.

Data is output from the printer port. The printer card is in slot 0 using the default address in the control program. Data is sent to the PLL device (U1) through the DIP Switch (S1), and 74HCT241 buffer (U5).

The MC145190/91 PLL has 3 output lines which are routed through a 74LS126 line driver (U2) to the computer.

U5 provides isolation, logic translation, and a turn on delay for PLL input lines. Logic translation is needed from the TTL levels on the printer port to the CMOS levels on the MC145190/91 inputs. Turn on delay is used to ensure the power on reset functions properly. The clock line to the PLL must be held low during power up.

Power passes from J6 to U3 (78L08 regulator) configured as a 8.5 V regulator.

D1 increases output voltage of the regulator by 0.6 V. On the MC145190EVK board, 8.5 V is routed through J3 to the charge pump supply and VCO. The MC145191EVK board uses 8.5 V only for VCO power. J3 is a cut–trace and jumper. If it is desired to power the 8.5 V circuits directly, the trace under J3 can be cut, J3's shorting plug is removed and 8.5 V is applied through J7 (2SV03 power connector).

Power for the 5 V logic is provided by U4 (78L05). U4 steps down the 8.5 V from U3. Output voltage from U4 passes through J4, the 5 V cut–trace and jumper. To supply separate power to logic, the trace under J4 is cut, J4 shorting plug is removed and 5 V is applied to J7. U3 and U4 are cascaded to lower their individual voltage drops. This lowers the power dissipated in the regulators.

The PLL loop itself is composed of the MC145190/91 device (U1), 741 – 751 MHz VCO (M1) and a passive loop filter (R11, R12, C4, C5, C6). A passive loop filter was used to keep the design simple, reduce noise, and reduce the quantity of traces susceptible to stray pickup.

A single VCO module was used for both the MC145190EVK and the MC145191EVK. Frequency range of operation is different for the MC145190 and MC145191 due to the lower charge pump supply voltage of the MC145191. The VCO is an internal Motorola part which is not sold for other applications. Power is fed to the PLL chip through a voltage divider, R14 and R10. Two resistors terminate the PLL chip in 50 ohms and provide isolation.

MC145190 and MC145191 PLLs are operated as shown in the enclosed data sheet. Reference frequency is 10 MHz and step size is 100 kHz default. Output tuning range is 741.6 – 751.6 MHz for the MC145190EVK and 733 – 743 MHz for the MC145191EVK. Phase Detector current is set at 2 mA on both boards. J1 and J2 are removable jumpers and cut traces. They are used as connection points for a current measurement of V_{PD} or V_{CC}. J10 is a wire jumper that selects 5 V or 8.5 V for V_{PD}. A trimmer Capacitor VC1 is used to set X1 (10 MHz, 18 pF, parallel resonance crystal) on frequency.

4.3 DATA TRANSFER FROM COMPUTER TO EVK

To control the serial input to the EVK with the parallel printer port, a conversion is done. Printer cards are designed to output 8 bits through 8 lines. A bit mask is used to obtain the bit combination for the three required output lines (Data, Clock, Load). As bytes are sent to the printer card in sequence, it appears to be a serial transfer. The printer port was used because data transfer using the serial port would have been much slower. A standard IBM PC can support a parallel port data rate of 4.77 MHz.

598 EXAMPLE OF WIRELESS SYNTHESIZERS USING COMMERCIAL ICs

IBM PCs and compatibles can accept up to three printer port cards. These ports are called LPT1, LPT2, and LPT3. Each printer card has jumpers or DIP switches on it to set a unique address. Two sets of addresses are in common use. One set applies to IBM PC XT, AT, and clones. The other is for the PS 2 line. To load data into the EVK, the correct address must be selected. The program default is $278, which is LPT1 in a clone. If $278 is not the address in use, it must be modified by entering the O menu item in the main menu. All allowed addresses given in hexadecimal are as follows:

Label	IBM PC and Clones	PS 2
LPT1	278	3BC
LPT2	378	378
LPT3	3BC	278

Up to three boards can operate independently from one printer port. All lines on the printer port are connected to every board. Even with three boards operating, only three output lines (Clock, Data, and Load) from the printer card are used. If two boards are controlled together, data for the second board is received from the Output A of the first. Output A is a configurable output on the MC145190/91 device, which in this case is used to shift data through chip 1 into chip 2. Output A and Data are connected using a printer port input line. This was done to avoid connecting extra wires. Fortunately not all port input lines are needed for computer input. Load and Clock are common to both boards.

A three board cascade is handled similarly to a two board cascade. Out A on the first board is fed to Data on the second. Out A on the second connects to Data on the third. Instructing the program on the quantity of boards connected together allow it to modify the number of bits sent.

The MC145190/91EVK has DIP switch S1 which gives each board a unique address. Switch positions for all possible addresses are given in the following diagram:

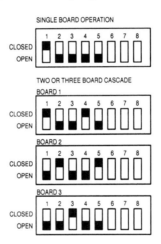

Figure 3. DIP Switch Positions

The control program is able to address any combination of '190 and '191 EVKs.

In Figure 3, DIP switch sections 6, 7, 8 allow the computer to read Out A, Out B, or Lock Detect from the MC145190/91 PLL device. Each of the inputs can only be read on one board at a time. But each item could be read on a different board. In a three board cascade, Out A could be read from the first, Out B from the second, and Lock Detect from the third board. There is no way to determine in software the board address of a particular input. The control program doesn't make use of these inputs, however source code could be modified as required.

Pin assignment on the printer port connector is referenced in the schematic in Appendix A.

600 EXAMPLE OF WIRELESS SYNTHESIZERS USING COMMERCIAL ICs

APPENDIX A
MC145190/91EVK SCHEMATIC

NOTE: Schematic continued on next page.

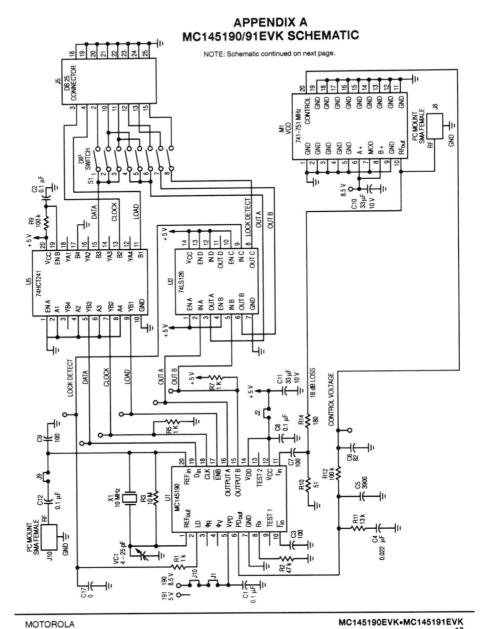

EXAMPLE OF WIRELESS SYNTHESIZERS USING COMMERCIAL ICs

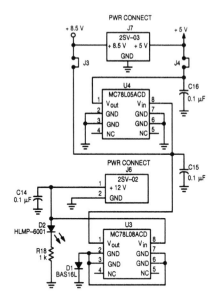

NOTES:
1. All ceramic capacitors and resistors are 1206 surface mount, ± 5% tolerance.
2. Default unit for capacitance is pF.
3. Default unit for resistance is ohms.
4. Board material is 1/16" thick G10.
5. Test points are 0.04" diameter plated through holes.
6. Board shown in '190 version. For '191, J10 is connected to 5 V, R2 is changed to 22 k and U1 is MC145191.
7. As shipped J9 isn't present and board operates in crystal mode.

EXAMPLE OF WIRELESS SYNTHESIZERS USING COMMERCIAL ICs

APPENDIX B
MC145190/91EVK PARTS LIST

Part	Type	Package	Manufacturer	Value	Quantity	References
2SV–02	Pwr Connect	Special	Augat		1	J6
2SV–03	Pwr Connect	Special	Augat		1	J7
50–651–0000–31	SMA Female	Special	ITT Sealectro		2	J8, J10
51R05279V17	VCO	Special	Special		1	M1
76SB08S	DIP Switch	16 Pin	Grayhill		1	S1
617Y–025SAJ120	DB–25 Female	Special	Amphenol		1	J5
BAS16L	0.2 A Diode	SOT–23	Motorola		1	D1
Capacitor	Ceramic	1206		0.022 µF	1	C4
Capacitor	Ceramic	1206		0.1 µF	7	C1, C2, C8, C12, C14, C15, C16
Capacitor	5% NPO	1206		82 pF	1	C6
Capacitor	5% NPO	1206		100 pF	3	C3, C7, C9
Capacitor	5% NPO	1206		3900 pF	1	C5
Crystal	18 pF, Parallel	HC–49/U	US Crystal	10 MHz	1	X1
HLMP–6001–011	LED	Special	HP		1	D2
Jumper	2 x 1 Header	0.1" Grid			4	J1, J2, J3, J4
MC74HCT241ADW	Octal Buffer	SO–20	Motorola		1	U5
MC78L05ACD	5 V Regulator	SOP–8	Motorola		1	U4
MC78L08ACD	8 V Regulator	SOP–8	Motorola		1	U3
MC145190	PLL IC	SOG	Motorola		1	U1
Resistor	5% CF	1206		1 kΩ	4	R1, R5, R7, R18
Resistor	5% CF	1206		10 MΩ	1	R3
Resistor	5% CF	1206		13 kΩ	1	R11
Resistor	5% CF	1206		47 kΩ	1	R2
Resistor	5% CF	1206		51 Ω	1	R10
Resistor	5% CF	1206		100 kΩ	2	R9, R12
Resistor	5% CF	1206		180 Ω	1	R14
SN74LS126D	Quad Buffer	SO–14	Motorola		1	U2
TDC–336K010NSF	Tant. Cap	Special	Mallory	33 µF, 10 V	2	C10, C11
TZB04Z250BB	Trim Cap	Special	Murata	4 – 25 pF	1	VC1
Wire Jumper	0.6", 24 AWG				1	J10

Motorola reserves the right to make changes without further notice to any products herein. Motorola makes no warranty, representation or guarantee regarding the suitability of its products for any particular purpose, nor does Motorola assume any liability arising out of the application or use of any product or circuit, and specifically disclaims any and all liability, including without limitation consequential or incidental damages. "Typical" parameters can and do vary in different applications. All operating parameters, including "Typicals" must be validated for each customer application by customer's technical experts. Motorola does not convey any license under its patent rights nor the rights of others. Motorola products are not designed, intended, or authorized for use as components in systems intended for surgical implant into the body, or other applications intended to support or sustain life, or for any other application in which the failure of the Motorola product could create a situation where personal injury or death may occur. Should Buyer purchase or use Motorola products for any such unintended or unauthorized application, Buyer shall indemnify and hold Motorola and its officers, employees, subsidiaries, affiliates, and distributors harmless against all claims, costs, damages, and expenses, and reasonable attorney fees arising out of, directly or indirectly, any claim of personal injury or death associated with such unintended or unauthorized use, even if such claim alleges that Motorola was negligent regarding the design or manufacture of the part. Motorola and Ⓜ are registered trademarks of Motorola, Inc. Motorola, Inc. is an Equal Opportunity/Affirmative Action Employer.

EXAMPLE OF WIRELESS SYNTHESIZERS USING COMMERCIAL ICs 603

APPENDIX C

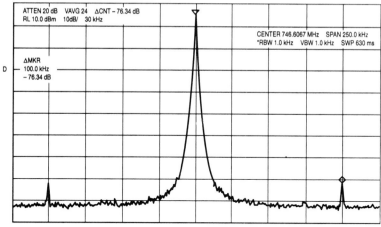

MC145190 EVK Typical Signal Plot

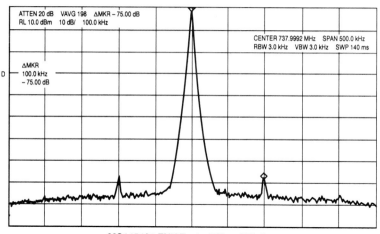

MC145191 EVK Typical Signal Plot

Literature Distribution Centers:
USA: Motorola Literature Distribution; P.O. Box 20912; Phoenix, Arizona 85036.
EUROPE: Motorola Ltd.; European Literature Centre; 88 Tanners Drive, Blakelands, Milton Keynes, MK14 5BP, England.
JAPAN: Nippon Motorola Ltd.; 4-32-1, Nishi-Gotanda, Shinagawa-ku, Tokyo 141, Japan.
ASIA PACIFIC: Motorola Semiconductors H.K. Ltd.; Silicon Harbour Center, No. 2 Dai King Street, Tai Po Industrial Estate, Tai Po, N.T., Hong Kong.

2GHz low-voltage Fractional-N synthesizer

SA8025A

Philips Semiconductors — Product specification

DESCRIPTION

The SA8025 is a monolithic low power, high performance dual frequency synthesizer fabricated in QUBiC BiCMOS technology. The SA8025A is an improved version of the SA8025, suitable for narrow band systems like the Japan Personal Digital Cellular (PDC) system. The new design improves the performance of the fractional spur compensation circuitry. The new version is pin-for-pin compatible with the previous version. Featuring Fractional-N division with selectable modulo 5 or 8 implemented in the Main synthesizer to allow the phase detector comparison frequency to be five or eight times the channel spacing. This feature reduces the overall division ratio yielding a lower noise floor and faster channel switching. The phase detectors and charge pumps are designed to achieve phase detector comparison frequencies up to 5MHz. A four modulus prescaler (divide by 64/65/68/73) is integrated on chip with a maximum input frequency of 1.8GHz at 3V. Programming and channel selection are realized by a high speed 3-wire serial interface. A 1GHz version (SA7025DK) is also available with the same pinout.

FEATURES

- Operation up to 1.8GHz at 3V
- Fast locking by "Fractional-N" divider
- Auxiliary synthesizer
- Digital phase comparator with proportional and integral charge pump output
- High speed serial input
- Low power consumption
- Programmable charge pump currents

PIN CONFIGURATION

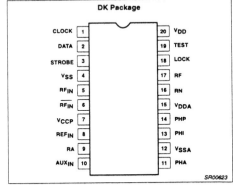

Figure 1. Pin Configuration

- Supply voltage range 2.7 to 5.5V
- Excellent input sensitivity: $V_{RF\_IN} = -20$dBm

APPLICATIONS

- PHS (Personal Handy-phone System)
- PDC (Personal Digital Cellular)
- PCS (Personal Communication Service)
- Portable communication systems

ORDERING INFORMATION

DESCRIPTION	TEMPERATURE RANGE	ORDER CODE	DWG #
20-Pin Plastic Shrink Small Outline Package (SSOP)	–40 to +85°C	SA8025ADK	SOT266-1

ABSOLUTE MAXIMUM RATINGS

SYMBOL	PARAMETER	RATING	UNITS
V	Supply voltage, V_{DD}, V_{DDA}, V_{CCP}	-0.3 to +6.0	V
V_{IN}	Voltage applied to any other pin	-0.3 to (V_{DD} + 0.3)	V
T_{STG}	Storage temperature range	-65 to +150	°C
T_A	Operating ambient temperature range	-40 to +85	°C

NOTE: Thermal impedance (θ_{JA}) = 117°C/W. This device is ESD sensitive.

Copyright of Philips Semiconductors. Used with permission.

2GHz low-voltage Fractional-N synthesizer SA8025A

PIN DESCRIPTIONS

Symbol	Pin	Description
CLOCK	1	Serial clock input
DATA	2	Serial data input
STROBE	3	Serial strobe input
V_{SS}	4	Digital ground
RF_{IN}	5	Prescaler positive input
$\overline{RF_{IN}}$	6	Prescaler negative input
V_{CCP}	7	Prescaler positive supply voltage. This pin supplies power to the prescaler and RF input buffer
REF_{IN}	8	Reference divider input
RA	9	Auxiliary current setting; resistor to V_{SSA}
AUX_{IN}	10	Auxiliary divider input
PHA	11	Auxiliary phase detector output
V_{SSA}	12	Analog ground
PHI	13	Integral phase detector output
PHP	14	Proportional phase detector output
V_{DDA}	15	Analog supply voltage. This pin supplies power to the charge pumps, Auxiliary prescaler, Auxiliary and Reference buffers.
RN	16	Main current setting; resistor to V_{SSA}
RF	17	Fractional compensation current setting; resistor to V_{SSA}
LOCK	18	Lock detector output
TEST	19	Test pin; connect to V_{DD}
V_{DD}	20	Digital supply voltage. This pin supplies power to the CMOS digital part of the device

1995 Feb 23

606 EXAMPLE OF WIRELESS SYNTHESIZERS USING COMMERCIAL ICs

Philips Semiconductors   Product specification

2GHz low-voltage Fractional-N synthesizer   SA8025A

BLOCK DIAGRAM

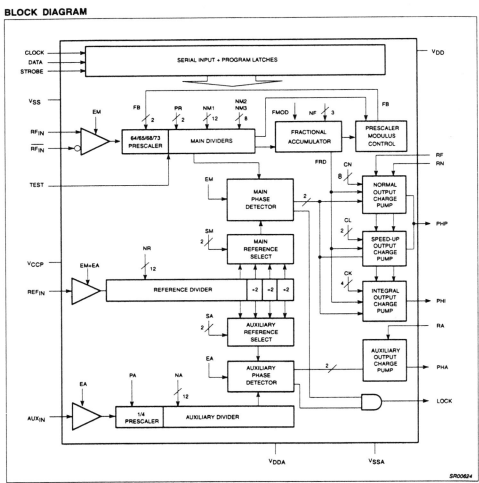

Figure 2. Block Diagram

EXAMPLE OF WIRELESS SYNTHESIZERS USING COMMERCIAL ICs

Philips Semiconductors

Product specification

2GHz low-voltage Fractional-N synthesizer — SA8025A

DC ELECTRICAL CHARACTERISTICS

$V_{DD} = V_{DDA} = V_{CCP} = 3V$; $T_A = 25°C$, unless otherwise specified.

SYMBOL	PARAMETER	TEST CONDITIONS	MIN	TYP	MAX	UNITS
V_{SUPPLY}	Recommended operating conditions	$V_{CCP} = V_{DD}$, $V_{DDA} \geq V_{DD}$	2.7		5.5	V
$I_{STANDBY}$	Total standby supply currents	EM = EA = 0, $I_{RN} = I_{RF} = I_{RA} = 0$		50	500	μA
Operational supply currents: $I = I_{DD} + I_{CCP} + I_{DDA}$; $I_{RN} = 25μA$, $I_{RA} = 25μA$, (see Note 5)						
I_{AUX}	Operational supply currents	EM = 0, EA = 1		3.5		mA
I_{MAIN}	Operational supply currents	EM = 1, EA = 0		9.5		mA
I_{TOTAL}	Operational supply currents	EM = EA = 1		12		mA
Digital inputs CLK, DATA, STROBE						
V_{IH}	High level input voltage range		$0.7 \times V_{DD}$		V_{DD}	V
V_{IL}	Low level input voltage range		0		$0.3 \times V_{DD}$	V
Digital outputs LOCK						
V_{OL}	Output voltage LOW	$I_O = 2mA$			0.4	V
V_{OH}	Output voltage HIGH	$I_O = -2mA$	$V_{DD}-0.4$			V
Charge pumps: $V_{DDA} = 3V / I_{RX} = 25μA$ or $V_{DDA} = 5V / I_{RX} = 62.5μA$, V_{PHX} in range, unless otherwise specified.						
$|I_{RX}|$	Setting current range for any setting resistor	$2.7V < V_{DDA} < 5.5V$		25		μA
		$4.5V < V_{DDA} < 5.5V$		62.5		
V_{PHOUT}	Output voltage range		0.7		$V_{DDA}-0.8$	V
Charge pump PHA						
$|I_{PHA}|$	Output current PHA	$I_{RN} = -62.5μA$; $V_{PHP} = V_{DDA}/2$ [13]	400	500	600	μA
		$I_{RN} = -25μA$; $V_{PHP} = V_{DDA}/2$	160	200	240	
$\frac{\Delta I_{PHP\_A}}{|I_{PHP\_A}|}$	Relative output current variation PHA	$I_{RA} = -62.5μA$ [2, 13]		2	6	%
ΔI_{PHA_M}	Output current matching PHA pump	$V_{DDA} = 3V$, $I_{RA} = 25μA$			±50	μA
		$V_{DDA} = 5V$, $I_{RA} = 62.5μA$			±65	
Charge pump PHP, normal mode [1, 4, 6] $V_{RF} = V_{DDA}$						
$|I_{PHP\_N}|$	Output current PHP	$I_{RN} = -62.5μA$; $V_{PHP} = V_{DDA}/2$ [13]	440	550	660	μA
		$I_{RN} = -25μA$; $V_{PHP} = V_{DDA}/2$	175	220	265	
$\frac{\Delta I_{PHP\_N}}{I_{PHP\_N}}$	Relative output current variation PHP	$I_{RN} = -62.5μA$ [2, 13]		2	6	%
$\Delta I_{PHP\_N\_M}$	Output current matching PHP normal mode	$V_{DDA} = 3V$, $I_{RA} = 25μA$			±50	μA
		$V_{DDA} = 5V$, $I_{RA} = 62.5μA$			±65	
Charge pump PHP, speed-up mode [1, 4, 7] $V_{RF} = V_{DDA}$						
$|I_{PHP\_S}|$	Output current PHP	$I_{RN} = -62.5μA$; $V_{PHP} = V_{DDA}/2$ [13]	2.20	2.75	3.30	mA
		$I_{RN} = -25μA$; $V_{PHP} = V_{DDA}/2$	0.85	1.1	1.35	
$\frac{\Delta I_{PHP\_S}}{I_{PHP\_S}}$	Relative output current variation PHP	$I_{RN} = -62.5μA$ [2, 13]		2	6	%
$\Delta I_{PHP\_S\_M}$	Output current matching PHP speed-up mode	$V_{DDA} = 3V$, $I_{RA} = 25μA$			±250	μA
		$V_{DDA} = 5V$, $I_{RA} = 62.5μA$			±300	
Charge pump PHI, speed-up mode [1, 4, 8] $V_{RF} = V_{DDA}$						
$|I_{PHI}|$	Output current PHI	$I_{RN} = -62.5μA$; $V_{PHI} = V_{DDA}/2$ [13]	4.4	5.5	6.6	mA
		$I_{RN} = -25μA$; $V_{PHI} = V_{DDA}/2$	1.75	2.2	2.65	
$\frac{\Delta I_{PHI}}{I_{PHI}}$	Relative output current variation PHI	$I_{RN} = -62.5μA$ [2, 13]		2	8	%
ΔI_{PHI_M}	Output current matching PHI pump	$V_{DDA} = 3V$, $I_{RA} = 25μA$			±500	μA
		$V_{DDA} = 5V$, $I_{RA} = 62.5μA$			±600	
Fractional compensation PHP, normal mode [1, 9] $V_{RN} = V_{DDA}$, $V_{PHP} = V_{DDA}/2$						
$I_{PHP\_F\_N}$	Fractional compensation output current PHP vs F_{RD} [3]	$I_{RF} = -62.5μA$; $F_{RD} = 1$ to 7 [13]	-625	-400	-250	nA
		$I_{RF} = -25μA$; $F_{RD} = 1$ to 7	-250	-180	-100	

1995 Feb 23

608 EXAMPLE OF WIRELESS SYNTHESIZERS USING COMMERCIAL ICs

Philips Semiconductors

Product specification

2GHz low-voltage Fractional-N synthesizer

SA8025A

DC ELECTRICAL CHARACTERISTICS (Continued)

SYMBOL	PARAMETER	TEST CONDITIONS	LIMITS MIN	LIMITS TYP	LIMITS MAX	UNITS
Fractional compensation PHP, speed up mode [1, 10] $V_{PHP} = V_{DDA}$, $V_{RN} = V_{DDA}$						
$I_{PHP\_F\_S}$	Fractional compensation output current PHP vs F_{RD} [3]	$I_{RF} = -62.5\mu A$; $F_{RD} = 1$ to 7 [13]	-3.35	-2	-1.1	µA
		$I_{RF} = -25\mu A$; $F_{RD} = 1$ to 7	-1.35	-1.0	-0.5	
	Pump leakage		-20		20	nA
Charge pump leakage currents, charge pump not active						
I_{PHP_L}	Output leakage current PHP; normal mode [1]	$V_{PHP} = 0.7$ to $V_{DDA} - 0.8$		0.1	20	nA
I_{PHI_L}	Output leakage current PHI; normal mode [1]	$V_{PHI} = 0.7$ to $V_{DDA} - 0.8$		0.1	20	nA
I_{PHA_L}	Output leakage current PHA	$V_{PHA} = 0.7$ to $V_{DDA} - 0.8$		0.1	20	nA

AC ELECTRICAL CHARACTERISTICS

$V_{DD} = V_{DDA} = V_{CCP} = 3V$; $T_A = 25°C$; unless otherwise specified. Test Circuit, Figure 4. The parameters listed below are tested using automatic test equipment to assure consistent electrical characteristics. The limits do not represent the ultimate performance limits of the device. Use of an optimized RF layout will improve many of the listed parameters.

SYMBOL	PARAMETER	TEST CONDITIONS	LIMITS MIN	LIMITS TYP	LIMITS MAX	UNITS
Main divider guaranteed and tested on an automatic tester. Some performance parameters may be improved by using optimized layout.						
f_{RF_IN}	Input signal frequency	Pin = -20dBm, Direct coupled input [14]	0	1.8		GHz
		Pin = -20dBm, 1000pF input coupling		1.8		
V_{RF_IN}	Input sensitivity	$f_{IN} = 1700MHz$	-20		0	dBm
Reference divider ($V_{DD} = V_{DDA} = 3V$ or $V_{DD} = 3V$ / $V_{DDA} = 5V$)						
f_{REF_IN}	Input signal frequency	$2.7 < V_{DD}$ and $V_{DDA} < 5.5V$			25	MHz
		$2.7 < V_{DD}$ and $V_{DDA} < 4.5V$			30	
V_{REF_IN}	Input signal range, AC coupled	$2.7 < V_{DD}$ and $V_{DDA} < 5.5V$	500			mV$_{P-P}$
		$2.7 < V_{DD}$ and $V_{DDA} < 4.5V$	300			
Z_{REF_IN}	Reference divider input impedance [15]			100		kΩ
				3		pF
Auxiliary divider						
f_{AUX_IN}	Input signal frequency		0		50	MHz
	PA = "0", prescaler enabled	$4.5V \leq V_{DDA} \leq 5.5V$	0		150	
	Input signal frequency		0		30	
	PA = "1", prescaler disabled	$4.5V \leq V_{DDA} \leq 5.5V$	0		40	
V_{AUX_IN}	Input signal range, AC coupled		200			mV$_{P-P}$
Z_{AUX_IN}	Auxiliary divider input impedance [15]			100		kΩ
				3		pF
Serial interface [15]						
f_{CLOCK}	Clock frequency				10	MHz
t_{SU}	Set-up time: DATA to CLOCK, CLOCK to STROBE		30			ns
t_H	Hold time; CLOCK to DATA		30			ns
t_W	Pulse width; CLOCK		30			ns
	Pulse width; STROBE	B, C, D, E words	30			

1995 Feb 23

EXAMPLE OF WIRELESS SYNTHESIZERS USING COMMERCIAL ICs

Philips Semiconductors

Product specification

2GHz low-voltage Fractional-N synthesizer

SA8025A

AC ELECTRICAL CHARACTERISTICS (Continued)

SYMBOL	PARAMETER	TEST CONDITIONS	LIMITS			UNITS
			MIN	TYP	MAX	
t_{SW}	Pulse width; STROBE	A word, PR = '01'		$\frac{1}{f_{VCO}} \cdot (NM2 \cdot 65) + t_W$		ns
		A word, PR = '10'		$\frac{1}{f_{VCO}} \cdot [(NM2 \cdot 65) + (NM3 + 1) \cdot 68] + t_W$		
		A word, PR = '11'		$\frac{1}{f_{VCO}} \cdot [(NM2 \cdot 65) + (NM3 + 1) \cdot 68 + (NM4 + 1) \cdot 73)] + t_W$		
		A word, PR = '00'		$\frac{1}{f_{VCO}} \cdot [(NM2 \cdot 65) + (NM4 + 1) \cdot 73] + t_W$		

NOTES:
1. When a serial input "A" word is programmed, the main charge pumps on PHP and PHI are in the "speed up mode" as long as STROBE = H. When this is not the case, the main charge pumps are in the "normal mode".
2. The relative output current variation is defined thus:
$$\frac{\Delta I_{OUT}}{I_{OUT}} = 2 \cdot \frac{(I_2 - I_1)}{|(I_2 + I_1)|} \text{ with } V_1 = 0.7V, V_2 = V_{DDA} - 0.8V \text{ (see Figure 3).}$$
3. F_{RD} is the value of the 3 bit fractional accumulator.
4. Monotonicity is guaranteed with C_N = 0 to 255.
5. Power supply current measured with f_{RF_IN} = 1667.4MHz, NM1 = 0, NM2 = 1, NM3 = 1, NM4 = 4, FMOD = 8, N = 694 6/8, main phase detector frequency = 2.4MHz, f_{REF_IN} = 19.2MHz, NR = 8, SM = 1, f_{AUX_IN} = 150MHz, NA = 125, SA = 1, PA = 0, auxiliary phase detector frequency = 300kHz, IRN = IRA = IRF = 25µA, CN = 160, CL = 0, CK = 0, lock condition, normal mode, $V_{CCP} = V_{DD} = V_{DDA}$ = 3V. Operational supply current = $I_{DDA} + I_{DD} + I_{CCP}$.
6. Specification condition: CN = 255
7. Specification conditions:
   1) CN = 255; CL = 1, or
   2) CN = 75; CL = 3
8. Typical output current $|I_{PHI}| = -I_{RN} \times CN \times 2^{(CL+1)} \times CK/32$:
   1) CN = 160; CL = 3; CK = 1, or
   2) CN = 160; CL = 2; CK = 2, or
   3) CN = 160; CL = 1; CK = 4, or
   4) CN = 160; CL = 0; CK = 8
9. Any RFD, CL = 1 for speed-up pump. The integral pump is intended for switching only and the fractional compensation is not guaranteed.
10. Specification conditions: F_{RD} = 1 to 7; CL = 1.
11. Specification conditions:
    1) F_{RD} = 1 to 7; CL = 1; CK = 2, or
    2) F_{RD} = 1 to 7; CL = 2; CK = 1.
12. The matching is defined by the sum of the P and the N pump for a given output voltage.
13. Limited analog supply voltage range 4.5 to 5.5V.
14. For f_{IN} < 50MHz, low frequency operation requires DC-coupling and a minimum input slew rate of 32V/µs.
15. Guaranteed by design.

1995 Feb 23

610 EXAMPLE OF WIRELESS SYNTHESIZERS USING COMMERCIAL ICs

Philips Semiconductors

Product specification

2GHz low-voltage Fractional-N synthesizer

SA8025A

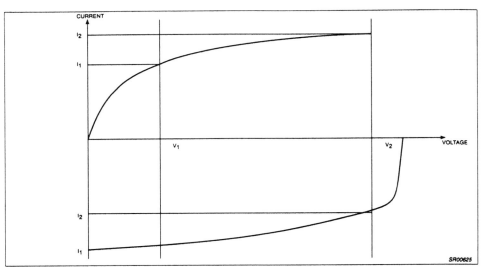

Figure 3. Relative Output Current Variation

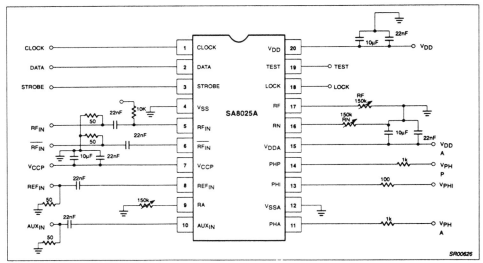

Figure 4. Test Circuit

1995 Feb 23

EXAMPLE OF WIRELESS SYNTHESIZERS USING COMMERCIAL ICs

Philips Semiconductors — Product specification

2GHz low-voltage Fractional-N synthesizer — SA8025A

AC TIMING CHARACTERISTICS

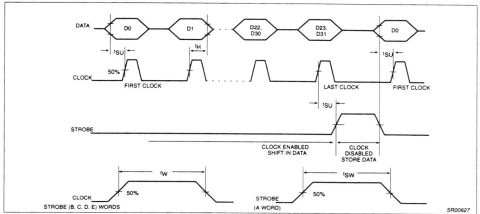

Figure 5. Serial Input Timing Sequence

FUNCTIONAL DESCRIPTION

Serial Input Programming

The serial input is a 3-wire input (CLOCK, STROBE, DATA) to program all counter ratios, DACs, selection and enable bits. The programming data is structured into 24 or 32 bit words; each word includes 1 or 4 address bits. Figure 5 shows the timing diagram of the serial input. When the STROBE = L, the clock driver is enabled and on the positive edges of the CLOCK the signal on DATA input is clocked into a shift register. When the STROBE = H, the clock is disabled and the data in the shift register remains stable. Depending on the 1 or 4 address bits the data is latched into different working registers or temporary registers. In order to fully program the synthesizer, 4 words must be sent: D, C, B and A. Figure 6 and Table 1 shows the format and the contents of each word. The E word is for testing purposes only. The E (test) word is reset when programming the D word. The data for CN and PR is stored by the B word in temporary registers. When the A word is loaded, the data of these temporary registers is loaded together with the A word into the work registers which avoids false temporary main divider input. CN is only loaded from the temporary registers when a short 24 bit A0 word is used. CN will be directly loaded by programming a long 32 bit A1 word. The flag LONG in the D word determines whether A0 (LONG = "0") or A1 (LONG = "1") format is applicable. The A word contains new data for the main divider.

Main Divider Synchronization

The A word is loaded only when a main divider synchronization signal is also active in order to avoid phase jumps when reprogramming the main divider. The synchronization signal is generated by the main divider. The signal is active while the NM1 divider is counting down from the programmed value. The new A word will be loaded after the NM1 divider has reached its terminal count; also, at this time a main divider output pulse will be sent to the main phase detector. The loading of the A word is disabled while the other dividers are counting up to their programmed values.

Therefore, the new A word will be correctly loaded provided that the STROBE signal has been at an active high value for at least a minimum number of VCO input cycles at RF_{IN} or $\overline{RF_{IN}}$.

For PR = '01'
$$t\_strobe\_min = \frac{1}{f_{VCO}}(NM2 \cdot 65) + t_w$$

For PR = '10'
$$t\_strobe\_min = \frac{1}{f_{VCO}}[NM2 \cdot 65 + (NM3 + 1) \cdot 68] + t_w$$

For PR = '11'
$$t\_strobe\_min = \frac{1}{f_{VCO}} \cdot [(NM2 \cdot 65 + (NM3 + 1) \cdot 68 + (NM4 + 1) \cdot 73)] + t_w$$

For PR = '00'
$$t\_strobe\_min = \frac{1}{f_{VCO}} \cdot [(NM2 \cdot 65) + (NM4 + 1) \cdot 73] + t_w$$

Programming the A word means also that the main charge pumps on output PHP and PHI are set into the speed-up mode as long as the STROBE is H.

Auxiliary Divider

The input signal on AUX_IN is amplified to logic level by a single-ended CMOS input buffer, which accepts low level AC coupled input signals. This input stage is enabled if the serial control bit EA = "1". Disabling means that all currents in the input stage are switched off. A fixed divide by 4 is enabled if PA = "0". This divider has been optimized to accept a high frequency input signal. If PA = "1", this divider is disabled and the input signal is fed directly to the second stage, which is a 12-bit programmable divider with standard input frequency (40MHz). The division ratio can be expressed as:

if PA = "0": N = 4 × NA

1995 Feb 23

2GHz low-voltage Fractional-N synthesizer SA8025A

if PA = "1": N = NA; with NA = 4 to 4095

Reference Divider
The input signal on REF_IN is amplified to logic level by a single-ended CMOS input buffer, which accepts low level AC coupled input signals. This input stage is enabled by the OR function of the serial input bits EA and EM. Disabling means that all currents in the input stage are switched off. The reference divider consists of a programmable divider by NR (NR = 4 to 4095) followed by a three bit binary counter. The 2 bit SM register (see Figure 7) determines which of the 4 output pulses is selected as the main phase detector input. The 2 bit SA register determines the selection of the auxiliary phase detector signal.

Main Divider
The differential inputs are amplified (to internal ECL logic levels) and provide excellent sensitivity (−20dBm at 1.7GHz) making the prescaler ideally suited to directly interface to a VCO as integrated on the Philips front-end devices including RF gain stage, VCO and mixer. The internal four modulus prescaler feedback loop FB controls the selection of the divide by ratios 64/65/68/73, and reduces the minimum system division ratio below the typical value required by standard dual modulus (64/65) devices.

This input stage is enabled when serial control bit EM = "1". Disabling means that all currents in the prescaler are switched off.

The main divider is built up by a 12 bit counter plus a sign bit. Depending on the serial input values NM1, NM2, NM3, NM4 and the prescaler select PR, the counter will select a prescaler ratio during a number of input cycles according to Table 2 and Table 3.

The loading of the work registers NM1, NM2, NM3, NM4 and PR is synchronized with the state of the main counter, to avoid extra phase disturbance when switching over to another main divider ratio as explained in the Serial Input Programming section.

At the completion of a main divider cycle, a main divider output pulse is generated which will drive the main phase comparator. Also, the fractional accumulator is incremented with NF. The accumulator works modulo Q. Q is preset by the serial control bit FMOD to 8 when FMOD = "1". Each time the accumulator overflows, the feedback to the prescaler will select one cycle using prescaler ratio R2 instead of R1.

As shown above, this will increase the overall division ratio by 1 if R2 = R1 + 1. The mean division ratio over Q main divider will then be

$$NQ = N + \frac{NF}{Q}$$

Programming a fraction means the prescaler with main divider will divide by N or N + 1. The output of the main divider will be modulated with a fractional phase ripple. This phase ripple is proportional to the contents of the fractional accumulator FRD, which is used for fractional current compensation.

EXAMPLE OF WIRELESS SYNTHESIZERS USING COMMERCIAL ICs

Philips Semiconductors

Product specification

2GHz low-voltage Fractional-N synthesizer

SA8025A

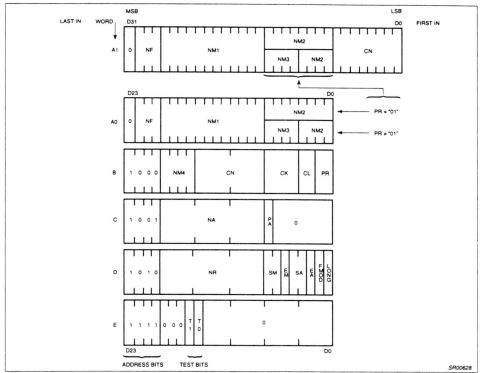

Figure 6. Serial Input Word Format

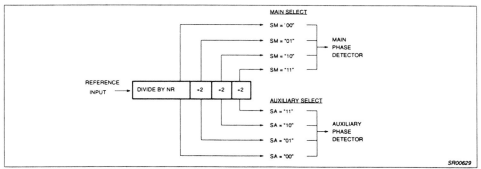

Figure 7. Reference Divider

1995 Feb 23

Philips Semiconductors

Product specification

2GHz low-voltage Fractional-N synthesizer

SA8025A

Table 1. Function Table

Symbol	Bits	Function
NM1	12	Number of main divider cycles when prescaler modulus = 64[*]
NM2	8 if PR = "01" 4 if PR = "10"	Number of main divider cycles when prescaler modulus = 65[*]
NM3	4 if PR = "10"	Number of main divider cycles when prescaler modulus = 68[*]
NM4	4 if PR = "11" or "00"	Number of main divider cycles when prescaler modulus = 73[*]
PR	2	Prescaler type in use PR = "01": modulus 2 prescaler (64/65) PR = "10": modulus 3 prescaler (64/65/68) PR = "11": modulus 4 prescaler (64/65/68/73) PR = "00": modulus 3 prescaler (64/65/73)
NF	3	Fractional-N increment
FMOD	1	Fractional-N modulus selection flag "1": modulo 8 "0": modulo 5
LONG	1	A word format selection flag "0": 24 bit A0 format "1": 32 bit A1 format
CN	8	Binary current setting factor for main charge pumps
CL	2	Binary acceleration factor for proportional charge pump current
CK	4	Binary acceleration factor for integral charge pump current
EM	1	Main divider enable flag
EA	1	Auxiliary divider enable flag
SM	2	Reference select for main phase detector
SA	2	Reference select for auxiliary phase detector
NR	12	Reference divider ratio
NA	12	Auxiliary divider ratio
PA	1	Auxiliary prescaler mode: PA = "0": divide by 4 PA = "1": divide by 1

[*]Not including reset cycles and Fractional-N effects.

Table 2. Prescaler Ratio

The total division ratio from prescaler to the phase detector may be expressed as:	
if PR = "01"	N = (NM1 + 2) x 64 + NM2 x 65
	N' = (NM1 + 1) x 64 + (NM2 + 1) x 65 (*)
if PR = "10"	N = (NM1 + 2) x 64 + NM2 x 65 + (NM3 + 1) x 68
	N' = (NM1 + 1) x 64 + (NM2 + 1) x 65 + (NM3 + 1) x 68 (*)
if PR = "11"	N = (NM1 + 2) x 64 + NM2 x 65 + (NM3 + 1) x 68 + (NM4 + 1) x 73
	N' = (NM1 + 1) x 64 + (NM2 + 1) x 65 + (NM3 + 1) x 68 + (NM4 + 1) x 73 (*)
if PR = "00"	N = (NM1 + 2) x 64 + NM2 x 65 + (NM4 + 1) x 73
	N' = (NM1 + 1) x 64 + (NM2 + 1) x 65 + (NM4 + 1) x 73 (*)
(*) When the fractional accumulator overflows the prescaler ratio = 65 (64 + 1) and the total division ratio N' = N + 1	

Table 3. PR Modulus

PR	Modulus Prescaler	Bit Capacity			
		NM1	NM2	NM3	NM4
01	2	12	8	–	–
10	3	12	4	4	–
11	4	12	4	4	4
00	3	12	8	–	4

1995 Feb 23

EXAMPLE OF WIRELESS SYNTHESIZERS USING COMMERCIAL ICs   615

Philips Semiconductors   Product specification

2GHz low-voltage Fractional-N synthesizer   SA8025A

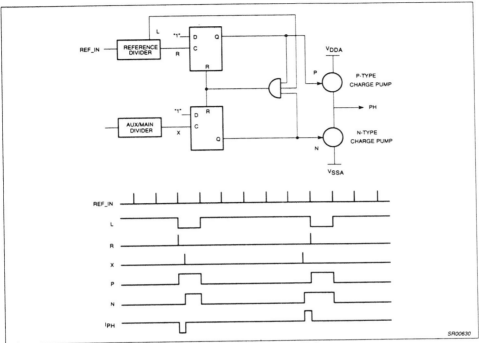

Figure 8. Phase Detector Structure with Timing

Phase Detectors
The auxiliary and main phase detectors are a two D-type flip-flop phase and frequency detector shown in Figure 8. The flip-flops are set by the negative edges of output signals of the dividers. The rising edge of the signal, L, will reset the flip-flops after both flip-flops have been set. Around zero phase error this has the effect of delaying the reset for 1 reference input cycle. This avoids non-linearity or deadband around zero phase error. The flip-flops drive on-chip charge pumps. A source current from the charge pump indicates the VCO frequency will be increased; a sink current indicates the VCO frequency will be decreased.

Current Settings
The SA8025A has 3 current setting pins: RA, RN and RF. The active charge pump currents and the fractional compensation currents are linearly dependent on the current connected between the current setting pin and V_{SS}. The typical value R (current setting resistor) can be calculated with the formula:

$$R = \frac{V_{DDA} - 0.9 - 150\sqrt{I_R}}{I_R}$$

The current can be set to zero by connecting the corresponding pin to V_{DDA}.

Auxiliary Output Charge Pumps
The auxiliary charge pumps on pin PHA are driven by the auxiliary phase detector and the current value is determined by the external resistor RA at pin RA. The active charge pump current is typically:

$$|I_{PHA}| = 8 \cdot I_{RA}$$

1995 Feb 23

2GHz low-voltage Fractional-N synthesizer

SA8025A

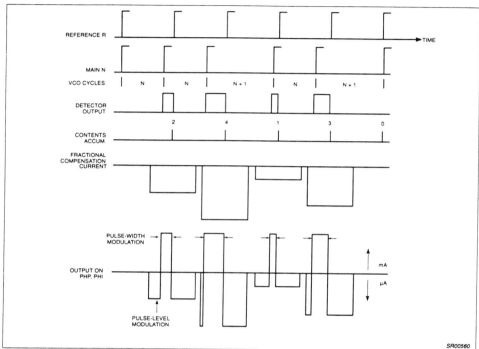

Figure 9. Waveforms for NF = 2, Fraction = 0.4

Main Output Charge Pumps and Fractional Compensation Currents

The main charge pumps on pin PHP and PHI are driven by the main phase detector and the current value is determined by the current at pin RN and via a number of DACs which are driven by registers of the serial input. The fractional compensation current is determined by the current at pin RF, the contents of the fractional accumulator FRD and a number of DACs driven by registers from the serial input. The timing for the fractional compensation is derived from the reference divider. The current is on during 1 input reference cycle before and 1 cycle after the output signal to the phase comparator. Figure 9 shows the waveforms for a typical case.

When the serial input A word is loaded, the output circuits are in the "speed-up mode" as long as the STROBE is H, else the "normal mode" is active. In the "normal mode" the current output PHP is:

$$I_{PHP\_N} = I_{PHP} + I_{PHP\_comp}$$

where:

$$|I_{PHP}| = \frac{CN \cdot I_{RN}}{32} \quad \text{:charge pump current}$$

$$|I_{PHP\_comp}| = FRD \cdot \frac{I_{RF}}{128} \quad \text{:fractional comp. current}$$

The current in PHI is zero in "normal mode".

In "speed-up mode" the current in output PHP is:

$$I_{PHP\_S} = I_{PHP} + I_{PHP\_comp}$$

$$|I_{PHP}| = \left(\frac{CN \cdot I_{RN}}{32}\right)(2^{CL+1} + 1)$$

$$|I_{PHP\_comp}| = \left(\frac{FRD \cdot I_{RF}}{128}\right)(2^{CL+1} + 1)$$

In "speed-up mode" the current in output PHI is:

$$I_{PHI\_S} = I_{PHI} + I_{PHI\_comp}$$

where:

$$|I_{PHI}| = \left(\frac{CN \cdot I_{RN}}{32}\right)(2^{CL+1}) \text{ CK}$$

1995 Feb 23

EXAMPLE OF WIRELESS SYNTHESIZERS USING COMMERCIAL ICs

Philips Semiconductors

Product specification

2GHz low-voltage Fractional-N synthesizer

SA8025A

$$|I_{PHI\_comp}| = \left(\frac{FRD \cdot I_{RN}}{128}\right)(2^{CL+1}) \text{ CK}$$

Figure 9 shows that for proper fractional compensation, the area of the fractional compensation current pulse must be equal to the area of the charge pump ripple output. This means that the current setting on the input RN, RF is approximately:

$$\frac{I_{RN}}{I_{RF}} = \frac{(Q \cdot f_{VCO})}{(3 \cdot CN \cdot F_{INR})}$$

where:
- Q = fractional-N modulus
- $f_{VCO} = f_{INM} \times N$, input frequency of the prescaler
- F_{INR} = input frequency of the reference divider

PHI pump is meant for switching only. Current and compensation are not as accurate as PHP.

Lock Detect

The output LOCK is H when the auxiliary phase detector AND the main phase detector indicates a lock condition. The lock condition is defined as a phase difference of less than ±1 cycle on the reference input REF_IN. The lock condition is also fulfilled when the relative counter is disabled (EM = "0" or respectively EA = "0") for the main, respectively auxiliary counter.

Test Modes

The lock output is selectable as f_{REF}, f_{AUX}, f_{MAIN} and lock. Bits T1 and T0 of the E word control the selection (see Figures 6 and 10).

If T1 = T0 = Low, or if the E-word is not sent, the lock output is configured as the normal lock output described in the Lock Detect section.

If T1 = Low and T0 = High, the lock output is configured as f_{REF}. The signal is the buffered output of the reference divider NR and the 3-bit binary counter SM. The f_{REF} signal appears as normally low and pulses high whenever the divider reaches terminal count from the value programmed into the NR and SM registers. The f_{REF} signal can be used to verify the divide ratio of the Reference divider.

If T1 = High and T0 = Low, the lock output is configured as f_{AUX}. The signal is normally high and pulses low whenever the divider reaches terminal count from the value programmed into the NA and PA registers. The f_{AUX} signal can be used to verify the divide ratio of the Auxiliary divider.

If T1 = High and T0 = High, the lock output is configured as f_{MAIN}. The signal is the buffered output of the MAIN divider. The f_{MAIN} signal appears as normally high and pulses low whenever the divider reaches terminal count from the value programmed into the NM1, NM2, NM3 or NM4 registers. The f_{MAIN} signal can be used to verify the divide ratio of the MAIN divider and the prescaler.

Test Pin

The Test pin, Pin 19, is a buffered logic input which is exclusively ORed with the output of the prescaler. The output of the XOR gate is the input to the MAIN divider. The Test pin must be connected to V_{DD} during normal operation as a synthesizer. This pin can be used as an input for verifying the divide ratio of the MAIN divider; while in this condition the input to the prescaler, RF_{IN}, may be connected to V_{CCP} through a 10kΩ resistor in order to place prescaler output into a known state.

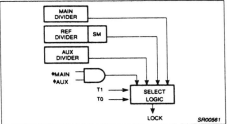

Figure 10. Test Mode Diagram

618 EXAMPLE OF WIRELESS SYNTHESIZERS USING COMMERCIAL ICs

Philips Semiconductors

Product specification

2GHz low-voltage Fractional-N synthesizer

SA8025A

PIN FUNCTIONS

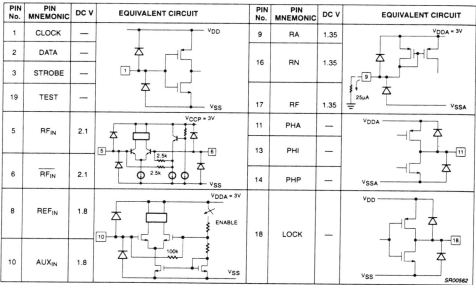

PIN No.	PIN MNEMONIC	DC V
1	CLOCK	—
2	DATA	—
3	STROBE	—
19	TEST	—
5	RF_{IN}	2.1
6	$\overline{RF_{IN}}$	2.1
8	REF_{IN}	1.8
10	AUX_{IN}	1.8

PIN No.	PIN MNEMONIC	DC V
9	RA	1.35
16	RN	1.35
17	RF	1.35
11	PHA	—
13	PHI	—
14	PHP	—
18	LOCK	—

Figure 11. Pin Functions

1995 Feb 23

777

EXAMPLE OF WIRELESS SYNTHESIZERS USING COMMERCIAL ICs

Philips Semiconductors

Product specification

2GHz low-voltage Fractional-N synthesizer

SA8025A

TYPICAL PERFORMANCE CHARACTERISTICS

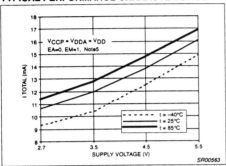

Figure 12. Operational Supply Current vs Supply Voltage and Temperature

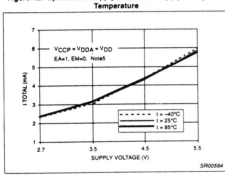

Figure 13. Auxiliary Operational Supply Current vs Supply Voltage and Temperature

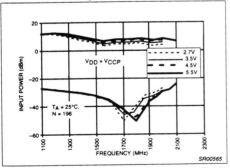

Figure 14. Main Divider Input Power vs Frequency and Supply

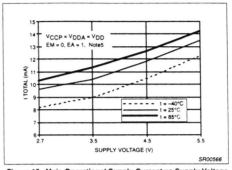

Figure 15. Main Operational Supply Current vs Supply Voltage and Temperature

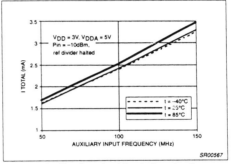

Figure 16. Auxiliary Operational Supply Current vs Frequency and Temperature

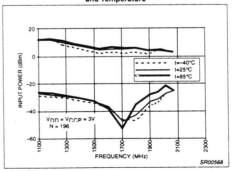

Figure 17. Main Divider Input Power vs Frequency and Temperature

2GHz low-voltage Fractional-N synthesizer — SA8025A

TYPICAL PERFORMANCE CHARACTERISTICS (Continued)

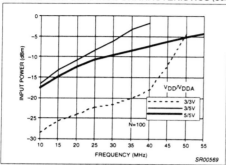

Figure 18. Reference Divider Minimum Input Power vs Frequency and Supply

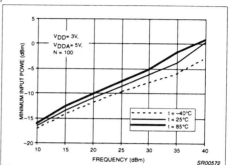

Figure 21. Reference Divider Minimum Input Power vs Frequency and Temperature

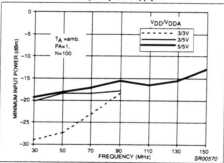

Figure 19. Auxiliary Divider Minimum Input Power vs Frequency and Supply

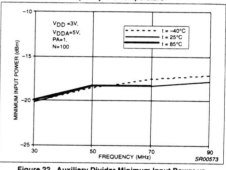

Figure 22. Auxiliary Divider Minimum Input Power vs Frequency and Temperature

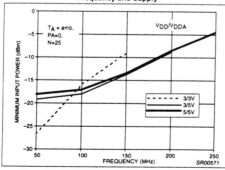

Figure 20. Auxiliary Divider Minimum Input Power vs Frequency and Supply

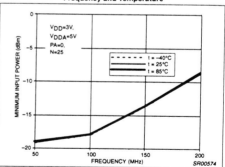

Figure 23. Auxiliary Divider Minumum Input Power vs Frequency and Temperature

1995 Feb 23

EXAMPLE OF WIRELESS SYNTHESIZERS USING COMMERCIAL ICs

Philips Semiconductors

Product specification

2GHz low-voltage Fractional-N synthesizer

SA8025A

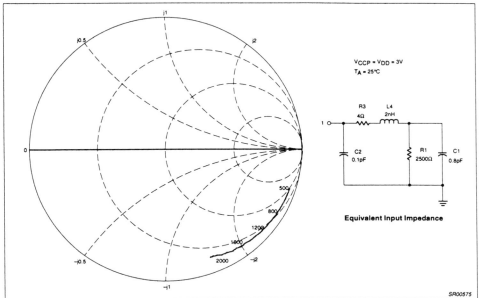

Figure 24. Typical RF$_{IN}$ Input Impedance

Philips Semiconductors   Product specification

2GHz low-voltage Fractional-N synthesizer   SA8025A

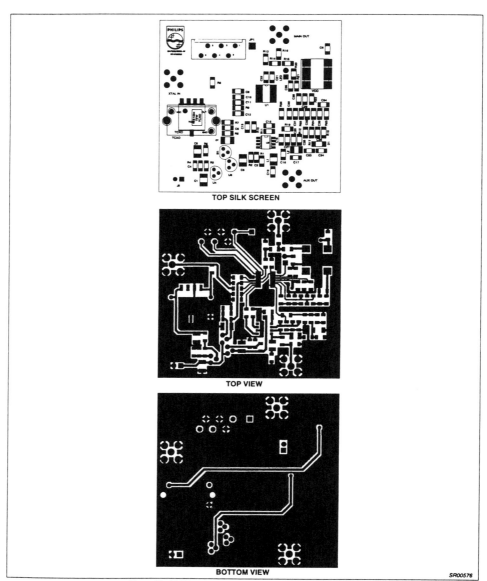

Figure 25. SA8025ADK Demoboard Layout (NOT ACTUAL SIZE)

EXAMPLE OF WIRELESS SYNTHESIZERS USING COMMERCIAL ICs

Philips Semiconductors Product specification

2GHz low-voltage Fractional-N synthesizer — SA8025A

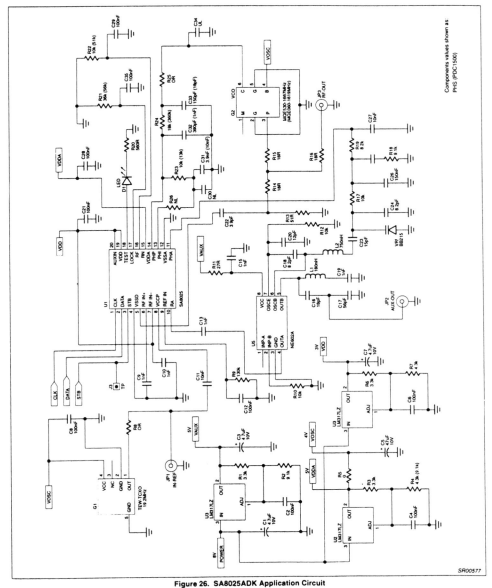

Figure 26. SA8025ADK Application Circuit

APPENDIX D

MMIC-BASED SYNTHESIZERS

The design of MMIC-based synthesizers is limited to companies equipped with GaAs foundries, but the synthesizers are interesting in the sense of seeing where the frequency limitations lie. As previously outlined, the millimeter-wave oscillators generally have a low Q and therefore are fairly noisy. This means they either have to be cleaned up by a wide loop bandwidth or special precautions must be taken to make them high performance.

Figure D-1 shows the component diagram of such a VCO. This particular VCO has been selected on the basis of common source, common gate, and common drain as outlined in Figure D-2.

Figure D-3 shows the measured phase noise of such a VCO at 18 GHz using a characteristic MMIC chip. In order to obtain the appropriate frequency division, a parametric type of frequency divider must be used, as shown in Figure D-4. The

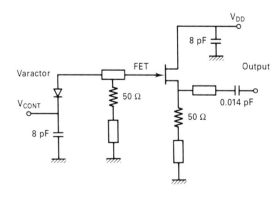

Figure D-1 VCO circuit diagram.

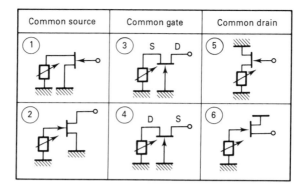

Figure D-2 VCO circuit configurations.

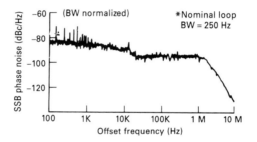

Figure D-3 PLO output SSB phase noise ($f_{osc} = 15.0$ GHz).

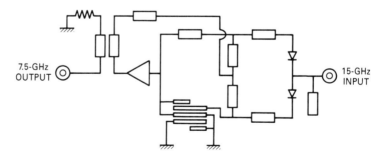

Figure D-4 MMIC frequency divider.

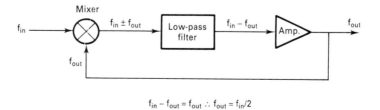

$$f_{in} - f_{out} = f_{out} \therefore f_{out} = f_{in}/2$$

Figure D-5 Principle of the regenerative frequency divider.

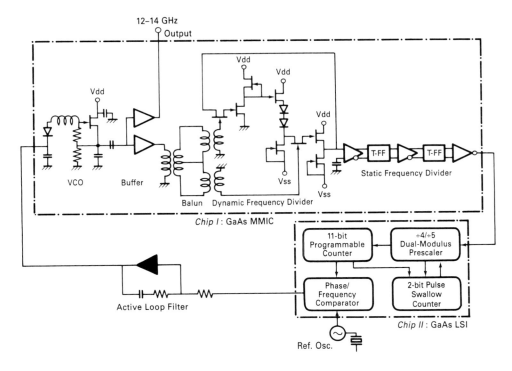

Figure D-6 Ku-band frequency synthesizer integrated in two GaAs monolithic chips.

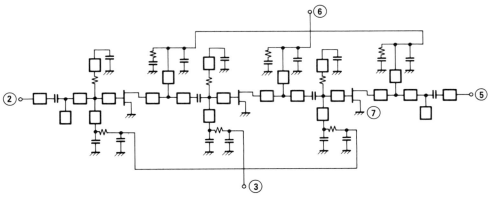

PRECAUTION
It is recommended to connect dc block capacitors to input and output terminal (②) and (⑤)) in order to prevent failure to surge

Figure D-7 Equivalent circuit of the IC.

dividers can be characterized as either regenerative, dynamic, or static devices. The dynamic frequency divider refers to a "flip-flop" type of frequency divider using a delay line between the input and output and is built on a heterojunction bipolar approach. These types of divider have been built up to 26 GHz. The

628 MMIC-BASED SYNTHESIZERS

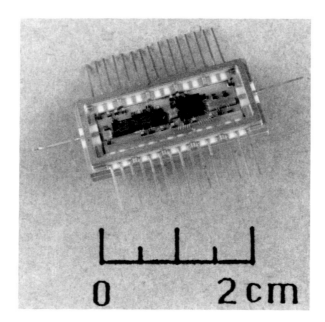

Figure D-8 Photograph of synthesizer.

regenerative frequency dividers are built on the popular principle shown in Figure D-5. A modification of this principle using a double-balanced mixer can also be used as a frequency doubler. This KU band frequency synthesizer is contained in two GaAs monolithic chips, as shown in Figure D-6.

Some of these integrated circuits can be obtained as discrete components. Figure D-7 shows a high-gain stage model MGF7201 GaAs monolithic microwave IC. At 14 GHz it exhibits 20-dB gain and approximately 20-dBm output power.

Figure D-8 shows a photograph of the MMIC synthesizer.

REFERENCES

1. T. Yamasaki and E. Nagata, *Microwave Synthesizer for Terrestrial and Satellite Communications*, NEC Corporation, June 1989.
2. M. Suzuki et al., "A 9 GHz Frequency Divider Using Si Bipolar Super Self-aligned Process Technology," *IEEE Electron Device Letters*, Vol. EDL-6, April 1985, pp. 181–183.
3. R. H. Derksen et al., "Monolithic Integration of a 5.3 GHz Regenerative Frequency Divider Using a Standard Bipolar Technology," *Electron Letters*, Vol. 21, October 1985, pp. 1037–1039.
4. T. Ohira et al., "14 GHz Band GaAs Monolithic Analogue Frequency Divider," *Electron Letters*, Vol. 21, October 1985, pp. 1057–1058.
5. K. Honjo et al., "Novel Design Approach for X-Band GaAs Monolithic Analog 1/4 Frequency Divider," *IEEE Transactions on Microwave Theory and Techniques*, Vol. MTT-34, No. 4, April 1986, pp. 436–441.

6. M. G. Stubbs et al., "A Single Stage Monolithic Regenerative 1/2 Analog Frequency Divider," *IEEE GaAs IC Symposium Digest*, October 1986, pp. 199–201.

7. K. Osafune et al., "An Ultra-High Speed GaAs Prescaler Using a Dynamic Frequency Divider," *IEEE Transactions on Microwave Theory and Techniques*, Vol. MTT-35, No. 1, January 1987, pp. 9–13.

8. P. Weger et al., "Static 7 GHz Frequency Divider IC based on a 2 μm Si Bipolar Technology," *Electronic Letters*, Vol. 23, February 1987, pp. 192–193.

9. Y. Yamauchi et al., "22 GHz 1/4 Frequency Divider Using AIGaAs/GaAs HBTs," *Electronic Letter*, Vol. 23, No. 17, August 1987, pp. 881–882.

10. K. C. Wang et al., "A 20 GHz Frequency Divider Implemented with Heterojunction Bipolar Transistors," *IEEE Electron Device Letters*, Vol. EDL-8, No. 9, September 1987, pp. 383–385.

11. J. F. Fensen et al., "26 GHz GaAs Room-Temperature Dynamic Divider Circuit," *IEEE GaAsIC Symposium Digest*, October 1987, pp. 201–204.

12. T. Mizutani et al., "A High-Speed Static Frequency Divider Employing $n^{\dagger}$ Ge Gate AIGaAs MISFET," *IEEE Proceedings, IEDM*, December 1987, pp. 603–606.

13. R. H. Derksen et al., "7.3 GHz Dynamic Frequency Dividers Monolithically Integrated in a Standard Bipolar Technology," *IEEE Transactions on Microwave Theory and Techniques*, Vol. MTT-36, No. 3, March 1988, pp. 537–541.

14. M. Shigaki et al., "High-Speed GaAs Dynamic Frequency Divider Using a Double-Loop Structure and Differential Amplifiers," *IEEE Transactions on Microwave Theory and Techniques*, Vol. MTT-36 No. 4, April 1988, pp. 772–774.

15. Y. Yamauchi et al., "AIGaAs HBT Dynamic Frequency Divider Constructed of a Single D-Type Flip-Flop," *Electron Letters*, Vol. 24, No. 17, August 1988, pp. 1109–1110.

16. H. Ichino et al., "Super Self-Aligned Process Technology (SST) and Its Applications," *IEEE Bipolar Circuits and Techniques Meeting Digest*, September 1988, pp. 15–18.

17. K. Morizuka et al., "AIGaAs/GaAs HBTs Fabricated by a Self-Alignment Technology Using Polyimide for Electrode Separation," *IEEE Electron Device Letters*, Vol. EDL-9, No. 11, November 1988, pp. 598–600.

18. K. Kanazawa et al., "A 15 GHz Single-Stage GaAs Dual-Gate FET Monolithic Analog Frequency Divider with Reduced Input Threshold Power," *IEEE Transactions on Microwave Theory and Techniques*, Vol. MTT-36, No. 12, December 1988, pp. 1908–1912.

19. M. Takahashi et al., "A 9.5 GHz Commercially Available 1/4 GaAs Dynamic Prescaler," *IEEE Transactions on Microwave Theory and Techniques*, Vol. MTT-36, No. 12, December 1988, pp. 1913–1919.

20. P. Weger et al., "A Si Bipolar 15 GHz Static Frequency Divider and 100 gB/s Multiplexer," *IEEE International Solid-State Circuits Conference Digest*, February 1989, pp. 222–223.

21. T. Ohira et al., "A Compact Full MMIC Module for Ku-Band Phase-Locked Oscillators," *IEEE Transactions on Microwave Theory and Techniques*, Vol. 37, No. 4, April 1989, pp. 723–728.

22. T. Ohira et al., "A Ku-Band MMIC PLL Frequency Synthesizer," *1989 IEEE MTT-S Digest*, pp. 1047–1050.

23. T. Ohira et al., "Development of Key Monolithic Circuits to Ka-Band Full MMIC Receivers," *IEEE 1987 Microwave and Millimeter-wave Monolithic Circuits Symposium*, pp. 69–74.

24. T. Ohira et al., "MMIC 14 GHz VCO and Miller Frequency Divider for Low-Noise Local Oscillators," *IEEE Transactions on Microwave Theory and Techniques*, Vol. MTT-35, No. 7, July 1987, pp. 657–662.
25. Shigeki Saito, "Low Power and Fast Switching Synthesizers for Mobile Roads," NTT Radio Communication Systems Labs.
26. A. N. Riddle, Avantek, Inc., and R. J. Trew, North Carolina State University, "A New Measurement System for Oscillator Noise Characterization," *1987 IEEE MTT-S Digest*, pp. 509–512.

INDEX

Acceleration, sensitivity of crystal resonators, 251–252
Accumulator, 179–180
Acquisition, 53–75
　lock-in time, 56
　loop stability, 64, 67–75
　pull-in range, 57, 60–62
　pull-out range, 56
　time, 9, 13
　VCO coarse steering, 62–64
Active integrator, 527, 530
Aliasing, 141
Allan variance, 84–85
Amplifier:
　feedback, 328
　noise, 129, 131
　phase noise characteristics, 101–102
　wideband high-gain, 319–330
　　differential limiters, 323–324
　　isolation amplifiers, 324–330
　　slew rate, 320
　　summation amplifiers, 319–323
Amplitude stability, oscillator, 211–212, 538–544
Analog Devices AD7008 DDS modulator, 491–492
Analog phase-locked loop, 3–4
API counters, 188
API current sources, 188
Asynchronous counters, 330–333
Autocorrelation method, 123
Autonomous circuits, noise analysis, 570
Avantek AT21 00 chip, equivalent circuits and measurement data, 91–92

Barkhausen criterion, 199
BFO synthesizer, single-loop, 496, 498
Bias voltages, drain-source, 560–561
Bipolar transistor, 101
　Gummel–Poon model, 552–553
B-mode, 250, 265–267
Bode diagram, 445, 518–524
　type 2 third-order loop, 499
Bridge oscillator, 274–276
Butler oscillator, 256–257, 259, 266–267
　fifth overtone, 267

Capacitances, connected in parallel or series with tuner diode, 230–232
Capture time, 9
CD4046 phase/frequency comparator, 308–309
Ceramic resonator oscillator, 453–454
Ceramic resonators, 405–409
Chain scattering parameters, 282
Charge pump, 5
Chebyshev approximation, conjugate roots, 381
Clapp circuit, 209
Clapp–Gouriet circuit, 209–211
Closed loop:
　feedback transfer function, 7
　formula, 6
　gain, 7
　transfer function, 10, 15, 44, 70, 74
CMOS, 50
Coarse-tuning loop, 435
Colpitts oscillator, 203–205, 255–264, 266, 269
　advantages and disadvantages, 259, 262
　with b-mode and fundamental trap, 266
　configurations, 255, 257
　modified, 262–264
　semi-isolated, 259–262
　with tuned collector, 269–270
Complex frequency plane, functions, 514–518
Complex impedance plane, 513
Complex planes, 513–518
Complex variable, functions, 507–512
Conversion noise analysis, 570–571
Counters, 366–367
　API, 188
　asynchronous, 330–333
　swallow/dual-modulus, 352, 356, 358–359, 361–364
　synchronous up/down, 333–360
Crosstalk, 102
Crystal power dissipation, 254
Curtice–Ettenberg model, 557
CV3594 microwave down-converter, 438–440

Damping factor, 381
Damping ratio, 70–71
　closed-circuit, 70
　peak overshoot as function, 71

631

Data registers, 185
Delay compensation, divider, 368–375
Dielectric resonator, 399–403
Dielectric resonator oscillator, 394–396
  with bipolar heterojunction transistor, 462, 464–466
  discriminator stabilized, 581–584
  frequency stabilization methods, 400
  L-band, phase noise, 452
  on microstrip, 400–401
  phase noise, 402, 572–573
  schematic, 402
  10-GHz, 463
Differential limiters, 323–324
Differentiation, 527
Digital/analog converter, 139
  data skew, 168
  ideal output, 140
  nonlinear transition effects, 167–168
  SNR, 145
Digital loops, 9–11
Digital recursion oscillator, 141–143, 152
Digital synthesis, 136–170
  digital waveform synthesizers, see Digital waveform synthesizers
  fractional division N synthesizers, see Fractional division N synthesizers
  future prospects, 169–170
  signal quality, 153, 167–169
  spurious sideband mechanisms, 167–169
Digital-to-analog converter, 62–63
Digital tri-state phase/frequency comparator, 305–319
  with antibacklash circuit, 312–314
  auxiliary circuits, 317–318
  controlled leakage, 311
  logic diagram, 306
  output voltage as function of frequency ratio, 307–308
  output waveforms, 306–307
  with pulse compensation, 312, 314–319
  quad-D circuit, 308
  transfer characteristic, 306–307
Digital waveform synthesizers, 139–153
  block diagram, 139
  digital recursion oscillator, 141–143, 152
  phase accumulator method, 143–147, 153
  RAM-based synthesis, 147–153
  summary of methods, 152–167
  systems concerns, 141
Diode, for frequency multiplication, 236, 238
Diode rings, 289–291
Diode switches, 233, 235–237
Direct digital synthesizers, 491–492
Direct frequency synthesis, 419–422
Divider:
  asynchronous or synchronous, 9
  phase noise characteristics, 102–105
  programmable, 330–375, 433
    asynchronous counters, 330–333
    divider chain, 358–359, 368
    look-ahead and delay compensation, 364–375
      delays, 368–375
      division by 584, 366–367
      10/11 counter, 367–368
    swallow counters/dual-modulus counters, 352, 356, 358–359, 361–364
    synchronous up/down counters, 333–360
      applications, 346
      block diagram, 344
      logic diagrams, 342–343
      timing table, 340–341
Double-balanced mixer, 287–288
Double-base transistor, 55
Double-mix-divide modular approach, 420–421
Duty cycle, 60–62

ECL phase/frequency comparator, 59
Edge-triggered JK master/slave flip-flop, 5, 302–305
Electronic band selection, diode switches, 235–237
Error function, 26
  equation, 8
  type 1 second-order loop, 21
Exclusive-OR gate, 5, 13, 291–294
External sweeping device, 55–56

Feedback amplifier, 328
Field-effect transistor, 101
  equivalent noise circuit, 556–557
  linear equivalent circuit, 554, 556
  microwave, 555–557
  noise, 131
  noiseless, 563
Fifth-order loop, transient response, 44–48
Filter, see also Loop filters
  active
    first order, 28–29
    transfer function, 31, 54
  passive
    first order, 29–30
    transfer function, 32, 54
Final value theorem, 527
Finite-pulse-width model, 297
First-order loop, see Type 1 first-order loop
Flicker frequency, noise sideband as function of, 97–98
Flicker noise, 564
Fluke synthesizer, delay-line-stabilized, 172, 177
FM:
  incidental, 33
  noise suppression, 43–44
  residual, 83
Fourier coefficients, normalized, 542
Fractional division N, 430

Fractional division N synthesizers, 137–139, 172, 175–193, 468–469
  advantages and drawbacks, 175
  average output frequency, 175
  block diagram, 138, 186
  divide-by-N loop with pulse remover block, 178–179
  frequency resolution, 175
  low-voltage, circuits, 604–623
  patents, 191–193
  phase-locked loop, *see* NF loop
Frequency, *see also* Offset frequency
  fundamental, 104
  instantaneous, 104
  stability, 79
    Allan variance, 84–85
Frequency agile synthesizers, 489
Frequency comparator, delay line and mixer as, 122–123
Frequency-conversion approach, limitations, 569–571
Frequency discriminator, phase noise measurement, 121–122
Frequency lock, 22, 57–60
  requirements for, 54–55
Frequency multiplier, 238
Frequency registers, 185, 187
Frequency standards, 236, 239–285
  examples of oscillator specifications, 240–242
  low-noise microwave synthesizers, 444, 446–448
  requirements, 236, 239
  resonators, crystal, 242–252
  specifying oscillators, 239–240
Frequency synthesizers:
  intelligence, 472–473
  multiloop, 473
Frequency West MS-70XCE-XX Microwave Source, 117–119
FRN synthesizer, 470, 472
Functions:
  complex frequency plane, 514–518
  complex variable, 507–512

Gummel–Poon bipolar transistor model, 552–553, 558

Harmonic-balance technique, 551, 564
Harmonic generation circuit, 458–459
Harmonic sampling, 107
Heterodyning techniques, 49–50, 52, 119–120
Hewlett-Packard:
  HP8662A, voltage-controlled oscillator, 218–219, 221
  HP2001 bipolar chip, 392
  HP8640 signal generator, 216, 219
Higher-order loops, 44–48, 71, 73
High-performance hybrid synthesizer, 488–504
  basic approach, 490–497

  loop filter design, 496–504
  single-sideband phase noise, 500–501
Hybrid coupler, 325

IF filter, phase shift, 51–52
Image-rejection mixer, 287
Initial value theorem, 527
Injection-lock oscillator, 107
Integration, 527
Intel devices, 474–475
Intermodulation distortion, two tone, 168–169
Isolation amplifiers, 324–330

Lag filter, transfer function, 530
Laplace transformation, 524–532
LC low-pass filter, passive, 382–384
Linearity theorem, 527
Line receiver, 323–324
Local oscillator synthesizer, 496–497
Lock-in function, fifth-order phase-locked loop, 445
Locking behavior, phase-locked loop, 530–532
Lock-in range, 9
Lock-in time, 56
Lockup time, 429
Loop filter, 375–384, 496–504
  active RC, 376–378
  active second-order low-pass, 378–382
  dual-time-constant, 57
  passive LC, 382–384
  passive RC, 375–376
Loop gain, linearizing, 64
Loops, with delay line as phase comparators, 171–172, 174
Lowpass filter, 140
  active second-order, 378–382
Lumped resonator oscillator, 393–394

Microprocessor, synthesizer applications, 471–481
Microstrip, dielectric resonator oscillator, phase noise calculation, 572–573
Microwave oscillator, 382, 384–397
  buffered, 391
  compressed Smith chart, 387–388
  conditions for oscillation, 384–385, 390
  design flowchart, 393
  dielectric resonator, 394–396
  equivalent circuits, 389
  lumped resonator, 393–394
  series or parallel resonance, 388–390
  specifications, 386
  Spice format, 396–397
  two-port, design, 390–397
  types, 382
Microwave resistors, parameters, 554–555
Microwave resonators, 397–409
  ceramic resonators, 405–409
  dielectric resonators, 399–403

Microwave resonators (*Continued*)
  SAW oscillators, 398–399
  varactor resonators, 403–405
  YIG oscillators, 402–405
Microwave synthesizers, low-noise, 437–473
  building blocks, 437–443
  commercial examples, 467–473
  early type, 438, 441
  frequency standards, 444, 446–448
  harmonic generators, 458–459
  interaction of frequency-determining modules, 441–442
  isolation stage, 457–458
  millimeter-wave oscillators, 459–463
  oscillators, 447–452
  other key components, 450, 452–463
  output loop response, 443–446
  simulator, predicting phase noise, 449, 451
  time domain analysis, 461–466
  wideband, 441
Millimeter-wave applications, phase noise calculation, 580–581
Mix-and-divide method, 420
Mixer, 284–285, 287–288
  circuits, noise figure, 565–567
  digital loops with, 48–53
  double-balanced, 5
MMIC-based synthesizers, 625–628
Modulation noise, 569–571
Motorola:
  MC12040 phase/frequency comparator, 309–310
  MC12012 universal dual-modulus counter, 352, 361
  MC68EC040, 475
  MC68060, 475
  MC1451XX, 490
  MC145156 synthesizer, 345–349
  MC145190/91EVK, 588–603
Multiloop synthesizer, 113–116, 422–427
  output loop, 425
  performance, 436
Multiple sampler loops, 170–171, 173
Multipliers, phase noise, 106–110

NEC NE42484A, data sheet, 93–95
Negative resistance oscillators, 517
NF loop:
  block diagram, 183–184
  integrated currents, 188–189
  integrator waveform, 188–189
  as modified divide-by-N loop, 185
  with pulse remove command section, 180–181
  structure, 187
Noise:
  analysis, oscillator, 567–568
  bandwidth, 8
  type 1 first-order loops, 11–12

  components at high-frequency deviations, 568
  contributions, practical results, 117–118
  ears, 117
  figure, mixer circuits, 565–567
  flicker, 564
  floor, 120
  generation in oscillator, 552
  from linear elements, 563–565
  modulation, 569–571
  oscillator, 80
  phase, 80–81
  from power supplies, 110–111
  power versus frequency, 86
  Van der Ziel model, 564
  at VCO output, 112
Noise-conversion nomograph, 85
Noise model, bias dependent, 552–562
Noise sideband, 79–82
  as function of flicker frequency, 97–98
  multipliers, 108–109
  oscillator, 97–99
  spurious mechanisms, 167–169
Noisy circuits, general concept, 562–565
Nyquist criterion, 198

Offset, 473–474
Offset frequency, 79
Open-loop:
  gain, 35, 38
  transfer function, 64, 67–68, 72
Oscillation, conditions for, 384–385, 390
Oscillator, 197–237. *See also* Microwave oscillator; Voltage-controlled oscillator; YIG oscillator
  with AGC and traps, 272–273
  amplifier model, 201
  amplitude stability, 211–212, 538–544
  analytical methods, 253–254
  bias current, 270
  block diagram, 198
  bridge, 274–276
  Butler, 256–257, 259
  Clapp–Gouriet, 209–211
  collector current, 272
  Colpitts, *see* Colpitts oscillator
  configuration, 254, 256–263
  covering range from 225 to 480 MHz, 219, 222, 224
  crystal power, 254
  differences between RF and microwave, 384
  differential mode gain, 275–276
  diodes for frequency multiplication, 236, 238
  diode switch use, 233, 235–237
  forward loop gain, 200
  frequency standards, *see* Frequency standards
  Hartley, 203
  high-performance capabilities, 277–278
  injection-lock, 107

large-signal model, 270–271
limiting amplitude, 272–279
low-noise
  design, 533–538
  LC, 213–219
  microwave synthesizers, 447–452
low phase noise, 448–449
  bipolar implementation, 456
  FET, phase noise calculation, 576–580
millimeter-wave, 459–463
mode selection, 263, 265–272
negative resistance, 207–209, 517
noise, 80, 90
  analysis, 567–568
  generation, 552
  sideband, 97–99
open-loop transfer function, 198
output voltage, 198
phase-controlled high-frequency oscillator, 315–316
phase noise, 273
  calculation, 86–99
  equivalent feedback models, 88–89
  floor, 274
  single-sideband, calculation, 498
  spectrum, 571–572
phase stability, 212–213
Pierce, 256–258
push–push, 546–548
quarter-wavelength, 533–536
  with grounded source electrode, 539
reference, 197
SAW, *see* Surface acoustic wave oscillator
series–parallel resonance, 254–256
single-sideband, 466
specifications, examples, 240–242
specifying, 239–240
stability, external influences, 276–277
steady-date loop equation, 207
switchable/tunable *LC*, 216, 218–227
TCXO, 240–242
  AT cut resonators, 248
tuning diode use, 226–233
very low phase noise, 455–456, 587–623
  MC145190/91 EVK, 588–603
  SA 8025A, 604–623
very wideband low phase noise, 457
wideband, 222, 225

Parallel resonant circuit, tuner diode, 228–230
Peak phase deviation, 81
Phase accumulator method, 143–147, 153
Phase accumulator synthesizer, modulation with, 147
Phase comparator, loops with delay line as, 171–172, 174
Phase detector, 1. *See also* Voltage-controlled oscillator

  divide-by-*N*, 176–179
  exclusive-OR, 291–292
  output voltage, 2
  output waveforms, 346–347
  quasi-digital, 4
  sawtooth output, 178–179
    requirements to approximate, 182
  sensitivity, 13
  sinusoidal, 3
  waveform and transient characteristic, 4
Phase detector scale factor, 291
Phase/frequency comparators:
  circuit diagrams and input and output waveforms, 12
  phase noise, 105–106
Phase-locked loop, 1–4
  acquisition, *see* Acquisition
  analog, 3–4
  block diagram, 2–3, 100, 519
  characteristics, 4–8
  classifications, 8
  closed-loop 3-dB bandwidth, 67
  defined, 2
  digital, 9–11, 27
    with mixers, 48–53
    pull-in performance, 60
  fifth-order, 443, 445
  filters, circuit and transfer characteristics, 18–19
  fractional *N*, *see* NF loop
  important formulas, 66
  linearized model, 112
  locking behavior, 530–532
  multiplier, 107
  natural loop frequency, 429
  noise contributions, 99–111
    amplifiers, 101–102
    contribution from power supplies, 110–111
    dividers, 102–105
    multipliers, 106–110
    phase/frequency comparators, 105–106
  noise suppression, 43
  normalized output response comparison, 40–42
  with open-loop transfer function, 72–73
  operating ranges, 53–54
  single-loop synthesizer, 492–494, 496
  stability, 64, 67–75
  transfer function, 3
  type of loop, 8
Phase margin, 68–73, 519
  closed-circuit damping ratio, 70
Phase noise, 80–81, 168–169
  amplifiers, 101–102
  dielectric resonator oscillators, 402
  dividers, 102–105
  measurement, 118–132

Phase noise, measurement (*Continued*)
    delay line and mixer as frequency comparator, 122–123
    frequency discriminator, 121–122
    heterodyne frequency measurement technique, 119–120
    spectrum analyzer, 120–121
    two sources and phase comparator, 123–132
        mathematical analysis, 126–132
        noise floor, 125
        noise spectrum, 127–129, 131
        potential problems, 125–126
        power spectrum, 128–129
        voltage at IF port, 124
    microwave resonators, 402–403
    minimizing, 91–93
    multipliers, 106–110
    oscillator, 273
        calculation, 86–99
        equivalent feedback models, 88–89
        single-sideband, calculation, 498
    overall performance of a system, 111–118
        practical results, 117–118
        rules for low noise operation, 112–114
    phase/frequency comparators, 105–106
    resonant oscillator, 215
    sideband, computations, 551–585
        bias-dependent noise model, 552–562
    spectral density, 86–89, 101
    spectrum, oscillator, 571–572
    surface acoustic wave oscillator, 277–279
Phase register, 179, 181
    purposes, 188
Phase shift, IF filter, 51–52
Phase stability, oscillator, 212–213
Phase truncation noise, 144
Phillips Semiconductors:
    HEF4750, 347, 350–351
    HEF4751 universal divider, 347, 352–358
    SA 8025A, 604–623
Pierce oscillator:
    advantages and disadvantages, 263
    modified, 262–263, 265
    third overtone, 265, 267–269
Power splitter, 283–284
Power supplies, noise contribution, 110–111
Prescaler, 364
Pull-in, 27–30
Pull-in range, 9, 57, 60–62
Pull-in time, 58
Pull-out range, 56

Q-factor, 231
Quad-D circuit, 308

RAM-based synthesis, 147–153
    applications, 152
    block diagram, 147–148
    components, 148–149
    frequency resolution, 151
    understanding design variables, 149–152
RC filters:
    active, 376–378
    lag, 515–516
    passive, 375–376
Reference frequency, 47–48
    standards, *see* Frequency standards
Reference oscillator, 197
Residual FM, 83
Resonator, 263, 265–266. *See also* Microwave resonators
    crystal, 242–252
        acceleration sensitivity, 251–252
        b-mode, 250
        crystal aging, 251
        crystal cuts, 248–249
        crystal specifications, 252
        dynamic thermal characteristics, 249–250
        equivalent circuit, 243–246
        with load capacitances, 244–245
        maximum obtainable Q, 250–251
        overtone crystals, 246–247
        quartz, temperature behavior, 247
    hair pin, 546–547
    quarter-wave, output impedance, 543–544
Rise time, 17
Rohde & Schwarz:
    EK47 receiver, 290–291
    EK070 receiver, 422–423
        voltage-controlled oscillator, 218, 220
    ESH2/ESH3 receiver, oscillator, 219, 222–223
    ESM500 receiver, voltage-controlled oscillator, 222–223, 226
    ESN/ESVN40, oscillator, 232–233
    ESV receiver, voltage-controlled oscillator, 223, 227
    model SMS synthesized generator, 476–480
        block diagram, 479
        frequency ranges, 478, 480
        specifications, 477
    multiloop synthesizer model SMK, 501–502
    signal generator SMP 22, 467
        single-sideband phase noise, 471, 473
    SMDU signal generator, 428–429
        oscillator, 216–218, 538
    SMHU85 signal generator, 469–470
    SMHU synthesizer, 470, 472
    spectrum analyzer Series FSEA-30, 469
    type XPC synthesizer, 113–116

Sample/hold comparator, 10
Sample/hold detectors, 295–302
    input and output waveform, 294–295
    open-loop gain, 298–299
    switch as pulse modulator, 294
    T-notch filter, 302
    zero-order data hold filter, 296, 298

Second-order loop, *see* Type 1 second-order loop; Type 2 second-order loop
Self-acquisition, 9–10
Sideband, *see* Noise sideband
Signal, quality, digital synthesis, 153, 167–169
Signal generators, requirements, 467–468
Signal-to-noise ratio, 96
Simulator, predicting phase noise, 449, 451
Single-loop synthesizer, requirements, 427
Single-sideband mixer, 285
Slew rate, 320
Small-signal theory, 538
Smith chart, compressed, 387–388
Spectral density:
   frequency fluctuations, 82–83
   phase noise, 86–89, 101
Spectrum analyzer, phase noise measurement, 120–121
Spice format, 396–397
s-plane, 513
Square wave, sine-wave contents, 524–525
STEL 2173, data sheet, 154–163
STEL 2273, data sheet, 164–166
Step function, 526
Step loop, 435
Summation amplifiers, 319–323
Surface acoustic wave oscillator, 277–284, 398–399
   chain matrix, 281–282
   design techniques, 279–285
   phase noise, 277–279
   power splitter, 283–284
   specifications, 279
   T parameters, 282–283
   Y parameters, 280–281
Swallow counters/dual-modulus counters, 352, 356, 358–359, 361–364
Synthesis, direct frequency, 419–422
Synthesizers, *see also* Microwave synthesizers, low-noise
   auxiliary, 431
   block diagram, 483
   direct digital, 491–492
   direct frequency synthesis, 419–422
   dual-loop frequency, 431–432
   high-performance hybrid, 488–504
   introduction of mixer, 433
   with large-scale integrated circuits, 474
   microprocessor applications, 471–481
   MMIC-based, 625–628
   multiloop, 106, 113–116, 422–427
   noise sideband, 431, 433
   one-loop, 136
   step loop, noise sideband, 435
   system analysis, 427–436
   transceiver applications, 481–484
   triple-loop, 434–435
   using heterodyne technique, 49–50

T-equivalent circuit, 552–553
Termination-insensitive mixer, 287–288
Texas Instrument 8132 VCO, 460
Thermal noise, 563
Third-order loops, *see* Type 2 third-order loop
Time domain analysis, low-noise microwave synthesizers, 461–466
T-notch filter, 302
Transceiver, synthesizer applications, 481–484
Transconductance, 540
Transfer function:
   active second-order low-pass filters, 379–380
   closed-loop, 70, 74
   fifth-order loop, 44
   lag filter, 530
   linearized loop, 111
   low-pass, 88
   open-loop, 64, 67–68, 72
   output characteristics, 168
   phase-locked loop, 3
   type 1 first-order loops, 11
   type 2 second-order loop, 25–26
   type 2 third-order loop, 35–42
Transistor, with lowest noise figure, 91
Triple-loop synthesizer, 434–435
Tri-state phase/frequency comparator, 5–6, 170, 289–319, 436
   digital, 305–319
   diode rings, 289–291
   edge-triggered JK master/slave flip-flops, 302–305
   exclusive-OR gate, 291–294
   gain, 27
   lock-in characteristic, 30–33
   pull-in characteristic, 27–30
   reference suppression, 48
   sample/hold detectors, 295–302
TS950, phase noise, 578–579
Tuned circuit, with negative resistor, 516–517
Tuning diode, 226–233, 454–455
   capacitances connected in parallel or series, 230–232
   capacitance/voltage characteristics, 227–228
   noise influence, 93
   noise sideband and, 97, 99
   noise voltage, 93, 96
   operating range, 228
   parallel resonant circuits, 228–230
   practical circuits, 232–233
   tuning range, 232
Tuning range, 9, 232
Type 1 first-order loop, 8, 11–15
   error function, 14–15
   transfer function, 11
Type 2 nth order loop, 523–524
Type 1 second-order loop, 15–24
   block diagram, 15
   error function, 17–18, 21–22

Type 1 second-order loop (*Continued*)
　filters, 18–19
　frequency response, 16
　noise bandwidth, 21, 23
　normalized output response, 41
　with phase compensation, 521
　rise time, 17
　with simple *RC* filter, 519–521
　transfer function, 15, 20
Type 2 second-order loop, 24–33, 521–523
　with active filter, 515
　advantages, 23
　error function, 26, 31
　lock-in characteristic, 30–33
　normalized output response, 42
　pull-in characteristic, 27–30
　pull-in time, 58
　transfer function, 25–26
Type 2 third-order loop, 34–44
　Bode diagram, 499
　closed-loop response, 42
　FM noise suppression, 43–44
　normalized output response, 41
　open-loop gain, 35, 38
　time constant, 40
　transfer function, 35–42

Unijunction transistor, 55

Van der Ziel noise model, 564
Varactor resonators, 403–405
VCXO, extremely low phase noise, 448–450
Voltage-controlled oscillator, 2–3
　ceramic resonator, phase noise, 453–454
　calculation, 574–576
　coarse steering, 62–64
　divide-by-N loop, 177
　gain, 13
　low phase noise cavity stabilized, 450, 453
　millimeter-wave, 580–581
　noise
　　sideband, 428, 430
　　suppression, 43
　output
　　advanced by a fraction of a cycle of phase, 181–182
　　frequency, 531
　　noise at, 112
　phase detector output frequencies with same offset, 182
　push-push, 547–548
　SAW delay line, phase noise, 452
　short-term frequency stability, 448
　that violates rules of good design, 579
　very low phase noise, 544–548
　wideband, with large number of tuning diodes, 576, 578

YIG oscillator, 402–405, 437
　coarse/fine steering, 447
　single-sideband phase noise, 443–444
Yttrium–iron–garnet oscillator, *see* YIG oscillators

Zero gain area, 105–106
Zero-order data hold filter, 296, 298
z-plane, 513